D1084329

Nanotechnology

GREGORY TIMP

EDITOR

Nanotechnology

Springer

Gregory Timp
Bell Laboratories
Lucent Technologies
Murray Hill, NJ 07974-0636
USA

Library of Congress Cataloging-in-Publication Data
Nanotechnology / [edited by] Gregory L. Timp.
    p. cm.
  Includes bibliographical references and index.
  ISBN 0-387-98334-1 (hardcover : alk. paper)
  1. Nanotechnology.   I. Timp, Gregory L.
T174.7.N373   1998
620´.5—dc21   98-4681

Printed on acid-free paper.

© 1999 Springer-Verlag New York, Inc.
AIP Press is an imprint of Springer-Verlag New York, Inc.

All rights reserved. This work may not be translated or copied in whole or in part without the written permission of the publisher (Springer-Verlag New York, Inc., 175 Fifth Avenue, New York, NY 10010, USA), except for brief excerpts in connection with reviews or scholarly analysis. Use in connection with any form of information storage and retrieval, electronic adaptation, computer software, or by similar or dissimilar methodology now known or hereafter developed is forbidden.
The use of general descriptive names, trade names, trademarks, etc., in this publication, even if the former are not especially identified, is not to be taken as a sign that such names, as understood by the Trade Marks and Merchandise Marks Act, may accordingly be used freely by anyone.

Production managed by Steven Pisano; manufacturing supervised by Jeffrey Taub.
Photocomposed pages prepared from the author's `troff` files.
Printed and bound by Maple-Vail Book Manufacturing Group, York, PA.
Printed in the United States of America.

9 8 7 6 5 4 3 2

ISBN 0-387-98334-1 Springer-Verlag New York Berlin Heidelberg SPIN 10767646

Contents

Contributors

David A. Awschalom, Department of Physics, University of California, Santa Barbara, CA 93106 USA

Harold U. Baranger, Bell Laboratories, Lucent Technologies, Murray Hill, NJ 07974 USA

Louis Brus, Chemistry Department, Columbia University, New York, NY 10027 USA

Vincent P. Conticello, Department of Chemistry, Emory University, Atlanta, GA 30322 USA

Timothy J. Deming, Department of Materials, University of California, Santa Barbara, CA 93106 USA

G. Dresselhaus, Francis Bitter National Magnet Laboratory, Massachusetts Institute of Technology, Cambridge, MA 02139 USA

M.S. Dresselhaus, Department of Electrical Engineering and Computer Science and Department of Physics, Massachusetts Institute of Technology, Cambridge, MA 02139 USA

Don Eigler, IBM Research Division, Almaden Research Center, San Jose, CA 95120 USA

R.E. Howard, Bell Laboratories, Lucent Technologies, Murray Hill, NJ 07974 USA

Evelyn Hu, Department of Electrical and Computer Engineering, University of California, Santa Barbara, CA 93106 USA

Leo P. Kouwenhoven, Department of Applied Physics, Delft University of Technology, 2600 GA Delft, The Netherlands

Herbert Kroemer, Department of Electrical and Computer Engineering, University of California, Santa Barbara, CA 93106 USA

N.C. MacDonald, Department of Electrical Engineering, Cornell University, Ithaca, NY 14853 USA

P.M. Mankiewich, Bell Laboratories, Lucent Technologies, Murray Hill, NJ 07974 USA

Jabez J. McClelland, Electron Physics Group, National Institute of Standards and Technology, Gaithersburg, MD 20899 USA

Paul L. McEuen, Department of Physics, University of California and Materials Science Division, Lawrence Berkeley Laboratory, Berkeley, CA 94720 USA

Stephan von Molnár, Department of Physics, Materials Research and Technology Center, Florida State University, Tallahassee, FL 32308 USA

Mara Prentiss, Physics Department, Harvard University, Cambridge, MA 02138 USA

R. Saito, Department of Electronics Engineering, University of Electro-communications, Chofugaoka, Chofu, 182 Tokyo, Japan

H. Sakaki, University of Tokyo (RCAST), 4-6-1 Komaba, Megoru-ku, Tokyo, Japan

D.M. Tennant, Bell Laboratories, Holmdel, NJ 07733 USA

G. Timp, Bell Laboratories, Lucent Technologies, Murray Hill, NJ 07974 USA

David A. Tirrell, Department of Polymer Science and Engineering, University of Massachusetts at Amherst, Amherst, MA 01003 USA

R.M. Westervelt, Division of Applied Sciences and Department of Physics, Harvard University, Cambridge, MA 02139 USA

George M. Whitesides, Department of Chemistry, Harvard University, Cambridge, MA 02138 USA

James L. Wilbur, Department of Chemistry, Harvard University, Cambridge, MA 02138 USA

Chapter 1

Nanotechnology

G. Timp

*Bell Laboratories, Lucent Technologies,
Murray Hill, N.J. 07974 USA*

I. INTRODUCTION

In A.D.1295, after a 24 year sojourn in the court of Kublai Khan, Marco Polo, his father, Nicolo, and his uncle, Maffeo, returned to Venice. They were no longer wearing the fine raiments of trusted advisors to the Khan of khans, however. Instead, they looked like vagrants. Their garments, of a Tartar cut, were torn and tattered by the rigors of the trip, and they could hardly speak their native tongue for lack of practice. Upon arriving home, they had to force their way into their family quarters. Their relatives hadn't recognized them – this trio had been given up for dead long ago.

Marco, Nicolo and Maffeo were both merchants and explorers. During their travels throughout Asia, they made careful observations and mapped the circuitous overland and sea routes from Russia to India, from Suez to Japan. But Marco had an eye for nature, too, and so the flora and fauna of the regions they transited were also documented. They did these things because they were curious and because they perceived an economic advantage to it. Their efforts on behalf of the Khan over the years had earned them the *chop*, a golden tablet with the Great Khan's royal seal, entitling them to safe passage throughout the Mongol empire and its tributaries. They used it to return home with incredible wealth - rubies, sapphires, emeralds, and diamonds sewn into their tattered clothing - and detailed knowledge of the East inculcated in their brains.

Eventually, their kinsmen came to be convinced that these suspicious looking characters were who they claimed to be, and they began to interrogate Marco about the East, especially after he revealed the jewels sewn into their garments. However, it seems that the jewels and silks could not buy credibility outright. Marco's fantastic tales of the marvels witnessed at Kublai's court – the Great Wall, coal, paper money, porcelain, movable type, water clocks, and the various great hunts – were not widely accepted by his contemporaries and were often dismissed as gross exaggerations or unmitigated lies.

It is natural to stay within sight of the shore, to tread a familiar path, and to drink from your own well. However, extraordinary people like Marco Polo can be provoked by curiosity and by the promise of wealth (either monetary or intellectual) to venture beyond the horizon, and they can be persuaded by vanity to come back and tell about what they have seen. Marco Polo made the world larger and secured the future of Western civilization by doing this. His travel log, *The Book of Ser Marco Polo Concerning the Kingdoms and Marvels of the East,* inspired Christopher Columbus to seek a westward route from Spain to the East, and Prince Rupert to found the Hudson Bay Company. And though his tales of the wonders of Asia seemed incredible to his contemporaries, we now know them to be true.

This book, *Nanotechnology,* is composed of fifteen chapters written by explorers of a new frontier; a frontier that exists on the head of pin, as incredible as that may seem. These explorers are from a variety of different disciplines: atomic physics, electrical engineering, chemistry, materials science, and numerical physics to name just a few. Like Marco Polo, they were motivated by curiosity, and the promise of intellectual and monetary rewards, to use science and their powers of observation to map the *terra icognita* of a microscopic world that extends from the 100 micrometer diameter of a human hair to a fraction of a nanometer, the size of a single atom. But this world is even more exotic than Marco Polo's Asia seemed to be, since the classical laws of physics that govern the mechanics of our common experience are suspended on this frontier.

As Rich Howard, Paul Mankiewich and I elucidate in **Chapter 2,** the exploration of this frontier was spawned by the integrated circuit (IC) revolution and the requirements of miniaturization that make ICs possible. The market for ICs has already generated a 150 billion dollars in sales in 1996 and it's growing. It is estimated that sales will reach 300 billion dollars by the year 2000. ICs now pervade our lives and have enabled space flight, the information age, and toasters that get it right every time. Moreover, the same techniques used to integrate and manufacture electronics have provided us with the opportunity to fabricate micro-machines and micro-robots with moving parts of the type Noel MacDonald describes in **Chapter 3.** The requirements of miniaturization, micro-machining and integration are becoming more stringent as the minimum feature size (MFS) shrinks, however, as Don Tennant reminds us in **Chapter 4.** For example, a Pentium® processor is comprised of about 4 million electronic switches, and each switch is only a few thousand atoms long. Devices this small are shorter than the wavelength of visible light, and consequently the conventional means for producing them, which employs optical lithography as a key element, cannot be inexpensively extended to much smaller scales. Following the evolutionary development of technology, there have been numerous forecasts of a *small wall* near an MFS of 100nm, beyond which conventional IC technology will stall because of the cost of fabrication[1,2]. Yet, there is still no consensus on a revolutionary, inexpensive route for producing smaller features to breach this wall.

An economical route to feature sizes 100 nm and smaller has been a primary motivation for research on the nanometer frontier. The next seven chapters of this book form a less than comprehensive list of some revolutionary routes to fabrication of features smaller than 100 nm. A fundamental dichotomy exists between the methodology for manufacturing proposed in these seven chapters, where nanometer-scale features are built up from their elemental constituents reminiscent of atomic or molecular architecture, and that represented in chapters 1, 2, and 3, where successive applications of planar lithography and etching are used to carve small features into a single crystal. For example, Hiro Sakaki annotates the remarkable progress toward the fabrication of atomically controlled structures using epitaxy in **Chapter 5**. The emergence of molecular beam epitaxy (MBE) and organo-metallic vapor phase epitaxy has allowed depositions of arbitrary thicknesses and composition with an accuracy of one atomic layer along the direction of growth, enabling the development of high performance devices such as quantum-well lasers, high-electron mobility transistors and resonant tunneling diodes. Although not yet as useful as MBE synthesis, Louis Brus demonstrates in **Chapter 6** the extraordinary potential for chemical synthesis or the spontaneous self-assembly of molecular clusters from simple reagents in solution for the production of three dimensional nanostructures or quantum dots of arbitrary diameter.

Two examples of molecular self assembly, which use the unusual chemistry of carbon, are reviewed by Millie Dresselhaus, Gene Dresselhaus and R. Saito in **Chapter 7,** and by George Whitesides and Jim Wilbur in **Chapter 8.** In particular, the mysteries of the carbon-based fullerene molecule, C_{60}, which is only 0.7nm in diameter and represents one of the most reproducible nanostructures currently available, are unraveled by Dresselhaus and co-authors, while Whitesides and Wilbur delineate how the characteristics of self-assembly make it especially promising as a technique for patterning nanostructures. Since the basic principles of self-assembly also underpin biology, e.g. protein folding and aggregation, the pairing of base pairs in DNA, etc., it seems only *natural* that David Tirrell and his co-authors have pursued the production of artificial protein sequences, synthesized with absolute uniformity of structure, by utilizing recombinant DNA technology and *Escherichia coli* to express the target proteins. As Tirrell describes in **Chapter 9,** a DNA template for the synthesis of a polypeptide chain, coupled with an intrinsic proofreading or self-correcting capacity, has enabled the production of macroscopic structures with nearly atomic precision.

Molecular self-assembly works so well, and reproducibly defines nanometer-scale structures spontaneously at room temperature because the fabrication occurs at a thermodynamic minima, which rejects defects according to the size of fluctuations about the minima. Unfortunately, the utter absence of manual control over the means of fabrication is construed as a pitfall for conventional applications. This limitation has prompted two atomic physicists, Jabez McClelland and Mara Prentiss, to exploit atom optics to manipulate, deflect, and focus atoms using forces that develop from spatial undulations in a light field. As they observe in **Chapter 10**, the principle advantages derived from manipulating atoms this way come from massive parallelism (which preserves one of the most attractive features of self-assembly) and the precise registration associated with the use of a laser. Yet, the same laser interferometer that is currently used to register one lithographic cell to another in conventional lithography can now be used to quickly produce features with nanometer-resolution in an arbitrary design within a cell. Finally, through serendipity, Don Eigler and his colleagues from I.B.M. have discovered a way to manipulate single atoms with exquisite precision into an arbitrary pattern. They reached this milestone in nanotechnology by using a scanning tunneling microscope to drag atoms across a very cold substrate, but the current implementation, which is tersely described in **Chapter 11,** is impractical from a manufacturing perspective because it requires an exorbitant amount of time to position atoms so precisely.

In addition to the economical manufacturing of nanostructures, a prerequisite for their implementation, and another motive for research on the nanometer frontier, is a detailed knowledge of how they work. On the nanometer frontier, magnetism, electricity, heat flow, friction and mechanics can perform contrary to our expectations based on experience in a macroscopic world. For example, in **Chapter 12**, David D. Awschalom and Stephan von Molnár reveal a variety of new and unexpected classical and quantum mechanical phenomena that occur when magnets become so small that only a few atomic spins are included in the particle. The obsequious pursuit of miniaturization has led Leo Kouwenhoven and Paul McEuen to a transistor that is so small that the quantum effects of electricity predominate over classical charge transport. In particular, the current through their *single electron transistor,* which is described in **Chapter 13,** depends crucially on quantum mechanical tunneling through a classically forbidden energy barrier and is regulated by the quantized charge of a single electron. And finally, the chaotic and regular behavior that develops from the propagation and interference of quantum mechanical electron waves formed in devices that are so small that scattering originates only at the boundaries of the device is illustrated by Harold Baranger and Robert Westervelt in **Chapter 14** and by Herb Kroemer and Evelyn Hu in **Chapter 15** respectively.

It took almost three years to cajole these authors into surrendering, and eventually contributing to this book on the nanometer frontier. In that interval, at a metabolic rate typical for an adult male from Illinois, a significant fraction of the self-replicating, error-correcting, DNA ladened nanostructures that comprise me have been rejuvenated and replaced. I am not the same person that I was when I started compiling this book, and this field of endeavor has already changed in subtle ways, too. Yet, I believe that the enthusiasm of these explorers for their subjects, which exudes from these chapters, transcends time just as Marco Polo's enthusiasm does from a travel log written seven centuries ago. The marvels of the nanometer frontier – i.e., the small wall, carbon fullerenes, self-assembled monolayers, atoms arranged by holograms, moving nanostructures, clocks that count electrons one-by-one, and the various experiments and theories used to discover the quantum nature of these wonders – are weirdly reminiscent of the marvels of Marco Polo's Asia – i.e., the Great Wall, coal, paper money, porcelain, movable type, water clocks, and the great hunts of the Khan. Despite my effort to focus this book on marvels that are demonstrably true, i.e., experimentally verified, the utility of these observations is still speculative. It may be years, if ever, before any of the ideas propounded in these chapters can be exploited for the benefit of mankind.

On his deathbed Marco's friends pleaded with him to save his soul and recant his fantastic prevarications about the East. In defiance he retorted, "I have not told half of what I saw"[3,4]. Certainly, the most profound reason I can think of for reading this book is to be inspired, just as Christopher Columbus was inspired by Marco Polo, to examine a new frontier that is only half explored.

REFERENCES

1.   Hutcheson, G.D. and Hutcheson, J.D., *Scientific American* **274**, 54–63 (1996).

2.   Stix, G., *Scientific American* **272**, 90–95 (1996).

3.   Polo, Marco, *The Travels of Marco Polo, the Venetian*, Komroff, M., ed., New York: Van Rees Press, 1930.

4.   Polo, Marco, *The Adventures of Marco Polo*, Walsh, R., ed., New York: John Day Co., 1948.

Chapter 2

Nano-electronics for Advanced Computation and Communication

G. Timp, R.E. Howard, and P.M. Mankiewich

Bell Laboratories, Lucent Technologies,
Murray Hill, N.J. 07974 USA

I. INTRODUCTION

The transistor was invented in 1947[1] for use as an amplifier and electronic switch. The invention grew in economic importance as it became a smaller, lower power, and more reliable alternative to the mature vacuum tube technology, but for many years it showed lower performance and often at higher cost than the earlier technology. So, the transition was slow. We often talk about the *revolution* caused by the invention of the transistor, but, in fact, there was a relatively gradual *evolution* that continued for decades: e.g., the development of silicon bipolar junction transistors (BJT) with a diffused emitter and base circa 1956[2], followed by the development of planar silicon transistors[3] and the metal-oxide-semiconductor field effect transistor (MOSFET) circa 1960, etc.[4,5].

A. The Integrated Circuit Revolution

The transistor became the cornerstone of modern computation and communications, not because it caused a revolution in electronics, but because it enabled the development of the integrated circuit. On the other hand, the invention of the integrated circuit (IC) by Kilby and Noyce[6] circa 1959 began an explosive revolution. In the last 30 years the cost to manufacture a silicon transistor along with the wires necessary to incorporate it into a useful circuit has decreased by a factor of 100,000. Figure 1 illustrates this precipitous drop as measured by the cost of a transistor in the densest memory possible on a single chip in volume manufacture – dynamic random-access memory (DRAM). The cost per transistor drops about 30% when the cumulative volume of DRAM bits doubles. The current cost per transistor for DRAM is only 0.017 millicents. While the costs to produce a transistor in an application specific integrated circuit (ASIC) and in a microprocessor are generally more expensive, about 0.3 millicents for an ASIC and about 1 millicent for logic, they are still a bargain, nevertheless. Over the next fifteen years, there is every indication that the trend shown in Figure 1 will continue, driving the cost down by another factor of 1000 and at the same time making

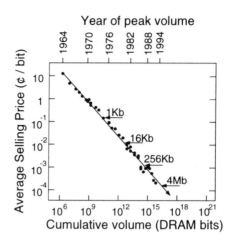

FIGURE 1. The development of the average selling price per bit with time and cumulative volume of (DRAM) bits. Prior to 1970 at least 3 transistors were typically used to represent a bit in DRAM. After 1973 only 1 transistor is used to store a bit.

electronics the largest industry in the world.

This unprecedented increase in productivity is the basis of one of the most astounding revolutions in history. Electronic equipment, once a novelty, is now ubiquitous. Computers run our cars, our washing machines, our entertainment equipment, and even our toasters. More importantly, cheap computation has enabled a global communications network that links the farthest corners of the world by TV, FAX, voice, and data in a way that has changed the nature of national boundaries and the relationships between countries. On a more personal level, it is also poised to dramatically change the relationship between individuals and the vast body of knowledge and information that has been and is being created by human civilization. The internet is just one example.

Figure 2 illustrates the growth in complexity of silicon IC technology as measured by the number of transistors in a DRAM. Gordon Moore of Intel astutely observed this astounding trend, now denoted as Moore's law, about two decades ago[7]. Almost since the inception of the IC, the number of transistors incorporated into a memory chip has increased by a factor of 4 every 3 years with unerring regularity[7]. In general, about half of this increase has come from a regular reduction of the size of features possible on the chip and the other half from factors like design improvements or an increase in chip die size.

The extrapolation of the curve in Figure 2 into the future is based on the National Technology Roadmap, which is the current blueprint for the US semiconductor industry[8]. It is not wishful thinking. It seems physically reasonable, based on devices that have been developed by research, and it will satisfy a

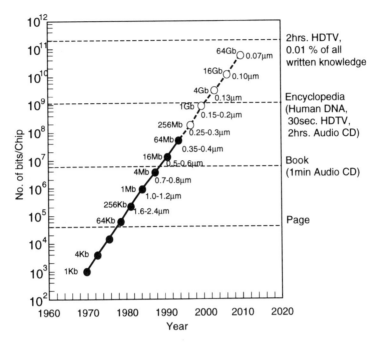

FIGURE 2. The complexity curve adapted from Moore[7] showing the development in time of the information storage capacity of silicon DRAM chips. The information was tabulated based on publications in the ISSCC Technical Digests. The extrapolations (indicated by the dotted line) are based on the SIA Roadmap. The approximate information storage capacity of an encyclopedia, human DNA, two hours of audio CD, 30 seconds of HDTV, a book, 1 minute of Audio CD, and a page are indicated for reference.

market need. The extrapolation is grounded in the assumption that the economics that currently drives integration will continue to do so. Higher levels of integration are economical because of incredibly inexpensive on-chip interconnections between transistors relative to chip-to-chip connections. An on-chip interconnection may cost as little as 1/10,000-th of a cent. When compared to the 1 cent cost associated with a connection between chips in a multi-chip module[9], there is little wonder why more transistors are being integrated into a single chip.

Despite the reduction in design rules by a factor of 50, and an increase in the die area by 150, the cost of a 1 cm^2 chip has not change over the last twenty years. The price is a little less than 4 dollars per cm^2, and is expected to remain so in the future, independent of the technology. To produce more transistors and wires at ever decreasing cost, however, the factories themselves are becoming ever more complex and expensive. The industry extrapolations shown in Figure 3 have the factory costs increasing along with the scaling or miniaturization of the IC technology, which we denote by the design rules used to implement them: i.e. 0.5 μm, 0.25 μm, 0.1 μm etc. Figure 3 shows that the economies of scale for the current generation of IC technology are now requiring factories that cost over 1 billion dollars to build. While such a factory can produce tremendous revenue and profit for its owners, it also represents a growing barrier to entry in the business and an enormous risk to manage. One cannot start out with a factory at half this size and scale up if the product is successful; it is necessary to start at an economical size just to stay in the game.

One of the main reasons for the escalating cost is associated with the stringent controls that must be maintained over the process used to produce ICs with high yield. Some factors that contribute to the cost are high resolution lithography steps, the complexity of deep submicron capacitors, multi-layer metallization, stringent defect, flatness and roughness control of the substrates, high purity chemicals, and testing facilities[12]. Yield in IC manufacturing is plagued by process fluctuations and contamination from impurities and particles, which together give rise to statistical variations from device to device. So there is a

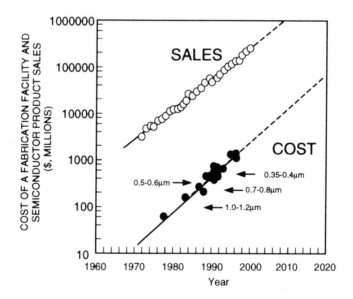

FIGURE 3. A compilation of estimates of the required capital investment for a large semiconductor fabrication line taken from the SEMATECH[165] and the estimated world-wide semiconductor electronics (IC) sales taken from DataQuest, Makimoto[10] and Bois[11]. The year for introduction of a particular technology, denoted by the design rule, e.g. 0.35 μm, is plotted versus millions of U.S. dollars.

chance that a few devices will not meet a design specification because of statistical variations, and there is a small probability that enough device failures will occur to affect circuit performance. Even the failure of a single device or interconnection can reduce a 500 dollar microprocessor to a piece of cheap jewelry.

Thus, yield, process control and circuit margin are all intertwined, and the efficacy of a factory is crucially dependent on them. To be feasible an IC design has to accommodate variations in a specification within a certain tolerance or error margin and still perform as expected. To illustrate the effect of process control and margin on circuit yield[13], consider a transistor with a drain current (I_D) designed to be I_{D0}. Assuming that the parameter I_D has a normal distribution with a standard deviation of $I_{D0}\sigma$ about the design value, if I_D for a device lies between the upper and lower margins $I_{D0}(1-M)$ and $I_{D0}(1+M)$ respectively, then the probability that a single component will fall within a margin of the nominal design specification is given by:

$$P_o = 2/\sqrt{\pi} \int_0^{M/\sqrt{2}\,\sigma} \exp[-x^2]\,dx = erf[M/\sqrt{2}\,\sigma]. \qquad (1)$$

It follows that for a circuit consisting of N nominally identical transistors, the circuit yield will be at most P_o^N.

As shown by Figure 4, the chip yield described by Eq. (1) has a process margin threshold that depends on the level of integration[13]. So, unless the normalized margin, M/σ, develops beyond a certain minimum, the yield crashes and production becomes unprofitable. Notice that for 50% yield the minimum margin occurs at $M/\sigma=2$ for 16 devices, while it occurs at $M/\sigma=6.7$ for 64 billion devices. Here lies an important distinction between research demonstrations that indicate the potential performance of a technology, but yield only ten to one hundred devices, and manufacturing that must produce millions or even billions of high performance devices. As the level of integration increases, it is expected that either the process control will become much more stringent (i.e. the process standard deviation must be smaller) or the process and circuit margins will increase to produce an adequate yield without increasing the cost of production per square centimeter of silicon. But, on the contrary, for fundamental reasons it is proving difficult just to maintain the same level of process control and circuit margin from one technology to the next rather than improve on it. An especially ominous indicator of the severity of this problem is the net probe yield in DRAM[12]. Probe yield, which was approximately 90% for 256 kb DRAM, has steadily declined to less than 12% for 256 Mb DRAM (although gross yield has been maintained through the use of redundant circuitry.)

It is not surprising then that in the pursuit of process control, the cost of a factory is increasing at a annual rate of 19%. However, according to Figure 3, the cost of a factory is increasing even faster than the 16% annual return from

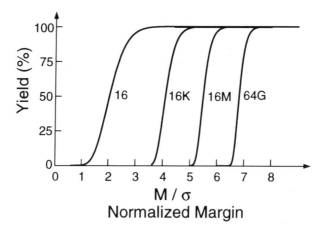

FIGURE 4. The effect of process control and uniformity on circuit yield. There is a process margin threshold that depends on the complexity. Adapted from Miller *et al*[13].

worldwide sales of semiconductor products. Soon only the largest companies (or even countries) will stay in business because of the diminishing financial returns. Thus, progress will continue as long as economically possible, i.e. as long as the cost per function in an IC continues to drop, driven by a collective industry-wide need to remain competitive through continuous improvement. Since the improvements in productivity and functionality are so great when moving from one technology generation to the next, any company that does not remain near the leading edge, where the profit margin is the largest, will not remain viable. The extrapolations of costs in Figure 3 suggest that some change must occur within the next 15 years if progress is to continue to be driven by economics[11]. But, even if the economics becomes unattractive because of increasing capital requirements, ICs are so central to the economy of a modern nation, that progress will probably continue afterward, but at a different rate than that indicated in Figure 1, driven just by political considerations, as use to be the case for steel.

B. Integrated Circuit Applications

To discover the impetus for new IC technologies, it is important to develop an appreciation for the human impact of the enormous numbers associated with them. For example, Figure 2 also shows what can be done with different amounts of memory. Right now, nearly an entire book can be stored on a single chip without any data compression; with compression that could increase by factors of 2 to 10. The extrapolation in Figure 2 suggests that just after the turn of the century, a full encyclopedia will be stored on a single chip. Furthermore, if we use memory to store audio, the same chip that can hold a full 24 volume encyclopedia, could also contain about 2 hours of CD quality music using compression.

While this may sound impressive, it's really not. When the use of video as a communication medium becomes more routine, a Gigabit of memory will look much less impressive. The same chip that could hold 2 hours of high quality sound can only contain 30 seconds of compressed high definition TV using current techniques. So, when people start to expect images as well as sound as part of their communications and information capabilities, even the massive amount of memory available when silicon ICs reach design-rules of 0.1 μm will still fall far short of easily imagined needs.

In addition to memory, it is also interesting to look at some of the applications that require computation and the demands they put on electronics technology. Table 1 enumerates the broad spectrum of computational processing power required for common numerical problems that are to be targeted by VLSI. Ten years ago, high performance computing was considered the realm of single processor supercomputers for scientific and engineering applications. These drove the computing technology and defined the leading edge. As Table 1 indicates, today a dichotomy exists between high performance designs. There are high performance processor designs that require low power consumption because they are intended for portable applications such as cell-phones, games or personal digital assistants. There are high performance, high volume consumer electronic designs that require low cost manufacturing; among these are multimedia applications such as motion picture expert group (MPEG2) encoders, and digital signal processors (DSPs) for real time, full-motion video signal processing. Full-motion video demands both high speed and large data processing capabilities, which is why multi-media consumes huge memory area and occupies a substantial fraction of the CPU time in today's personal computers. And finally, problems such as the human genome project require extremely high performance, Tera- and Peta-FLOPS computers (i.e. computers capable of 10^{12} or 10^{15} floating point operations per second.) The genome project should produce the entire sequence of all 3×10^9 base pairs of human DNA with the next few decades. Implicit in the project is an opportunity for advanced computation to directly influence human evolution through the prediction of high-order structure within the one-dimensional gene sequence. For example, secondary structure predictions for sequences of RNA (with approximately 10,000 nucleotides) can be done currently on an 8-processor CRAY Y-MP in about 6 hours, but there are hundreds of thousands of sequences ranging in size from 50 to 10,000 nucleotides[14].

Today, there are only about 1000 extremely high performance machines in the world, but nearly 60 million (60 M) personal computers were shipped in 1995 alone. These PCs are no longer toys; the most powerful of them can execute hundreds of millions of instructions per second (MIPS), but cost only a few thousand dollars. The performance gap between the low cost PC and the multi-million dollar supercomputers is shrinking so fast that many large problems are now being

TABLE 1. Numerical Problems to be targeted and estimated speed required in millions of instructions per second (MIPS).

TASK:	MIPS
Speech Recognition (numbers)	≈ 50
Digital Signal Processor (audio)	≈ 50
Cellular Phone	≈ 50
Personal Computer	≈ 100
Video Games	≈ 200
HDTV	≈ 1000
Color FAX	≈ 7000
Speech Recognition (5000 words)	$\approx 10,000$
MPEG2 Encoder	$\approx 20,000$
multimedia DSP	$\approx 50,000$
Video Graphics (Ray Trace)	$\approx 10^9$
Secondary Structure of genome	$\approx 10^{12}$

tackled with networks of low cost computers rather than single large machines. For example, two avenues that are currently being explored for the development of TeraFLOPS computers are: (1) scaling up a GigaFLOP computer through miniaturization of the components; and (2) networking of 32,000,000 processors together[14]. Both approaches are expected to produce a product in about 4 years for the same price: 50 million dollars. Thus, the whole character of high performance computation is changing as networks of small, inexpensive computers replace the larger centralized facilities. With parallelism, pipelining, and redundancy, the available computational power to solve a problem can grow almost indefinitely, while physical limits place severe constraints on the advances in higher speed single processors.

C. Prospectus

With a start like this and the impetus behind it, the integrated circuit revolution will be a hard act to follow. Yet, as we examine the extrapolations of Figures 1, 2, 3, and 4 into the future, we find that there are ample reasons to believe that this phenomenal record of growth and progress will be stymied by practical limits to manufacturing. Instead of following Moore's law, we anticipate that the rate of growth will slow in the near future because of economics. The central issue, threatening the extension of the curves in Figures 1 and 2 is that practical, economical device scaling limits are not accounted for by Moore's law and are not in the SIA (Semiconductor Industry Association) roadmap. As manufacturing approaches a hierarchy of limits set by circuits, devices, materials, physics and even the size of atoms, we expect that the margins and parameter spreads required to economically produce high performance Giga-scale integrated circuits may not be achievable. In what follows below, we will illustrate some of the issues that encroach on circuit performance, margin and parameter uniformity as the technology approaches these limits. Various aspects of these limits have already been reviewed in the literature; elements of which we will highlight and further refine below[15–24].

In Section II we examine some of the rudimentary elements of complementary metal oxide semiconductor (CMOS) integrated circuit technology, which is based on that ubiquitous VLSI (very large scale integration) switch: the MOSFET. CMOS technology dominates the market. It is so successful because there are not just one, but a number of different aspects of its performance that recommend it over other technologies such as bipolar or *GaAs*. Thus, we discover four of the prerequisites for a successful technology: gain, speed, low power and reliability, which all derive from a single overarching proposition to produce a circuit with "not one but a billion" components. Using analytical models we will show how these prerequisites are intertwined, and especially how the sensitivity to process uniformity affects them as the complexity of the circuit increases. In Section III, we attempt to infer the improvements still possible in CMOS through miniaturization and growing the die. The optimum (economical) size of the die is determined by the defect density and processing cost, and the effectiveness of miniaturization is deduced from (constant field) scaling. Scaling is a method in which the classical transport equations, describing the operating characteristics of an existing, successful, circuit, are scaled to smaller dimensions to predict the characteristics of a smaller design. Eventually, scaling will be frustrated by fluctuations in the energy barriers obstructing current in a MOSFET, and by the size and number of interconnections. The fluctuations develop from the choice of materials, the atomic-scale variations in the device and quantum effects such as tunneling. In this section we use the limitations of the MOSFET and wires to define the limiting IC performance targets for new technologies to exceed.

When the economic advantage associated with integration via miniaturization and the growth of the die are exhausted, the obvious question to ask is, "what's next?" Over the last 20 years or so, many alternatives have been proposed either to replace silicon transistors or to be their successor after existing technology reaches its limitations. As we learn more about the nature of the IC revolution and the underlying physics of alternative technologies, it is interesting to examine some of these alternatives and try to see if any have the potential to enable change on the same scale as the transistor. By looking at the nature of the successful technology and understanding the origin of its strengths, and the source of weakness, we hope to guide the path to the next revolution or at least identify

technologies which offer no fundamental improvement in the important characteristics and are not candidates for the next great technology change. Even this
modest achievement can be useful in focusing efforts and allocating resources.

Some of the same effects, which prove ruinous to the operation of a MOS-
FET, have formed the basis of and give a technological edge to alternative
devices. Section IV provides a comparison with three novel technologies
represented by the single electron transistor, the electron waveguide, and the
Josephson junction. These three respectively use the charge quantization, conductance quantization and magnetic flux quantization as a basis for operation and
promise enhanced performance as "drop-in" replacements using extensions of the
silicon fabrication technology. All three alternatives are expected to operate only
at very low voltage and temperature, though, because the energy barriers to
current are much smaller than that found in a MOSFET.

II. THE VLSI PROPOSITION, "not one, but a billion"

A very large scale integrated circuit (VLSI) is a conglomerate of a large
number of practically identical, very reliable switches interconnected to express a
computing function. Both the number of switches, and the number, layout, and
extent of the interconnections figure into the functionality and processing power
of a circuit. Understanding and predicting the emergent behavior of such systems
with millions of components is a challenge as daunting as producing the components themselves. Because of the expense and complexity involved, a new
design is usually developed heuristically, for instance, from a scaled version of a
previous product. This is because the number of decisions required to achieve an
optimum design is proportional to N!, where N denotes the number of switches,
whereas a heuristic design requires between $N \times \log N$ and N^3. As the complexity of a circuit is extended, it is anticipated that the margin between an optimum
and heuristic design will continue to grow. This margin is manifested in the
resources required for design and testing. For example, currently a certain
switching circuit can be implemented in 0.9 μm CMOS either by using standard
cells to generate a design in 3 months that incorporates 600,000 transistors in
300 mm^2 of silicon, operating at 46 MHz, dissipating 24 W, and requiring 6
months to verify; or by using a single custom chip designed in 9 months with
about half the number of transistors (350,000) in 1/4 of the area (70 mm^2) operating at 4× the frequency (195 MHz), but dissipating only a third of the power
(7.5 W) and requiring 9 months to verify.[166]

We can easily discriminate at least two separate functions performed by ICs:
memory and logic; the characteristics of each depend on the implementation.
Further distinctions can be drawn within memory between the different capacities, data rates and volatility (or memory retention when the power is turned off.)
For example, static random access memory (SRAM) has a 256k–1 Mb capacity, a

100 Mb/s data rate, but is volatile; DRAM has a 4 Mb capacity, a 10–100 Mb/s data rate, but is not so volatile; and flash electrically-erasable programmable read-only memory (EEPROM) has a 16 Mb capacity, a 1–10 Mb/s rate, and is considered nonvolatile. Similarly, within logic, distinctions can be drawn between microprocessors which are performance driven, and application specific integrated circuits such as digital signal processors (DSP).

The different functions or applications place disparate requirements on the design that ultimately translate into a dichotomy in characteristics such as the length of the interconnections, power dissipation, the number of pins on the package etc., even if it is implemented in the same technology. Table 2 enumerates some of the characteristics of DRAM and a microprocessor implemented in 0.35 μm CMOS as of 1995. A DRAM design emphasizes a regular layout and increases memory cell density by using multi-layer polysilicon stacked capacitors to store more charge in less area, and by using a package that takes advantage of the low power dissipation and a low pin count. Only two metal layers are needed to interconnect 64 M transistors. In contrast, a microprocessor design has a very complex layout which uses the example set by a previous generation to guide the design of the next. To minimize the time delay and to sustain high currents, a microprocessor uses the addition of multiple metal layers, from 4–7, for interconnections between about 4 M transistors A microprocessor package usually has a large I/O pin count which is delicately balanced to minimize noise and lead inductance but, at the same time, permits high power dissipation[25,26].

The peculiarities of the implementation of the wires and switches (e.g. whether the switches are implemented using BJTs, N-MOSFETs or heterojunction bipolar transistors (HBT)) are just as important in determining the performance of a circuit. Figure 5 illustrates this, showing the trade-off between performance measured by the frequency of operation, and power dissipation made in different implementations of a phase-lock loop/clock recovery circuit. Notice that

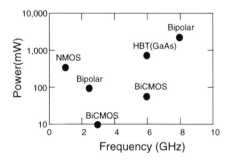

FIGURE 5. An illustration of the trade-off between power dissipation and operating frequency for a particular type of circuit, a phase-locked loop/clock recovery circuit.

TABLE 2. Characteristics typical of CMOS memory and logic applications.

	MEMORY	LOGIC
	High volume DRAM/Flash	*High volume: Microprocessor*
transistors/cm^2	64 M	4 M
On-Chip Frequency (MHz)	150	150–300
Chip size (mm^2)	190	250
Number of Defects	560	425
Number of Package Pins/Balls	30–60	512
Interconnect Density (Metal 1) (m/cm^2/level)	60	35
Number of Metal Interconnect Levels	2	4–5
Interconnect Length (m)	–	380
Maximum power dissipation (W/die)	3–5	5–80

a *Si* bipolar transistors and *GaAs* HBTs give a high operating frequency because of the large transconductance, but they also dissipate about 1–2 W because of a high power supply voltage. In contrast, the operating frequency of the N-MOS implementation is limited ultimately by the transit time across the channel, which does not compare favorably to the base transit time of a bipolar transistor. The hybrid bipolar-BICMOS technology may be a harbinger of the future of technology. BICMOS delivers 3.2 GHz operation at only 10 mW dissipation. This hybrid takes advantage of both the speed and threshold control associated with a bipolar technology, as well as the low power derived from CMOS using a MOSFET as a current source with a low saturation voltage.

A. MOS Technology

The quintessential IC design is implemented using complementary pairs of silicon P- and N-MOSFETs as the switches, and either polysilicon, or polysilicon strapped with a silicide, or aluminum wires purposefully contaminated with copper and silicon usually provide communication. Consequently, we review the salient features of a MOSFET and the associated interconnections below. From a manufacturing viewpoint, this choice may seem odd especially because of the process sensitivity of the MOSFET and the resistivity of aluminum and its susceptibility to electromigration. But, on the other hand, the relative simplicity and reliability of the fabrication process, which hinges on the use of SiO_2 as an insulator and polycrystalline silicon as a barrier to contamination, and the compatibility of the materials recommend it.

(1) The MOSFET. An N-MOSFET is represented in cross-section in Figure 6(a)[27]. The MOSFET is a four-terminal device incorporating two junction diodes and a MOS capacitor. The source and drain are represented by the two n^+ contacts in the p-Si substrate. The p-Si substrate contact is usually held at a fixed value. The gate electrode, usually heavily doped polycrystalline silicon with a silicide strap, is the "metal" electrode of a MOS capacitor, and is separated from the substrate by a thin dielectric layer. The dielectric is typically SiO_2. In operation, the source and substrate are usually grounded, and the drain is raised to a positive potential. Under these circumstances, the MOSFET consists of a reverse-biased (drain) diode and a forward-biased (source) diode. A positive gate bias induces a positive potential in the channel of the MOSFET, lowering the potential barrier between the source and drain. With a positive gate bias, the source acts like an electron emitter, while the drain behaves as an electron collector. The critical gate bias which lowers the surface potential within the silicon sufficiently, and enables electron flow from the source to the drain, is denoted by the threshold voltage, V_t, which is given by Gauss's law:

$$V_t = V_{FB} + 2\Psi_B + \frac{t_{ox}}{\varepsilon_{ox}} Q_B \approx 0.4 V, \qquad (2)$$

where V_{FB} is the flat-band voltage, $Q_B = qN_AD = q\sqrt{2\varepsilon_s qN_A(2\Psi_B)}$ is the charge within the surface depletion region and D is the depletion width, ε_{ox}, ε_s are, respectively, the dielectric permittivity of the oxide and the semiconductor, and Ψ_B is the bulk potential. $V_t \approx 0.45$ V for the device of Figure 6(b).

The relationship between the drain-to-source current, I_D, the drain-to-source voltage, V_{DS}, and the gate-to-source voltage V_{GS} measured in the device of Figure 6(a) is represented graphically in Figure 6(b). For small drain voltages, the drain current is proportional to the voltage: $I_D \approx (W/L)\mu C_{ox}(V_{GS} - V_t)V_{DS}$ where $\mu(V_{DS}, t_{ox})$ is the channel mobility, W and L are the channel width and length respectively, and $C_{ox} = \varepsilon_{ox}/t_{ox}$ where t_{ox} is the electrical thickness of the gate dielectric. As the drain voltage increases beyond $(V_{GS} - V_t)$, a high field region develops adjacent to the drain in which carriers are rapidly swept out of the channel, and the current saturates. This condition, which is generally referred to as pinch-off, is identified in text books as the point when the inversion charge at the drain is totally depleted, and it follows that $I_D \approx \frac{1}{2}(W/L)\mu C_{ox}(V_{GS} - V_t)^2$, but this is inaccurate and not physical. Actually, the pinch-off condition does not occur (even in long channel devices) because of carrier velocity saturation[28,29]. When the velocity of the carriers at the drain becomes saturated, the drain current is instead given by the product of the carrier density and drift velocity:

G. Timp, R.E. Howard, and P.M. Mankiewich

$$I_D = v_{sat} WC_{ox}(V_{GS} - V_t - V_{Dsat}) \qquad (3)$$

where the saturation drain voltage is $V_{Dsat} = F_{sat}L(V_G - V_t)/(F_{sat}L + V_G - V_t)$
and F_{sat} is the drain field at which the velocity saturates. For n-type silicon,

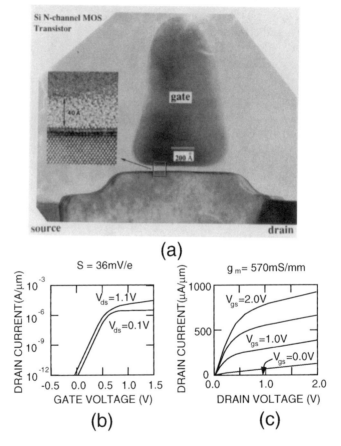

FIGURE 6. (a) A high-resolution transmission electron micrograph of a cross-section of
an N-MOSFET with a gate length of 0.13 µm adapted from[27] and through the courtesy
of Y. Kim. The channel is less than 400 atoms long. The inset shows a lattice image of
the channel region of this device. (b) and (c) represent the measured subthreshold and
drain characteristics found at room temperature for an N-MOS transistor like that shown
in (a). From these measurement it can be inferred that the transconductance is approximately 570 µS/µm, and the subthreshold slope, $S = 36$ mV per e-fold change in I_D or
$84mV$/decade, and the threshold voltage is $V_t = 0.45$ V.

$F_{sat} \approx 1.5 \times 10^4$ *V/cm* and $v_{sat} \approx 10^7$ *cm/s*. In velocity saturation, the drain current is independent of the gate length. This follows since both the transit time and the total mobile charge are proportional to L. Notice that for $F_{sat}L \gg V_{GS} - V_t$ the conventional form for the saturation current is recovered: i.e. $I_D \approx \frac{1}{2}(W/L)\mu C_{ox}(V_{GS} - V_t)^2$, while for a short channel device: $I_D \approx v_{sat} W C_{ox}(V_{GS} - V_t)$. Velocity saturation can be avoided by reducing the source-drain field.

As shown in the figure, a drain current generally exists even for voltages below threshold. The subthreshold current is governed by the physics of a forward-biased diode and vanishes exponentially with gate voltage: i.e.

$$ I_{DS} \propto (W/L)\mu C_{ox}(1 - \exp[-qV_{DS}/kT])\exp[(V_G - V_t)/S]. \quad (4) $$

This exponential nonlinearity is due to the Boltzmann factor associated with activation over the energy barrier between the source and drain. It is a universal feature of electronic switches, even neurons exhibit an ion conductance that depends exponentially on the membrane potential at low currents[30]. The gate voltage swing, S, which takes into account the charge associated with ionized donors or acceptors under the gate electrode, diminishes the effectiveness of the gate at controlling the barrier energy. The subthreshold swing is given by:

$$ S = \frac{kT}{q}(1 + \frac{C_B}{C_{ox}})/(1 + \frac{C_B}{C_{ox}}\frac{dV_B}{dV_G}), \quad (5) $$

where C_B represents the depletion layer capacitance and V_B represents the body voltage. $S \approx 37$ *mV* per *e*-fold change in the drain connect or 85 *mV* per decade at 25C in a MOSFET, but is 20% larger or 44 *mV* per *e*-fold change at the normal operating temperature of 85C. The subthreshold swing is the key to understanding many of the limitations inherent in CMOS technology.

(2) The Interconnections. At low levels of integration, the MOSFET may determine the circuit speed, packing density and yield, but as the level of integration increases, the role played by the interconnections predominates because the distributed wiring capacitance becomes larger than the gate capacitance. An interconnection can be approximated simply by its capacitance, while the resistance and inductance are ignored, so long as the rise time is much longer than the signal propagation time along the line, and so long as the output resistance of the MOSFET is much larger than the characteristic impedance of the transmission line.

The delay, which essentially results from the time required to charge the interconnection capacitance, increases as the circuit density increases, because the

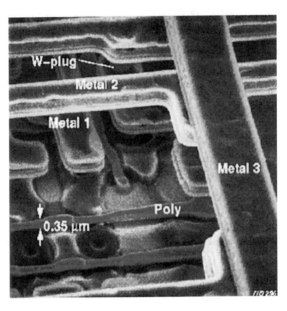

FIGURE 7. Multilevel interconnections in an SRAM produced in 0.35 μm technology, courtesy of F. Baumann. (See color plate.)

current delivered by the driving circuit is reduced in proportion to the shrinking area of the MOSFET, while the capacitance increases in proportion to the square root of the increasing area of the die. Alternatively, the delay specification may not include the entire chip. Following this paradigm, complex high speed systems are comprised of smaller, faster subsystems where the communication between subsystems is made to be less demanding. For faster performance, the subsystem can be made with smaller design rules. Thus, many layers of interconnections, each layer with different design rules, are utilized, with different layers isolated from one another by SiO_2 to reduce interference or cross-talk[31]. Local interconnects are limited by the minimum feature size, while the global interconnects are constrained by the charging time. The dimensions of the intermediate levels of wiring are determined by the chip size. Figure 7 illustrates the complex and dense chip wiring for high performance 0.35 μm technology in an SRAM, which utilizes three separate levels of metal lithography for global interconnection and contacts, and tungsten plugs and polysilicon wires as local interconnects. To guarantee that the output resistance of the MOSFET dominates the line resistance, there is a critical wire length that depends on the resistivity of the material and wire geometry. So, for example, using 0.35 μm line rules, a polysilicon runner can be only 4 μm long and the silicide interconnect can be only 30 μm long, while an aluminum wire can be 1.4 mm long.

The multi-level interconnection scheme leads to disparity in the number and length of the wires comprising a circuit. The wire length distribution can be inferred using Rent's rule[32,33], a well established empirical relationship between the number of I/O to the number of circuits: i.e. $PO = KN^p$, where PO is the number of I/Os, K is the average number of I/Os per circuit and N is the number of circuits, and p is the Rent exponent[34] which is supposed to reflect the organization of the circuit. Usually $0.5 < p < 1$ and $3 < K < 5$. Essentially, Rent's rule is a measure of information flow in a circuit[35]. It has implications for the line length distribution because with short lines only devices in an annulus near the circuit periphery can contribute to I/O, whereas with long lines devices deep within the circuit can contribute. Assuming that the entire chip area is randomly interconnected with wire, the stochastic distribution of wire lengths has the following form according to Meindl and Davis[36]. For $1 \le l \le \sqrt{N}$

$$f(l) = \Gamma \frac{FO}{FO+1} \frac{k}{2} \frac{l^3}{3} - 2\sqrt{N} l^2 + 2Nl^{2p-4}, \qquad (6a)$$

while for $\sqrt{N} \le l \le 2\sqrt{N}$

$$f(l) = \Gamma \frac{FO}{FO+1} \frac{k}{6} (2\sqrt{N} - l)^3 l^{2p-4}, \qquad (6b)$$

where

$$\Gamma = \frac{2N(1 - N^{p-1})}{-N^p \dfrac{1+2p-2^{2p-1}}{p(2p-1)(p-1)(2p-3)} - \dfrac{1}{6p} + \dfrac{2\sqrt{N}}{2p-1} - \dfrac{N}{p-1}},$$

and where l is the wire length in gate pitches and FO is the average circuit fanout. For large N and $0.5 < p < 0.67$, the total interconnection length, L, approaches $L = N/(1-p)(k/2)(FO/FO+1)$ in gate pitches. So, for example, if $N = 4$ million, $p = 0.6$, $K = 4$, and $FO = 3$, the total interconnection length asymptotically approaches $L \approx 2N$ in gate pitches or about 70 m. That this is a reasonable guide to the wiring length on a real logic chip as can be ascertained by examining Table 2.

The interconnect capacitance represents the dominant load in VLSI. To minimize the load, the line capacitance can be reduced by reducing the length and width of the wire, by increasing the thickness of the dielectric, or by increasing the wire-to-wire spacing. To improve packing density, it is desirable to keep constant the thickness of the conductor and the insulator while reducing the wire width and pitch, but this increases the capacitance between lines and increases cross-talk. Furthermore, fringing fields diminish the effectiveness of increasing the thickness of the dielectric beyond a width-to-thickness ratio of one. Thus, the distributed capacitance depends only on the ratios of the interconnection dimensions and is independent of feature size. The minimum total capacitance per unit length of a line is about $c_{dist} = 0.1$-$0.2 fF/$ μm for densely wired chips with the width and thickness comparable to the insulator thickness. Using this value for c_{dist}, we can translate Eq. (6) into an estimate of the load capacitance. The total capacitance per node is approximately $C_L = c_{dist} \bar{L} \approx 50$-$100\lambda fF$, where $\bar{L} \approx (2p+3)(p+1)N^{p-0.5}/6p(4p^2-1)(1-p))\times(10\lambda)$ is the average interconnection length assuming that $p > 0.5$, and 10λ is the circuit pitch with a minimum feature size of λ given in micrometers. C_L depends on the technology only through the minimum feature size, λ. Currently, the capacitance associated with chip-to-chip wires is at least an order of magnitude larger than this number. This difference encourages higher levels of integration, and so the growth of the die and miniaturization are the keys to reducing capacitance associated with wiring.

B. CMOS Design

Complementary MOSFETs and aluminum interconnects are used so prevalently in VLSI, despite the inadequate transconductance, the process sensitivity, and the expense associated with the fabrication of complementary transistors, because they satisfy a number of practical prerequisites for integration. Each prerequisite has to do with the overarching proposition to produce a circuit with "not one, but a billion" components that subject information, represented by voltages stored as electric charge on well-isolated circuit capacitances[37], to a large number of logical operations. After von Neumann[38] and Keyes[16], we recognize that in every computing engine: (1) the information must be standardized regularly[16,39]; (2) the manipulations must be performed quickly[16]; (3) the average power dissipated per operation must be minuscule, and (4) the IC complex comprised of the gates, the architecture as well as the programming must be highly reliable and produced economically with high yield. These four prerequisites are mutually dependent and, taken all together, they have far-reaching implications for the design, performance, and manufacture of any IC, which supersede Moore's empirical law.

(1) Gain. The first prerequisite develops from the observation that small cumulative errors, propagating through a large number of computations, will irreversibly destroy information. For a logic signal to propagate through a large number of circuits, despite noise, it is necessary that the noise not propagate from one stage to the next[16]. Thus, digital circuits have become the bulwark of computation and communication. If a digital circuit rejects noise at the input because of the logic threshold, then the output will be accurate provided there is sufficient gain. But this type of standardization requires a nonlinear transfer characteristic in each step: i.e. the output must reach an accepted value for a tolerable range of input values to accommodate small errors in the input or variability in the circuit elements[16,39]. In other words, the slope of the transfer characteristic between input and output must be less than unity over the range of valid logic levels, while the slope must be greater than unity in the regime between the valid logic levels.

The MOSFET, in conjunction with the power supply, provides the nonlinearity, gain, and noise immunity required for VLSI. To illustrate the suitability of the MOSFET to this task, we choose to evaluate the gain of a MOSFET configured as a CMOS inverter, motivated by its widespread use in VLSI. The inverter of Figure 8 utilizes two matched enhancement-type MOSFETs, one N-MOSFET, and one P-MOSFET. In an IC, one logic gate usually drives another. Consequently, the output of the inverter of Figure 8 is connected to a capacitor, C_L, which is a lumped element representation of the load presented by the next stage. The input-output transfer characteristic of a CMOS inverter shown in Figure 8 satisfies the criterion for gain outlined above. It exhibits three distinct regions: the relatively flat low-input region where $V_{in} < V_{IL}$; the high gain, sharp transition region where $V_{IL} \leq V_{in} \leq V_{IH}$; and the flat high-input regime where $V_{in} > V_{IH}$. V_{IL} is the maximum allowable logic-0 value: i.e. input voltages less than V_{IL} are acknowledged by the switch as representing logic 0. Similarly, V_{IH} is the minimum allowable logic-1 value. A low input signal turns the P-MOS device ON while turning the N-MOS device OFF, and so the output is pulled to $V_{OH} = V_{DD}$. A high input signal has the opposite effect of turning the N-MOS device ON and leaving the P-MOS device OFF so $V_{OH} \approx V_{Dsat}$.

At minimum each stage must be able to drive the input of another similar stage (fan-out equals one) with the associated interconnection. Thus, the energy required to change the output must be as large as that required to change the input. Since switching must be accomplished within a finite interval of time, the region between valid logic levels must therefore exhibit power gain. Since a circuit usually provides more signal to the input of the next stage than it was supplied with, the circuit must obtain power from the power supply, separate from the actual signal path[37].

While a power supply voltage of 3.3 V currently prevails in CMOS VLSI, it is anticipated that the supply voltage will be reduced in future designs to lower the electric fields, maintain reliability and reduce power consumption (as shown below). Following Swanson and Meindl[40], we can evaluate the minimum power supply voltage required for operating an inverter with greater than unity gain by equating the drain currents due to the P- and N-MOSFETs and differentiating with respect to V_{in}. The available voltage gain is highest in the subthreshold regime where the drain current responds exponentially to the input voltage and the output conductance is minimized. If the threshold voltage associated with the P-MOSFET is exactly the opposite of that of the N-MOSFET, then the maximum gain is found at the logic threshold in Figure 8 where $V_{out} = V_{DD}/2$:

$$A_v = \frac{\partial V_{out}}{\partial V_{in}}\Big|_{V_{out}=V_{DD}/2}$$

$$= \frac{g_{mn}+g_{mp}}{g_{Dn}+g_{Dp}+2\frac{\beta}{kT}\exp[\frac{V_{DD}-2V_t}{2S}]\exp[-\frac{V_{DD}}{2kT}]\cosh[\frac{V_{DD}-2V_{in}}{2S}-\frac{V_{DD}-2V_{out}}{2kT}]}. \tag{7}$$

This expression differs from that derived by Swanson and Meindl because we take into account the output conductances, g_{Dn} and g_{Dp}, associated with the N-MOSFET and P-MOSFET respectively. One source of output conductance is derived from the channel-length modulation induced by the source and drain voltages. The widths of the depletion regions surrounding the source and drain increase as the bias voltage increases and so the channel length decreases.

Swanson and Meindl found that for a voltage gain $A_v \approx 10$, the supply voltage must be at least $V_{DD} = 3$ to $4kT/q$. The same conclusion can be deduced from Eq.(7) if the output conductance is ignored. It is not a coincidence that, to achieve sufficient nonlinearity in the transfer characteristic, the voltage must be larger than the uncertainty associated with thermal fluctuations in the charged particle energy[16]. The difference in the rates of diffusion of electrons comprising the subthreshold currents flowing from drain to source and vice versa only become significant for energies larger than a few kT. However, ignoring the output conductance in the denominator of Eq. (7) for the voltage gain is not always justified in practice; it can be the dominant term. The gain, which is determined by the finite slope of the transfer characteristic at the logic threshold in Figure 8, is related to the output conductance, and its effect on the slope is exaggerated as the power supply is reduced. The voltage gain is approximately the ratio of the sum of the transconductances and the sum of the output conductances, $A_v = (g_{mn}+g_{mp})/(g_{Dn}+g_{Dp})$. Typically, $1nS < g_D < 100$ μS, so that the minimum supply should be at least $V_{DD} \approx 0.32$ V for unity gain or $V_{DD} \approx 0.45$ V to guarantee a gain of 10, accounting for an output conductance of 1 μS.

Fundamentally, the gain must also be sufficient to overcome the deleterious effects of noise on the input. Generally, the noise margin determines the minimum gain which, in turn, determines the minimum supply voltage. The noise margin in an inverter can be evaluated assuming that another identical inverter with the output HIGH drives it with a minimum input logic level. The difference between these two values represents the noise margin, NM. The driven gate will not make an error provided the noise voltage is lower than NM. Applying this definition with reference to Figure 8:

$$NM \equiv V_{IL} - V_{OL}. \tag{8}$$

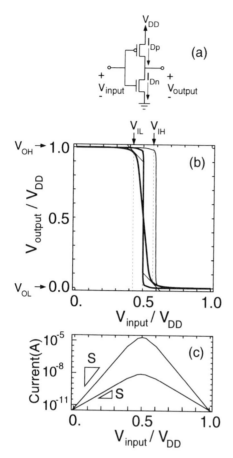

FIGURE 8. The transfer characteristic of the CMOS inverter is robust with respect to variations in the power supply and threshold voltage. Shown in the figure are a schematic of a CMOS inverter circuit and a theoretical inverter transfer function, where the input and output voltages are normalized to the power supply V_{DD}, and the corresponding current available at the output. The red lines represent the characteristic expected for a power supply voltage of $V_{DD} = 0.6$ V, while $V_{DD} = 1.2$ V is shown in black for the same threshold voltages. The output conductance is $g_D = 50$ nS and $|V_{tn,tp}| = 0.6$ V for each case. The blue curve represents the effect of a 0.2 V variation in V_{tn} on the transfer characteristic for a 1.2 V power supply. As Figure 8(b) illustrates, the choice of power supply is not arbitrary. The crux of the problem is associated with the inability to scale the subthreshold characteristic. A factor >10,000 decrease is expected in ON current, when the power supply is reduced from 1.2 V to 0.6 V if the threshold voltage remains at 0.4 V.

Thus, the sharper the voltage transfer characteristic or the higher the gain, the larger the noise margin. Since the transfer characteristics are usually smooth, it is customary to estimate the noise margin by identifying the voltages, V_{IH} and V_{IL}, with the input voltages for which the voltage gain is unity: i.e. $A_v = -1$; as indicated in Figure 8(b)[41]. The margin associated with an archetypical CMOS inverter is given approximately by: $NM \approx 8(3V_{DD} + 2V_t)/8$.

If variations in the signal level are considered to be noise, then noise can develop not only from resistive, inductive or capacitive cross-talk between switches and interconnections[31], but also from process variations. The variability of the threshold voltage across the chip, or the MOSFET drain current, or W/L ratios therefore become factors influencing the supply voltage. For example, the resilience of the inverter characteristic to a change in threshold voltage is indicated by the dashed curve in Figure 8. A shift of $\delta V_t = 0.2$ V in the N-MOSFET threshold voltage induces a comparable change in the logic threshold V_{inv} from $V_{DD}/2$. For a CMOS inverter operating in velocity saturation, the logic threshold varies with the threshold voltage and the drain current of the N-MOSFET, I_{Dn}, and the drain current of the P-MOSFET, I_{Dp}, according to:

$$V_{inv} = \frac{V_{tn} I_{Dn} + (V_{DD} - V_{tp}) I_{Dp}}{I_{Dn} + I_{Dp}}.$$

Thus, the logic threshold is acutely sensitive to changes in the threshold voltage or the drain current that disrupt the balance between the N-MOSFET and P-MOSFET.

In particular, the control of the threshold voltage and the subthreshold slope in a MOSFET are two of the most demanding elements defining VLSI processes and circuit design. Uncertainties in the threshold voltage associated with the fabrication process and the temperature across a chip during operation may be as large as 0.1–0.2 V (6σ)[42], and usually a MOSFET will be designed so that the sensitivity of the threshold voltage to the definition of the gate electrode by lithography and etching is less than 1 $V/\mu m$. Accounting for such variations the threshold voltage is mandated to be between 0.4–0.6 V to minimize the subthreshold leakage current and adequately shut OFF the transistor.

In a MOSFET, the OFF current and the ON current are related through the exponential subthreshold characteristic (EQ. 4): $I_{OFF} \approx I_{ON} exp[(V_G - V_t)/S]$. As Eq. (5) indicates, the subthreshold swing S can be reduced either by reducing the temperature, or the ratio of the depletion capacitance to the oxide capacitance[43], but room temperature operation, S is theoretically limited to $kT/q \approx 26$ meV per e-fold change in current. Consequently, either the leakage current or the ON current are limited by the voltage swing $-V_t < (V_{GS} - V_t) < (V_{DD} - V_t)$. For example, if $I_{ON} \approx 10$ μA is required to drive the capacitance associated with 100 μm long interconnection at a clock frequency of 1 GHz, then a threshold voltage of at least 0.4 V will be necessary to achieve a leakage current of 1 pA, the mandated SIA leakage tolerance. But if the threshold voltage is fixed at 0.4 V then, as Figure 8(c) indicates, the ON current is reduced by a factor of about 10,000 when the power supply voltage is reduce from 1.2 V to 0.6 V, a change that cannot be mitigated by increasing the W/L ratio.

A compromise is generally made in the circuit design to diminish the sensitivity of the noise margin to the threshold voltage and to guarantee sufficient ON current. The supply voltage is chosen to be at least $V_{DD} = (4\text{-}5) \times V_t$[44]. However, as the power supply voltage is reduced, it becomes increasingly difficult to maintain the noise margin and current drive capability, and simultaneously satisfy the criterion on the leakage current because the subthreshold characteristic does not change. Based on these considerations, the minimum power supply voltage is expected to be in the range: 1.2 $V < V_{DD} < 2.5$ V.

(2) **Speed.** The second prerequisite develops from the size of the numerical problem tackled by the circuit, and the compatibility with the time scale for human interaction[16]. As Table 1 enumerates, speech recognition is an example of a numerical problem that must be solved in a way compatible with human thought. A substantial computational investment, approximately $10^4 MIPS$, must be dedicated to recognize even a limited vocabulary of words *in real time*. Thus, to be practical the MOSFET must provide sufficient gain to act like a switch at high frequency. The current gain, i.e. the ratio of the drain current to the gate current in a MOSFET, decreases from an extremely large value at low frequency in proportion to reciprocal frequency. A figure of merit for the high frequency performance is the frequency at which the current gain decreases to unity: $f_T = g_m / 2\pi(C_{GS} + C_{GB})$, where C_{GS} and C_{GB} represent the capacitances between the gate to source and gate to body, respectively. Thus, the higher the value of f_T, the more effective a MOSFET is as a high frequency amplifier. $f_T = 118$ GHz for 0.08 μm N-MOSFET, while $f_T = 81$ GHz for a P-MOSFET with the same dimensions[27].

The frequency f_T is a figure of merit for the device that is independent of the amplifier circuit. However, in a VLSI circuit there are many elements, transistors and wires, collected into logic gates and cascaded to represent a complex logic function. In this conglomerate, a critical signal path can usually be identified which defines the circuit delay. The total delay is just the sum of the time-constants and delays through the critical path of components[45]. To study the propagation of signal through an indefinite number of cascaded logic circuits, we imagine a chain of archetypical logic circuits, identical inverters. Once again, we assume that load presented by the next inverter and the interconnections can be represented by a lumped capacitor. The dynamic operation of a capacitively loaded CMOS inverter circuit is delineated in Figure 9.

Ignoring the rise-time associated with the input, the delay is essentially the time required for the inverter to discharge the load capacitance to ½ V_{DD}. At time

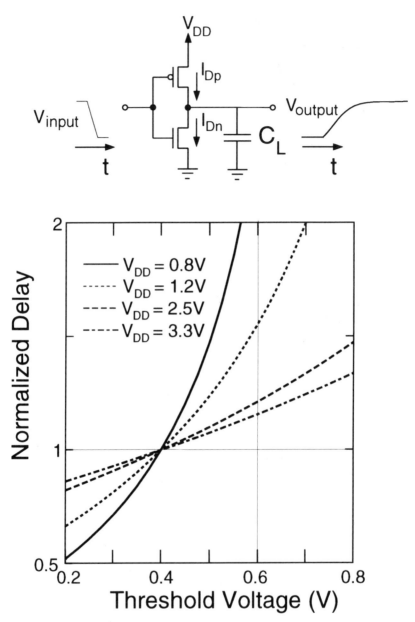

FIGURE 9. One of the major obstacles encountered in the design of practical VLSI is the excessive margin needed to accommodate deviations in the circuit delay. The delay variations expected as a function of the threshold voltage and power supply normalized to the delay found at $V_t = 0.4$ V are shown for an inverter operated in velocity saturation. Adapted from references [43,44].

t=0, the input voltage abruptly rises 0 to V_{DD}. We assume that the capacitor is initially charged to the output voltage $V_{out} = V_{DD}$ just prior to the leading edge of the input pulse. Then, at t=0 the rise of V_{in} to V_{DD} causes the P-MOSFET to immediately turn OFF and the capacitor to discharge through the N-MOSFET. It can be shown that the total delay time, corresponding to the 50% point where $V_{out} = V_{DD}/2$ is:

$$\tau^{50\%} \approx \frac{1}{4} C_L V_{DD} \left[\frac{1}{I_{Dnsat}} + \frac{1}{I_{Dpsat}} \right] \propto \frac{C_L V_{DD}}{W \, v_{sat} \, C_{ox} (V_{DD} - V_t - V_{Dsat})}. \qquad (10)$$

According to Eq. (10), the delay is critically dependent upon the saturation current, and decreases when the current increases. For technology generations beyond 0.5 μm, the drain current does not necessarily scale simply as the reciprocal channel length because of velocity saturation, but instead is proportional to the saturation velocity and the gate capacitance as indicated in Eq. 10. The exact value of the saturation velocity depends on the high energy band structure of the material because of the high electric fields employed in short-channel MOS-FETs[46]. However, the similarity in the nearly free electron-like bands found at high energy in all semiconductors does not promise a dramatic improvement in v_{sat}. To circumvent this limitation, it has been proposed that devices operate in velocity over-shoot, where ballistic transport permits the velocity to exceed v_{sat},[167] but velocity overshoot has been elusive, so far. Thus, thinning the gate oxide has become an economical way to increase the current and shorten the delay.

Equation (10) indicates that speed will be compromised if the power supply is reduced since the delay is a monotonically increasing function of the ratio of the threshold voltage to the power supply. (Note: we expect that $V_{Dsat} \approx 0.35 V_{DD}$ for N-MOSFETs and $V_{Dsat} \approx 0.55 V_{DD}$ for P-MOSFETs.) The increase in delay can be minimized as the power supply is reduced, however, provided that the threshold voltage is scaled accordingly. If the supply and threshold voltage are not reduced at the same rate[47–49], the sensitivity of the delay to threshold voltage distribution due to temperature or process variations is aggravated. This sensitivity is implicit in Eq. (10) since $\partial \tau / \partial V_t \propto [1 - V_t/V_{DD}]^{-2}$ in velocity saturation. (If the drain field is reduced to avoid velocity saturation, the sensitivity of the delay to variations in the threshold voltage gets worse since $\partial \tau / \partial V_t \propto [1 - V_t/V_{DD}]^{-3}$.) Because of the concomitant variations in the drain current, variations develop in the delay across a die[50]. Figure 9 illustrates the expected range in the delay associated with an inverter operating in velocity saturation. According to the figure, with $V_{DD} = 0.8 \, V$ and a worst case threshold voltage of $V_t = 0.4 \, V$, the delay can vary by a factor of 2 when the the threshold changes by less than 0.15 V. This should be compared with the 10% variations in the delay (corresponding to a 10% drain current margin) that is tolerated in

0.5 μm technology using a 3.3 V power supply with the same threshold voltage.

Variations in the delay affect the sequence of events in a computation and are detrimental to timing whether the system is synchronous or self-timed. In synchronous systems, which are by far the most prevalent, the sequence and time are related by a clock signal that permeates the entire system. This means that information must move from gate to gate within a single clock period. Consequently, the timing has to account for the worst-case delay or errors will occur. On the other hand, in a self-timed system (see Section IV.C) the computation is initiated by signal events at the input of an element and completed when signals on the output indicate so. There is no system-wide timing. The time required to do a computation is related to the delay imposed by the various elements and the associated interconnections and tends toward the average delay[37]. The computed answer is correct provided that the interconnections between the various self-timed elements is correct. The design of such systems is problematical, however, because of the combinatorics associated with finding legal interconnections for large scale circuits[37], so it is not widely used yet.

Changes in the design of a synchronous system can be implemented to accommodate variations in the delay associated with an increase in the drain current margin, but there are associated costs because it means that the design will not scale, and performance will suffer. For example, the clock frequency, f_{clk}, can be lowered to avoid synchronization errors and guarantee that the relaxation to the desired state in a circuit is completed every clock cycle. For this reason, the maximum clocking frequency is usually chosen to be to be about 100× smaller than f_T or about 10× longer than the gate-delay This is because the (unloaded) inverter and logic delays at each stage contribute ($\approx 2\tau^{50\%}$) to the total delay and, because at least another $\approx 2\tau^{50\%}$ is contributed by parasitic and line capacities. If one clock phase is $\approx 4\tau^{50\%}$, then the two clock phases required per cycle give a minimum clocking period of $t_c \approx 8\tau^{50\%}$. This is the minimum clock period. If the single phase clock period is doubled to accommodate a factor of two variation in the gate delay, the speed is at least halved, but such variations in the delay would have an even more deleterious effect if a more complicated timing scheme is utilized or if the FO=3 instead of 1.

Since each step of a sequential computation generates an intermediate result that must be stored at least for a clock cycle, the second prerequisite for a successful technology also tacitly recognizes that the results of a computation must be stored reliably and hence preserved against decay in time. However, if the information in the switched state relaxes or decays with time due to leakage, an error can occur. The relative importance of the this error and an error due to incomplete relaxation to the desired state, depends on the clock period[51]. The longer time allowed for switching, the more complete the relaxation to the final state, but the greater the chance of relaxation of the stored information. Thus, the volatility of charge stored on a node can restrict the interval of time required between clock cycles[37]. For example, the refresh cycle in a DRAM cell is sensitive to OFF state leakage current through a pass transistor[52]. Figure 10(a) is an illustration of the single-transistor DRAM cell that is used to store one bit of information. A leakage current of 1 pA through the pass transistor, which connects the capacitor $C \approx 50 fF$ storing information to the bit line for read and write operations, will discharge 1 V in about 50 ms. Thus, the minimum refresh time must be shorter than this. In current technology, an OFF device may be below threshold by $20kT/q \approx 0.50$ V to achieve a corresponding subthreshold current of approximately $I_{OFF} = (1 \ \mu A/\mu m) \times \exp[-V_t/S] \approx 1 \ pA/\mu$m, where $S = 37$ mV. If it takes time τ for charge to pass through an ON transistor; then in the OFF state, it will take approximately $\tau_{eff} \approx \exp[V_t/S] = 5 \times 10^4 \tau (\approx 50 ms)$, or a factor of 5×10^4 longer for charge to decay through the energy barrier presented by the transistor. However, if we attempt to reduce the threshold voltage to $V_t = 4kT/q \approx 0.10$ V, then the OFF state current is decreased by a factor of only 15 from the ON state, and consequently, it would take only $15\tau (\approx 15 \ \mu s)$ for the charge to leak through. Moreover, even a threshold voltage of 0.5 V would not be sufficient to guarantee a 50 ms refresh cycle if the chip is operating at 80C since $S = 44$ mV. At this elevated temperature $V_t \approx 0.64$ V to preserve the stored information.

While the data in a DRAM cell must be periodically refreshed because of leakage currents, static memory uses amplification and positive feedback to restore and maintain information indefinitely, as long as the power supply is available. Figure 10(b) shows a six-transistor SRAM cell. This static RAM cell consists of a flip-flop formed by two cross-coupled inverters, and two access transistors, T5 and T6. If the access transistors, T5 and T6, are turned OFF then the cell has two stable states that can be used to store a bit of information, a LOW state with T1 ON and T2 OFF and a HIGH state with T1 OFF and T2 ON. The cell is stable in either of these configurations as long as the node voltage is less than the logic threshold voltage. While SRAM is faster and more reliable, it costs more, consumes more power, the storage capacity is typically about a factor of 4 smaller than DRAM, and it is still susceptible to errors. Especially for low supply voltages and minimum cell areas, mismatches that occur between the transistors T1 and T2, or T5 and T6, e.g. due to geometry or threshold voltage differences, cosmic rays, alpha particles or 1/f noise may cause the cell to inadvertently switch states when the mismatch opposes the stored state[53,54]. For example, a 100mV mismatch in threshold voltage between T1 and T2 can cause a 0.9 V increase in the minimum supply voltage required for reliable operation[55].

Alternatively it may be possible to lower the threshold voltage to improve performance and yet suppress the subthreshold current at the same time by using an appropriate circuit design[56]. For example, there are proposals for circuits that can be incorporated into a DRAM design such as that represented in Figure 10(c) that use a switched supply which supplies power only during the active period and employs a level holder to maintain the output voltage when it is inactive. These schemes involve using a lower threshold voltage for speed and

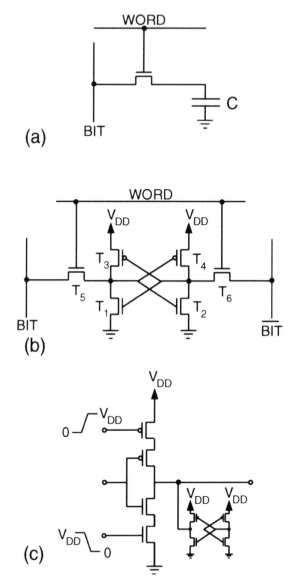

FIGURE 10. A schematic illustration of (a) a single-transistor DRAM cell formed by a pass gate and a trench or fin capacitor, and (b) a six-transistor CMOS SRAM cell which uses a static latch. While SRAM is faster and more reliable, it costs more, consumes more power and the storage capacity is typically about a factor of 4 smaller than DRAM. (c) A subthreshold current reduction circuit after Sakata *et al.*[56], that switches the power supply and stores an intermediate result using a latch.

then shortening the period in which a large subthreshold current flows and shutting-off the subthreshold current otherwise by using another switch. The other switch can be a MOSFET with an especially narrow width channel that limits the leakage or a MOSFET with an especially high threshold voltage. In this way the subthreshold current becomes independent of the threshold voltage and so the active current can be reduced, but there is an area penalty and additional cost associated with the increased circuit complexity just as there is for an SRAM over DRAM.

(3) **Power Dissipation.** The third prerequisite develops from the constraints imposed on VLSI by an upper bound on the operating temperature[57], and (for portable applications) by the lifetime of the battery. In the former case, the temperature reached by a chip dissipating a power P_{total} is determined by the thermal resistance of the package, i.e. $T = \theta P_{total}$, which depends on the ambient (air, water or immersion cooling), and the distribution of the heat production throughout the chip. Lowering the overall circuit density improves Θ and allows more power to be delivered to the component circuits which, in turn, permits faster operation. Similarly, more effective cooling by water instead of air, for instance, results in less heating. For example, $\theta < 50 C/W$ is a typical case to air thermal resistance for a $1\,cm^2$ chip, while $\theta \approx 20 C/W$ for a water cooled technology. On the other hand, if even the lowest thermal resistance – that associated with immersion cooling at boiling, $\theta \approx 0.3\text{-}1\,C/W$ – cannot satisfy the temperature specification, then the power dissipation must be lowered[58]. For battery power, not only must the overall power dissipation be lowered to accommodate the energy available from rechargeable batteries, but the rate at which the power is withdrawn must be budgeted, as well[59].

CMOS design strategy accounts for more than 50% of the IC market, primarily, because of one feature of the technology – low power dissipation. The total dissipated power can be written:

$$P_{total} = P_{dynamic} + P_{sc} + P_{static} + P_{passive}, \qquad (11)$$

where $P_{dynamic}$ is the dynamic power calculated assuming that the rise and fall times of the circuit are much less than the repetition rate, P_{sc} is the power dissipated during abrupt switching transient when both devices are ON for a short period of time, P_{static} is the power dissipated in sub-threshold MOSFETs leakage, and $P_{passive}$ represents the power dissipated by passive elements in the circuit, which is usually negligible and will be ignored.

To ascertain the relative significance of the various components to the dissipation let us consider the dynamical operation of the inverter shown in Figure 9. When the input is HIGH, i.e. $V_{in} \approx V_{DD}$, the N-MOS device is biased ON, the P-MOS device is OFF, and the output voltage is LOW, i.e. $V_{output} \approx 0$. If the input is switched abruptly LOW then the N-MOSFET is turned OFF, the P-MOSFET is ON, and the output voltage goes HIGH. During the transient, when the input is switched abruptly LOW, both the N-MOS and P-MOS transistors are ON for a short period of time, resulting in a short current pulse from V_{DD} to ground. This "short-circuit" dissipation is dependent on the rise and fall time, the load, whether or not the device is in velocity saturation, and the ratio of the threshold to the supply voltage, but for typical loads it is not appreciable. The dominant term in the dissipation, is usually associated with the current that charges and discharges the load capacitance. Charging the capacitor requires an energy $\delta E = C_L V_{DD}^2$ to extract the charge $Q = C_L V_{DD}$ from the power supply through the P-MOSFET. Initially, because of the abrupt application of V_{DD}, most of the voltage appears across the P-MOSFET. Ultimately, only the energy $\frac{1}{2} C_L V_{DD}^2$ is actually stored on the load capacitor, however. Consequently, about $\frac{1}{2} C_L V_{DD}^2$ of the energy is dissipated in the channel resistance of the ON P-MOS transistor. If the cycle is completed and the N-MOSFET is used to abruptly discharge the capacitor to return it to its initial state, and drive the node LOW again, that stored energy is dissipated similarly in the channel resistance of the N-MOSFET.

A total energy of $E_{sw} = C_L V_{DD}^2$ is finally dissipated during a complete switching cycle: half in charging and half in discharging the node capacitance, while no appreciable current flows through the gate except during the switching transient. There is no direct connection between the power supply and ground. Therefore, only the leakage current between reverse biased contacts and the substrate, and the subthreshold current contribute to the power dissipation in the time interval outside the transient. It is even possible to avoid dissipating this much energy during switching by implementing a scheme in which a linear or stepped increase in the input voltage is used in place of the abrupt switching transient[60–63]. For example, a linear increase in voltage over a time T where $T >> V_{DD} C_L / I_{ON}$ yields $E_{sw} = \int VI dt = \int V_{DD} (C dV/dt) dt = (V_{DD} C_L / T I_{ON}) \, C_L V_{DD}^2$ or a linear decrease in energy dissipation during the transient. While the number of computations per Joule is increased, the timing and redesign required to adopt such a scheme is complicated and the number of computations per second is compromised because of the slow voltage ramp required.

The dynamic power dissipation is related to the product of the energy dissipated per switching event and the cycling frequency, i.e. $P_{dynamic} = f_{clk} C_L V_{DD}^2$, for an inverter switched ON and OFF at a frequency f_{clk}. It is the dominate term in Eq. (11). Accordingly, the faster a circuit works, the more power it dissipates. Thus, a compromise between the speed and dissipation is fundamental to circuit design, and so the product of *delay* × *power dissipation* has become a common measure of performance. If the power-delay product is written as: $P_{total} \tau^{50\%} \approx C_L V_{DD}^2 \times (\tau^{50\%} f_{clk})$ the main advantage derived from CMOS becomes transparent; it is the combination of the low duty factor $\kappa = \tau^{50\%} f_{clk}$, and negligible static power dissipation[19]. See that for 0.35 μm technology, $\tau^{50\%} \approx 30 - 50 \ ps$, but the clock frequency is only projected to be 150–300 MHz, so that $\kappa = \tau^{50\%} f_{clk} < 0.01$. In contrast, for bipolar, emitter coupled logic, $\kappa \approx 1$, and for N-MOS $\kappa \approx 0.5$.[95]   Since power dissipation limits the scale of integration, and since the dissipation can be reformulated as follows, $P_{total} = f_{clk}(C_L V_{DD}/I_{Dsat}) V_{DD} I_{Dsat} \approx \kappa V_{DD} I_{Dsat}$, it becomes apparent that the activity is an important factor enabling VLSI.

To improve performance optimally, all the components of the dissipated power should be reduced. All four components in Eq. (11) can be reduced by lowering the power supply voltage, but supply voltage, is determined by the device technology, and cannot be lowered arbitrarily. The main impediments are the reluctance to scale the threshold voltage in a MOSFET due to process nonuniformities and the inability to scale sub-threshold swing at room temperature, and a reduction in the drive current which increases the time required to charge a capacitive load. Figure 11 illustrates that even if the power supply voltage and the saturation current are reduced according to the constant field program proposed by the SIA, the scale of integration is restricted to less than one million

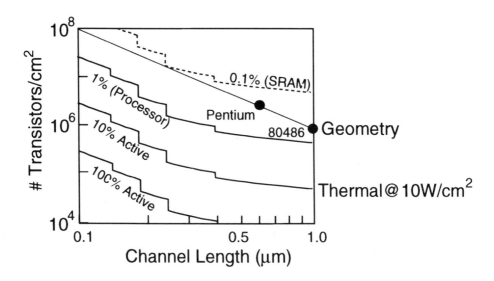

FIGURE 11. The thermal limitation associated with air-cooled technology (10 W/ cm²) on the scale of integration. We assume that each cell contains one transistor and that the number of lithographic units per cell is approximately 100 in accordance with microprocessor designs.

transistors by an air-cooled technology. This vexing limitation imposed by a maximum dissipation of 10 W/cm^2 is relaxed, however, when the devices on a chip are not all operating concurrently, i.e. when $\kappa < 1$, as is the case for a large logic circuit with lots of memory. Thus, with an activity factor between 0.1 and 1%, it is possible to produce a Pentium® microprocessor with about 4 million transistors per chip. But a successful design requires careful management of the chip area as well as the timing. In particular, notice that the integration scale of the Pentium® and 486 processors are estimated to be limited by the area consumed by the layout for κ in this range.

If the problem of dynamic power dissipation is resolved by using a complementary architecture with a low activity factor, then static power dissipation and OFF current become key measures of performance. Therefore, the size of the fluctuations in the threshold voltage in conjunction with the OFF-state subthreshold leakage specification become crucial to performance. In particular, the low voltage portion of the threshold voltage distribution is especially important since it disproportionally affects the total subthreshold leakage current and dissipation. For example, consider the stand-by power dissipation in DRAM which ranges from $P_{stand-by}^{o} \approx V_{DD} I_{OFF} = 0.25{-}10$ mW. Accounting for fluctuations in the threshold voltage increases the stand-by power because

$$P_{stand-by} = V_{DD} I_{ON}$$

$$\sum_{V_t} \frac{1}{\sqrt{2\pi}\,\sigma_{Vt}} \exp[-(V_{GS}-V_t)^2/2\sigma_{Vt}^2]\exp[(V_{GS}-V_t)/S],$$

$$(12a)$$

so that,

$$P_{stand-by} = P_{stand-by}^{o} \exp[\sigma_{Vt}^2/2S^2]. \qquad (12b)$$

Thus, devices with a threshold voltage about σ below the nominal threshold excessively contribute to the sum. If the worst case threshold is to be 0.4 V, but there are 0.1 V fluctuations, then the design threshold must be no less than 0.5 V to satisfy the same stand-by power specification, which has serious implications for reducing the supply voltage.

(4) Reliability. The fourth prerequisite develops from the desire to produce an answer, with a small chance of error, out of a large number of logical operations using a circuit constructed from components subject to probabilistic failures. The failures might be associated with the fabrication process, design mistakes, random component defects or external disturbances such as alpha particles or cosmic rays. Especially with VLSI, some failures may be accommodated by design through the use of redundancy techniques such as duplication logic, self-checking circuits, reconfigurable arrays and yield enhancement[64], but these techniques are not 100% effective, unfortunately. Nevertheless, provided that the chance of failure for a gate is sufficiently small, and that the failures are statistically independent, von Neumann guarantees that it is possible to reliably realize any function[65].

The chance of failure for communications is given by the bit error rate which is typically $BER \approx 10^{-15}$ per clock cycle, while for computations, $BER \approx 10^{-30}$ per clock cycle. Superficially, this may appear to be a very stringent reliability criterion, but it is not. Suppose that we have a computing system of 20 million identical devices with a BER of 10^{-30} switching at a frequency 1 GHz, and suppose that we build 1000 such systems expecting a lifetime of 10 years for each. Furthermore, if we assume that system failures occur when a single device fails, and that the device errors are independent and are not corrected by the architecture of programming, then the probability of zero system failures is: $P_{0 failures} \approx \exp[-BER \times 1 GHz \times 20M \times 1M \times 10 \times 3 \times 10^{7}] = 0.9940$, which is marginal performance for a system with a functional throughput rate, $FTR = 20M \times 1 GHz = 2 \times 10^{16} gate - Hz$. Thus, in practice, it is required that $BER \rightarrow 0$ or that the reliability approaches 100%, i.e. $1 - BER \rightarrow 1$, for a device, faster than $FTR \rightarrow \infty$ for the system.

To be economical, the materials and processes used in manufacturing must be designed at the outset to rectify reliability problems[66]. Defects in the aluminum used as interconnections and dielectric breakdown in insulators are just two examples. The failure rate of an aluminum wire due to electromigration-induced mass transport depends on the line dimensions[67,68]. Empirically, the lifetime is found to be proportional to the cube of the line-width, and to either the square or cube of the thickness, depending on the grain structure. Right now, for 0.35 µm rules with 4–5 levels of metal, the lines are 0.4 µm wide and thick and there is about 60 m of wiring per cm^2, but for 0.1 µm designs, we anticipate 6–7 levels of metal with lines 0.12 µm wide and thick, and 300 m of wiring per cm^2. The smaller dimensions of the lines coupled with the larger extent suggests that miniaturization and increasing levels of integration will lead to a collapse of interconnect reliability.

The impetus to scale the gate-oxide thickness in order to deliver the current drive and enhance circuit performance for low supply voltage implies a degradation in reliability[69], as well. The principal failure mechanisms of the oxide are time-dependent dielectric breakdown (TDDB)[70,71] and tunneling. These two modes are interrelated and are especially sensitive to defects introduced at the Si/SiO_2 interface during processing. The minimum oxide thickness for a technology is given by the maximum field in the oxide:

$$t_{ox}^{min} = (V_t - V_{FB} - 2V_B)/F_{ox}^{max} . \qquad (13)$$

Conversely, the maximum supply voltage is determined by the thickness of the oxide in a technology. Fields larger than $F_{ox}^{max} = 13.5$ MV/cm have been measured for oxides approximately 10 nm thick, while current state-of-the-art limits the maximum field in production to values between 5–8 MV/cm with $t_{ox} \approx$ 8–10 nm. This limitation is presently dictated by the TDDB phenomenon.

Breakdown eventually occurs as a result of current injection into the oxide. The gate voltage needed to maintain a constant current in the oxide increases with time until breakdown, at which point the voltage required collapses catastrophically. The gate voltage increase comes about because injected carriers fill traps that either are already in the oxide or induced by the stress. The charge injected into the oxide, in turn, causes shifts in the threshold voltage[72]. An especially sensitive parameter that is used to characterize oxide breakdown is the charge-to-breakdown,

$$Q_{bd} = \int_0^{t_{BD}} J dt \approx J t_{bd},$$

where J is the current density in the oxide and t_{bd} is the time necessary for breakdown to occur.

For voltages larger than 4 V and $t_{ox} < 5$ nm, the current is described by Fowler-Nordheim tunneling[73–75] while direct tunneling is observed for voltages < 4 V; the physical distinction between the two is illustrated by Figure 12. Direct tunneling (DT) presents a trapezoidal barrier to tunneling with a field independent thickness, whereas Fowler-Nordheim tunneling (FN) presents a triangular barrier with a thickness that depends on the applied voltage. If we ignore, for mathematical economy, the depletion of the polysilicon gate and the quantum mechanical thickness of the inversion layer, and assume that the gate voltage is entirely dropped across the insulator, the current density in the DT regime can be written as:

$$J = \frac{AF_{ox}^2}{[1-(1-qV_G/\phi)^{1/2}]^2} \exp[\frac{8\pi\sqrt{2m_{ox}}}{3hqV_G}\phi^{3/2}((1-qV_G/\phi)^{3/2}-1)t_{ox}] \quad (14)$$

$$= \frac{AF_{ox}^2}{[1-(1-qV_G/\phi)^{1/2}]^2} \exp[\frac{\phi((1-qV_G/\phi)^{3/2}-1)}{qV_g}\frac{t_{ox}}{\Lambda}],$$

where A is a constant, $\Lambda = 4\sqrt{2m_{ox}\phi}/3\hbar \approx 0.1-0.2 nm$ is the evanescent decay length of the electron, $F_{ox} = V_G/t_{ox}$ is the field in the insulator, ϕ is the barrier height, and m_{ox} is the effective mass in the oxide bandgap. The barrier height and the effective mass are usually determined empirically by a fit of Eq. (14) to experimental data; typically for thick barriers $\phi \approx 3.1-3.2$ eV and

$m_{ox} \approx 0.42 - 0.38 m_e$, where m_e is the free electron mass. For high fields $qV_G = \phi$, the familiar Fowler-Nordheim (FN) equation for tunneling is recovered from Eq. (14), but in the DT regime the current is enhanced over the FN prediction, owing to the seemingly insignificant change in the barrier from triangular to trapezoidal[71]. The current enhancement still shows an exponential dependence on

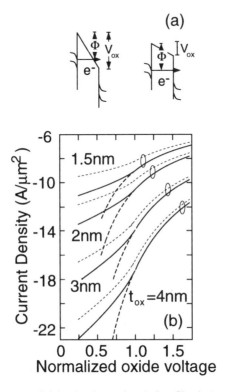

FIGURE 12. (a) The potential barrier in an insulating film between two electrodes for $V \approx 0$, $V < \phi_o/q$; and $V > \phi_o/q$. (b) The tunneling current through a large (≈ 100 μm $\times 100$ μm) MOS capacitor structure as a function of the applied voltage normalized by the barrier height, ϕ for t_{ox} = 1.5, 2, 3, and 4 nm. No accounting is made for the depletion of the polysilicon gate or the quantum mechanics of the inversion layer. The solid lines result from a calculation in which interfaces are atomically smooth, while the dotted lines represent the same calculation assuming Gaussian roughness of 0.25 nm-rms at one interface. The dashed lines indicate the current due to Fowler-Nordheim tunneling alone without direct tunneling. The same prefactor for the tunneling current is used for all of the calculations and it is made to be consistent with experimental data obtained from a 3nm oxide.

the insulator thickness, but the exponential gate voltage dependence is weaker.

An established correlation between the tunneling current and TDDB lifetimes for $t_{ox} \geq 4$ nm dictates[71] an acceptable current level to be at $J < 0.1\ pA\mu m^{-2}$. If we assume that the charge necessary for breakdown is independent of the gate voltage, then the TDDB lifetime will be proportional to the reciprocal of the tunneling current. For example, Hu[76] found that: $t_{TDDB} = \tau(T) \exp[G(T) t_{ox}/V_G] \approx 10^{-11} \exp[350(MV/cm) t_{ox}/V_G]$ at 300K. So, a 5 nm thick oxide has a 10 year lifetime as long as the operating gate voltage is less than about 4 V. If the above mentioned criterion on the leakage current in the FN regime still applies in the DT regime, then the onset of direct tunneling essentially determines the thinnest practical oxide. We estimate that the minimum oxide thickness for 1.5 $V < V_{DD} < 2.2\ V$ to be $t_{ox} \approx 3$ nm or else the supply voltage must be derated[77].

Beyond these guidelines set by reliability, there is no well accepted theory of the intrinsic breakdown strength of gate oxides, and so the extent of improvement that is still possible is unknown. For example, it has been observed that small regions of a 10 nm thick oxide can support 40 MV/cm[78]. Moreover, while it is assumed that the onset of direct tunneling near 3 nm establishes the minimum SiO_2 thickness that can be tolerated, it does not necessarily represent a limitation on circuit design. For example, with an exponentially increasing risk of a timing failure due to stored charge leaking through the gate capacitor, a polynomial improvement in performance, measured by the increase in saturation current for example, can be achieved by thinning the oxide. If the criterion for leakage is 1 $pA/\mu m$, similar to that discussed in connection with DRAM in Section II.A.2, then the minimum oxide thickness could be less than 2.5 nm at 1.2 V, according to Figure 12, or in the case of a high performance processor where the criterion is 1$nA/\mu m$, the thinnest oxide would be 2 nm at 1.2 V. In particular, Momose et al.[79] recently demonstrated that 0.1 μm MOSFETs with an oxide 1.5 nm thick are practical and enhance I_{Dsat} over more conventional designs.

For such thin oxides it is expected that asperities that arise from defects in the oxide, from contamination prior or subsequent to the growth, from subsequent processing, or from non-stoichiometric or rough interfaces will affect the reliability and yield. Asperities that are smaller in amplitude than t_{ox}, or fluctuations in the barrier height associated with defects behave effectively like punctures and lead to an enhanced tunneling current. It has been demonstrated that a Si surface roughened by cleaning produces a decrease in the breakdown yield of MOS capacitors[80], while preliminary studies indicate that smooth interfaces $h_{ox} \approx 0.15$ nm-rms enhance gate oxide integrity in the thickness range 2.3 − 6.4 nm[81,82].

The intrinsic roughness of the silicon surface prior to gate oxide growth is determined by factors such as the production method (epitaxy, CZ, Float Zone), chemo-mechanical polishing, and pre-gate oxide cleaning. The interface roughness, once the oxide is grown, can be affected by factors such as particulate

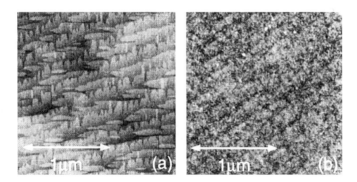

FIGURE 13. AFM images of an epitaxial silicon surface area $2\mu m \times 2\mu m$ before (a) and after (b) growth of a 2 nm thick oxide (courtesy of B.J. Sapjeta). Prior to the rapid thermal oxidation at 1050C, the surface of the silicon was cleaned with NH_4OH and H_2O_2 at room temperature, followed by an HF dip in 15:1 $HF:H_2O$. The rms roughness of the epi-silicon prior to oxidation is 0.06 nm, while subsequent to oxidation and an HF dip to remove the oxide, we found the silicon to be rougher, 0.16 nm-rms. (See color plate.)

organic and metallic contaminants and the termination of the silicon. Figure 13 exemplifies the perfection of the silicon surface, which is currently routinely achieved before and after oxidation. While the root mean square (rms) roughness after oxidation is only about 0.2 nm, the range of variations observed in the 4 μm^2 area shown is about 1–2 nm. X-ray scattering[83], STM, and AFM[81] have provided similar estimates for the interface roughness, $0.14 < h_o < 0.28$ nm, on thinner oxides (2.3 nm $< t_{ox} <$ 10 nm). However, the surface roughness inferred from transport measurements in long channel MOSFETs gives a standard deviation of $h_o \approx 0.43$ nm with a lateral extent of 1.5 nm[84–86].

A realistic appraisal of the lower bound on t_{ox} should incorporate the effects of interface roughness on the reliability. Following Krylov and Suris[87] we can refine our estimate of the minimum oxide thickness in the presence of asperities assuming that a single rough interface determines the tunneling transmittance. If we consider a random interface profile, $h_{ox}(\rho)$ in a MOS structure, where ρ is a two-dimensional coordinate and where $<h_{ox}(\rho)>=0$, then Eq.(14) can be reformulated as follows:

$$J_{tot} = \frac{AF_{ox}^2}{[1-(1-qV_G/\phi)^{1/2}]^2} \exp[((1-qV_G/\phi)^{3/2}-1)(t_{ox}-h_{ox}(\rho))/\Lambda]$$

$$= J\exp[\frac{-\phi((1-qV_G/\phi)^{3/2}-1)}{qV_g} \frac{h_{ox}(\rho)}{\Lambda}], \tag{15}$$

where J is the tunneling current through a smooth film given by Eq. (14). If $h_{ox}(\rho)/\Lambda \gg 1$, then the spread in the local current density through the oxide is exponentially large[88] and there exists local minima in the film thickness which effectively resemble punctures. The current through such protrusions is determined, not only by the tunneling probability at the thinnest point which we identify as $\rho = 0$, but also by the probability to find such a protrusion in that area. This optimum protrusion and tunneling in the immediate neighborhood of this optimum, essentially determine the average current. If $h_{ox}(\rho)$ obeys Gaussian statistics, the largest contribution to the current is derived from the maximum of the tunneling exponent which can be shown, using the method of steepest descent, to be:

$$J_{tot} = J \exp \left[\left(\frac{\phi((1 - qV_G/\phi)^{3/2} - 1)}{qV_g} \right)^2 \frac{h_{ox}^2}{2\Lambda^2} \right]. \qquad (16)$$

with exponential precision.

According to Eq. (16), the roughness gives rise to an exponential enhancement of the leakage current through the oxide. Another way to look at the roughness is effectively as a reduction of the average thickness of the oxide by an amount proportional to h_{ox}^2/Λ. Using the approximations in Eq. (16) for the leakage current, we can estimate the effect of interface roughness on the current by substituting for Λ whether we are in the FN or DT regimes. The results, using a rms roughness of 0.25 nm, are plotted in Figure 12 alongside the calculations which ignore interface roughness altogether. This difference, which can be more than a factor of 10, is only an indication of the range of improvement still possible in the oxide associated with a large area device. Since shrinking the area of the device to a size comparable to the lateral extent of the protrusions would affect the statistical variations observed from device to device[88], Figure 12 represents a lower bound on the possible improvement since the polysilicon-SiO_2 interface is rougher still.

III. Limiting Performance of CMOS: The Target to Exceed

These prerequisites allow us to infer the limiting performance of CMOS technology. An accurate forecast of the limiting performance is necessary not only to identify the main impediments, but to spur the development of alternative technologies, as well[23]. Currently, the primary means for improving performance and increasing the scale of integration on a chip are: (1) growing the size of the die; and (2) miniaturization of the individual electronic devices and the wires which interconnect them. The SIA roadmap projects that in the future gains will continue to accrue from both of these two approaches, implicitly assuming that each

TABLE 3. Overall Roadmap of Technology characteristics adopted from the SIA including a threshold voltage inferred from a worst-case scenario including manufacturing tolerances, using the leakage criterion given by the SIA and stated in the table and assuming a subthreshold slope of 44 mV per e consistent with 100C.

	1995	2001	2007	2010
Feature size (nm)	350	180	100	70
Gate CD control (nm)	35	18	10	7
Overlay control (nm)	100	50	30	20
t_{ox}(nm)	7–12	3–5	<4	<4
−front end roughness(nm−rms)	0.2	0.1	tbd	tbd
$V_t(V)$	>0.45	>0.43	>0.4	>0.4
$I_{OFF}(pA/\mu m$ at 100C)	5–10	20	30	40
short−wire pitch (μm)	0.6–1.05	0.33–0.54	0.16–0.30	0.12–0.25
Interconnects				
−R(Ω/μm)	0.15	0.29	1.34	1.34
−C(fF/μm)	0.17	0.21	0.27	0.27
Gates/chip				
−logic	4 M	13 M	50 M	90 M
−DRAM	64 M	1 G	16 G	64 G
Wafer processing cost (*dollars*)	3.90	3.70	3.50	3.50
Chip size (cm^2)				
−Logic	2.50	3.60	5.20	6.20
−ASIC	4.50	7.50	11.0	14.0
−DRAM	1.90	4.20	9.60	14.0
Electrical Defect Density (*defects*/cm^2)	2.40	1.40	1.00	0.25
Maximum power (W/die)				
−high performance	80	120	160	180
−Logic without sink	5	10	10	10
−portable	2.5	3	4	4.5
Power supply (V_{DD})				
−desktop	3.3	1.8	1.2	0.9
−portable	2.5	0.9–1.8	0.9	0.9
Clock Frequency(MHz)				
− on chip	300	600	1000	1100

will continue to provide the same economic advantages as before. Table 3 represents the proposed scaling scenarios for four technologies taken from the SIA roadmap, and it also indicates the respective performance improvements expected from these extrapolations.

There is no compelling physical reason why these performance targets cannot be reached and, in fact, the performance might continue to improve even beyond a die size of 14 cm^2 and a minimum feature size of 0.07 μm projected for 2010. For example, the effective defect density is expected to continue to improve beyond 25 m^{-2} using yield enchancement strategies, and correspondingly we naively expect that the die area will grow in proportion to the reciprocal of D_o. Moreover, room temperature operation of MOSFETs with a channel length of 40 nm has already been demonstrated, and the utility of gate oxides 1.5 nm thick[79] and p-n junctions 10 nm deep[89] has been shown. However, there are compelling economic reasons why the performance targets in Table 3 might not be attained. Below we follow an argument presented by Warwick et al.[23] which indicates that the growth of the die will be economically exhausted at 10 cm^2 because of the cost of processing. Moreover, looming problems with circuit margin and process control, reliability, and power dissipation indicate that economical miniaturization may be exhausted beyond 0.1 μm. Thus, the euphoria surrounding the research into extreme sub-micron devices should be tempered by the requirements of manufacturing.

A. Growing an IC or Die

According to Warwick et al.[23], three competing costs determine the optimal die size: the cost of packaging, processing, and increasing yield[90]. The cost of packaging increases with the number of pin-outs which empirically is inversely proportional to the perimeter, $\sqrt{A}$ according to Rent's rule[91,92]. The cost of processing is area independent, while the cost of waste depends linearly on the area because the yield is related to the product of the defect density and the critical area in the following way: i.e. $Y = \exp[-D_o A]$ where D_o is the defect density. Thus, a chip with an area many times larger than $1/D_o$ will probably be flawed, or instead the area must be kept smaller than a $3/D_o$ if a single flaw will kill the design. Figure 14 shows the cost per die and its constituent elements versus the area of the die for an optimistic representation of the yield. With an increase in the size of the die, packaging costs (a) are reduced until defects (b) cause yield to fall (and cost to rise) precipitously.

If the optimal die size is the most economical one, then currently the total cost of a finished chip is about twice the processing cost (to the user) or 20 dollars/ cm^{-2}, which corresponds to a maximum die area of 1.3 cm^2, but if the defect density is reduced as anticipated, the cost of packaging and waste will be subsumed into the total cost, which is then expected to approach the cost of

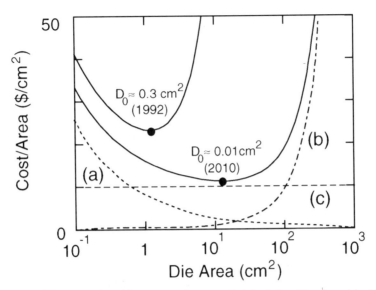

FIGURE 14. The cost of a chip versus die area at defect densities found in 1992 and anticipated in 2010. The cost increases dramatically as the size departs from the optimum value either because of packaging costs (a) or waste (b). In 2010, the cost is dominated by processing (c). Adapted from C. Warwick *et al.*[27].

processing the silicon (c) as indicated in the figure. Notice that the die area increases more slowly than the reciprocal of the defect density[23], D_o^{-1}, however. Currently, $D_o \approx 0.3$ cm^{-2} while the effective D_o^{eff} is 0.0240 m^{-2} given in the SIA projections which takes into account the effect of redundancy and repair. By the year 2010, D_o should approach 0.01 cm^{-2}, while $D_o^{eff} \to 0.001$ cm^{-2}. The former value for D_o translates to an economical die size around $A \approx 10$ cm^2 as shown in the figure. This is only about a factor of 4 larger than a microprocessor die in current technology, but it requires a thirty fold improvement in D_o[23,93]. Thus, unless the cost of processing silicon falls precipitously, owing to a breakthrough in technology, the economic advantage to growing the die will be exhausted with 10–15 years.

B. Scaling VLSI and the Tyranny of Large Numbers

The guiding principal for the performance improvements expected through miniaturization has been scaling. The classical equations, like Poisson's equation and the current continuity equation, which govern the design of a MOSFET are analytically intractable, especially for deep submicron devices, principally because the problem is three dimensional and can range over nine orders of magnitude in current. Solving for the electromagnetic fields surrounding a multi-level wiring scheme on a chip by using Maxwell's equations is a computationally intensive problem that is just as formidable. Scaling is a method in which the classical transport equations, which describe the operating characteristics of an existing device or wire, are scaled to smaller dimensions to predict the characteristics of a smaller device or wire with minimal computational and design investment.

For example, according to constant-field (CF) scaling[94], the shape and magnitude of the internal electric field in the existing device or wire remains the same in the scaled device, while the horizontal and vertical dimensions of the miniaturized version are scaled by $\kappa^{-1} < 1$. Consequently, the oxide thickness t_{ox}, the junction depth, r_j, the channel length and width are all supposed to scale by the factor κ^{-1} in a MOSFET. And, if the depletion width $D = \sqrt{2\varepsilon_s[V + \Psi_B]/qN_A}$ is to be scaled by κ^{-1} then all the operating voltages as well as the threshold voltages and the currents must be scaled by κ^{-1} as well. On the other hand, the doping density N_A must be scaled by κ. Table 4 lists some of the relevant physical quantities with the corresponding scaling factor according to the CF rule.

TABLE 4. Constant field scaling after Dennard *et al.*[94].

Quantity:	Scaling factor
Device and wire dimensions: e.g. L, W, t_{ox}, r_j	$1/\kappa$
Packing density: e.g. number of devices per die	κ^2
Doping concentrations: e.g. N_A	κ
Bias voltages: e.g. V_t, V_{DD}	$1/\kappa$
Fluctuations in voltages: e.g. σ_{Vt}[108]	$\kappa^{-3/4}$
Bias currents: e.g. I_D	$1/\kappa$
Power dissipation for a circuit: e.g. inverter	$1/\kappa^2$
Power dissipation per area:	1
Capacitances: e.g. C_{ox}	$1/\kappa$
Charge: e.g. Q_i	$1/\kappa^2$
Electric field: F	1
Transistor transit time: e.g. τ	$1/\kappa$
E_{sw}	$1/\kappa^3$
FTR: e.g. gate-Hertz	κ^3

The advantages of CF scaling are that field-induced breakdown in the oxide or semiconductor can be avoided, while the power-delay product improves. Both the current and voltage scale like κ^{-1}, consequently, the power dissipation scales like κ^{-2}. Although the capacitors and the current driving them scale by κ^{-1}, the rate at which capacitors are charged will not change, and so the time needed to charge them will be scaled down by κ because of the lower final voltage and much lower final charge. Consequently, the power-delay product scales by κ^{-3}. However, the device area scales by κ^{-2}, so the dissipation per unit area remains constant. Thus, it can be shown that through scaling the performance improves between technologies. The improvement is measured by a reduction of total energy dissipation with a concomitant increase in speed, density and functionality on a chip.

The SIA roadmap is similar to a CF rule. For example, the power density is approximately constant as the feature size is reduced from 0.35 μm to 0.1 μm, because the the power supply is reduced from 3.3 V to 1.2 V, and the clock frequency is increased from 300 MHz to 1 GHz. The functional throughput follows the CF rule, as well, since the anticipated improvement is 43× by 2007. Today, using 0.35 μm line-rules in VLSI we have microprocessors with 4 M MOSFETs and 64 M DRAM operating essentially flawlessly at a clock frequency of 150–300 MHz. By the year 2007, using 0.10 μm line-rules, we anticipate Giga-scale integration (GSI) in microprocessors with 50 M logic gates/chip and memory with approximately 16 billion transistors on a chip operating at 0.5–1 GHz with improved system reliability.

(1) The Wires. Scaling the interconnects into the 0.1 μm regime and beyond has been examined theoretically[95]. To improve packing density, the interconnection dimensions are usually scaled down by the same factors as the transistors, but the effect of the interconnections on the time delay cannot be reduced by simply reducing the scale according to a CF rule. It can be shown[95] that the delay can be represented approximately as: $\tau = R_w C_w + 2.3(R_{ch} C_w + R_{ch} C_L + R_w C_L)$ where R_{ch} denotes the transistor channel resistance which is approximately $(L/W)/\mu C_g (V_{DD} - V_t)$, and where R_w and C_w denote the wiring resistance and capacitance respectively. As the width of the wire decreases, the wiring capacitance decreases, but the resistance increases by the same factor and eventually becomes comparable to or larger than the transistor channel resistance as shown in Figure 15. Although the resistance of the polysilicon used for local interconnections is large relative to aluminum, the lengths are small so that the RC delay associated with local wiring is not really a limitation. However, as the chip area increases, the global interconnection length increases and so the distributed RC delay of the longest lines degrades. Thus, for each new technology an optimal layout is chosen in which both the horizontal and vertical pitches for global interconnections increase, *opposite to the scaled reductions in the local interconnections,* while the number of wiring levels increases. The reverse scaling of the

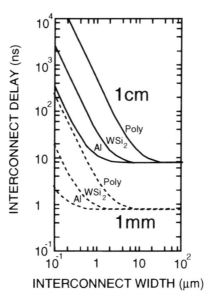

FIGURE 15. The interconnection delay for 1 cm and 1 mm long lines as a function of line width for three materials Al, WSi_2, and polysilicon assuming a channel resistance of $R_{ch} \approx 1 k\Omega$. It is assumed that the spacing between lines is equal to the width, that the interconnection thickness is a third of the width, and that the dielectric thickness is about a fifth of the width. Constant field scaling is applied and we assume that $\rho_{Al} = 3\mu\Omega cm$, $\rho_{WSi_2} = 30\mu\Omega cm$, and $\rho_{Poly} = 500\mu\Omega cm$, for the respective resistivities. Adapted from Bakoglu[95].

horizontal and vertical pitch is required to minimizing the RC delay associated with the interconnections.[96]

(2) **The MOSFET.** Scaling the MOSFET to the 0.1 μm regime and beyond has been examined theoretically down to a 0.03 μm channel length[46,97–101] and accomplished experimentally to the proof-of-concept stage down to a channel length of 0.04 μm[89,102–104]. From these studies, it has become apparent that the characteristics of future technologies cannot follow a CF rule because subthreshold leakage prevents us from lowering the threshold voltage. Yet, future technologies cannot follow the roadmap given in Table 3 either, and improve performance at the same time. In particular, the primary motivation for shrinking design rules, which is to improve performance for constant power consumption at the same cost, is diminished for channel lengths less than 0.1 μm because of velocity saturation and the resultant limitation on I_{Dsat} at low power supply[105]. If we impose on the gate delay the constraint that the dynamic power dissipation per area remains constant: i.e. $P_{dynamic}/(\kappa_{active} V_{sat} C_{ox}) = V_{DD}(V_{DD} - V_t - V_{Dsat})$ is

constant then there is a minimum delay that occurs for a power supply of $V_{DD} \approx 0.9 V$ associated with technologies below 0.1 μm. So, unless the power supply and threshold voltage both decrease at the same rate, the dissipated power density must increase to improve speed. However, the maximum power supply voltage has to be derated to preserve hot carrier and gate oxide reliability, while the minimum power supply, determined by the noise margin and the immutable threshold voltage, remains at 1.2–1.5 V. A reduction in the threshold voltage is therefore central to the preservation of CF scaling and improving performance.

To reduce the worst case threshold voltage, the variations in the threshold voltage must be reduced or the tolerance of subthreshold leakage should increase. Currently, for 0.5 μm logic, the 3σ specification on a 0.5 V threshold voltage with a 3.3 V supply is approximately 10%[49] or about 50 mV in order to keep the ratio of variations in gate delay at about 10%. But for the lower supply voltages required by the CF rule and the larger normalized margin required for 0.1 μm technology, the threshold voltage control must improve fantastically. For 0.1 μm VLSI, we must consider at least 6.7σ variations and, according to the results of Figure 9, for $V_{DD} = 1.2$ V, the 6.7σ variation in V_t must be about 8% or $\sigma_{V_t} \approx 6$ mV if the threshold voltage is kept at 0.5 V.

This extrapolation represents an upper bound on the distribution of the threshold voltage, because the model assumes the transistor is operating in velocity saturation[107]. Yet, the required process control is Homeric, especially when compared with the statistical fluctuations observed in the threshold voltage of deep submicron MOSFETs due to fluctuations in the channel length, and in the number and distribution of dopants in the active region of the device[108–110]. Fluctuations in the channel length develop from variations in the exposed resist that are transferred and amplified by the gate stack etch. The fluctuations in the number of dopants arise from both random ion diffusion and implantation processes. Figure 16 shows the threshold voltage distribution measured by Mizuno et al.[106] in 8,000 nominally identical $L = 0.1$ μm and $L = 0.5$ μm channel length MOSFETs with a peak channel doping of about 10^{18} cm^{-3}, an oxide thickness of $t_{ox} = 4$ nm, and a channel width of $W = 1.3$ μm. These fluctuations, approximately $\sigma_{V_t} = 10 mV$ for 0.5 μm technology and $\sigma_{V_t} = 35 mV$ for 0.1 μm technology, were observed in devices on the same chip manufactured simultaneously. It is important to understand the significance of such a small number to the yield of a low voltage circuit. Out of 1 billion 0.1 μm transistors comprising a chip operating at 1.2V with a 0.4V threshold voltage, 630,000 devices would be outside the $\pm 3\sigma$ range, even if σ were taken to be 10% of the threshold voltage.

The distribution of threshold voltages found in the $L = 0.5$ μm MOSFETs has been attributed to statistical fluctuations in the number of channel dopants[111]. This is especially worrisome because these fluctuations are considered to be intrinsic to the device. According to Eq. (2), the variation in the threshold voltage, δV_t, is the square root of the sum of the squares of the variations due to the flat-band voltage, the surface potential, the gate capacitance, and the depletion

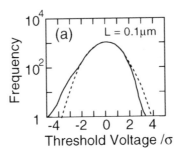

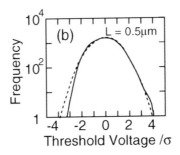

FIGURE 16. The V_t distribution of 8000 MOSFETs with $t_{ox} = 4$ nm, $W = 1.3$ µm, a peak channel doping of 1×10^{18} cm^{-3}, and gate lengths of L = 0.10 µm and (b) 0.50 µm obtained from Mizuno *et al.*[106] In (a) the origin is associated with $\bar{V}_t = 0.082\,V$ with $\sigma_{Vt} = 35\,mV$, while in (b) the origin occurs at $\bar{V}_t = 0.21\,V$ with $\sigma_{Vt} = 9.6\,mV$.

charge layer. Mizuno, Okamura and Toriumi[112,113] have shown experimentally that, in the devices they have fabricated[108], the dominant term is the latter one associated with the depletion charge. Consequently, based on the one-dimensional model for the threshold voltage given in Eq. (2), we expect that:

$$\sigma_{Vt} = \frac{Q_B t_{ox}}{2\varepsilon_{ox}} \frac{\delta N_A}{N_A} = q^{3/4} \frac{\varepsilon^{1/4} \Psi_B^{1/4}}{\sqrt{2}\,\varepsilon_{ox}} t_{ox} N_A^{1/4} W^{1/2} (L-D)^{1/2} \quad (17)$$

and so the width of distribution should scale linearly with the oxide thickness (as observed[108]), but should show a much weaker $N_A^{1/4}$ dependence on doping concentration (also as observed[108]), a $\sqrt{W}$ dependence on transistor width, and $1/\sqrt{L}$ dependence on gate length. Following a CF scaling rule, the width of the threshold voltage distribution is expected to scale as $\kappa^{-3/4}$. Consequently, if $\sigma_{Vt} \approx 10\ mV$ can be attributed to dopant fluctuations in a 0.5 µm design, then

$\sigma_{Vt} \approx 3$ mV for 0.1 µm, which is already comparable to the process control necessary to guarantee a 10% delay margin, but still smaller than the observation made by Mizuno *et al.* and represented in Figure 16(b).

Notice that both of the threshold voltage distributions in Figure 16 are asymmetrical and deviate from a Gaussian; the frequency of lower threshold voltages is much higher for the 0.1 µm MOSFETs, whereas the frequency of the lower threshold voltages is much smaller for the 0.5 µm transistors. Moreover, the width of the 0.1 µm distribution, $\sigma_{Vt} \approx 46$ mV, is 2.3× larger than that extrapolated using a $\sigma_{Vt} \propto 1/\sqrt{L}$ rule and the results obtained from measurements using $L = 0.5$ µm MOSFETs. The voltage distribution function, that characterizes the $L = 0.5$ µm measurements, is derived from the observed Gaussian distribution of channel dopants[108]:

$$f(V_t) = \frac{1}{\sqrt{2\pi}\,\sigma_{Vt}} \exp[-(V_t^2 - \overline{V}_t^2)/8\,V_t^2\,\sigma_{Vt}^2] \qquad (18)$$

where $\overline{V}_t$ is the mean threshold. When $\sigma_{Vt}/V_t \ll 1$, Eq. (18) is approximately Gaussian, but if either V_t is small or σ_{Vt} is large, the distribution becomes asymmetrical with a *lower* probability of a lower V_t compared to a Gaussian. This trend is the opposite to the observations made for $L = 0.1$ µm MOSFETs, however. Mizuno *et al.*[106] attribute the fluctuations observed for $L = 0.1$ µm predominately to a short channel effect that derives from fluctuations in the channel length, even though the channel length distribution was characterized by $\sigma_L \approx 6.5$ nm.

Currently, 0.5 µm technology, which has been sold for revenue, has gate CD control at post etch of 50 nm with overlay control of 150 nm. The 0.35 µm and 0.25 µm technologies, that are currently being used for product development, require 35 nm and 25 nm gate CD control and 100/75 nm overlay respectively. Thus, it is anticipated that the 0.1 µm and 0.07 µm generations will require 10 nm/7 nm CD control and 30 nm/20 nm overlay to be profitable. But, the above data suggests that the naive extrapolation of the 0.5 µm criterion for CD control may not be enough to guarantee yield at 0.1 µm. In particular, if $\sigma_{Vt} = 46$ mV, then we expect a range of thresholds at least 0.25 V wide which leads to a factor of 4 variation in gate delay with a 0.4 V threshold voltage and a 1.2 V power supply. Instead of 10 nm CD control, $\sigma_L \approx 1 - 2$ nm is required to reduce the magnitude of σ_{Vt} comparable to that due to statistical dopant fluctuations[106]. Consequently, even though the effects of fluctuations in channel length are considered to be extrinsic to MOSFET operation, the cost of the lithography, which currently represents about 35% of the manufacturing cost, may become onerous if the required CD control must improve by a factor of 10. On the other hand, if cost is not a consideration, then even "intrinsic," atomistic fluctuations in the dopant distribution might be eliminated through a better device design. For example, the variance in the threshold voltage due to dopant distribution could be reduced by removing the dopants altogether and using another gate to control the workfunction instead[97] or by removing the charge needed to induce the electric field at the interface to a depletion layer about a delta-doped region, well separated from the interface (see Section IV.B below.)

Variations in the threshold voltage are likewise susceptible to the circuit design choices. For example, the analysis that gives rise to Eq. (2) for the threshold voltage was based on Gauss's law, assuming that the potential of the source and drain were the same. However, if the drain potential is large, it can act as a second gate and modulate the amount of charge Q_B in the channel. The charge modulation affects the barrier potential between the source and drain causing an apparent lowering of the threshold voltage and a punchthrough current[114]. Drain-induced-barrier-lowering (DIBL) is supposed to depend exponentially on the channel length, depletion charge and oxide thickness according to:

$$\delta V_t = \frac{2\dfrac{\varepsilon_s}{\varepsilon_{ox}}t_{ox}}{D + \dfrac{\varepsilon_s}{\varepsilon_{ox}}t_{ox}} \exp\left[-\pi L/4\left(D + \frac{\varepsilon_s}{\varepsilon_{ox}}t_{ox}\right)\right]\delta V_{DS}. \quad (19)$$

Thus, the barrier lowering and the concomitant change in the threshold voltage become especially important as the channel length decreases or as the supply voltage increases.

According to most scaling scenarios, as the channel length decreases, the substrate doping concentration is supposed to increase to prevent barrier lowering effects such as DIBL or punchthrough, and the oxide thickness is reduced so that the field required to invert the substrate is maintained. For example, it has been proposed that, to mitigate the effects of DIBL[115], the substrate concentration scale as κ^2, while the oxide thickness and channel length scale as κ^{-1}. This scenario minimizes DIBL, but the standard deviation of the threshold voltage, due to statistical variations in the doping, increases to 39 mV for a 0.07 μm channel length with $W/L = 4$ and $t_{ox} = 3$ nm which, in turn, gives rise to intolerable (100%) variations in the saturation current and in the delay for a power supply voltage of 0.9 V[115]. Ultimately, there is no room for compromise. As we have already seen, there is a minimum oxide thickness, which corresponds to the maximum electric field that the oxide can support, before breakdown occurs and reliability degrades. Moreover, there is a maximum substrate doping, which correspond to the maximum band-bending, before tunneling occurs in the space charge region of the reverse drain p-n junction which comprises the drain contact. Therefore, the suppression of DIBL and punchthrough eventually lead to either an increase in the variations of device parameters such as the threshold voltage, or breakdown of the reverse-biased drain junction or breakdown of the gate oxide.

Junction breakdown occurs via tunneling to the substrate. Tunneling occurs through the barrier in the space charge region of a reverse biased p-n diode. The tunneling transmittance depends exponentially on the barrier height, E_g, and width. Following Kane and Blount[116,117] and ignoring variations in the tunneling length due to statistical fluctuations in the doping, the tunneling current can be written:

$$J_t = \frac{\sqrt{2m^*}\, q^3 E_j V}{4\pi^3 \hbar^2 \sqrt{E_g}} \exp[-\frac{4\sqrt{2m^* E_g^3}}{3qE_j\hbar}] \tag{20}$$

$$= 0.741\, V\sqrt{N_A(V+V_{bi})}\, \exp[-\frac{32.5}{\sqrt{N_A(V+V_{bi})}}]\, A/\mu m^2$$

where the doping concentration is measured in $10^{18}\, cm^{-3}$, the electric field is $E_j = (V+V_{bi})/D$ for an abrupt junction, and V is the applied bias in Volts while $V_{bi} \approx 1.0\, V$ is the built-in voltage. Equation 20 indicates that, while the tunneling current will increase dramatically as the doping concentration increases to avert punchthrough, it is still manageable at a channel length of 0.1 μm. For example, for $V_{DD} = 1\, V$, the leakage current is $< 1 fA/\mu m$ for a square $W/L=1$ MOSFET with a doping density of $N_A = 5\times10^{17}\, cm^{-3}$ which, according to a CF rule, corresponds to a channel length of about 0.10 μm, while the leakage current is $1 nA/\mu m$ or about 400A per chip for an array of square MOSFETs with a channel doping density of $N_A = 5\times10^{18}\, cm^{-3}$, which correspond to a 0.05 μm channel length.

The minimum oxide thickness imposes the most severe constraint on scaling through a cascade of effects. A thinner oxide promotes a higher drain saturation current which is required to improve speed and to charge parasitic capacitances. It can also be used to effectively mitigate the variations in the threshold voltage due to DIBL or the dopant distribution, and to minimize the subthreshold swing. However, as we have already seen, preliminary work on the reliability of ultra-thin oxides indicates that the maximum operating voltage collapses[77] from 3.3 V at $t_{ox} < 3–4$ nm to $< 1\ V$ at 2.4 nm at the onset of direct tunneling as shown in Figure 17. While this assessment may appear to be overly pessimistic, it is really not, especially if we consider that a tunneling current of 0.1 $pA/\mu m^2$ through a 3.2 nm oxide with 1.5 V applied to the gate will induce a 2 mV shift in V_t in about 1 year under constant stress[77]. A 2 mV threshold shift is comparable to the standard deviation required to keep gate delay variations below 10% with a $V_{DD} = 1.2\ V$.

A minimum oxide thickness of about 3 nm, in conjunction with a minimum subthreshold swing of about $S = 44\ mV$ per e-fold change in the drain current means that technologies smaller than 0.1 μm may not be viable. Because of the subthreshold slope, the worst case threshold voltage must be at least 0.4 V to avoid leakage. On the other hand, the power supply voltage must be reduced according to a CF rule to at least $V_{DD} = 1.2$ V to guarantee reliability. But at this voltage, excessive circuit margins ruin the performance. For example, a factor of two variation in the drain current, associated with $6.7\sigma = 0.25$ V variations in the threshold voltage, requires at least a factor of two reduction in the clock frequency from the scaled estimate of 1 GHz to 500 MHz. Consequently, technologies beyond 0.1 μm will require both an aggressively scaled insulator thickness in order to avoid short channel effects (such as an increase in the subthreshold swing), and a lower supply voltage (which will further aggravate circuit margins and further reduce speed.) Figure 17 indicates the limited room for improvement beyond 0.1 μm technology imposed by reliability concerns. The maximum operating voltage is an abrupt function of the oxide thickness near the onset of direct tunneling, and it will collapse below 3nm unless dramatic improvements can be made in the gate insulator or in the circuit margin. For example, for 0.2nm–rms interface roughness, the minimum oxide thickness assuming 1.5 V operation is calculated to be about 2.5 nm, provided that the maximum tunneling current allowed is 0.1 $pA/\mu m^2$. With an atomically flat surface the minimum oxide thickness can be improved to about 2.1 nm. However, if the maximum

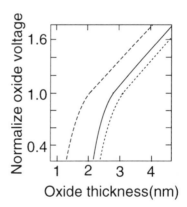

FIGURE 17. A calculation of the maximum supply voltage that is consistent with a leakage current less than 0.1 $pA/\mu m^2$ (dotted line) and 1 $nA/\mu m^2$ (dashed line) assuming that the interface micro-roughness is about 0.2 nm-rms. The supply voltage which is normalized to the barrier height, φ, is severely derated near the onset of direct tunneling leading to a minimum oxide thicknesses of 2.4 nm and 1.4 nm respectively. The minimum oxide thickness is also sensitive to the interface roughness. The solid line represents the maximum supply voltage consistent with a leakage current of 0.1 $pA/\mu m^2$ calculated with an atomically flat interface. Adapted from Schuegraf et al.[71].

tunneling current is stipulated to be 1 $nA/\mu m^2$ (consistent with the OFF-state leakage in a microprocessor) instead of $0.1\ pA/\mu m^2$, then the minimum oxide thickness could be as thin as 1.5 nm.

IV. REPLACEMENTS ON THE HORIZON AND COMPARISONS WITH CMOS

Based on what we know about the CMOS IC and its limitations, what characteristics should a replacement have and what new capabilities should it provide that CMOS will never be able to achieve? A replacement technology should fill the existing VLSI applications and yet be able to expand into new areas that would otherwise be inaccessible. Clearly, "drop in" material replacements such as the compound semiconductor *GaAs* fall into this category. There is no dispute that for high speed digital and analog applications at room temperature, *GaAs* ICs have a speed advantage over *Si* and have penetrated this market to a limited extent. Unfortunately, the absence of a circuit configuration like CMOS, which has very low static power, has made the transition of *GaAs* to VLSI difficult. Silicon can continue to provide much of the digital needs through parallelism or pipelining at an adequate throughput because more computation elements can be put on a single chip economically than with *GaAs*. For the same reason, large memories using acceptably low power are also now impossible in *GaAs*. Since a large processor needs enormous quantities of memory to operate efficiently, this is a serious handicap for an alternative technology that wants to get on the cost learning curve of Figure 1.

Another familiar "drop-in" replacement is the silicon bipolar transistor, which operates on a different principle than a MOSFET. The operation of a bipolar transistor features the direct control of the emitter-base potential height by the emitter-base voltage, V_{BE}, which is different than the capacitively coupled control of the barrier employed in a MOSFET[19]. The intimate control exercised over the barrier potential permits exponentially large currents to be transmitted using small voltage drops: i.e. $I_C = I_o(exp[qV_{BE}/kT] - 1)$ where I_C is the collector current and I_o is the base-emitter reverse saturation current, and is manifested in superior device performance. For example, the transconductance of a bipolar transistor is proportional to the collector current, $g_m = I_C/kT$, while it is proportional to the square root of the drain current in a long channel MOSFET, $g_m = \sqrt{2}\,\mu(\varepsilon_{ox}/t_{ox})(W/L)\sqrt{I_D}$. Consequently, for the same operating current, the latter has about ten times larger transconductance. The device mismatch, characterized by the difference between the base-emitter voltage in bipolar $V_{BE} = (kT/q)\ln[I_C/A_e I_o]$ where A_e is the emitter area, is usually within 1 mV for minimum area transistors, which is much smaller than the $\approx 5-15\ mV$ that typifies the mismatch in a pair of N-MOSFETs. And finally, the speed of a bipolar switch, measured by $f_T = (1/2\pi)\,g_m/(g_m\tau_b + C_{BE} + C_{BC})$, is very high because the base transit time, τ_b is so short, and because the parasitic base-emitter and base-collector capacitances C_{BE} and C_{BC} are usually negligible. The base transit time is short because the base width may be only 10–20 nm. But despite the higher intrinsic speed, the reduced process sensitivity and the lower dynamic power, a bipolar transistor is not yet a serious competitor in the electronics market. One reason is the higher cost required to produce a bipolar technology which develops from the extra steps needed for an epitaxial growth step and the incorporation of 4–6 more lithography levels into the silicon process flow. The increase in processing complexity might be mitigated in high volume VLSI production, but the absence of a low static power circuit configuration which makes VLSI impossible. Consequently, system designers are forced to seek alternatives.

Below, we explore the operation and limitations of three, more tentative technologies: the single electron transistor (SET)[118–121]; the electron waveguide (EW)[122,123]; and the Josephson junction (JJ)[124], which superficially seem to promise fantastic performance through the use of fundamentally different materials and operating principles than those employed in the MOSFET. All of the proposed alternatives operate at extremely low voltages to make effective use of quantum mechanics. While quantum mechanics is also manifested in the room temperature operation of the MOSFET indirectly through the silicon bandgap and the quantization of the inversion layer, its operation has been successfully described by classical transport equations such as Poisson's equation, and the current continuity equation, so far. This is because the thermal deBroglie wavelength, $\lambda_T = (\hbar^2/2m^*kT)^{1/2}$, is only about 1–2 nm at room temperature. On the other hand, the operation of an SET is based on quantum mechanical tunneling of single electrons and mutual Coulombic repulsion, relying on devices which are on the nanometer scale. The operation of an electron waveguide is based on energy quantization and the coherence of an electron wave, while the operation of a Josephson junction develops from the coherence of the superconducting ground state wavefunction and flux expulsion as well as tunneling. Because of their size, these three alternatives can utilize a basic quantum to represent a bit of information instead of the voltage on a capacitor. The SET uses the quantum of charge on a single electron, q; the electron waveguide uses the quantum of resistance, h/q^2; and the JJ uses a quantum of magnetic flux, $hc/2q$.

With the exception of JJ technology, these alternative technologies are not yet developed to the point that circuit yield and process control are considerations. Generally, their functionality has not been explored beyond the level of elementary circuits and devices and so the performance projections are much more tentative than the SIA roadmap, which is based on evolutionary improvements to well established technologies. Nevertheless, it is possible to identify some of the advantages and limitations inherent in these more tentative alternatives, especially as they are juxtaposed with CMOS technology.

In principle, the first two alternatives can be implemented in a complementary architecture that is functionally analogous to CMOS, but all of the alternatives improve over the power dissipation found in CMOS because of the extremely low voltage operation mandated by the physics. Consequently, the level of integration envisioned is much higher than that achieved by CMOS. While low voltage swing minimizes power dissipation, it sacrifices voltage gain, fan-out and increases leakage current, unless it is employed at low temperature. Thus, not one of the proposed alternatives will operate at room temperature reliably. The chance of an error derived from thermal fluctuations suppresses the operating temperature to at most 77K, even if state-of-the-art fabrication techniques are used to enhance the relevant energy scales. Thus, additional costs are expected from the need for refrigeration and from problems associated with differential thermal expansion of the variety of materials on a chip and in a package. At present, cooling a single superconducting computer chip in a closed-cycle cryo-cooler with a footprint of 2 sq. ft. costs about 20 thousand dollars. Unless the cost of refrigeration drops drastically, and the footprint is reduced, there will be no mass market applications.

Low temperature operation might seem advantageous with the discovery of high T_c (>77K) superconductors, especially since the high parasitic resistance encountered with wire dimensions less than 0.1 μm adversely affects the speed and noise performance[125]. However, the resistance of a superconductor is strictly zero only at d.c. and increases as the square of the frequency because the reactance associated with the inertia of the Cooper pairs causes the current to flow through the normal channel resistance of the superconductor. Consequently, the height-to-width aspect ratio required to reduce the resistance of a narrow superconducting wire becomes prohibitive (i.e. > 100 for W = 40 nm) at 100 GHz[126]. This number is comparable to that which can be achieved using a normal metal T-gate structure, and worse than the aspect ratio required if multiple normal metal wires are used in parallel[126]. Yet, low temperature operation can be beneficial especially for high performance computing and communications. Cooling in a low temperature bath allows the direct and efficient transfer of heat dissipated by the chip into the coolant. Low temperature also affects the rates of thermal diffusion, corrosion and electromigration so that, at least, the potential exists for improved reliability.

Finally, it is noteworthy that low temperature (77K) operation of CMOS[127,128], is suppose to extend the utility and reliability of a viable room temperature design, primarily because of the reduction in the subthreshold swing, with added benefits such as an increase in the channel mobility and thermal conductivity, and a decrease in the interconnection resistance[129]. Provided that the worst-case threshold voltage is not determined by other manufacturing tolerances, a collateral decrease in the threshold voltage will accompany the reduced subthreshold swing and temperature variations across the wafer. Corresponding to this change, it is expected that the speed will improve by a factor of 2–3. Moreover, since the doping levels will already be so high in a deep submicron technology, no special fabrication allowance is mandated. However, for purposes of comparison, we assume that the room temperature 0.10 μm technology is the limiting technology.

A. The Single-Electron Transistor (SET)

The number of carriers in the active region of MOSFET is expected to decrease to about 10 when the gate length is 20 nm. If it could be fabricated, numerical simulations[130,131] have indicated that single electron charging effects could be observed in a MOSFET this small. From this perspective then, an SET is a MOSFET contrived to exaggerate single electron charging effects.

An SET operates by controlling the tunneling of single charges into an isolated dot[119,132]. Its operation is thoroughly reviewed in Chapter 14 (by P. McEuen and L. Kouwenhoven) in this book. The dot can be either semiconducting, metallic or superconducting. To change the number of electrons on the dot, charge must tunnel through a junction comprised of an insulating barrier sandwiched between the dot, and the drain or source contacts. If the insulator is thin enough and the voltage is small compared to the height of the barrier, then tunneling current at zero temperature is given by an extension of Eq. (14):

$$J_t = V \frac{\sqrt{2m^* \phi} \, q^2}{4\pi^2 \hbar^2 t_{ox}} \exp[-\frac{2\sqrt{2m^* \phi} \, t_{ox}}{\hbar}] \equiv V/R_T. \qquad (21)$$

Thus, the current is a linear function of V, i.e. the barrier appears "Ohmic," although the nature of the transport through a junction is different from that of a generic Ohmic resistor. This resistance is denoted by R_T.

In an SET, two junctions, configured as illustrated in Figure 18, are used to control tunneling into the dot. When the area of the dot is extremely small, an electron cannot always tunnel through the insulator. During a tunneling event, the charge must discontinuously change by an elementary charge, q. At zero temperature, the energy of the circuit as a function of the number of charges added to the dot, δN, is given by:

$$E_{eq}(N) = \sum_{p=1}^{N} E_p + \frac{(-q\delta N + C_G V_G + Q_o)^2}{2C_\Sigma} - \delta Nq[\frac{C_S}{C_\Sigma} V_S + \frac{C_D}{C_\Sigma} V_D] \qquad (22)$$

where E_p represents the single particle energy, δN is the number of charges added to the dot, C_G represents the capacitance between the dot and the gate, $C_\Sigma = C_S + C_D + C_G$ represents the sum of the capacitances between the source, the drain and the gate, V_G, V_D and V_S denote the gate, drain and source voltages respectively, and Q_o represents the sum of the charge induced on the dot by doping (intentional or otherwise), Q_i, and the differences in electrode work functions, i.e.

$$Q_o = Q_i + \frac{1}{q}[(\phi_D - \phi_G)C_D - (\phi_S - \phi_G)C_S].$$

While this charge, Q_o, can be almost compensated for by δN, there will be a residue charge difference because of charge quantization. Consequently, we assume the residue $|Q_o - q\delta N|$ is equally likely to be distributed over the interval $<q/2$ and vanishes elsewhere.

The number of electrons in the dot is an integer and can only be changed by an integer. By definition, the chemical potential is the minimum energy required to add another electron: i.e. $\mu_o(N+1) = E_{eq}(N+1) - E_{eq}(N)$. Consequently,

$$\mu_o(N+1) = E_{N+1} + \frac{(q\delta N - qC_G V_G - Q_o)}{C_\Sigma} + \frac{q^2}{C_\Sigma}.$$

Thus, if the number of electrons on the dot changes by one and the gate voltage is fixed, then $\delta\mu_o = \mu_o(N+1) - \mu_o(N) = E_{N+1} - E_N + q^2/C_\Sigma$. This energy gap, $\delta\mu_{N+1,N}$, leads to a blockade for tunneling of electrons into or out of the dot. So essentially, energy conservation prevents tunneling because the concomitant charging of the dot would increase the electrostatic energy of the system. The charge quantization and the Coulomb repulsion between electrons give rise to a Coulomb blockade and the peculiar current-voltage characteristics shown in Figure 18.

This description of single electron tunneling, which is originally due to Averin and Likharev[133] applies provided the barrier resistance greatly exceeds the resistance quantum, $R_T \gg h/q^2$, so that quantum fluctuations of the charge can be avoided and an excess electron can be localized on the dot with confidence. If the separation between energy levels becomes comparable to the width because the lifetime is short or because of disorder, then each energy level cannot be treated independently. Another necessary condition for operation is that the dot must be small enough to guarantee that the energy required to add a charge to the dot at least exceeds that of thermal fluctuations, i.e. $q^2/2C_\Sigma \gg kT$ in order to enhance the probability that only N electrons are on the dot. With these assumptions, the device characteristics are determined solely by the capacitance and the electrostatic potentials of the source, drain and gate electrodes.

The Coulomb blockade can be intentionally eliminated by changing the voltage, V_G on the central gate electrode, which is capacitively coupled to the dot. For example, if the chemical potential in the dot, μ_o, is lowered below the

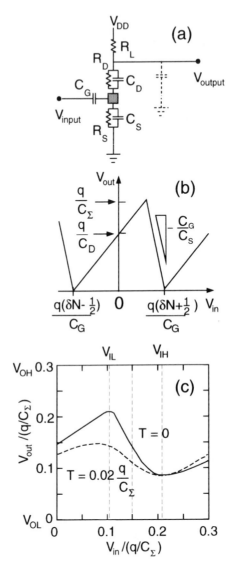

FIGURE 18. (a)Schematic illustration of a single electron transistor in an inverter circuit configuration. The dot is isolated by the source and drain tunnel junctions, each represented by a parallel connection through a capacitor and resistor, but is capacitively coupled to the input voltage through C_G. A change in the input potential removes the Coulomb blockade and permits electrons to tunnel from the source one-by-one. (b)An idealized transfer characteristic for the SET inverter shown in (a). (c) A calculation of an inverter characteristic adapted from Korotkov *et al.*[139] showing the effect of different temperatures on the gain.

chemical potential of the source and drain contacts, $\mu_S, \mu_D > \mu_o(N)$ where μ_S and μ_D denote the chemical potentials associated with the source and drain respectively, then the electron can tunnel from the source to the dot increasing the chemical potential in the dot by $\delta\mu_o \approx q^2/C_\Sigma$ (ignoring the change in the single particle energy.) Because of a corresponding increase in the chemical potential caused by the additional electron, $\mu_S > \mu_o(N+1) > \mu_D$, an electron can now tunnel from the dot to the drain which will return the chemical potential of the dot to $\mu_o(N)$. Thus, as a function of V_G the conductance through the dot will oscillate by approximately q^2/C_Σ where, corresponding to the transmission of a single electron, the period of oscillation in the gate voltage is[134]:

$$\delta V_G = \frac{C_\Sigma}{C_G} \frac{(E_{N+1} - E_N)}{q} + \frac{q}{C_G} \approx \frac{q}{C_G}. \tag{23}$$

Associated with dependence of the blockade on gate voltage, it has been shown experimentally that greater than unity voltage and current gains are possible for capacitive coupling to the dot[135,136]. For example, for the inverter configuration shown in Figure 18(a), with $C_S = C_D$, $C_G = 8C_D$, $R_S = R_D \equiv R_T$, and a supply voltage of the order of $q/2C_\Sigma$, the transfer characteristic resembles the sawtooth shown in Figure 18(b). The maximum current gain is approximately $1/\omega C_G R_T$ which can be very large, depending on the frequency. The maximum voltage gain, found on the falling portion of the sawtooth, is $A_v = -C_G/C_S = -8$. (The gain on the rising portion is $C_G/(C_D + C_G) < 1$.) In practice, the largest voltage gain achieved[137], so far, has been $A_v \approx 2.8$ with FO=1 using a gate with $C_G = 1.2 fF$ and tunnel junctions with capacitances near 0.2fF, and a tunneling resistance of the order of $R_T \approx 10^{12} \Omega$ in a prototype circuit operated at a temperature of 100mK $(=0.1q^2/C)$. An extrapolation to zero temperature yields a maximum gain of 8 for the same device.

SETs have been effectively operated only at low temperature (T<<1K), so far, although there have been indications that high temperature operation is possible[138]. Calculations by Korotkov et al.[139] have shown that, with a realistic load, the voltage gain is less than unity for temperatures as low as $0.015q^2/C \approx 27.9K/C(aF)$. Thus, the unity gain condition places a very stringent requirement on either the operating temperature or the size of the capacitance C. For high temperature operation $T_o = 300K = 0.015q^2/kC$, we find that $C < 0.1 aF$, whereas presently most experiments involve $C \approx 0.1 fF$. To appreciate the size of this capacitor, suppose that the junction can be represented by a parallel-plate geometry, then the capacitance will be given by: $C = \varepsilon A/d$. So, if $A = (10\ nm)^2$, and we choose a dielectric material with a high barrier energy and low dielectric constant such as SiO_2 with $\varepsilon/\varepsilon_o = 4$, then the thickness of the insulator would have to be smaller than the diameter of an atom, $d < 0.035$ nm, to obtain $C = 1 aF$. The smallest features achieved by electron beam lithography[140] are about 2 nm on a side. Using this feature size to define the area of the capacitor, instead of 10 nm, we find that the corresponding minimum is now $d = 0.87$ nm for the same type of capacitor, which is comparable to the Debye screening length in a metal. The latter observation suggests that minimum effective barrier thickness in a tunnel junction will be determined by the Debye screening length, even if an insulator with a smaller dielectric constant is found. Of course, this estimate for the minimum insulator thickness ignores reliability issues (which currently determine the minimum oxide thickness in a MOSFET.) In particular, it is expected that higher order tunneling processes[141], such as macroscopic quantum tunneling, will require that single junctions be replaced by stacks of 8–12 junctions to provide stability against charge inadvertently tunneling through the Coulomb blockade presented by an array of junctions, thus increasing the junction count per function well beyond the ideal represented in Figure 18.

The prospect of low gain translates into a small noise margin. Even at zero temperature it can be shown that the noise margin, defined by Eq. (8), is no larger than: $NM = (q/C_G)[1/2 - (C_S/C_\Sigma)] \approx q/2C_G$. This margin cannot easily accommodate the residue charge variations that originate either with random charged impurities lying in or near the tunnel junctions, or in the workfunction difference between the electrodes. The poor noise margin relative to the charge fluctuations probably represents the most serious obstacle for practical digital single electronics[139]. An appraisal of the margin for fluctuations in the background charge, given by Korotkov et al.[139], has shown that for a realistic load, δQ_o must be less than 0.03q at an operating temperature of $T \approx 0.01 q^2/C_\Sigma$. Thus, a small background charge renders digital logic impossible unless the bias of each device can be adjusted separately.

Even if this problem disappears due to charge relaxation[141], we anticipate work function fluctuations, even for similar metals, due to statistical variations in the small size of the dot required for high temperature operation. Following Brodie[142], we can estimate the effect of an atomistic change in the size of the dot on the work function. Assuming that the surface of a metal sphere behaves as a classical conductor to subatomic tolerances, for the purposes of calculating electric fields and the image force on an external charge, it can be shown that the work function associated with a metal dot of radius r is given by: $\phi = (q/8\pi\varepsilon) r/[(r+\chi)^2 - r^2]$, where χ is distance the between the surface of the sphere and the electron. If we identify χ as the uncertainty in the position of an electron inside the metal (typically, $\chi \approx 0.1 - 0.2$ nm), then a variation in r from 1 nm to 1.25 nm results in a change in the workfunction of 33 meV which corresponds to about one quarter of the power supply voltage.

Finally, although the gain and noise immunity do not depend on the transparency of the junction, large scale integration of SETs will still require excessive circuit margin because of the exponential sensitivity to microscopic fluctuations in the insulator thickness. For example, consider the fantastic speed promised by recent experiments using a single electron turnstile[143]. We assume that the delay required to charge the next gate in a string of SET inverters[120] is indicative of the speed, analogous to the calculation for a chain of CMOS inverters performed in Section II.A.2. The input capacitance of the next gate will be C_G and it will be charged through the source and drain resistances. If the load capacitance, including the interconnection capacitance is much greater than C_Σ, then the link will not form another blocking dot and the gate will therefore charge and discharge with a characteristic time constant of $\tau_c \ll 2R_S C_\Sigma \approx [R_T(1M\Omega) \times C(aF)]ps$. Thus, the minimum delay will be given the thinnest d consistent with the condition that $R \gg 4h/q^2$. If $R_T = 1M\Omega$ and $C \approx 1aF$ then $\tau_c \approx 1ps$, which is a only factor of 10 smaller than the gate delay measured in a 0.1 μm CMOS inverter with FI=FO=1 at a supply voltage of 1.8 V. However, the thickness required to produce $R_T = 1M\Omega$ and for $C = 1aF$ is $d = 0.2$–0.4 nm which is comparable to the interface roughness. Resistance fluctuations due to interface roughness can randomly decrease R by a factor of $exp[-2h_o^2/\Lambda]$ however, as shown in Section II.A.4 above. For $h_o \approx 0.1$nm-rms, and using $Al_x O_y$ for the tunneling barrier, we find that the resistance could decrease by a factor of about 1000. Consequently, the clock frequency must be below 1 GHz to avoid timing errors due to a worst case delay in a synchronous system, which is worse than the performance anticipated from 0.1 μm VLSI.

Thus, realistic junction parameters seem to preclude high temperature (77K) operation and limit the performance of an SET technology. The potential for high speed is compromised by sensitivity of the transparency to junction parameters, while the potential for high density is offset by the large junction count per functional group required for stability against thermal fluctuations and macroscopic quantum tunneling. A large junction count is also necessary because the small fanout that develops from low voltage gain requires more stages. Nevertheless, SETs may still come into use in a mainstream technology, for example as a sense amplifier in memory, because of their exquisite sensitivity and noise performance.

B. The Electron Waveguide

An electron waveguide can be construed as a MOSFET with a channel width comparable to the electron wavelength, and a channel length smaller than the elastic and inelastic mean free path. Ideally, in an electron waveguide there is no scattering in the channel at all. The transport is ballistic (which seems to obviate the limitations imposed by velocity saturation.) Under these conditions, the contacts completely determine the resistance of the device. An electron waveguide is operated by modulating the transmission of an electron wave, analogous to its electromagnetic counterpart, but with the difference that the wavefunction can be controlled by external voltages[144]. This provides the opportunity to duplicate with electron waveguides many of the concepts that are employed in integrated optics.

The insets to Figure 19 show one way of implementing an electron waveguide, using split-gate electrodes on a modulation-doped *GaAs/AlGaAs* heterostructure. In these devices, the two-dimensional electron gas (2DEG), confined to the *GaAs/AlGaAs* interface by the heterostructure band offset, is depleted immediately beneath the gate electrodes by applying a negative bias less

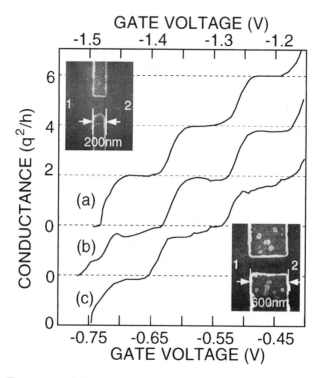

FIGURE 19. The two terminal conductance of an electron waveguide at T=280mK as a function of gate voltage (or the width of the constriction). The inset at the top of the figure shows a top view of 200 nm long split-gate electrodes with a 300 nm gap between them placed on a high mobility *GaAs/AlGaAs* heterostructure. The bottom inset shows a similar device on the same heterostructure with a 600 nm lithographic length. The quantization of the conductance ($\delta G = (1 \pm 0.01) 2q^2/h$) of the 200 nm long constriction shown in (a) deteriorates after cycling to room temperature, as shown in (b). We attribute the deterioration to a difference in the configuration of depletion charges corresponding variations in the width of the constriction. The poor quantization of the conductance of a 600 nm long constriction, shown in (c), is also supposed to develop from fluctuations in the width. (See color plate.)

than -0.325 V. As the gate voltage decreases further, the constriction within the gap between the electrodes narrows; the carrier density within the constriction decreases; and steps or plateaus are observed in the conductance. Figures 19 (a),(b) and (c) show the two terminal conductances measured as a function of the voltage applied to the split-gate electrodes. The *average* conductance on a plateau (minus the series resistance) is approximately $2q^2N/h$ with N an integer ranging from 1 to about 10 typically, and is evidently quantized in steps of $2q^2/h$ with about 1% accuracy. While the gate voltage does not correspond directly to either the carrier density, the width of the constriction or the chemical potential, experiments and numerical simulations have shown that the width of the constriction measured relative to λ_F changes as a function of V_G[122].

The essential theory of the quantized conductance was given in the first experimental papers[145,146]. Consider a one-dimensional (1D) channel of constant cross-section between two 'reservoirs', whose Fermi energies differ by the (small) applied bias qV. The width of the channel is so constricted that it becomes comparable to the Fermi wavelength of the electron. Consequently, the transverse energy is quantized. For example, for a square constriction $E = \pi^2 \hbar^2 N^2/2mW(x)$ where x extends along the length of the channel. The reservoirs are simply regions where the cross-section is much wider than in the channel and the density of states is large, while the current density is negligible. The reservoirs provide a Fermi distribution of electrons, with a well-defined Fermi energy, E_F, that impinges on the 1D constriction from each end. At zero temperature, the net current is carried by the range of energies between the Fermi energies, where the occupation functions of the two reservoirs differ. Thus,

$$I = qv(E_F)T(E_F) \; [\frac{1}{2}n_{1D}(E_F)qV] \qquad (24)$$

where $v(E_F)$ is the velocity of the electrons, $T(E_F)$ is the (flux) transmission probability, and the quantity in brackets is the linear density of electrons with a factor of half because we only include motion along one direction. The density of states in energy, for both spins, is given by:

$$n_{1D}(E) = \frac{2}{\pi\hbar v(E)} \qquad (25)$$

The crucial result is that the velocity cancels when Eq. (25) is substituted into Eq. (24), leaving

$$G = \frac{I}{V} = \frac{2q^2}{h} T(E_F), \tag{26a}$$

which is known as the Landauer formula. This formula relates the quantum mechanical transmission coefficient to the conductance directly[147,148]; the factor of 2 in Eq. (26) accounts for the spin degeneracy. A perfect 1D ($N=1$) constriction has T=1, so that the maximum conductance is $2q^2/h \approx 77.4 mS$ per spin. If we assume that: (1) the temperature is lower than the separation between transverse energy levels at the narrowest point in the constriction; (2) that the length of the constriction is longer than the decay length of evanescent modes; and (3) there is no scattering then:

$$G \equiv \frac{2q^2}{h} N_{min} (k_F W), \tag{26b}$$

where N_{min} is the number of 1D channels or modes propagating through the constriction. For $N > N_{min}$ the transmission coefficient is exponentially small and classically forbidden.

Using the Landauer formula, we can obtain an estimate of the maximum gain attributable to an electron waveguide. According to Eq. (26) the drain current is given by $I_D = 2(q^2/h) TV_D$, so that the transconductance is given by:

$$g_m = 2 \frac{q^2}{h} \frac{\partial T}{\partial V_G} V_D = 2 \frac{q^2}{h} \frac{\partial T}{\partial W} \frac{\partial W}{\partial V_G} V_D. \tag{27}$$

Incorporating tunneling and above-barrier reflections leads to a spreading of the sharp-edged steps implicit in Eq. (26). For example, according to Glazman et al.[149] for an adiabatic constriction the step shape depends on the curvature at the center of the constriction, R, according to:

$$T(W) = [1 + \exp(-(\frac{k_F W}{\pi} - N) \pi^2 \sqrt{2R/W})]^{-1},$$

where k_F is the Fermi wavevector, so that $\partial T/\partial W < 2k_F = 0.35(n/10^{12} cm^{-2})^{-1/2} nm^{-1}$, where n is the electron density in cm^{-2}. Following Sols et al.[123] we can estimate the quantity $\partial W/\partial V_G$, which appears in Eq. (27) for g_m, using the fact that it is simply the negative derivative of the depletion length: i.e.

$$\frac{\partial W}{\partial V_G} \approx \frac{\partial D}{\partial V_G} = -\frac{1}{2}(\frac{2\varepsilon}{qN_D})^{1/2}\frac{1}{(V_G+V_{bi})} \approx \frac{20}{\sqrt{V_G+V_{bi}}}(nm/V),$$

where the voltages are given in Volts[123]. For $V_{bi} \approx 1$ V and $N_D = 10^{18}$ cm^{-3}, we find that $g_m < (2q^2/h) V_D$. The current drive capability is so low fundamentally because the maximum conductance of a 1D constriction is so small. To remedy this deficit, the use of multiple parallel wires has been proposed by Sakaki and others (see chapter 5). The maximum voltage gain, estimated from the ratio of maximum transconductance and the minimum output conductance,

$$A_v^{max} = \frac{g_m}{g_o} \leq \frac{2q^2/h}{2Nq^2/h} = \frac{1}{N},$$

cannot exceed unity. Unity voltage gain can only be achieved in single mode operation, i.e. when $N = 1$. Without voltage gain, the implementation of voltage restoring logic using only electron waveguides is impossible. There is power gain, however. The low frequency current gain can be tremendous, which is reflected in the unity current gain frequency:

$$f_T = \frac{g_m}{2\pi C_G} \approx (\frac{2q^2}{h}) V_D (\frac{\partial T}{\partial W}) \frac{1}{(2\pi qnL)}$$

The numerical estimates of Sols et al[123], given in Table 5, indicate that a 10 nm wide structure could have $f_T \approx 3 THz$. This high (theoretical) performance develops from the modulation of quantum mechanical interference as much as from the small gate capacitance. It corresponds to the short transit time[123] necessary for an electron to travel across the width of the constriction ($2W/v_F \approx 0.3 ps$).

Ignoring multi-mode correlations[150], it is desirable to operate an electron waveguide in the fundamental mode to maximize the swing in transmittance and minimize the current carried in other modes. Single mode operation is not so easily achieved, however. Table 5 lists parameters which characterize a waveguide with a square well confinement potential after Sols *et al.*[123]. The energy difference between the thresholds for the first and second subband permits room temperature operation only for W = 10 nm and $m^* = 0.01 m_o$. We are not aware of a material system that simultaneously provides such a light effective mass and a band-offset large enough to support an intersubband separation of 400 *meV*. Furthermore, it has been observed that an electron waveguide cannot operate single mode over distances longer than about 1 μm in conventional devices even though the elastic mean free path is an order of magnitude larger.

TABLE 5. Estimated parameters of an electron waveguide for various widths, W, electron effective mass, m, and carrier densities, n, adapted from[123]. ΔE_o is the energy difference between first and second subbands, T_o is the operating temperature chosen to guarantee sufficient monochromicity of the electrons, and f_T is an estimate of the unity current gain frequency.

W	m/m_0	ΔE_0	$n \times 10^{10}$	T_0	v_F	f_T
(nm)		(meV)	(cm^{-2})	(K)	$(10^7 cm/s)$	(THz)
10	0.01	380	150	330	55	2.9
10	0.05	75	150	65	11	0.6
20	.01	94	38	82	27	0.7
50	0.01	15	6	13	11	0.11

Figure 19(c) illustrates the deterioration in the accuracy of the quantization in a constriction about 0.6 μm long[151] that is fabricated on the same heterostructure as Figure 19(a). The quantization deteriorates to less than 90% of the expected value even though the mean free path in the 2DEG is about 7–10 μm implying scattering within the constriction. Still more insidious is the deterioration in the accuracy of the quantization shown in Figure 19(b) which is observed, using the same constriction that produced the results of 19(a), after cycling it to room temperature. In each case the poor quantization has been attributed to fluctuations in the width of the constriction.

Fluctuations in the width ruin the prospects for single mode operation because of resonant scattering. This assertion has been corroborated by numerical simulations of the structures shown in Figure 19 by Davies et al.[152]. A realistic model for transport through a 1D constriction must include the shape of the constriction and the effects of coherent scattering from the random potential associated with impurities[153,154]. Davies has evaluated the effect of the electrostatic confinement by the split-gate electrodes on transport in the 2DEG numerically, including the effect of donors placed at random in a doped layer, their respective Coulomb energies, with their images in a self-consistent way. The resultant density of electrons in the 2DEG for a constriction similar to that shown in the top inset of Figure 19 is given in the top inset to Figure 20. In the model, the donors have a real density, $N_D^{(2D)}$, and are separated from the surface by a distance c ('cap'). The electrons are a distance s ('spacer') below the donors. This material is assumed to be 'δ-doped', with the donors restricted to an atomic plane. (A similar doping technique has been proposed as a remedy for threshold voltage variations and short channel effects in MOSFET.)

The top inset in Figure 20 shows the electron density contours associated with a disordered 200 nm long constriction. Although there are nearly 60,000 donors

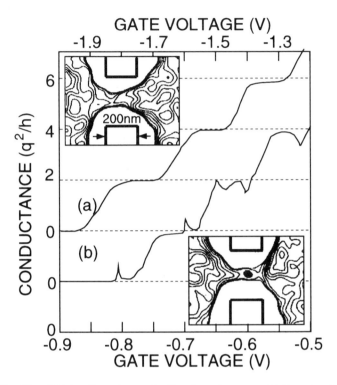

FIGURE 20. The deterioration of quantization in a long constriction due to resonant scattering. The top inset indicates the electron density variations in a 200 nm long constriction with a 300 nm gap and about 60,000 donors, corresponding to the structure in the top inset of Figure 19. The electron density changes by $4.2 \times 10^{14} m^{-2}$ between different contours. The conductance, calculated as a function of gate voltage V_G for this constriction, is shown in (a) and corresponds closely with the experimental result shown in Figure 19(a). (b)The calculated conductance as a function of gate voltage for the similar constriction, but with a different impurity configuration. In the corresponding inset there is a bulge in the electron density between 400 and 500 nm which gives rise to a resonance scattering and the deterioration in the accuracy of the quantization.

in the numerical sample, the number of features shown in the inset is much smaller than that. This emphasizes the long length scale of the potential fluctuations in the vicinity of the 2DEG, and that they do not arise from individual impurities. The transport mean free path for electrons away from the constriction is a few micrometers, several times the size of the sample, despite the fluctuations. The reason that the random potential is very inefficient at scattering these electrons is because its length scale, about 150 nm, is larger than the Fermi wavelength of 50 nm. This changes within the constriction, where the kinetic energy of the electrons falls and they become more susceptible to scattering. The potential gets rougher as the channel is squeezed further and the density of electrons is reduced.

The calculated two terminal conductance for the constriction of the inset is plotted in Figure 20(a) and reveals a well-quantized conductance. Why does the conductance remain well quantized, despite these fluctuations? The vital feature for preserving quantization is that *forward* scattering is dominant, which in turn follows from the slowly-varying spatial nature of the random potential. Forward scattering allows a 'compensating' process to occur to the left of the constriction. Electrons that scatter out of one of the conducting modes into a higher mode can therefore be exactly balanced by electrons scattering via the inverse process. This is not true if back-scattering is important, because the backward-going modes are not fully occupied and the two rates do not balance.

To investigate the deterioration of the quantization, Davies analyzed another device which differs from the previous one only in the arrangement of the random impurities. The electron density distribution and the corresponding poorly quantized conductance are illustrated in the lower inset to Figure 20. Notice that the electron density in the inset shows a bulge near the center, to the right of the narrowest region. Electrons are forward scattered from one of the conducting modes into the extra mode that propagates within the bulge. They are reflected when this mode cuts off at the ends of the bulge to form the resonance. Some electrons are 'forward' scattered out of the resonance into the backward-going conducting modes, giving an indirect back-scattering process which lowers the conductance. This result suggests that the maximum length of a constriction is set by the correlation length of the random potential, rather than a conventional mean free path, if resonances are the dominant back-scattering mechanism. This length is about $0.3\mu m$ in these devices.

Fluctuations in the width of the waveguide, and the associated resonance scattering are significant in conventional structures because the random potential is larger than the separation between 1D subbands. According to Davies[155] the variance of the random potential fluctuations due to dopants asymptotically scales like $\sigma_{scr} \approx (e^2/4\pi\varepsilon)(N_D^{2D}/q_{TF}^2)(1/s^2)$ which is about 2 *meV* for the devices in Figure 20 and is comparable to the intersubband spacing. σ_{scr} cannot be reduced much further because of the corresponding decrease in the interface field and reduction in the 2DEG density. From the electrostatics, it can be shown[155] that the optimum structure, based on the same design as shown in Figure 19, has $W \approx 1.6(s+c)$, which is rather narrow for typical layers, and has a threshold voltage of $V_W \approx 1.7 \ V_T$ where $V_T = (q/\varepsilon)(s+c)n_{2D}$ is the threshold voltage of a two-dimensional MODFET. Taking the confining potential to be parabolic gives an optimum separation between energy levels of:

$$\Delta E \approx 0.8 \sqrt{\frac{\hbar^2}{m} \frac{\Delta E_c}{s(s+c)}} . \tag{28}$$

where $\Delta E_c \approx 200 \ meV$ denotes the change in the conduction band minimum. Equation (28) shows that the spacing between the energy levels at threshold falls off as $\Delta \varepsilon \propto 1/s$ for large spacers in these optimized wires whereas the screened potential falls off as $1/s^2$ and therefore becomes much smaller than ΔE for thick spacers. The other important quantity, the width of the wire, is proportional to $1/\sqrt{E}$ and therefore to $\sqrt{s}$. So, the wire is more tightly confined on the length scale of the fluctuations ($\propto s$) as s increases. While there are preliminary indications that this recipe developed by Davies for increasing the length of a single mode electron waveguide is effective for reducing fluctuations in the wire width[156], the stringent requirements presented in Table 5, the expense associated with epitaxial growth, and the absence of voltage gain altogether preclude the development of an electron waveguide as a "drop-in" replacement for the MOS-FET.

C. The Josephson-junction

According to Likharev[157], there are three features of Josephson-junction integrated circuits that especially recommend them for high performance VLSI over CMOS technology. First, Josephson-junctions are high-speed, low impedance switches that dissipate very little heat. Even though the JJ impedance is small, the power consumption is still low because the voltages employed cannot exceed twice the superconducting energy gap $2\Delta/q \approx 2 \ mV$. Second, superconducting transmission lines can be used as the interconnections. If the JJ impedance is matched to a superconducting micro-strip, picosecond waveforms can propagate down this type of transmission line at nearly the speed of light with low attenuation and low dispersion. A dense layout pattern can be used with little interference and cross-talk, as well, because of the small dielectric thickness required[158]. Finally, the fabrication technology is simpler than that used for silicon VLSI. Yet, the physical limits on the junction size appear at about the lithographic feature size where limitations of silicon CMOS technology develop, i.e. near 0.1 μm.

There are two aspects of the operation of a Josephson-junction that differentiate it from a MOSFET: tunneling and superconductivity. A Josephson-junction consists of a sandwich of two superconducting electrodes separated by an insulator thin enough to permit tunneling, as shown in Figure 21(a). Usually the insulator is an oxide layer grown on the lower base electrode. Above the superconducting transition, the current through the structure is characteristic of a leaky capacitor:

$$I = \frac{V}{R_T} + C\frac{dV}{dt} \tag{29}$$

where R_T is the tunneling resistance defined by Eq. (21), and C is the sandwich capacitance. As the junction is cooled below the superconducting transition temperature, a superconducting energy gap develops which produces a nonlinearity in the tunneling component of the current.

Another contribution to the current develops from the Josephson effect[159]. The ground state of the superconductor is parameterized in terms of a macroscopic wavefunction. Whenever the superconductivity between two points is "weakened", e.g. through the introduction of an insulating gap, the current becomes a periodic function of the phase difference between the wavefunction across the junction which is termed the Josephson effect: i.e.

$$I = AJ_c \sin\phi \tag{30}$$

where J_c is the maximum current density, A is the area of the junction, and ϕ is the phase difference. This current depends on the electric field through the relation:

$$\frac{d\phi}{dt} = \frac{2q}{\hbar}V. \tag{31}$$

The I-V characteristic of the junction is derived from a partial differential equation that describes the time development of the phase of the superconducting wavefunction:

$$I = C\frac{\partial^2\phi}{dt^2} + \frac{1}{R_T}\frac{\partial\phi}{\partial t} + \frac{2qA}{\hbar}\sin\phi.$$

Since it is nonlinear, the solution to this equation can only be obtained numerically. When the spatial variations are neglected, the time response takes an especially simple form characterized by the R_TC time constant and the Josephson angular frequency $\omega_j = (2q/\hbar)(J_c AR_T)$.

There are two ways to use a JJ as a switch, which depend on the electrical characteristics of the junction through the McCumber-Stewart parameter, $\beta \equiv \omega_j/CR_T$. Using an underdamped junction with $\beta > 1$, a JJ can be used as a latch to produce voltage state logic. Voltage-state logic has been used to demonstrated impressive performance with a 1 GHz clock, but requires a slow reset operation and an a.c. power supply and cyrogenic temperatures. It was discarded

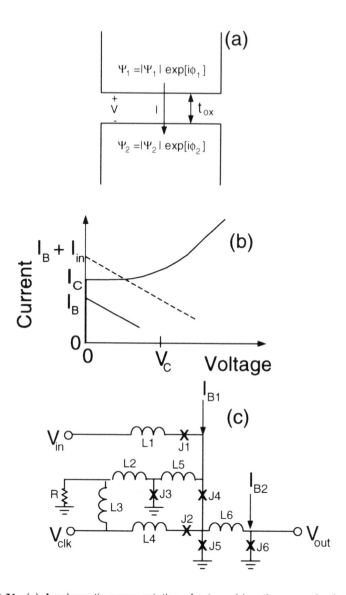

FIGURE 21. (a) A schematic representation of a tunnel junction comprised of an oxide layer sandwiched between two superconducting electrodes. The voltage that develops across the electrodes arises from the change in the phase of the wavefunction across the gap. (b) The I-V characteristic of an overdamped Josephson-junction, the building block of RFSQ logic. (c) Schematic representation of an inverter realized with JJ technology. The device count required to realize a simple function may preclude the use of JJ technology.

by IBM[160] in the 1980's as impractical because of the instability of the tunneling resistance, associated with the ultra-thin (2–3 nm) oxide layers used in conjunction with superconducting lead alloys, after cycling the temperature, and because the performance was already comparable to that obtained from *GaAs* circuits without cryogenics. Using an overdamped junction with $\beta < 1$ results in the I-V characteristic shown in Figure 21(b). A JJ with this characteristic can be used in non-latching, flux shifting logic that is based on the generation and transmission of voltage pulses with a time integral equal to a flux quantum, $\delta V \delta t = \Phi_o = 2.07 pH - mA$. The overdamped I-V curve is single valued, and there exists a broad range of amplitude and duration for I_{in} that triggers a quantized jump of the Josephson phase of the junction by 2π. According to the Josephson relation given in Eq. (31), a jump in phase corresponds to the generation of a voltage pulse across the junction a few picoseconds in duration. Provided that the d.c. bias current I_B is near the critical value I_C, this voltage pulse can be triggered by an incoming pulse of less than the nominal value, which means that the junction exhibits some voltage gain. This flux-shifting technology, denoted as Rapid Single Flux Quantum Logic (RSFQ), is just now being explored because of the potential for clock frequencies beyond the 300 GHz. So far, RSFQ gates operating up to 370 GHz have been demonstrated[161] at the proof-of-concept stage. Thus, JJ technology offers the prospect of one or two orders of magnitude improvement in clock frequency over extrapolations of the competing CMOS technology.

Because of the low energy content of the flux quantum ($\approx 1 aJ$), fabrication margins are especially important in circuit design[162]. RSFQ logic typically has margins of 19–50% at 4.2K and are practical within current processing technology that utilizes $Nb/AlO_x - Al/Nb$ Josephson-junctions instead of *Pb* alloys. For example, a critical margin of 19% for I_{C2} (the critical current through junction J2) has been identified in the inverter circuit of Figure 21(c). RSFQ logic requires, at minimum, a single flux to be trapped in the current loop described by 2 junctions and an inductance. Figure 21 shows an equivalent circuit for an RSFQ inverter. The main quantizing loop is defined by the junctions J3, J4, J5, and the inductance L5, where the critical current of J5 is larger than that of J4. Without a signal pulse during the operating window, the inverter is LOW and the persistent current is flowing counterclockwise through J2, J4, L5 and J3 with J5 in its subcritical state. When a clock pulse T is applied, it induces a 2π phase slip in J5 and a SFQ pulse is produced at the output. On the other hand, if is a signal is applied to the input during the operating window, it induces a 2π slip in J4 rather than J5, but no signal at the output. The application of a clock pulse then induces a 2π slip in J2, but there is still no output signal. However, a delayed version of the the clock pulse arrives at node B and resets the cell to its LOW state by inducing a phase slip in J3.

Simplistically, the relationship between the trapped flux and the current loop formed by the junctions and inductance can be summarized by: $\Phi_o = I_C L$, where

$\Phi_o = hc/2q$ and the inductance L is approximately 1pH at 77K. As noted by Likharev and Semenov[157], the ratio of the Josephson coupling energy $\hbar\omega_c$ to the thermal energy kT must be large enough to avoid inadvertently triggering the junction under these circumstances. If the bias current is chosen to lower the barrier by a factor of 10, then a $BER \approx 10^{-30}$ imposes the condition that $(1/10)(\hbar\omega_c/kT) \approx 70$. This condition means that $I_C > 200$ μA for liquid helium temperature operation or $I_C > 3$ mA for operation at 77K. For 0.1 μm^2 area junction the latter value for the critical current corresponds to the largest critical current density ever achieved in research. The need for a high current level in RFSQ logic for reliable 77K operation not only minimizes the amount of scaling that is possible, but also requires incredibly small inductances. For a high T_c superconductor with a penetration depth of 200 nm, this inductance represents about 3 lithographic squares from which it is impossible to fabricate a loop. Since the penetration depth changes little as a function of temperature below 77K, the only way to reduce the required current is to reduce the temperature to suppress thermal fluctuations, which results in an operating temperature as low as 30K.

In order to make a fair comparison to CMOS technology it is also necessary to analyze the device count per circuit. The problem is already apparent from the the schematic of a JJ inverter shown in Figure 21(c). After examining three different circuit functions, an inverter, a NAND gate, and a 1-bit full adder, we find that an inverter requires only 2 transistors in CMOS, but 6 junctions and 6 inductors in RSFQ; a NAND gate requires only 4 transistors but 19 junctions in RSFQ, and finally the adder requires only 17 transistors, but 80 junctions are required for RSFQ logic. Thus, as the complexity of the circuit grows the number of Josephson junctions required to implement it grows faster than the number of MOSFETs and even faster than the number of p-n junctions. The increased number is not only a reflection of the two-terminal character of the junction, but also indicates the need for buffers to isolate the output signal from the input, and the need for splitters in logic with low voltage gain, to reproduce pulses without a decrease in voltage amplitude when the fan-out has to be greater than one.

Finally, one of the most serious limitations of the RSFQ logic family is the relatively low density superconducting memory available. A von-Neumann-type computer relies on frequent exchanges between the processor and memory. For example, a state of the art parallel multi-processor comprised of a thousand high end microprocessors operating at clock frequency of 100 MHz with 1–4 instructions issued per cycle requires about 32 Mb of memory per processor. In general, the memory required for a specified level of GigaFLOPS performance scales as: *Memory* = $(\#ofFLOPS/1\,GigaFLOPS)^n$ *GB* where $n < 0.75$, depending on the allotment of simulation time[14]. A parallel-pipelined JJ RSFQ logic can produce 100 GFLOPS with a 1 ns latency in 1 μm design rules, yet superconducting memory is not expected to exceed 64Kb – 1 Mb with 0.1 ns access times. The memory density is low because of the required device count and the minimum design rules allowed by the superconducting penetration depth. Thus, the increase in speed that might be accomplished using JJ technology is useless because of much slower intra-chip communication between the microprocessor and the main memory. And even if room temperature silicon memory is employed, the extremely low voltage (mV) interface to the superconducting processor would be problematical.

V. Conclusion

The economic forces that have facilitated the integrated circuit revolution and the dramatic increase in performance that we have witnessed over the last thirty years are expected to diminish within the next two decades. The main problem with room temperature operation of CMOS is the subthreshold slope. To turn OFF a transistor the threshold voltage must be relatively high, 0.5 V or so, which in turn requires a high supply voltage of 1.2–1.5 V for noise immunity and circuit margin. It is possible to lower the threshold voltage somewhat without improving on the subthreshold current by reducing the variations in the threshold. For example, by using an undoped layer immediately beneath the channel[97] and removing the channel doping to a delta-doped layer a distance s away from the interface, the "intrinsic" fluctuations associated with the dopants can be reduced as s^{-2} as shown in Section IV.C in connection with the operation of an electron waveguide. So far, epitaxial growth has proven to be too expensive, however.

Because of the variations in the delay encountered in circumstances with a high threshold voltage, excessive circuit margins are required at low voltage so that performance may be compromised. Consequently, the two principal means for improving performance, growing the die and miniaturization, are effectively nullified for feature sizes at or below 0.1 μm in a constant field scaling scenario. Despite the advent of exciting prospects, such as single electronics, the electron waveguide and Josephson-junction technology, most of the proposed alternatives still depend on lithography, dry etching, ion implantation, clean rooms, metal interconnections, small line rules, etc. When coupled with a whole new set of reliability and fabrication problems that accompany a new material or device structure, the potential replacements are expected to come in at a greatly increased cost without a corresponding increase in performance.

What would it take for a new technology to replace the CMOS IC? The simple answer is an opportunity to enter production, and better performance at lower cost. After invention, an alternative technology has to start its way down the curve of Figure 1 and compete with the CMOS/IC solutions. Improved performance will let it command a higher price, but for it to replace the transistor/IC, it has to do so in the mass market. Here, there is little premium for improved performance and cost is everything. The only hope for a new technology is to be able to get on the curve at a much lower initial point than the MOSFET. It could do this if it either had a greatly simplified fabrication process or if it were a "drop-in" replacement into the existing IC processing fabrication lines.

It is possible that the silicon IC business will go the way of the steel industry becoming an inexpensive, mature commodity while other technologies (e.g. plastics or aluminum) arise and compete with it in specific applications where they offer a cost or performance advantage. Even with these alternatives, steel is still the core of our mechanical culture and coexists with these other technologies. The equivalent situation would be the growth of alternative memory or processing technologies which offer improved performance or cost.

An example of such a killer technology might be the recent advances in plastic transistors[163,164], which may eventually be fabricated according to a simpler scheme analogous to how a printing press prints newspapers. It is easy to imagine a low cost, high resolution printing technique that would give competitive densities at dramatically lower costs. If plastic transistors and wires could be made by an additive printing technology that uses fewer steps than a conventional IC process, the potential cost per device could be substantially lower than that of a conventional IC. Currently, the design rules for a plastic transistor are limited to tens of microns, however. Another down side to this particular technology is performance. Right now, these devices are factors of ten thousand or more slower than bulk silicon transistors, far too low to compete in the bulk of the applications. Even now, they could compete, though, in very low cost memories for audio or even video applications with the right parallel access architectures. But to make in-roads into the rest of the electronics applications, performance would have to be improved dramatically. While this may not be precluded by any obvious physical principle, even after decades of trying, we have no good ideas for getting high carrier mobilities in deposited materials as opposed to highly ordered single crystals.

REFERENCES

1. Bardeen, J., and Brattain, W.H., *Phys. Rev.* **74**, 230–231 (1948); Schockley, W., *Bell Syst. Tech. J.* **28**, 435–439 (1949); Schockley, W., *Proc. IRE* **41**, 970 (1953).

2. Tanebaum, M., and Thomas, D.E., *Bell Syst. Tech. J.* **35**, 1 (1956).

3. Hoerni, J.A., "Planar Silicon Transistor and Diodes," *IRE Electron Devices Meeting*, Washington, D.C. 1960.

4. Atalla, M.M., U.S. Patent 3,206,670 (filed in 1960 issued in 1965).

5. Kahng, D., and Atalla, M.M., *IRE-IEEE Solid-State Device Research Conference*, Carnegie Institute of Technology, Pittsburgh, PA, 1960; Kahng, D., U.S. Patent 3,102,230 (filed in 1960, issued in 1963).

6. Several patents were filed in 1959 on different aspects of the integrated circuit. The historical development of the integrated circuit is annotated in: Kilby, J.S., *IEEE Trans. Electron Devices* **ED-23**, 648 (1976).

7. Moore, G.E., *IEEE IEDM Tech. Dig.* 11–13, (1975).

8. *The National Technology Roadmap for Semiconductors,* Semiconductor Industry Association, San Jose, CA (1994).

9. Mayo, J.S., *Scientific American* **255**, 4, 58–65 (1986).

10. Makimoto, T., "Market and Technology Trends in the Nomadic Age," *Symp. on VLSI Tech.*, 1996, pp. 6–9.

11. Bois, D., *L'Onde Electrique* **73(6)**, 4–10 (1993).

12. Ogirima, M., "Process Innovation for Future Semiconductor Industry," *IEEE 1993 Symp. on VLSI Tech.* pp. 1–5.

13. Miller, D.L., Przybysz, J.X., and Kang, J.H., *IEEE Trans. Appl. Superconductivity* **3**, 2728–2731 (1993).

14. Cray, Seymour *Enabling Technologies for PetaFLOPS Computing,* Sterling, T., Messina, P., and Smith, P.H., eds., Cambridge MA: MIT Press, 1995.

15. Hoeneisen, B., and Mead, C.A., *Solid State Electron.* **15**, 819–829 (1972).

16. Keyes, R.W., "Physical Limits in Digital Electronics," *Proc. IEEE 63,* 1975, pp. 740–767.

17. Feynman, R., *Optics News,* 11–20 (1985).

18. Keyes, R.W., *VLSI Electronics: Microstructure Science*, Einspruch, Norman G., ed., New York: Academic Press, 1981, vol. 1, ch. 5, pp. 185–229.

19. Solomon, P.M., "A Comparison of Semiconductor Devices for High-Speed Logic," *Proc. IEEE,* **70** 1982, pp. 489–509.

20. Davidson, A., and Beasley, M.R., *IEEE J. Solid-State Cir.* **SC-14**, 758–762 (1979).

21. Bate, R.T., *VLSI Electronics: Microstructure Science* **5**, 359–386 (1982).

22. "Low Power Electronics," Terman, L.M., and Yan, R.H., eds., *Proc. IEEE* **83**, 1995, pp. 495–697.

23. Warwick, C.A., and Ourmazd, A., *IEEE Trans. Semicond. Manufacturing,* **6, 3** , 284–289 (1993).

24. Chatterjee, Pallab K., in *VLSI Electronics: Microstructure Science,* Einspruch, Norman G., and Huff, Howard eds., New York: Academic Press, 1985, vol. 12, ch. 7, pp. 307–383.

25. Kohyama, S., "Semiconductor Technology Crisis and Challenges Towards the Year 2000," in *IEEE 1994 Symp. VLSI Tech.,* pp. 5–8.

26. Barrett, C.R., "Microprocessor Evolution and Technology Impact," *IEEE 1993 Symp. on VLSI Tech.*, pp. 7–10.

27. Warwick, Colin A., *et al.*, *AT&T. Tech. Journ.* **72 (5)**, 50–59 (1993).

28. Ko, P.K., in *VLSI Electronics: Microstructure Science,* Einspruch, Norman G., and Huff, Howard, eds., New York: Academic Press, 1989, vol. 18, ch.1, pp. 1–37.

29. Tsividis, Y.P., *The Operation and Modeling of the MOS Transistor,* New York: McGraw-Hill, 1987.

30. Mead, C., *Analog VLSI and Neural Systems, Addison-Wesley VLSI System Series,* Reading, MA (1989) p. 54.

31. Brews, J.R., in *Submicron Intergrated Circuits*, Watts, R.K., ed., New York: John Wiley & Sons, 1989, pp. 269–331.

32. Christie, P., "A Fractal Analysis of Interconnection Complexity," *Proc. IEEE 81,* 1993, pp. 1492–1499.

33. Donath, W.E., *IEEE Trans. Circuits System* **CAS-26**, 272–277 (1979).

34. Davis, J.A., De, V., Meindl, J., *1996 IEEE Symp. VLSI Tech. Digest* 78–79.

35. Ferry, D.K., Grondin, R.O., and Akers, L.A., in *Submicron Intergrated Circuits*, Watts, R.K., ed., New York: John Wiley & Sons, 1988, pp. 377–412.

36. Meindl, J.D., and Davis, J., *Mat. Chem. and Phys.* **41**, 161–166 (1995).

37. Mead, C., and Conway, L., *Introduction to VLSI Systems,* Reading MA: Addison-Wesley, 1980.

38. von Neumann, J., *The Computer and the Brain*, New Haven, CT: Yale Univ. Press, 1958.

39. Lo, A.W., in *Micropower Electronics*, Keonjian, E., ed., New York: McMillian, 1964, pp. 19–39.

40. Swanson, Richard M., and Meindl, James D., *IEEE Journ. of Solid State Cir.,* **SC-7**, 14–153 (1972).

41. Glasser, Lance A., and Dobberpuhl, Daniel W., *The Design and Analysis of VLSI Circuits,* Reading, MA: Addison-Wesley Publishing Co., 1985, pp. 208–210.

42. Pimbley, J.M., *et al.*, in *VLSI Electronics Microstructure Science*, New York: Academic Press, 1989, vol. 19.

43. Yoshimura, H., *et al.*, *IEDM Tech. Dig.* 909–912, Dec. (1992).

44. Andoh, T., Furukawa, A., and Kunio, T., *IEEE 1994 IEDM,* 79–83.

45. Glasser, L.A., and Hoyet, L.P.J., "Delay and Power Optimization in VLSI Circuits," in *21st Design Automation Conf.*, Albuquerque, NM, pp. 529–535 1984.

46. Fischetii, M.V., and Laux, S.E., *Phys. Rev B* **48**, 2244–2274 (1993) and Fischetii, M.V., and Laux, S.E., *Phys. Rev B* **88**, 9721–9745 (1988).

47. Stork, J.M.C., *Proc. IEEE 83* (**4**), 607–618 (1995).

48. Kobayashi, T., and Sakurai, T., "Self-Adjusting threshold voltage scheme for low-voltage high speed operation," in *IEEE 1994 Custom Integrated Circ. Conf.*, p. 271.

49. Sun, S.W., and Tsui, P.G.Y., "Limitations of CMOS supply voltage scaling by MOSFET threshold voltage variation," in *IEEE 1994 Custom Integrated Circ. Conf.*, p. 267.

50. Mii, Y., *et al.*, "An Ultra-Low Power 0.1 μm CMOS," in *1994 IEEE Symposium on VLSI Tech. Digest of Technical Papers*, p. 9 (1994).

51. Landauer, R., *IBM Journ. Res. Dev.*, **5**, 183 (1961).

52. Nishizawa, J., *et al.*, *IEEE Trans. Electron Devices* **27**, 1640–1649 (1980).

53. Coones, M., *et al.*, *SPIE Microelectronics Manufacturing and Reliability* **1802**, 10–23.

54. Seevinck, E., List, F.J., and Lohstroh, J., *IEEE Journ. Solid-State Cir.* **SC-22**, 748–754 (1987).

55. Burnett, D., *et al.*, "Implication of Fundamental Threshlold voltage variations for High-Density SRAM and Logic Circuits," in *1994 Symp. VLSI Tech. Digest*, pp. 15–16.

56. Sakata, T., *et al.*, *IEEE J. Solid State Cir.* **29**, 761–769 (1994).

57. Kanai, H., *IEEE Trans. Components, Hybrids, Manuf. Tech.* **CHMT-4(2)**, 173 (1981).

58. Bar-Cohen, A., *IEEE Trans. Comp. Hybrids, Manuf. Tech.* **CHMT-19**, 159 (1987).

59. Powers, R.A., *Proc. IEEE* **83**, 687 (1995).

60. Kramer, A., *et al.*, "Adiabatic Computing with the 2N-2N2D Logic Family", in *1994 IEEE Symp. on VLSI Circuits Digest*, pp. 25–26.

61. Younis, S.G., and Knight, T.F., "Practical implementation of charge recovering asymptotically zero power CMOS," in *Research on Integrated Systems: Proceedings of the 1993 Symposium*, Cambridge, MA: MIT Press, 1993.

62. Koller, J.G., and Athas, W.C., "Adiabatic switching, low energy computing, and the physics of storing and erasing information," in *Proceedings of Physics of Computation Workshop*, Dallas, TX, October 1992.

63. Dickinson, A.G., and Denker, J.S., "Adiabatic Dynamic Logic," AT&T Bell Laboratories, Internal Memorandum, January (1993).

64. Johnson, Barry W., *Design and Analysis of Fault-Tolerant Digital Systems,* New York: Addison Wesley, 1989.

65. von Neumann, J., in *Automata Studies*, Shannon, C.E., and McCarthy, J., eds., Princeton, NJ: Princeton University Press, 1956, pp. 329–378.

66. Takeda, E., *Physics World* 48–52 (March 1993).

67. Fu, K.Y., *Appl. Phys. Lett.* **65**, 833–835 (1994).

68. Woods, M.H., "The implications of scaling on VLSI reliability," in *Proc. 22nd Annual International Reliability Physics Symposium,* 1984.

69. Sofield, C.J., and Stoneham, A.M., *Semicond. Sci. Technol.* **10**, 215–244 (1995).

70. Moazzami, R., and Hu, C., *IEEE Trans. Elec. Dev.* **TED-37**, 1643 (1990).

71. Schuegraf, K.F., King, C.C., and Hu, C., "Ultra-thin Silicon disoxide Leakage Current and Scaling Limit," in *1992 Symposium on VLSI Technology Digest of Technical Papers,* 1992, p. 18.

72. DiMaria, D.J., *Appl. Phys. Letter* **51**, 655–658 (1987).

73. Itsumi, M., and Muramoto, S., "Gate Oxide Thinning Limit Influenced by Gate Materials," *1985 Symposium on VLSI Technology,* Japan Society of Applied Physics, p.22.

74. Maserjian, J., *J. Vac. Sci. Technol.* **11**, 996–1003 (1974).

75. Lenzlinger, M., and Snow, E.H., *J. Appl. Phys.* **40**, 278 (1969).

76. Hu, C., *J. Vac. Sci. Technol. B* **12**, 3237–3241 (1994).

77. Schuegraf, K.F., Park, D., and Hu, C., "Reliability of Thin SiO_2 at Direct-Tunneling Voltages," *IEEE 1994 IEDM,* pp. 609–613.

78. Murrell, M.P., *et al., Appl. Phys. Lett.* **62**, 786 (1993).

79. Momose, H.S., *et al., IEEE IEDM Tech. Dig.* 593–596, (1994).

80. Heyns, M., *et al.,* 1992 Extended Abstracts, *Int. Conf. on Solid State Devices and Materials,* Tsukuba, Japan, p. 187.

81. M. Hirose, *et al., Journ. Vac. Sci. Tech. A* **12**, 1864 (1994).

82. Depas, M., *et al., Proc. 2nd Int. Symp on Ultra Clean Processing of Silicon Surfaces,* Heyns, H., *et al.* ed., (Lueven: Uitgeverij Acco), 1994, p. 319.

83. Tang, M., *et al., Appl. Phys. Lett.* **64**, 748–750 (1994).

84. Hartstein, A., Ning, T.H., and Folwer, A.B., *Surf. Sci.* **58**, 178–181 (1976).

85. Tang, Mau-Tsu, *et al., Appl. Phys. Lett.* **62 (24)**, 3144–3146 (1993).

86. Hahn, P.O., and Henzler, M., *J. Vac. Sci. Tech. A* **2**, 574 (1984).

87. Krylov, M.V., and Suris, R.A., *Sov. Phys. JETP* **61 (6)**, 1303 (1985).

88. Raikh, M.E., and Ruzin, I.M., in *Mesoscopic Phenomena in Solids,* Altshuler, B.L., Lee, P.A., and Webb, R.A., eds., Elsevier Science, 1991, pp. 315–368.

84   G. Timp, R.E. Howard, and P.M. Mankiewich

89. Ono, M., *et al.*, *IEEE IEDM Tech. Dig.* 119–122 (1993).

90. Meindl, J.D., *IEEE Trans. on Electronic Devices* **ED-31, 11** 1555–1561 (1984).

91. Landman, B.S., and Russo, R.L., *IEEE Trans. Computers,* **C-20**, 1469–1479 (1971).

92. Also see: Chiba, T., *IEEE Trans. Comput.* **C-27**, 319 (1975).

93. Semiconductor Technology Workshop Working Group Reports, Semiconductor Industry Association, San Jose, CA (1992).

94. Dennard, R.H., *et al.*, *IEEE Journ. Solid State Circuits* **SC-9, No. 5**, 256 (1974).

95. Bakoglu, H.B., *Circuits, Interconnections and Packaging for VLSI,* New York:Addison-Wesley, 1990.

96. Rahmat, K. *et al.*, *IEEE 1995 IEDM* 245–248.

97. Frank, D.J., Laux, S.E., and Fischetii, M.V., *IEEE IEDM Tech. Digest* 553–556, 1992.

98. Brews, J.R., *et al.*, *IEEE Elec. Dev. Lett.* **EDL-1, (1)**, 2–4, (1980).

99. Yan, R.H., *et al.*, *IEEE Trans. Elec. Dev.* **ED-39**, 1704 (1992).

100. Yan, R.H., *et al.*, *Appl. Phys. Lett.* **59**, 3315 (1991).

101. Aoki, M., *et al.*, *IEEE IEDM Tech. Dig.* 939–943 (1990).

102. Hori, A., *et al.*, "A 0.05 μm-CMOS with Ultra Shallow Source/Drain Junctions Fabricated by 5 keV Ion Implantation and Rapid Thermal Annealing," *IEEE 1994 IEDM*, pp. 94–97.

103. Yan, R.H., *et al.*, *IEEE Elec. Dev. Lett.* **13, (5)**, 256–258 (1992).

104. Taur, Y., *et al.*, *IEEE IEDM Tech. Digest* 127–130 (1993).

105. Nowak, E., *IEEE IEDM Tech. Digest* 115–118 (1993).

106. Mizuno, T., *et al.*, "Performance Fluctuations of 0.1 μm MOSFETs – Limitations of 0.1 μm ULSI," *IEEE 1994 Symposium on VLSI Technology*, p. 13–15.

107. Sakurai, T., and Newton, A.R., *IEEE J. Solid-State Cir.* **25**, 584–594 (1990).

108. Wong, Hon-Sum, and Taur, Yuan, *IEEE IEDM Tech. Dig.* 705 (1993).

109. Mizuno, T., Okamura, Jun-ichi, and Toriumi, Akira *VLSI Symp.* 1993, p. 41.

110. Nishinohara, K., *et al.*, *IEEE Trans. Electron Devices* **ED-39**, 634 (1992).

111. Keyes, R.W., *IEEE J. Solid State Circuits* 245–247 (1975).

112. Mizuno, T., Okamura, J., and Toriumi, A., *IEEE Trans. Elec. Dev.* **41**, 2216–2221 (1994).

113. Mizuno, T., Toriumi, A., *J. Appl. Phys.* **77**, 3538–3540 (1995).

114. Ratnakumar, K.N., Meindl, J.D., and Scharfetter, D.L., "New IGFET Short-Channel Threshold Voltage Model," *IEEE Internal Electron Device Meeting*, pp. 204–206, Washington,

D.C. 1981.

115. De, V.K., Tang, X., Meindl, J.D., *54th Device Research Conf. Digest* 114–115 (1996).

116. Kane, E.O., and Blount, E.I., in *Tunneling Phenomena in Solids*, Burstein, E., and Lundqvist, S., eds., New York: Plenum Press, 1969, pp. 79–91.

117. Fair, R.B., and Wivell, H.W., *IEEE Trans. Electron Devices* **ED-23**, 512 (1976).

118. Fulton, T.A., and Dolan, G.J., *Phys. Rev. Lett.* **59**, 109 (1987).

119. *Single Charge Tunneling: Coulomb Blockade Phenomena in Nanostructures,* Grabert, H., and Devoret, M.H., NATO ASI series, Plenum Press, 1991, vol. 294.

120. Tucker, J.R., *J. Appl. Phys.* **72**, 4399–4413 (1992).

121. Lutwyche, M.I., and Wada, Y., *J. Appl. Phys.* **75**, 3654–3661 (1994).

122. van Houten, H., Beenakker, C.W.J., and van Wees, B.J., in *Semiconductors and Semimetals,* Reed, M.A., volume ed., New York: Academic Press, 1991.

123. Sols, F., *et al., Journ. Appl. Phys.* **66 (8)**, 3892–3906 (1989).

124. *The New Superconducting Electronics,* Weinstock, H., and Ralston, R.W., eds., Boston: Kluwer Academic Publishers, 1993.

125. Allee, D.R., *et al., J. Vac. Sci. Technol. B,* **6**, 328–332 (1988).

126. Allee, D.R., Broers, A.N., and Pease, R.F.W., "Limits of Nano-Gate Fabrication," *Proc. IEEE. 79,* 1991, pp. 1093–1105.

127. Yokoyama, M., *et al., IEEE Electron Dev. Lett.* **15**, 202–205 (1994).

128. Dike, R.S.U., *Int. Journ. Electronics* **76**, 403–415 (1994).

129. Gildenblat, G., in *VLSI Electronics Microstructure Science,* New York: Academic Press, 1989, vol. 18, pp. 191–236.

130. Yano, K., and Ferry, D.K., *Superlatt. Microstruc.* **11**, 61 (1992).

131. Yano, K., and Ferry, D.K., *Phys. Rev. B* **46**, 3865 (1992).

132. Averin, D.V., Korotkov, A.N., and Likharev, K.K., *Phys. Rev. B* **44**, 6199 (1991).

133. Averin, D.V., and Likharev, K.K., *Mesoscopic Phenomena in Solids,* Altshuler, B.L., Lee, P.A., and Webb, R.A., eds., Elsevier Science 1991, pp. 173–271.

134. van Houten, H., Beenakker, C.W.J., and Staring, A.A.M., in *Single Charge Tunneling: Coulomb Blockade Phenomena in Nanostructures,* Grabert H., and Devoret, M.H., eds., NATO ASI Series B: Physics Vol. 294, New York: Plenum Press, 1992, p. 167.

135. Likharev, K.K., *IEEE Trans. Magnetic,* **MAG-23**, 1142–1145 (1987).

136. Delsing, P., *et al.*, *IEEE Trans. Mag.* **27, (2)**, 2581 (1991); and Likharev, K.K., *IEEE Trans. Magn.* **23**, 1142 (1987).

137. Zimmerli, G., Kautz, R.L., and Martinis, J.M., *Appl. Phys. Lett.* **61 (21)**, 2616 (1992).

138. Schönenberger, C., and van Houten, H., *1992 International Conference on Solid State Devices and Materials*, 1992, p. 726.

139. Korotkov, A.N., Chen, R.H., and Likharev, K.K., *J. Appl. Phys.* **78**, 2520–2530 (1995).

140. Liu, H.I., *et al.*, *Appl. Phys. Lett.* **64**, 2010 (1994).

141. Averin, D.V., and Likharev, K.K., in *Single Charge Tunneling,* Brabert, H., and Devoret, M.H., eds., New York: Plenum Press, 1992, ch. 9, pp. 311–332.

142. Brodie, I., *Phys. Rev. B* **51**, 660 (1995).

143. Geerligs, L.J., *et al.*, *Phys. Rev. Lett.* **64**, 1691 (1990).

144. Washburn, S., and Webb, R.A., *Advance Phys.* **35**, 375–422 (1986).

145. van Wees, B.J., *et al.*, *Phys. Rev. Lett.* **60**, 848 (1988).

146. Wharam, D.A., *et al.*, *J. Phys. C* **21**, L209 (1988).

147. Landauer, R., *Localization, Interaction, and Transport Phenomena*, Bergmann, G., and Bruynseraede, Y., eds., New York: Springer-Verlag, 1985, p.38–50.

148. Economou, E.N., and Soukoulis, C.M., *Phys. Rev. Lett.* **46**, 618 (1981).

149. Glazman, L.I., *et al.*, *JETP Lett.* **48**, 238 (1988); [*Pis'ma Zh. Teor. Fiz.* **48**, 218–220 (1988)].

150. Datta, S., *Superlattices and Microstructures* **6**, 83–93 (1989).

151. Timp, G., *et al.*, *Proc. Int. Symposium on Nanostructure Physics and Fabrication,* Reed, M.A., and Kirk, W.P., eds., Academic Press, 1989, p. 331.

152. Davies, John H., and Timp, Gregory, *Heterostructures and Quantum Devices,* Academic Press, 1994, pp. 385–418.

153. Nixon, J.A., Davies, J.H., and Baranger, H.U., *Phys. Rev. B* **43**, 12638 (1991).

154. Nixon, J.A., and Davies, J.H., *Phys. Rev. B* **41**, 7929 (1990).

155. J. Davies, private communication.

156. Koester, S.J., *et al.*, *Phys. Rev. B* **49**, 8514 (1994).

157. Likharev, K.K., and Semenov, V.K., *IEEE Trans. Appl. Supercond.* **1**, 3–28 (1991).

158. Kwon, O.K., *et al.*, *IEEE Electron Dev. Lett.* **EDL-8**, 582–585 (1987).

159. Matisoo, J., *IBM J. Res. Develop.* **24**, 113–129 (1980).

160. Anacker, W., *IBM J. Res. Develop.* **24**, 107–112 (1980).

161. Bunyk, P.I., *et al.*, *Appl Phys. Lett.* **66**, 646–648 (1995).

162. Hamilton, C.A., and Gilbert, K.C., *IEEE Trans. Appl. Superconduct.* **1**, 157–162 (1991).

163. Garnier, F., *et al.*, *Science* **265**, 1684–1686 (1994).

164. Garnier, F., *Recherche,* **26**, 76–77 (1995).

165. SEMATECH, Technology Transfer #91080669E-GEN (1993).

166. B.D. Ackland, private communication.

167. D.A. Antoniadas and J.E. Chung, *IEEE IEDM Tech. Dig.* 21 (1991).

Chapter 3

Nanostructures in Motion: Micro-Instruments for Moving Nanometer-Scale Objects

N.C. MacDonald

Department of Electrical Engineering
Cornell University,
Ithaca, New York 14853

"I would like to describe a field, in which little has been done, but in which an enormous amount can be done in principle."

"What I want to talk about is the problem of manipulating and controlling things on a small scale."

Richard P. Feynman, "There's Plenty of Room at the Bottom", December 26, 1959[1].

I. INTRODUCTION

This chapter describes the integration of micro-actuators and nanometer-scale tips to manipulate and control things on a small scale. The micro-instruments or micro-robots are made using micro-machining technology. The micro-machining field has been given the generic name micro-electromechanical systems or MEMS. This field is quite broad and includes integrated micro-sensors, micro-actuators, micro-instruments, micro-optics and micro-fluidics, and includes applications ranging from accelerometers[2] to deploy an automobile airbag, ink jet printer heads[3,4]; an array of movable mirrors for color projection displays[5]; to atom probes for imaging and transporting atoms[6].

Figure 1 shows two versions of atom probes or micro-scanning tunneling microscopes[6] (micro-STMs) that were micro-machined from single crystal silicon (SCS). One micro-STM measures a few millimeters on a side [Figure 1(a)], and a smaller micro-STM [Figure 1(b)] measures a few hundred micrometers on a side. Each micro-STM includes integrated x-y capacitive micro-actuators, [the comb-like structures in Figure 1(c)]; an integrated tunneling tip [Figure 1(d)] mounted on a 'teeter-totter' torsional z motion-micro-actuator [Figure 1(e)]; microstructural supports and springs [Figures 1(a-b)]; and wiring integrated on

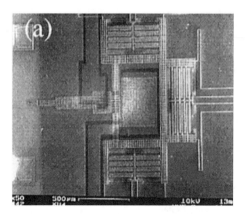

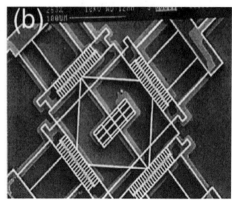

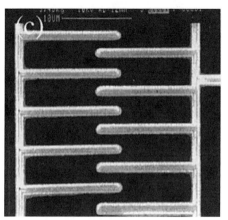

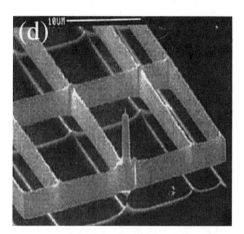

FIGURE 1. (a) SEM micrograph of the 2 mm × 2 mm micro-STM. Scanning in x and y is generated by interdigitated capacitor micro-actuators. The integrated 20 nm diameter tip is moved in the vertical (z) direction using a torsional micro-actuator. (b) SEM micrograph of a 200 μm × 200 μm micro-STM. (c) Interdigitated or comb drive micro-actuator. The maximum available displacement left-to-right is 10 μm. (d) SEM micrograph showing the integrated single crystal silicon tip for the micro-STM in (b). (e) Parallel-plate, torsional micro-actuator with integrated tip for the micro-STM in (c).

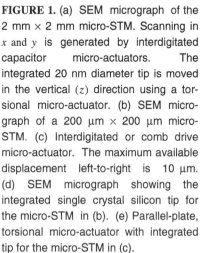

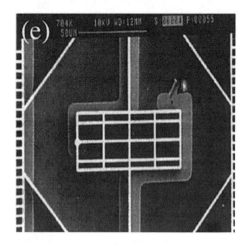

the suspended microstructures to supply power to the tip and the micro-actuators. The larger STM [Figure 1(a)] was used to image 300 nm metal lines on a 'silicon-chip-sample' placed on top of the micro-STM and supported by integrated SCS posts – seen on the left side of Figure 1(a). The building blocks used to make these 'micro-STMs' – the micro-actuators, the tips and probes, the silicon micro-machining processes, the micro-system architecture, and the design methods – are the basic components required to build micromanipulators and micro-robots for nanometer-scale manipulation. A discussion of the theory, architecture, design, fabrication and characterization of these micro-machined components forms the basic content of this chapter.

A. Introduction to MEMS

MEMS can be defined as the fabrication or micro-machining of materials to make stationary and moving structures, devices and systems of a nominal size-scale from 'a few centimeters to a few micrometers.' One major paradigm for MEMS is the integration of micro-machines with microelectronics such as silicon and gallium arsenide devices and integrated circuits. This is facilitated by the fact that many of the techniques used to fabricate MEMS are similar to or based on the batch fabrication techniques used in the fabrication of microelectronic circuits.

Although many unique MEMS fabrication processes exist, with new techniques being researched and developed, almost all the processes fall into the two general categories of surface micro-machining and bulk micro-machining.

Surface micro-machining consists of those processes that are used to fabricate micro-electromechanical structures from films deposited on the surface of a substrate. The most widely used surface micro-machining technique is the thin film polysilicon process. In this process, released and movable structures are fabricated on a single crystal silicon substrate from thin layers of polysilicon deposited on a sacrificial layer of silicon dioxide. A diagram of the process used to make a movable comb or interdigitated electrode micro-actuator is shown in Figure 2[7-10]. The comb drive structure is partially released by etching the silicon dioxide beneath the polysilicon layer. The released structure is supported on the periphery and insulated from the substrate with the remaining silicon dioxide or silicon nitride.

Many devices have been fabricated using the thin film polysilicon process, and a MEMS foundry has been established[11] to provide a service for fabricating prototype surface micro-machines and LIGA MEMS devices. Devices fabricated using the polysilicon on a sacrificial layer process include rotating micro-motors[9], an array of thermal, polysilicon infrared light sources, a room-temperature infrared camera[12], and accelerometers[2] to name just a few.

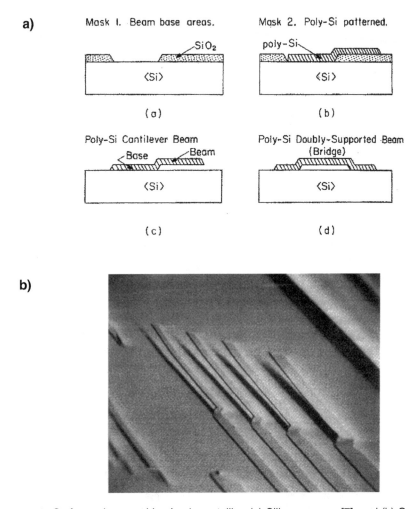

FIGURE 2. Surface micro-machined polycrystalline (a) Silicon process[7], and (b) Cantilever beams.

For the most part, the integration of released thin film structures and actuators on silicon integrated circuits has been quite successful. The main disadvantage of using thin films for structural materials is that the internal stress of the films limits the thickness of the layer to a nominal thickness of a few micrometers. The film thickness severely limits the size of thin film-based microstructures and the force produced by thin film micro-actuators ($\approx 100 \ \mu N$). Also, during the release step care must be taken not to allow permanent sticking and attachment of the released structures to the substrate or to other released structures. Fluid forces generated during the release process and structural sticking[13] due to van der Waals forces present significant challenges in MEMS manufacturing. For all of these reasons, the main applications – and very important applications – of polysilicon are for small micro-sensors and arrays of micro-sensors: accelerometers, pressure sensors, mirror arrays, micro-mechanical filters and many others. When compared with SCS, polysilicon has inferior electronic properties and many of its mechanical properties depend on the grain boundaries including size, structure, and internal stress. Because of the variation in the thin film properties, there is a saying in the MEMS community that "no two groups make the same polysilicon". Thus, characterization and standardization of the mechanical and electrical properties of polysilicon is an ongoing challenge. All thin films and thin film processes require routine characterization to certify that their physical and chemical properties are stable and predictable.

Thin films of metals and other materials, with and without a sacrificial layer, have been used instead of polysilicon for many applications. These thin film structures are used to achieve specific or selective properties[8,14,15] including high reflectivity (e.g. *Al, Au*); high mass density[16] (e.g. *W, Au, Pt*); low stiffness (e.g. polymers); specific adsorption and adhesion characteristics (e.g. *Pd, Ir, Au, Pt*, polymers and complex biological molecules such as enzymes and proteins)[8]; and stable damping characteristics.

An additive metal process using the selective chemical vapor deposition of tungsten on silicon or a metal seed layer [Figure 3(a)] can be used with or without a sacrificial layer of silicon dioxide to make multiple-level MEMS microstructures on fixed or moving substrates[16]. The tungsten films can be mechanically polished before the release-step to obtain optically flat surfaces. Such multiple layer tungsten structures are used to make mirrors, to add mass to structures, to make compact micro-actuators [Figure 3(b)], and to make electrodes and electrostatic lenses to focus charged particles. In addition, the tungsten process and other similar metal processes can be used to make microelectron/ion optics [Figure 3(c)] for micro-instruments with integrated tips[17]. The tungsten process can be integrated with the SCS processes outlined in Section V. The integration of thin films on larger moving microstructures is an important paradigm for MEMS.

Another approach to micro-machining involves an additive process of plating metal structures from a patterned seed-layer[18-21]. A number of very high aspect ratio structures and mechanical actuators have been fabricated using plating technologies. Examples of plated MEMS devices and structures include magnetic actuators and motors, inductors, gears and gear trains. Acronyms for the plating processes are LIGA[18] and SLIGA. (SLIGA is LIGA with sacrificial layers[19].)

The second category of MEMS processes uses the bulk material or substrate, single crystal silicon (SCS) for example, to form movable microstructures by etching directly into the bulk material. This approach to micro-machining is

(a)

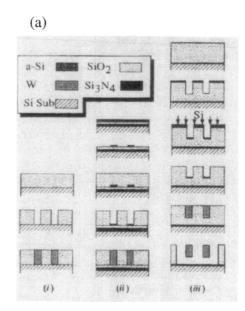

(b)   (c)

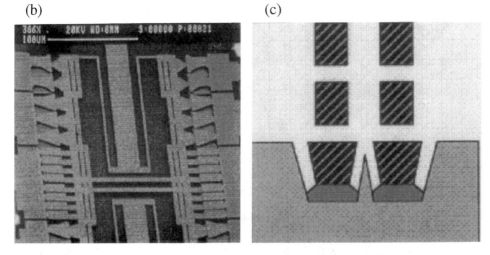

FIGURE 3. Tungsten surface micro-machining process. (a) Process sequences for selective chemical vapor deposition of tungsten, for seeding and growth of tungsten (i) exposed substrate silicon; (ii) thin film polysilicon or metals; and (iii) silicon dioxide implanted with silicon[16]. (b) SEM micrograph of a three level surface micro-machined tungsten interdigitated electrode micro-actuator[17], and (c) schematic of a three element electrostatic tungsten lens and field emitter tip.

called bulk micro-machining[8,14]. The structures can be formed by wet chemical etching[22-24] or by reactive ion etching. Wet chemical etching, using a boron etch-stop layer[25], is another technique to make large MEMS structures. Many suspended microstructures have been fabricated using wet chemical micro-machining including neural probes[26], actuators and sensors. The advantage of bulk micro-machining and chemical etching is that excellent substrate materials such as quartz and SCS are readily available, and relatively large scale (milli-meter to centimeter-scale), high aspect-ratio structures can be fabricated. Bulk micro-machining of SCS allows the integration of active devices and the use of integrated circuit technologies. However, the disadvantages of wet chemical bulk micro-machining include significant pattern and structure sensitivity and pattern distortion due to different selective etch rates on different crystallographic planes[22-24]; the use of both front side and back side processing; severe limits

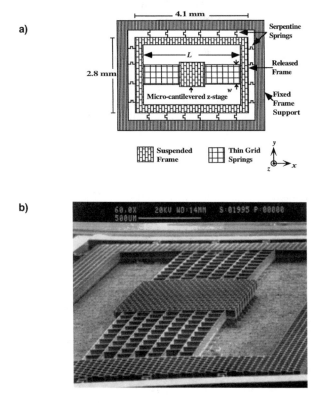

FIGURE 4. High aspect-ratio single crystal silicon structure. (a) Schematic, (b) SEM micrograph. The structure height is 85 μm and the minimum feature size is 1.5 μm[32].

FIGURE 5. SEM micrograph of a single crystal silicon accelerometer fabricated on a pre-processed, commercial wafer with operational amplifiers.

and constraints on the minimum feature size (MFS) and minimum feature spacing due to the wet chemical etching process; and the industry trend to reduce or eliminate wet chemical processing.

Bulk micro-machining is also accomplished by reactive ion etching (RIE) of the substrate material[27-31]. With RIE, it is possible to obtain sub-micron MFS structures and spacing and to produce very high aspect-ratio structures [Figure 4] of greater than 50:1 vertical height (depth into the substrate) to width in SCS[32]. High aspect-ratio, micro-machined SCS structures produce very high force micro-actuators due to the large surface area of the sidewalls (Section III); produce large area suspended structures with sub-micron features; facilitate the decoupling of vibrational modes; provide for electrical and thermal isolation through the thermal oxidation of sections of the silicon structures; and provide the capability to fabricate transistors and electronics on the suspended structures. Also, RIE-produced MEMS can be added to wafers with prefabricated integrated circuits[33] using a low temperature RIE process[28] as shown in Figure 5; the moving SCS accelerometer was integrated after the operational amplifiers were made by a commercial supplier. This approach to the integration of MEMS and microelectronics partitions and separates a complex, multiple mask, planar silicon integrated circuit process from the non-planar, high aspect-ratio MEMS processes.

The RIE approach to bulk micro-machining is covered in detail in the sections that follow.

B. Introduction to the MEMS Literature

A number of MEMS journals and conferences are now available. MEMS journals include the Journal of Micro-Electromechanical Systems[34], the Journal of Micro-Machining and Micro-Engineering[35], Sensors and Actuators[36], and Micro-System Technologies[37]. MEMS conferences include the IEEE Micro-Electromechanical Systems Workshop[38], the International Conference on Solid-State Sensors and Actuators (Transducers)[39], and The Solid State Sensor and Actuator Workshop[40]. In addition, sessions on MEMS are included in the Conference on Smart Structures and Materials[41], and SPIE's International Symposium on Optoelectronic and Laser Technologies[42]. For optical MEMS technologies, *GaAs* is sometimes considered the material of choice. Two papers provide an introduction to *GaAs* for MEMS[29,43].

Two books devoted to MEMS and sensors were published in 1994: *Semiconductor Sensors,* edited by S. M. Sze[8]; and *Micro-Sensors, Principles and Applications,* by Julian W. Gardener[14]. Both books discuss MEMS processing, devices and systems and include substantial references to the MEMS and sensor literature up to 1993. A third book, *Silicon Sensors,* by S. Middlehoek and S. A. Audet, addresses issues related to sensors and MEMS[15]. Early MEMS research and a concise listing of MEMS applications through 1983 are presented in a classic paper: "Silicon as a Mechanical Material", by Kurt Petersen[44]. For an early vision of micro-machines and nanometer-scale manipulation, the two Feynman lectures[1,45] are full of ideas and concepts, and are highly recommended reading. MEMS paradigms, concepts and components are discussed in a recent paper[46]: "Engineering Microscopic Machines", by K. Gabriel. For a review of the generation of force and motion using electrical energy refer to a three volume set of books: *Electromechanical Dynamics*, by Woodson and Melcher[47]. These volumes are packed with information and ideas on capacitive and magnetic motors and actuators, on energy conversion and electromagnetic principles, and on elastic and fluid media – a required reference set for the MEMS researcher.

C. This Chapter and the Book

"Nanotechnology" encompasses nanometer-scale processes, materials, structures, devices and systems, and involves nanometer-scale lithography, and nanometer-scale information storage, nanometer-scale motion, nanometer-scale probing and characterization, and nanometer-scale manipulation of objects such as atoms and molecules, and nanometer-scale information storage. All these subjects are addressed to some degree in the chapters of this book. This chapter deals in particular with SCS micro-machines, micro-instruments and micromanipulators used to move, position, probe, pattern, and characterize nanometer-scale objects and nanometer-scale features. Figure 1 shows such micro-instruments:

micro-STMs made using silicon micro-machining technology. We consider micro-system concepts and architectures, and the micro-machining processes used to fabricate micro-systems and micro-actuators with integrated tips that translate and rotate in three dimensions. Thus, the chapters on Precision Atom Manipulation (Chapter 11), and Physical Properties of Nanometer-scale Magnets (Chapter 12) address aspects of moving nanometer-scale probes and tips. The topics covered in those chapters are related to and can benefit from the material discussed in this chapter.

Conventional lithography, microelectronics, and conventional information storage technologies will experience major difficulties in reaching the nanometer-scale (Chapters 2 & 4). Consequently, there is interest in the possibility of building high-throughput, massively parallel, scanned-tip MEMS or micro-machined electron beam arrays[17] to pattern surfaces at the nanometer-scale, to store information at the nanometer-scale, and to manipulate and image atoms and molecules in a massively parallel fashion. Such an approach to nanometer-scale patterning will compliment and extend the applications for nanometer-scale materials and processes (Chapters 5-8). Here we discuss how arrays of micro-machines or micromanipulators made using silicon-based micro-machining can be used to produce a massively parallel array of electron beams or nanometer-scale tips to pattern and probe surfaces.

D. Chapter Organization

The chapter begins with an introduction to micro-machines (Section II) for the manipulation of nanometer-scale objects, and identifies the need and the challenge of making integrated, high force micro-actuators that can translate and rotate an object such as a nanometer-scale tip. Massively parallel micro-system concepts and architectures are introduced; and we identify the challenges of making 'scaled-down', compact (50 µm on a side) micro-actuators, and the challenge of wiring and connecting a large number of micro-actuators on suspended and moving microstructures.

The following sections of the chapter address these challenges in some detail, beginning with Section III which deals with the force and torque produced by capacitive micro-actuators. Micro-actuators that produce translation and torsional rotation are analyzed, and a scaling law is derived for each micro-actuator. The scaling law for the force or torque produced by each micro-actuator is related to the minimum feature size (MFS), a, and the height, b, of the capacitor plate or beam, the maximum displacement or rotation produced by the micro-actuator, and the area of the silicon-chip. Therefore, the key processing parameters: lithography and deep-etch RIE pattern transfer are identified and highlighted.

In Section IV, we return to discuss some microstructural issues related to supporting large array architectures of suspended and nested micro-actuators for manipulating microstructures and nanometer-scale tips. We identify structural stiffness, structural planarity and internal structural stress as important design and process related challenges in making micro-robotic systems. The need for high aspect-ratio, stiff microstructures is again identified as a key process requirement.

Section V discusses SCS as a material for MEMS and describes the SCS single crystal reactive etching and metalization (SCREAM) process. SCREAM meets the process requirements needed to build array architectures of micro-actuators and micro-robots with integrated nanometer-scale tips which are discussed in Sections II–IV, and IX.

The frequency response and frequency scaling of micro-actuators and MEMS are discussed in Section VI. The nonlinear properties of bending beams, microstructures and electric field-stiffened structures are discussed in Section VII. The force vs. displacement of a bending beam has both a linear term – Hooke's law, $F = -k_H x$, and a cubic term, $F \propto x^3$. These nonlinearities are used to design systems – some are chaotic – with deep potential wells with adjustable barriers, and are used to make voltage programmable nonlinear and linear resonators and other nonlinear devices.

Section VIII treats the integration of high aspect-ratio lateral and vertical probes and tips on moving, torsional cantilevers. The use of an integrated stiff cantilever beam attached to a torsional beam is identified as a new approach for making high sensitivity cantilevers for atomic force measurements and scanned-probe instruments. The concept of minimal detectable force for the new torsional cantilever is introduced.

Section IX, *Array Architectures and Nanometer-scale Manipulators,* 'puts the pieces together' by describing array architecture designs using the scaling concepts and microstructures presented in the preceeding sections. Then, we describe micro-robots that can translate and rotate nanometer-scale tips in three dimensions using the micro-actuators, microstructures and SCS processes discussed in this chapter.

Section X discusses the scaling of the actuators and structures to 50 nm – 100 nm minimum feature size using electron beam lithography and applying the scaling laws developed in the chapter. We show orders of magnitude improvements in important parameters for nanometer-scale actuators.

The chapter ends (Section XI) with a few comments on nanometer-scale SCS robotics and micro-instruments for nanometer-scale manipulation, nanometer-scale characterization, nanometer-scale lithography, and nanometer-scale information storage. The question 'why do it?' is addressed.

II. MICRO-INSTRUMENTS AND MICRO-SYSTEM ARCHITECTURES

System Architecture

The MEMS system architecture needed to move a single tip or probe in three dimensions is very similar to that of a macroscopic STM. Figure 6 illustrates the system concept: a tip attached to three piezoelectric actuators is used to scan a surface in x-y-z. The tunneling current is used as a feedback signal to control the z actuator. The actuators for the macroscopic STMs are usually piezoelectric scanners of various designs to which a tip is attached; the most popular actuator being a piezoelectric tube scanner[48]. Piezoelectric thin film drives have also been used to make millimeter-scale STMs[49,50] using ZnO piezoelectric thin films and an array of four force microscopy probes.

The micro-STMs shown in Figure 1 use capacitive micro-actuators to scan an integrated tip in x-y-z. A parallel-plate, capacitive micro-actuator (Section III.A) consists of two capacitor plates that support a potential gradient between them [Figure 7(a)]. The potential gradient or electric field produces an attractive force between the two oppositely charged plates. When the capacitive micro-actuators are fabricated by micro-machining the plates from silicon or other materials as part of a suspended structure, the attractive force is used to move the structure. Figure 8 shows an example of a compact set of parallel-plate micro-actuators that move a suspended structure in x-y; the 'white plates' are biased negative with respect to the adjacent plates. The resulting attractive force moves the structure in

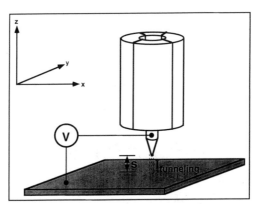

TUNNELING CURRENT: $I_{tunneling} \propto V \exp(-\kappa s)$; $\kappa \approx 10^{10}$ m^{-1}

FIGURE 6. Schematic of piezo tube scanner used in scanning tunneling microscopes. The tube scanner is used to position the tip in x, y, and z.

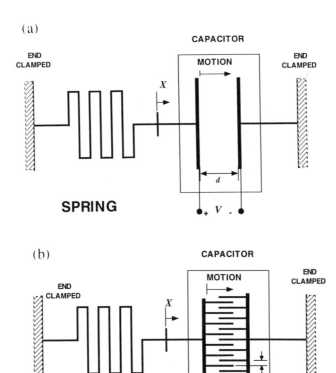

FIGURE 7. Capacitor micro-actuators (a) Parallel-plate micro-actuator with maximum displacement of d. (b) Interdigitated electrode or comb drive capacitor micro-actuator with maximum displacement proportional to l_p – not d.

the direction of the biased plates. Another capacitive micro-actuator, the interdigitated electrode micro-actuator (Section III.B), has one moving electrode (±) positioned between two (∓) fixed electrodes [Figure 7(b)]. The center electrode (±) moves parallel to the two fixed electrodes (∓) due to the fringing electric fields at the ends of the three plates. Figure 9 shows – on the same structure – the two types of SCS, high aspect-ratio capacitive micro-actuators: an interdigitated electrode capacitor [A] and a parallel-plate capacitive micro-actuator [D-E]. Both micro-actuators are connected to and supported by a SCS spring, [B]. Capacitive micro-actuators can be nested to produce x-y motion as shown in Figures 1 and 8.

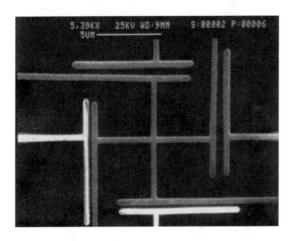

FIGURE 8. SEM micrograph of four parallel-plate capacitors suspended by springs. The cross in the center has been displaced downward and to the left by actuation of two sets of plates – white plates are negative. The single crystal silicon plates are 10 μm × 200 nm.

Figures 1 (a,b) illustrate the concept: two interdigitated electrode x micro-actuators move [Figure 7(b), 1(c)] the tip $\pm x$ while two y micro-actuators move the tip $\pm y$ and a torsional z micro-actuator [Figure 1(e)](Section III.C) moves the tip $\pm z$ respectively.

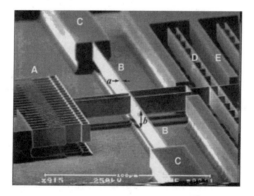

FIGURE 9. A high aspect-ratio (b/a) single crystal silicon device made using SCREAM processes[63]. [A]-Interdigitated electrode micro-actuator that moves the structure to the left when a voltage is applied to the two sets of electrodes. [B]-Suspended spring with supports-[C]; [D]-Moving suspended plate of a parallel-plate capacitor attached to the spring and a fixed plate [E]. The parallel-plate capacitor is used to sense displacement.

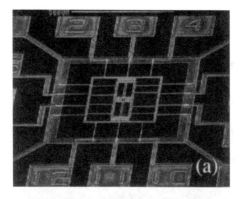

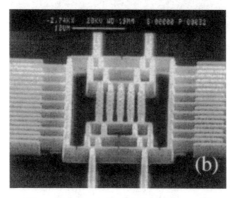

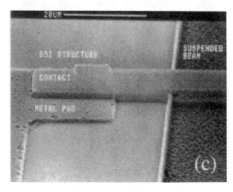

FIGURE 10. Micrograph of an *x-y-z* single crystal silicon stage-within-a-stage with integrated capacitive micro-actuators and a tip array. (a) The complete device shows bonding pads for electronic connections to the support springs. (b) One actuator with an array of twenty-five field emitter tips with apertures. (c) Detail of electrical isolation of a suspended beam.

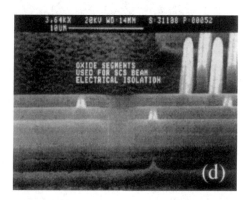

FIGURE 10. (d) Detail showing silicon dioxide segments for electrical isolation along the beam.

Another example of nested x-y actuators supporting an array of 25 field emitter tips[30] is shown in Figure 10. The larger interdigitated, comb-like micro-actuators [Figure 10 (a)] move the tip-array vertically in the y direction, and the smaller micro-actuators [Figure 10 (b)] in the center move the tip array horizontally in the x direction. For this structure the z motion is only inward toward the silicon substrate. Note all the actuators are attached to a rigid, suspended frame structure supported by fourteen suspended beams or springs that terminate on electrical contacts or wire bond pads [Figure 10 (c)]. Each of the fourteen SCS springs also acts as a suspended conductor to distribute power to the actuators and provide electrical connections to the tip and the tip-gate electrode. Electrical isolation is achieved by the insertion of silicon dioxide insulating sections along the SCS beams [Figure 10 (d)].

The devices shown in Figure 1 and Figure 10 highlight some of the important features related to MEMS architectures for micromanipulators and micro-instruments.

1. The x-y-z actuators are all suspended on a common, rigid backbone structure and operate independently.

2. Nesting of the actuators (actuator-within-an-actuator) helps to isolate the motions in the three coordinates – for example isolating the x motion from motions that occur in y and z.

3.   Electrical power and signal distribution is routed to the suspended devices by patterned conductors covering the suspended springs and support structures [Figure 10(c)].

4.   The micrometer-scale thermal isolation is also used to thermally isolate suspended devices and elements of an array and to isolate micro-heater elements.

5.   MEMS tip-array architectures can be used to move many tips (nanometer-scale objects) with one set of x-y or x micro-actuators while precise adjustment of each tip along the z axis would be accomplished by a compact, small deflection amplitude z axis micro-actuator – examples include a parallel-plate capacitive (Section III.A) or torsional micro-actuators (Section III.C).

Figure 11 shows a schematic of an array architecture. The two large arrays of interdigitated electrode micro-actuators (refer to Figure 9[A] for a 3D (three dimensional) perspective of the interdigitated electrode micro-actuator) located at each end of the array structure move the entire [10×10] tip-array ±25 μm in x. Each tip has an integrated z micro-actuator (Section III.C) like the one shown in Figure 1(e), which occupies an area of 50μm × 50μm. One 50 μm scan of the x micro-actuators generates 100 line scans in parallel, one for each tip. Each z micro-actuator 'writes a line' and adjusts for local topography. The backbone

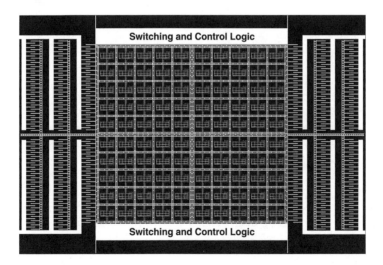

FIGURE 11. Schematic of an array architecture of 100 torsional, vertical actuators each with an integrated silicon tip. The entire array is moved horizontally with the interdigitated electrode micro-actuators positioned at each end.

structure is used to support the array and channel electrical energy to each array element. Electronic circuits for addressing are included on the moving structure to reduce the number of electrical connections required to be routed to the (stationary) silicon substrate via the support springs [Figure 10(d)].

An x-y scan is accomplished by the integration of a set of y micro-actuators (refer to Figure 10) to the array shown in Figure 11. Other 'two-chip' architectures are discussed in Section IX.

Micromanipulation architectures benefit from many of the concepts important in the design of integrated circuits and microcomputer systems. Array architectures and massively parallel architectures offer increased throughput – objects, pixels or signals handled per second – and take full advantage of 'on-chip' microelectronic circuits. As in the design of batch-fabricated integrated circuits, liberal use of 'on-chip' transistors and amplifiers, and the integration of sub-micrometer-scale electrical isolation, contacts [Figure 10(c)], and wiring are required to increase functional density of the circuits 'on-the-chip' (refer to Chapters 2 & 4). Likewise, massively parallel micromanipulators should be totally integrated and not require assembly of individual actuators, tips, or other moving structures. Integration, particularly with silicon processing, is fundamental to making large numbers of manipulators with acceptable yield and cost per micro-system.

The major component of the micromanipulator is the micro-actuator. The micro-actuator should be small, modular and capable of nesting to obtain independent motions along each axis. From the architectures described above and shown in Figures 10 and 12, an array of tunneling tips can be moved in parallel in the x-y plane, but each tip includes an individual, small amplitude z actuator to perform precision positioning of the tip along the z axis to adjust to the local surface topography [Figure 11 and 1(b,d)]. Thus, the spacing between adjacent tips in an array must be large enough to integrate a z micro-actuator with each tip. The area scanned by the x-y micro-actuators is then equal to the area of the z micro-actuator. Therefore, the x-y micro-actuators should produce enough force and displacement in the smallest 'chip-area' (volume) to move the structure, the array, over this area and to perform the desired function with the required force at the required speed (frequency).

Array architectures, particularly the massively parallel micromanipulator arrays, require compact (1-50 μm on a side) micro-actuators that can be integrated easily on suspended or sliding structures which are moved by larger micro-actuators. As more tips or micromanipulators are packed on the array, wiring, communication, and control become major issues. In the example shown in Figure 10, sixteen wires are connected to electrical bonding pads for wire bonding, but larger arrays will require on-the-structure decoding and routing of electrical signals like that shown in Figure 11. Figure 12 shows an example of the integration of a metal-oxide-semiconductor transistor[51] on a suspended SCS beam. The source and drain contacts and conductors are suspended nickel silicide

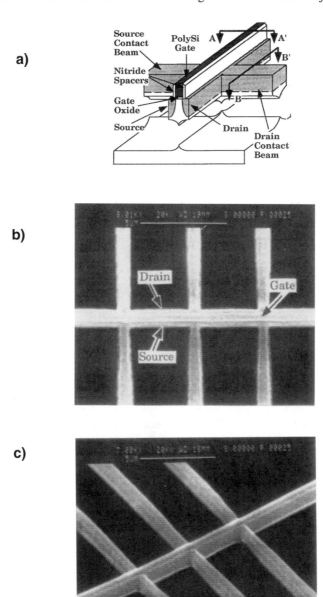

FIGURE 12. Field-effect or metal-oxide-semiconductor transistor fabricated on a suspended single crystal silicon beam; (a) Schematic drawing of device; (b) Top view of actual device; (c) Rotated view of actual device[51].

beams. For these architectures, the integration of transistors on the suspended structures and the integration of new, high bandwidth optical or millimeter wave links to moving microstructures for micro-actuator control are needed.

III. COMPACT, CAPACITIVE MICRO-ACTUATORS AND SCALING RULES

Capacitive micro-actuators use the energy stored in an electric field to produce motion. The capacitive micro-actuators are easily fabricated using silicon technology [Figures 1,7,8 and 9]; can be made very small for low force, small deflection ($\leq$ 100 nm) applications [Figure 1(d) and (e)]; can be scaled-up in 'chip area' to produce a force of greater than 1 N for a 1 cm^2 'silicon chip'; and can operate at low power and low voltage: V $\leq$ 50 Volts. Plus, the force generated by a capacitive micro-actuator scales as the square of the electric field – an advantage for sub-micrometer-scale MFS structures and micro-mechanical systems.

Like a macroscopic electromechanical system or robot, the capability to integrate, 'nest', and support micro-actuators on a larger, moving structure or stage allows for many degrees of freedom – for example, x-y-z displacements and θ rotation. In addition, small micro-actuators integrated and nested on a larger x-y-z stage [Figure 10] provide additional precision displacements at higher operating frequencies (Section VI) and provide access to high frequency vertical ('tapping mode') and lateral oscillation ('frictional force mode') of probes. Thus, the integration of multiple degree of freedom micro-actuators on a millimeter-scale [Figure 1(a)] and a micrometer-scale [Figures 1(c), 8 and 10] is a key paradigm for making micro-robots for nanometer-scale manipulation. Integrated, compact, high force micro-actuators are important building blocks for micro-machines and micro-robots.

A. Parallel-plate Capacitive Micro-actuator

The force and displacement of a capacitive micro-actuator[47] is obtained by calculating the gradient of the energy function for an ideal capacitor. The attractive force on the plates of the parallel-plate capacitor shown in Figure 7(a) is given by the equation:

$$F_p(x) = -\frac{\partial U}{\partial x}, \qquad (1)$$

where $F_p(x)$ is the magnitude of the attractive force normal to the capacitor plates, and U is the energy stored in the parallel-plate capacitor. The energy density u or the energy stored per unit volume in an electric field is[52]:

$$u = \frac{1}{2}\varepsilon_r\varepsilon_0 E^2 ,$$

(2)

where ε_0 is the vacuum permittivity and ε_r is the relative dielectric constant and E is the electric field. The stored energy in a volume v is:

$$U = \int_V u dv.$$

(3)

For the capacitor, $dv = A_p dx$ where A_p is the area of the capacitor plate. The attractive force, $F_p(x)$, on the capacitor plate for the parallel-plate micro-actuator is obtained by denoting the plate spacing as $(d-x)$, the plate area as $A_p = bl_p$, and $E = [V/(d-x)]$. After substitution in Eqs. (2) and (3) with $\varepsilon_r = 1$, and differentiating Eq. (3) we find Eq. (4) for $F_p(x)$, the attractive force on the movable capacitor plate as a function of x or the magnitude of the attractive force on the fixed plate positioned at $x = d$ [Figures 7, 8 and 9]:

$$F_p(x) = \frac{1}{2}\varepsilon_0 \left[\frac{V^2}{(d-x)^2}\right] bl_p.$$

(4)

The normal stress, $\sigma_p(x)$, on the capacitor plate is

$$\sigma_p(x) = \frac{1}{2}\varepsilon_0 \left[\frac{V^2}{(d-x)^2}\right].$$

(5)

An instructive exercise is the calculation of the magnitude of the *force* that is generated by an array of N parallel-plate capacitors [Figure 9] *per unit of (substrate) silicon surface area*, A_s. The variable a is the MFS produced by the lithography and pattern transfer processes. Here, the capacitor array is simple: modular parallel-plate capacitors [Figure 9] are connected in parallel with one set of plates fixed [Figure 9E] and electrically connected to the silicon substrate (or backbone structure) and the second set of plates [Figure 9D] supported mechanically on a nested, rigid, and movable support or backbone structure. The force produced by one capacitor module is given by Eq. (4). For a (substrate) silicon surface area of $A_s = L_S^2$ and for a capacitor plate area for each module of $A_p = bl_p$, then the maximum number, N, of capacitor modules or units that can occupy A_s is:

$$N \approx \frac{L_S^2}{(l_p + 2a)(a + 2d)}, \qquad (6)$$

where a and $2a$ in the denominator represent the minimum width of the plate and minimum plate spacing at the ends of the electrodes respectively. The plate pitch is $a + 2d$.

Equations (5) and (6) are used to determine the total force on the the mechanical support or backbone structure which is equal to the force, $F_{Array}(x)$, produced by the array of parallel-plate capacitors:

$$F_{Array}(x) = \frac{1}{2} \varepsilon_0 \left[\frac{V^2}{(d - x)^2} \right] \times \left[\frac{L_S^2 (bl_p)}{(l_p + 2a)(a + 2d)} \right]. \qquad (7)$$

For $l_p \gg (2a)$ and $d = a$, the force equation for the parallel-plate capacitor array reduces to the expression:

$$F_{Array}(x) = \frac{1}{6} \varepsilon_0 \left[\frac{V^2}{(1 - x/a)^2} \right] \times \left[\frac{L_S^2 b}{a^3} \right], \qquad (8)$$

which shows that $F_{Array}(x) \propto ba^{-3}$. For micro-actuator design, Eq. (8) highlights the importance of two key process related variables, the MFS, a, produced by the lithography and the pattern transfer processes, and the maximum etch depth which is $\propto b$ as shown in Figure 9 (refer to Chapters 2 and 4 and Section V). Also note that for micro-actuator arrays it is important to achieve the MFS on suspended, planar structures over the entire area of the 'chip', L_S^2, since the force scales as L_S^2 / a^3. In the structural section, Section IV, we show that the vertical stiffness of the suspended structures scales as b^3 / l^3 where l is the span-length for the structure. Therefore, maintaining a high aspect-ratio (b/a) is important for achieving both a large force and a rigid, large area micro-actuator or suspended microstructure.

The force, $F_{Array}(x)$, can be increased by coating the capacitor plates with a material with $\varepsilon_r \gg 1$ [Eq. (2)] such as silicon nitride, but the maximum displacement is then reduced by two times the thickness of the material[31].

The parallel-plate capacitor has the advantage that it is a compact, high force actuator, but the maximum nonlinear (plate-to-plate) displacement of the parallel-plate capacitor is limited to the length of the gap d which is typically of the order of 1-10 μm. For a capacitor attached to a linear spring where $F_{sp} = -kx$, the system becomes unstable when the nonlinear micro-actuator array force, $F_{Array} \propto (1/(1-x/d)^2)$, exceeds the linear spring restoring force $(-kx)$. A phenomenon known as pull-in occurs[53]. The moving plate clamps to the fixed plate and the voltage must be reduced well below the value of the pull-in voltage for the plates to release. This electric field-induced instability affects the operation of all MEMS and must be handled by proper system design (refer to Sections III.D, IV and VIII). For some applications, such as a micro-relay or an optical switch, this nonlinear pull-in, operation is advantageous.

For small displacements, $x/d \ll 1$, the displacement, x, is proportional to V^2. Small area (≈ 10 μm on a side) parallel-plate capacitor actuators[31] can be used for precision, small amplitude displacements of tips and structures [Figure 8] and for generating torsional motion [Figure 1(d), Section III.C.1].

B. Interdigitated-electrode Capacitive Micro-actuator

The interdigitated electrode or comb-drive capacitive micro-actuator [Figures 1(c), 7(b) and 9] can generate a large amplitude displacement parallel to the capacitor plate of length l_p. The force is independent of displacement and is proportional to V^2. The interdigitated electrode capacitive micro-actuator was first applied to MEMS by Tang and Howe[7]. The maximum displacement is l_p, the length of the suspended, movable center electrode in Figure 7(b). Using Eqs. (1) through (3) with $dv = 2bdx$, we obtain the force, $F_{Array}(x)$, produced by an array of interdigitated electrode capacitors:

$$F_{Array} = \frac{1}{4} \varepsilon_0 \left[\frac{V^2}{d} \right] bN_p, \tag{9}$$

where N_p is the maximum number of interdigitated electrode capacitors that occupy a silicon surface area $A_s = L_S^2$ and is given by the expression:

$$N_p = \left[\frac{L_S^2}{(2l_p + 2a)(2a + 2d)} \right]. \tag{10}$$

The $2a$ terms in the denominator of Eq. (10) set the minimum spacing required at the ends of each plate and the minimum pitch ($d = 2a + 2d$) for each unit.

For $l_p \gg (a + d)$ and $d = a$ the substitution of Eq. (10) into Eq. (9) yields Eq. (11) which highlights the importance of the process-controlled variables a and b in the scaling of the force $F_{Array}(x)$:

$$F_{Array}(x) = \frac{1}{16} \varepsilon_0 V^2 \left[\frac{L_S^2 b}{l_p a^2} \right]. \tag{11}$$

The array force, $F_{array}(x)$ [Eq. (11)], is independent of the displacement x and is proportional to V^2 not $V^2 / (1 - (x/a))^2$ which is the case for the parallel-plate capacitor [Eq. (8)]. Note that the a^3 term in the denominator of Eq. (8) is replaced by $(l_p a^2)$ in Eq. (11).

For a fixed array force, $F_{Array}(x)$, equation (11) highlights a trade-off between the maximum displacement l_p and the square of the MFS, a^2, in the term $(l_p a^2)$; and Eq. (11) also highlights a trade-off between the chip area, L_S^2, and the height of the capacitor plate, b, in the term $(L_S^2 b)$. Thus, a sub-micrometer dimension for a and a large b are very important for high force micro-actuators that consume the minimum silicon area, L_S^2.

C. Torsional Capacitive Micro-actuators

Rotation about an axis is most often required for object manipulation, grasping, and other robotic motions. Torsion or twisting of a thin SCS beam provides a method to achieve the rotation of microstructures [Figure 1 (a,b,d)][6]. Here we discuss the rotation of a microstructure out-of-plane (in the z direction) by generating a rotation about an axis parallel to the plane of the surface of the 'silicon chip' as shown in Figure 13. Rotation and twisting of microstructures about an axis perpendicular to the plane of the 'silicon chip' is also possible using similar micro-actuators and SCS processes. Such multiple axis rotation enriches the possibilities for MEMS-based micro/nano-robotics when building in the 'flat world' of silicon substrates. Rotation about an axis can be performed with torsional capacitive micro-actuators [Figure 1 (a,d)]. The large micro-STM shown in Figure 1(a) has a long cantilever (≈ 1 mm) attached to a parallel-plate (large square structures) torsional micro-actuator.

1. Torsional Capacitive Micro-actuator:
Parallel-Plate Electrodes

The torsional 'teeter-totter-like' structure [Figure 13] of length $2L_C$ is driven by an integrated capacitive micro-actuator. A simple torsional micro-actuator [Figure 1(b,e)] is made with a parallel-plate capacitive actuator in which one plate is the torsional microstructure and the other fixed plate is an electrode on the silicon substrate at a distance, d, below the microstructure. The capacitive micro-actuator force, $F_p(z)$, is given by Eq. (4) ($x \rightarrow z$) with bl_p the area of the capacitor plate. The torsional bar rotates or twists the capacitor plates out-of-plane by an angle θ, where $\theta(rad) \propto \left[F_p(z)/k_\phi\right]L_C$, and the z displacement is given by:

$$z = \phi L_C \propto \left[F_p(z)/k_\phi\right]L_C^2. \qquad (12)$$

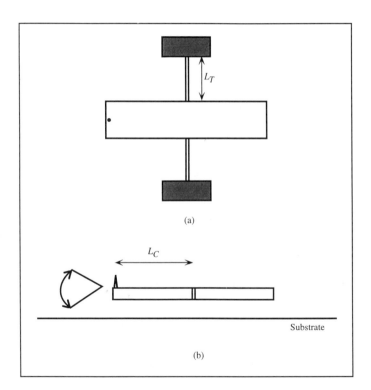

(a)

(b)

FIGURE 13. Schematic of a torsional or 'teeter-totter-like' structure with integrated tip, showing variables L_T and L_C.

The torsional rigidity, K_ϕ (*Nm/rad*) of a beam with a rectangular cross-section is given by the equation:

$$K_\phi = \beta G \left[\frac{a^3 b}{L_T} \right], \tag{13}$$

where G is the shear modulus of rigidity ($\cong$ 80 GPa for single crystal silicon), b is the height, a is the width of the beam, L_T is the length of the torsion bar, and β is a constant depends on the aspect ratio, b/a, of the bar. For $(b/a) > 10$, $\beta = 1/3$[54]. For the small STM[6] shown in Figure 1(b), the torsional z micro-actuator has parameters: $L_T = 58$ μm, $a = 1$ μm,, $b = 6$ μm, $L_C = 50$ μm, $l_p = 20$ μm $d = 2\mu$m, $\beta = 1/3$, and $k_\phi = 5.5 \times 10^{-9}$ (*Nm/rad*). The magnitude of the z transfer function [Eq. (12)] for $(z/d) \ll 1$, is given by the expression:

$$z \approx \left[\frac{V^2}{33} \right] \text{nm/V}^2. \tag{14}$$

or 48 nm for $V = 40$ V with a micro-actuator of area $L_S^2 = 100$ μm $\times 100$ μm.

2. *Torsional Capacitive Micro-actuator: Interdigitated electrodes*

The force produced by a torsional capacitative micro-actuator can be increased by integrating an interdigitated electrode microstructure with the torsional bar[55]. Figure 14 shows an interdigitated electrode torsional micro-actuator. The plate structure looks like a parallel-plate capacitor, but for the torsional micro-actuator the moving center plate rotates upward in the z direction as shown in Figure 15. This design produces nearly a constant force with the angle of rotation and eliminates the difficult process steps associated with depositing a capacitor plate on an insulator underneath the torsional microstructure – the gray pads shown in Figure 1(e). The torsional micro-actuator with interdigitated electrodes can be nested or attached to a 'robot arm' or cantilever-microstructure which is discussed in Section VIII. As shown in Figure 15(c) the asymmetry in the electric field at the top and the bottom of the electrodes drives the torsional micro-actuator in one direction[55,56]. The 'teeter-totter' torsional microstructure with interdigitated electrodes on each side of the 'teeter-totter' produces out-of-plane rotation and $\pm z$ displacements of microstructures and tips.

Eqs. (1), (2) and (3) are used to estimate the improvement in the magnitude of the transfer function for a torsional micro-actuator with interdigitated electrodes.

The energy stored in the electric field vs. ϕ produced by rotating a rectangular plate positioned between two fixed rectangular plates [Figure 14] is obtained by the integration of Eq. (3) with $\partial v \approx \partial \left[2d(l_p b - (b^2 \phi/4) - l_p^2 \phi) \right]$ for $\phi \leq \tan^{-1}(b/l_p)$ to yield:

$$U = \int_v \left[\frac{\varepsilon_0 V^2}{2d^2} \right] \partial \left[(2d) \times (l_p b - (b^2 \phi/4) - l_p^2 \phi) \right]. \qquad (15)$$

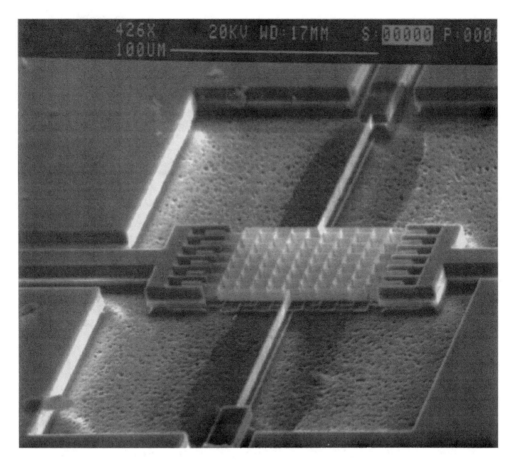

FIGURE 14. An SEM micrograph of a single crystal silicon, torsional micro-actuator with interdigitated electrode drives[55].

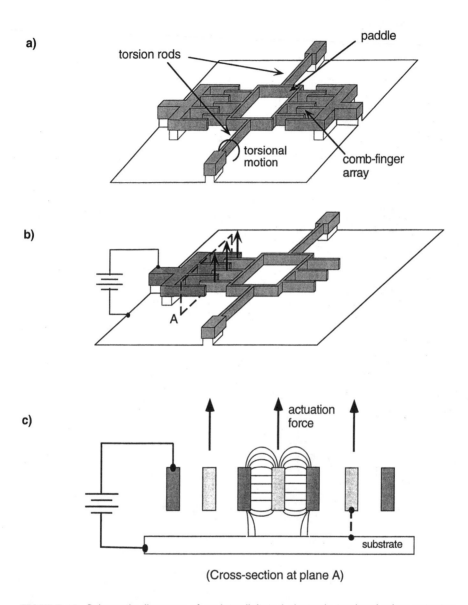

a)

paddle

torsion rods

torsional
motion

comb-finger
array

b)

A

c)

actuation
force

substrate

(Cross-section at plane A)

FIGURE 15. Schematic diagrams of an interdigitated electrode torsional micro-actuator.
(a) Overview showing torsion rods and comb-finger or interdigitated electrode array,
(b) Force vector generated by applying a potential difference between the two electrode-
pairs, and (c) Electric field distribution showing asymmetries at top and bottom of the
electrodes[55].

The torque, T is obtained by differentiation of Eq. (15) with respect to ϕ, so:

$$T = -\frac{\partial U}{\partial \phi} \approx \frac{1}{2}\varepsilon_0 \left[\frac{V^2}{a}\right]\left[(b^2/4)+l_p^2\right]N_p, \qquad (16)$$

where N_p is the number of micro-actuator (plate) modules that fill the area L_S^2, and is given by the expression:

$$N_P = \left[\frac{L_S^2}{(l_p+2a)(2a+2d)}\right]. \qquad (17)$$

Figure 16 shows four micro-actuators where the interdigitated electrodes fill the available area of 100 μm on a side. Using Eqs. (13), (16) and (17) with $d=a$, and $l_p \gg a$, we obtain expression (18) for the angle of rotation or twist, ϕ, for the N_p rectangular capacitor plates where $l_p^* = l_p\left[1+(b/2l_p)^2\right]$ and $\phi \ll 1$:

$$\phi(rad) = \left[\frac{\varepsilon_0 V^2}{8\beta G}\right]\left[\frac{l_p^* L_S^2 L_r}{a^5 b}\right]. \qquad (18)$$

The torque twists the rectangular torsion bar and rotates the 'teeter-totter' out-of-plane by ϕ. Note $\phi \propto a^{-5}$, so the reduction in the MFS is very important for producing a large rotation with compact, integrated actuators.

    We use Eq. (18) to calculate the characteristics of a 50 μm on a side, SCS torsional micro-actuator for an array architecture similar to that shown in Figure 16 with dimensions: $L_T=20$ μm, $L_C=l_p=23$ μm, $L_S=50$ μm, $a=1$ μm, $b=46$ μm, $\beta=(1/3)$, and $K_\phi=1.2\times10^{-7}(Nm/rad)$. The magnitude of the z transfer function [Eq. (19)] for the eight capacitor modules per micro-actuator that fill the area − (four modules on each side of the 'teeter-totter' structure) and for $z/L_C \ll 1$ (so $\phi(rad) \ll 1$) is given by the expression:

$$z \approx \pm \ \phi L_C = \left[\frac{\varepsilon_0 V^2}{8\beta G}\right]\left[\frac{l_p^* L_S^2 L_T L_C}{a^5 b}\right] \approx \left[\frac{V^2}{35}\right]nm/V^2. \qquad (19)$$

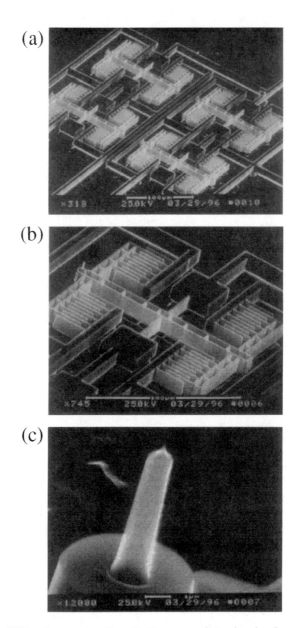

FIGURE 16. SEM micrograph of a (2×2) array of torsional micro-actuators with integrated 20 nm diameter silicon tips. Each compact micro-actuator produces vertical displacement of the tips. (a) The 2×2 array; (b) One array element, and (c) A high aspect-ratio tip.

Eq. (19) shows that the transfer function is about the same as that of Eq. (14) which is for a much larger torsional micro-actuator. Furthermore, the resonant frequency for the smaller torsional micro-actuator is substantially higher than that of the larger torsional micro-actuator shown in Figure 1(d) (refer to Section VI).

A scaling equation can be derived for the z displacement produced by a compact, torsional micro-actuator [Figure 16] using Eqs. (13) and (18). If we assume that the eight capacitor modules and torsion bar fill the area of the 'teeter-totter' structure with dimensions $l_p^* \approx l_p, L_C = L_T = l_p, L_S \approx 21_p$, then we obtain Eq. (20) for the maximum displacement, z, with $z/L_C \ll 1$, for a micro-actuator that occupies an area L_S^2:

$$z \approx \pm \phi L_C = \pm \left(\frac{\varepsilon_0 V^2}{32 \beta G} \right) \left(\frac{L_S^4 L_C}{a^5 b} \right)$$

$$\approx \left(10^{-14} V^2 \right) \left[\frac{L_S^4 L_C}{a^5 b} \right] nm/V^2. \qquad (20)$$

Eq. (20) clearly shows the advantage of reducing the MFS, a, when designing compact micro-actuators for an array architecture, since the displacement scales as $(L_S^4 L_C/a^5)$! We return to this scaling result [Eq. (20)] in Section X and let $a \rightarrow 50$ nm.

If we use a similar design for a micro-atomic force microscope (micro-AFM) with an interdigitated electrode torsional micro-actuator and a long (large L_C) stiff cantilever, the torsional micro-actuator [Eq. (20)] can be used to measure very small forces (refer to Section VIII).

Torsional, SCS micro-actuators with highly asymmetrical electrodes can rotate objects out-of-plane of the 'chip' by greater then 60°; the movable electrode would, for example, be made of a high aspect-ratio silicon dioxide segment[57,58] attached to a thin silicon electrode at the bottom. The fixed electrodes would be made of silicon. The torsional actuator would be attached to an SCS torsional rod or bar. SCS rods and bars have high fracture strength[59-61] and can be twisted to large angles. Micro-robotic applications for these torsional devices are discussed in Section IX.B.

D. Micro-actuator Design Issues

The forces calculated in Eqs. (8), (9) and (11) are based on finite-size and infinitely rigid, suspended capacitor plates that fill the entire chip surface area L_S^2. However, for large silicon chip-area actuators, the capacitor plates require additional support structures to increase the rigidity in both the plane of the chip (x-y) and in the vertical axis (z). Figure 17 shows the support structure – the backbone – that supports an array of 9000 interdigitated electrode or comb-drive capacitive micro-actuators that produce > 1.5 mN of force at 50 V. The micro-actuators are part of a micro-loading machine [Figure 17(b)] used to characterize buckling and fracture of microstructures[59,60]. Thus, the combination of dividing the capacitors into subgroups, and designing stiffer beams or frames to support and connect the subgroups, consumes additional silicon surface area [Figure 17 (b)]. Consequently, the force generated per unit area of silicon-chip would be multiplied by an area filling factor for the capacitor plates of 0.5 to 0.8 of the force predicted by Eqs. (8), (9) and (11).

The force produced by an array of capacitive micro-actuators is proportional to the geometrical factor $[L_S^2 b / Da^2]$ where D is the maximum displacement of the micro-actuator for both the parallel-plate ($D = a$) micro-actuator [Eq. (8)] and the interdigitated electrode ($D = l_p$) micro-actuator (Eq. 11). The force is essentially determined by the beam width, the plate spacing or gap, the beam height, and the maximum span that supports cantilevered, stress-controlled, and undistorted (planar) capacitor plates and the mechanical support (backbone) structures[59,61]. The beam width, a, and the beam aspect-ratio, (b/a), are process related dimensions determined by both pattern exposure (lithography) and pattern transfer (etching). The maximum span is determined by the rigidity of the beams. Thus, it is important to use high aspect-ratio beam structures (and deep etch

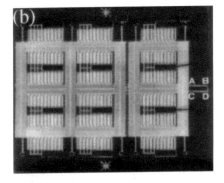

FIGURE 17. SEM micrograph of (a) A suspended and planar backbone structure with integrated micro-actuators made of 12 μm high SCS beams. (b) A micro-loading machine using the backbone structure to support 9000 interdigitated electrode micro-actuators. The structure spans an area of 4×5 mm^2.

processes) that produce released beams with beam widths and beam spacings that scale to a MFS compatible with state-of-the-art production lithography; and to use processes that produce high aspect-ratio, released and cantilevered beam structures that are rigid, planar, and stress free or stress controlled. In Section V we describe an SCS process that meets these criteria and we elaborate on structure planarity issues in Section IV.

IV. MICROSTRUCTURE ISSUES FOR 3D MICRO-ACTUATORS

In the design of array structures and array architectures we must constantly be concerned with internal stress, structural stiffness, and structural planarity during operation. Also, we must design the micro-actuators and support structures such that dynamic or static displacement in one dimension does not couple to the other dimensions or modes and become unstable as shown in Figure 18 for structural rotation [Figure 18(a)], structural lifting [Figure 18(b)], and capacitor plate bending and contact [Figure 18(c)]. For high frequency operation, we need to keep the modal frequencies high and separated.

In Sections III.A and III.B we derived the force generated by the two types of capacitor micro-actuators. These derivations were for an ideal 'one dimensional' capacitor and stress free microstructures. However the field distribution is three dimensional, and there are forces due to fringing electric fields. Also, internal bending stress and bowing can act on the suspended capacitor plates. For example, an unwanted vertical displacement, Δz [Figure 18(b)], of a suspended plate due to an asymmetric distributed load, $\sigma_z a$, produced by fringing electric fields acting on the top and/or bottom of the plates is given by:

$$\Delta z = \left[\frac{3\sigma_z}{16E} \right] \left[\frac{l_p^4}{b^3} \right], \tag{21}$$

where E is the modulus of elasticity which is 130 GPa for (100) silicon, b is the beam height, and l_p is the length of the cantilevered capacitor plate. In Eqs. (8) and (11), the force generated by the capacitor actuators is proportional to the height, b. Fortunately, the structural rigidity of the cantilevered capacitor plate or beam in the z direction is proportional to b^{-3} for a clamped suspended beam. Thus, as b is increased to increase the amount of lateral, in-plane force generated by the actuator, the out-of-plane stiffness of the structure increases as b^3, and the Δz displacement due to the vertical fringing fields Eq. (21) decreases as $(1/b^3)$.

Moreover, the capacitor plates must also be very stiff normal to the plane because of the large normal stress and bowing experienced by large span,

N.C. MacDonald

a)

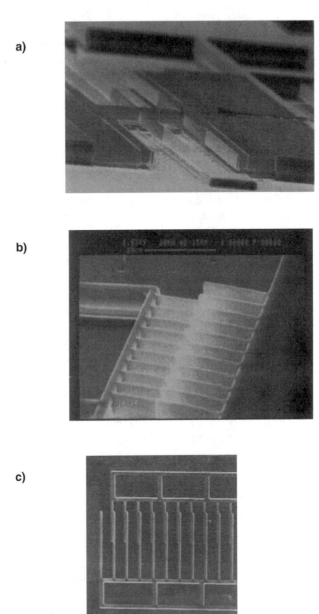

b)

c)

FIGURE 18. SEM micrographs showing structural defects and instabilities. (a) Rotation of plates out-of-plane due to process and design defects; (b) Non-planar misalignment of suspended and fixed plates; and (c) Structural instabilities showing clamping of interdigitated electrode micro-actuator plates.

(a)

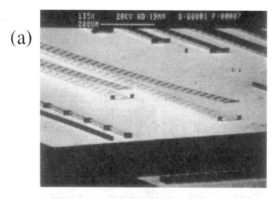

(b)

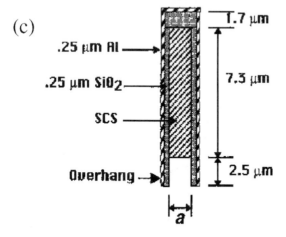

(c)

FIGURE 19. SEM micrographs showing the bending of composite-silicon beams. (a) Bending upward, (b) Bending downward and (c) Schematic of the beam showing SCS core and thin film coatings.

cantilevered structures. Figure 19(a,b) shows the bowing of beams out-of-plane due to internal stress in the structures. A cross-section of the structure is shown in Figure 19(c). The control of stress in the thin films is very important to achieve planarity over a large surface area. A complete analysis of these SCREAM structures[62] shows that the planarity or radius of curvature ρ_c of the structure also depends on the height of the beam or capacitor plate as shown in Figure 20.

In addition to remaining in the plane, the capacitor plates must remain parallel so that the capacitor gap remains constant along the beam during the operation of the actuator. For instance, for the comb drive, any displacement of the capacitor finger(s) along the $\pm y$ direction from vertical symmetry in the x-z plane will drive the center capacitor plate toward the opposite, 'fixed' plate [Figure 18(c)] as described in Section III.A in relation to the pull-in instability[53]. The suspended capacitor plates now act as two parallel-plate capacitors where the $\pm y$ forces on the central beam (plate) are equal in magnitude and opposite in direction with zero net force in the y direction. However, if the beam is displaced Δy, the force increases following Eq. (8) with $x \rightarrow y$; the beam becomes unstable unless the stiffness, k_y, of the suspended beam provides a restoring force greater than the net force produced by voltage applied to the actuator. The maximum voltage, V_{l_p}, on the plate occurs for the maximum displacement $x = l_p$; the suspended beams of stiffness, k_y, must supply the required restoring force in the $\pm y$ direction to keep

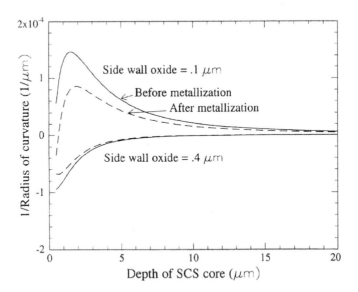

FIGURE 20. Plot of 1/radius of curvature, $1/\rho_c$, vs. the depth of the SCS core, b, for different composite-beam structures.

the plate (beam) centered between the two 'fixed' plates.

In array architectures, the N_p movable plates are connected in parallel on a strong backbone structure [Figure 17] which is supported by a spring of stiffness k_x where $|F_x| = k_x|x|$. Both plates are usually cantilevered, so the required effective lateral, plate stiffness, is k_y:

$$k_y \geq \left| \frac{\partial F(y)}{\partial y} \right|_{y=0;x=l_p} = \frac{1}{2} \left| \frac{\partial}{\partial y} \left[\frac{\varepsilon_0 bx}{(d-y)^2} \right] \left[\frac{N_p V_{l_p}^2}{2} \right] \right|_{y=0;x=l_p}$$

$$= \left| \left[\left[\frac{\varepsilon_0 bx}{(d-y)^3} \right] \left[\frac{N_p V_{l_p}^2}{2} \right] \right] \right|_{y=0;x=l_p}. \tag{22}$$

If $k_x l_p^2 = \dfrac{\varepsilon_0 bx}{d} \dfrac{N_p V_{l_p}^2}{2}$, then $k_y \geq \dfrac{k_x l_p^2}{d^2}$. As $d \to a$, $k_y a^2 \to k_x l_p^2$, and the plates do not bend and clamp together [Figure 18(c)] for the highest operating voltage, V_{l_p}, and for the maximum displacement, $x = l_p$.

From the previous force analysis (Section III.D), the force per unit surface area increases with smaller a, and for the interdigitated electrode capacitor, l_p determines the maximum displacement. For large force and large displacement, we want a small − an infinitely thin capacitor plate − and l_p large, but then the capacitor plate can bend (Δy), become unstable, move, and clamp to the opposite plate [Figure 18(c)]. The normal method used to stiffen the capacitor plate is to use the lattice design shown in Figure 21 for a capacitive accelerometer. Stiff capacitor plates that meet the requirements of Figure 22 are made by connecting two or more beams of width a with crossbars [Figure 21(b)] to form a beam of equivalent width $a \to 3a$ in Eq. (22). The stiffness, k_y, of the capacitor plates in Eq. (22) is increased by $3^3 = 27$, and additional chip area is consumed. A second approach used to increase k_y is to connect the moving plates together with a 'bridge-stiffener' near the end of the cantilever; this approach requires additional processing steps and significant redesign of the actuators. For compact, small displacement micro-actuators, the lattice-structured plates are not required. For large arrays of interdigitated electrode capacitors a very stiff lattice structure is required to support the actuator array so that all the capacitor plates only move parallel to the fixed plates. Examples of the backbone and plate structures are shown in Figures 9, 17, 19 and 21.

(a)   (b)

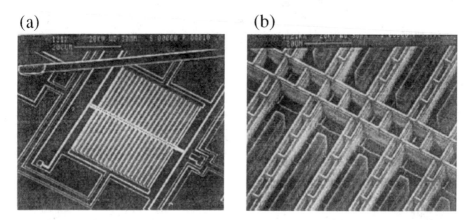

FIGURE 21. SEM micrograph of an accelerometer with 'stiffened' suspended interdigitated electrodes. (a) Accelerometer showing suspended electrode structure (white structure) supported by a spring on each end. (b) A higher magnification image of the stiff capacitor plates.

In-plane stiffness and vertical stiffness are the most critical design parameters for realizing stable, large displacement and multiple degrees of freedom, manipulators for nanometer-scale robotics. The design methods and the micromachining processes used to make these robots must support the scaling requirements for smaller minimum features (a) of high aspect-ratio (b/a) microstructures that span a large 'chip-area' (L_S^2). Single crystal silicon structures together with thin film structures offer such critical scaling for MEMS.

V. SINGLE CRYSTAL SILICON PROCESSES

A. Single Crystal Silicon: An Electronic and Mechanical Material

Single crystal silicon is the dominant electronic material; is relatively inexpensive for a single crystal material; and is the material of choice for microprocessor chips, memory chips, DSP chips (refer to Chapter 2), and, in the future, for sensor chips[8,14,15,46] and MEMS chips. In addition to its excellent electronic properties, SCS also has excellent mechanical properties[44] with a density of 2300 kg/m^3 and a modulus of elasticity, $E{\approx}130$ GPa – about the same as steel. SCS can be processed at temperatures of greater than 1100C, and in an oxidizing atmosphere forms silicon dioxide which is an excellent electrical and thermal insulator. Many compatible thin film deposition processes have been developed including physical vapor deposition, chemical vapor deposition, sputter deposition, plating, and epitaxial growth. These thin film processes provide a reasonable suite of materials compatible with SCS including silicon dioxide, polysilicon, silicon nitride, aluminum, many silicides, and heavy metals such as tungsten. Thus, extensive process technology is available to address the complex, electromechanical system requirements for sensor, 3D micro-actuator, and micromanipulator arrays.

Micromanipulator systems require contacts, electrical and thermal isolation[30,31], and integrated transistors[31,51] on fixed and moving SCS structures (Section II.A) to deal with system integration, architecture and conflicting system requirements. Scaling of integrated circuits and silicon-based processing to sub-micrometer and nanometer-scale minimum feature sizes continues unabated. MEMS scaling will follow. Lithography which determines a and high throughput reactive ion etching (RIE) pattern transfer which determines b, are very important for the future progress in building micro-instrumentation and massively parallel micromanipulators.

We have discussed how high aspect-ratio SCS capacitor plates with a submicrometer MFS generate large forces per unit surface area, L_S^2, of the SCS substrate. The challenge is to develop and implement an SCS process that meets the requirements needed to make integrated, high force, high torque micro-actuators and micromanipulators having both rotation and translation in all three dimensions. Also, we need a process that builds on the future trends in the semiconductor industry (refer to Chapters 2 & 4) and leverages future silicon technology growth so that smaller, smarter and higher density 3D micro-actuators and micro-sensors can be realized. The description of such a process, the SCREAM process, is presented in the following section.

B. Single Crystal Reactive Etching and Metallization (SCREAM) Process.

The micro-STM shown in Figure 1, the micro-loading machine shown in Figure 17(b), and the accelerometers shown in Figure 21 were fabricated using a process called SCREAM. The SCREAM process has been used to fabricate *released and moving structures* in both single crystal *GaAs*[29] and SCS[27,28,30]. Though the SCREAM process continues to be refined and extended, the generic SCS process exhibits the following key features:

1. The SCS structures are RIE etched into the silicon substrate with, for example, Cl_2, to form high aspect-ratio (etch-depth/MFS) features with the MFS determined substantially by the lithographic pattern. Aspect ratios of greater than 70:1 have been made for a 1.25 μm MFS [Figure 4] using new etch mask processes[57,58].

N.C. MacDonald

(a)   (c)

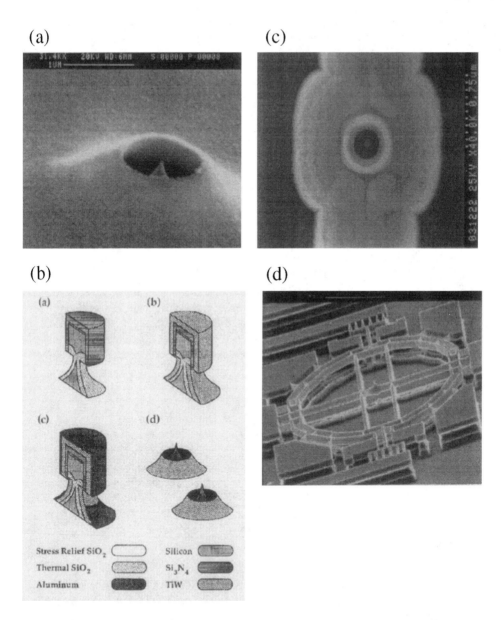

(b)   (d)

FIGURE 22. (a) Tip with aperture; (b) Process for tip with aperture[84]; (c) Tip with aperture[30]; (d) SCS heater with integrated tip.

2. The shapes of the suspended SCS structures are independent of the crystal-lographic planes[27] [Figure 22(d)].

3. Structures to remain after the release step are protected on the top and sidewalls by a slowly etching layer during an isotropic release etch (SF_6) that undercuts [Figure 19] the bottom of the protected microstructures.

4. The suspended structures are SCS and usually include with thin layers of insulators, metal or both [Figure 19(c)].

5. Electrical isolation on a micrometer-scale is achieved through the thermal oxidation of silicon structures[30] [Figure 10(c)] and/or by coating the structures[28] with CVD deposited, ceramic insulators – silicon dioxide or silicon nitride.

6. The SCS micro-machining processes are compatible with the integration of metal-oxide-semiconductor field effect [Figure 12] and bipolar transistors fabricated on the suspended SCS beams.

7. A method is included to both anchor and electrically isolate suspended SCS beams or to create moats to isolate the metal bonding pads from the SCS substrate. An 'overhanging' silicon dioxide layer[28] after the release etch is used to electrically isolate the thin film metal deposited on the insulator [refer to Figure 19(b)].

8. A method is included to integrate SCREAM MEMS on completed wafers with integrated circuits[33]. Thus the complex, multiple mask process used to make the integrated circuits is separated from the process to make the MEMS. Methods to electrically connect the MEMS to the integrated circuit are included in the process sequence.

9. A method is included to form and integrate a dense array of high aspect-ratio SCS tips [Figure 10(b)] on moving beams [Figure 1(e)]. The tip-support shaft is typically 0.5 µm in diameter and 6 µm long with a nominal 10-20 nm diameter silicon tip at the end. The silicon tip is coated with metal, or it can be converted to nickel silicide[62] and then coated with metal to produce a highly conductive tunneling tip.

10. A method to form tips with self-aligned electrodes to control the field at the tip are used for field emitters [Figure 22] and unique scan-probe tips. The tips can be integrated with SCREAM microstructures. Field emitter tips with integrated tungsten lens structures [Figure 3(b,c)] can be fabricated using the tungsten process[17]. Small heaters with an integrated tip [Figure 22(d)][62] can be used to heat individual tips in an array of tips.

A detailed process diagram is illustrated in Figure 23 which shows the SCREAM process steps required to produce released SCS structures with

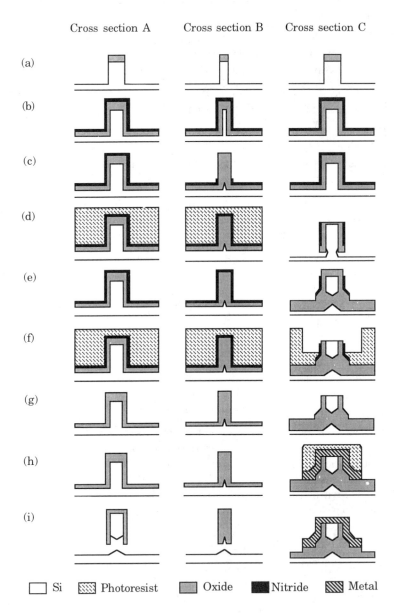

FIGURE 23. High temperature version of SCREAM process. The diagram shows the SCREAM process sequence for fabricating suspended silicon segments (cross-section A), suspended high temperature thermal silicon dioxide segments (cross-section B), and silicon-on-insulated structures with metal contacts (cross-section C).

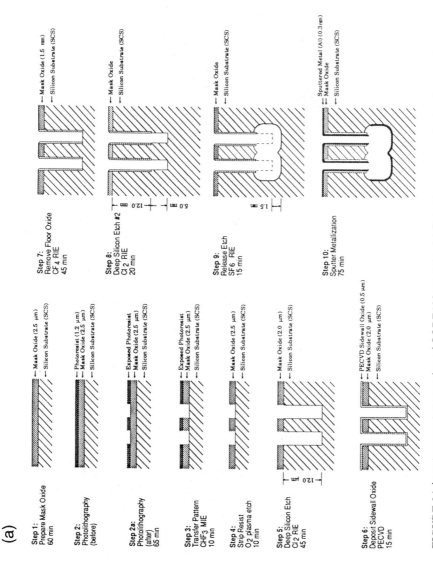

FIGURE 24. Low temperature version of SCREAM process. (a) Process sequence for low temperature version of SCREAM that uses isolated MESA structures for an anchor and bonding pads.

(b)

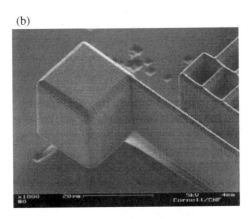

FIGURE 24. (b) SEM micrograph of SCREAM MESA structures.

insulated or conducting anchors and contacts, and silicon dioxide insulating segments on the released beams. The low temperature version of SCREAM[29] is outlined in Figure 24. This process uses the mesas shown in Figure 24(b) and an 'overhang' of silicon dioxide near the bottom of the SCS beams and the contact pads to insulate the metal electrodes from the conducting SCS beams and the silicon substrate [Figure 19(b)]. This silicon dioxide overhang-structure provides electrical isolation of metal conductors from the suspended beams and silicon substrate and allows low temperature MEMS processing on completed wafers with integrated circuits [Figure 5]. A recent review paper, "SCREAM Micro-Electromechanical Systems"[63], provides more detail on the different versions of the SCREAM process.

VI. FREQUENCY SCALING AND NESTED MICRO-ACTUATORS

A simple scaling law for the resonant frequency for the transverse (lateral) mode of a suspended clamped-clamped beam is that of a spring-mass system:

$$f_o = \frac{1}{2\pi} \left[\frac{stiffness}{effective\ mass} \right]^{1/2} \propto \left[\frac{E}{\rho} \right]^{1/2} \left[\frac{b}{l^2} \right] \tag{23}$$

where E is the modulus of elasticity, ρ is the density of the material, b and l are the vertical height and length of the beam respectively. The effective mass is the equivalent moving mass of the beam. An exact solution for the frequency is obtained by solving the time dependent Euler equation for a bending beam with the appropriate boundary conditions. The solution for the transverse modal frequencies for a bending beam is available in many text books[64] where the modal frequencies for the n-th mode are tabulated[65]. For the discussion that follows the expression (23) provides the necessary first order scaling results.

Consider a one dimensional (x) comb drive actuator (Section III.B) driving a spring of stiffness $k_x = 200$ N/m. If the maximum displacement $x = l_p$ is 25 μm, the actuator must generate a force of 5 mN. Using Eq. (11) with nominal values of $a = 1$ μm and $b = 50$ μm, and $V = 40$ Volts, we can calculate the minimum silicon area (mass) required to produce 5 mN of force with a 0.5 area fill factor which is $L_s^2 = 1.25 \times 10^{-6} m^2$. The resonant frequency is calculated using Eq. (23) which for the y micro-actuator is:

$$f_{ox} = \frac{1}{2\pi} \left(\frac{k_x}{m_x} \right)^{1/2} = \frac{1}{2\pi} \left(\frac{200}{2300 \cdot 50 \cdot 10^{-6} \cdot 1.25 \cdot 10^{-6}} \right)^{1/2} = 37 \, kHz$$

$$(24)$$

The interdigitated electrode x actuator is nested inside the y micro-actuator with $k_x = 200$ N/m and is supported by a backbone structure of mass $m_s = m_x$. If the y micro-actuator must produce a maximum displacement $y = l_p$ of ±25 μm, then the y micro-actuator must also generate a force of 5 mN. Since the mass of the y micro-actuator includes the masses m_x, m_y, and m_s, the approximate resonant frequency of the y micro-actuator is:

$$f_{oy} = \frac{1}{2\pi} \left(\frac{k_x}{m_x + m_y + m_s} \right)^{1/2} = \frac{1}{2\pi} \left(\frac{200}{3m_x} \right)^{1/2} = 21 \, kHz \qquad (25)$$

The z or vertical displacement of the tip is accomplished with a small area, high frequency micro-actuator. Vertical displacement of each tip in an array is accomplished with a torsional micro-actuator attached to each tip. The size of the torsional micro-actuator depends on the application. For an array architecture of torsional micro-actuators, the size of the torsional micro-actuator should be comparable to the maximum displacements of the x-y support structure or stage. A pair of interdigitated electrode x actuators and a pair of y interdigitated electrode actuators produce an x-y scan amplitude of ±25 μm in both x-y according to Eq. 11. So we choose a compact torsional actuator of 50 μm on a side like the one described in Section III.C.2. The approximate resonant frequency of a torsional resonator is given by the expression:

$$f_\phi = \frac{1}{2\pi} \left[\frac{k_\phi}{I_\phi} \right]^{1/2} , \qquad (26)$$

where k_ϕ is the torsional rigidity of a bar [Eq.(13)] and I_ϕ is the moment of inertia. The 50 μm on a side torsional micro-actuator shown in Figure 1 has a
resonant frequency of $f_\phi \approx 1\,MHz$. The first resonant frequency for torsional actuators of this scale are in the range of $f_\phi \approx 1 \rightarrow 20$MHz.

Array architectures of torsional micro-actuators scanned with x-y linear actuators offer Giga-pixel/second scanning using actuators with the above frequency
characteristics. These high throughput architectures are discussed in Section IX.

VII. ELECTROMECHANICAL TUNING, NONLINEAR MEMS AND Q

Electromechanical tuning of MEMS offers unique opportunities for making
tunable filters and resonators; for feedback control of the electromechanical system or sensor; for real time control of the dynamic range and sensitivity of a sensor; and for real time control of the response of micro-robotic systems. In addition, nonlinearities in the mechanical stiffness can be electromechanically
tuned[66,67,68] to realize nonlinear control functions and chaotic MEMS.

An example of a tunable resonator[69] in which both the linear and nonlinear
stiffness coefficients can be electromechanically tuned is shown in Figure 25.
The schematic diagram of the structure shows two sets of closely-spaced fingers.
The moving fingers (black) are attached to a backbone structure while the fixed
fingers (gray) are located near, but not interleaved with, the moving fingers. The
fringing electric fields produced by applying a tuning voltage, V $_{tune}$, to the fixed
fingers can either augment or reduce the electromechanical stiffness of the structure which is supported by two mechanical springs (black) – one at each end of
the moving structure. The interdigitated micro-actuators located at each end of
the suspended structure are used to drive or excite the tunable, electromechanical
resonator.

Figure 26 shows the detail alignment of the structure for both the augmentation and reduction fringing field actuators. When the fingers are aligned, as in
Figure 26(a), or misaligned, as in Figure 26(b), the top half of the actuator does
not experience a net force in the "motion" direction. This result is clear from
symmetry. In Figure 26(a), if the device deflects over a small region to the left or
to the right, the top half of the actuator will experience a restoring force that is
proportional to the deflection. This behavior is modeled as an electrostatic spring.

a)

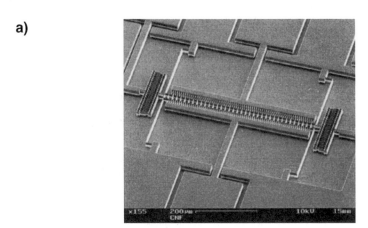

b)

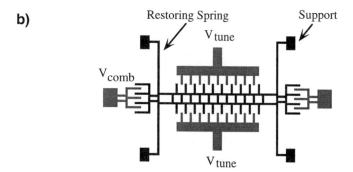

FIGURE 25. Voltage tunable resonator. (a) SEM micrograph, and (b) Diagram of resonator identifying key components.

In Figure 26(b), the behavior is similar, except that the electrostatic spring constant is negative. For large deflections of the mechanical spring the deflection is nonlinear. Thus, the mechanical springs exhibit both linear and nonlinear behaviors and are modeled as

$$F_{spring} = -(kx + \eta x^3). \qquad (27)$$

Also, the combination of an augmentation and reduction fringing field actuators can produce linear and nonlinear responses. The integrated fringing field actuators are modeled by the following set of equations for the tunable, nonlinear oscillator.

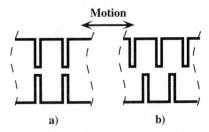

FIGURE 26. Diagram of voltage or electric field tuning actuators. (a) Augmentation actuator and (b) Reduction actuator. The actuators augment or reduce the stiffness, k, of the system.

$$F_{Aug.}(x,V_1)=-(a_1 x+a_3 x^3)V_1^2 \qquad (28a)$$

$$F_{Red.}(x,V_2)=-(r_1 x+r_3 x^3)V_2^2 \qquad (28b)$$

Eqs. (28 a-b) are functionally equivalent to the nonlinear mechanical restoring force given in Eq. (27).

An integrated device with both augmentation and reduction fringing field actuators experiences the sum of the two forces in Eq. (28):

$$F_{sum}(x,V_1,V_2)=-(a_1 V_1^2+r_1 V_2^2)x-(a_3 V_1^2+r_3 V_2^2)x^3 \qquad (29)$$

A micro-mechanical oscillator with a combination of augmentation and reduction actuators is governed by the equation of motion:

$$m\ddot{x}+k_1 x+k_3 x^3=F_{ex}(t) \qquad (30a)$$

$$k_1=k+(a_1 V_1^2+r_1 V_2^2) \qquad (30b)$$

$$k_3=\eta+(a_3 V_1^2+r_3 V_2^2) \qquad (30c)$$

where k and η, are the linear and cubic stiffness coefficients of the mechanical system respectively given in Eq. (28), and $F_{ex}(t)$ is an external forcing function generated by the interdigitated micro-actuators located at the ends of the structure in Figure 25(b). Equation (30a) is a form of the Duffing equation[64].

Figure 27 shows a graph of the resonant frequency response of a resonator with integrated augmentation and reduction actuators as a function of the square of the tuning voltage. The amplitude frequency responses for the integrated device are shown in Figures 28 and 29.

From Eq. (29), if the following tuning-voltage ratios hold:

$$\frac{V_1^2}{V_2^2} = -\frac{r_1}{a_1}, \tag{31}$$

the linear term cancels. This equality is possible if a_1 and r_1 have opposite signs. Ideally, a_1 and r_1 would be equal in magnitude but opposite in sign and the actuator pair is balanced. For the same voltage applied to each actuator, the linear contributions cancel, and the cubic term can be tuned. In either case, as long as the ratio a_1/a_3 is not equal to r_1/r_3, the electrostatic cubic term can be scaled independently of the linear term by scaling V_1 and V_2 while maintaining the equality in Eq. (30). For tuning-voltage ratios:

$$\frac{V_1^2}{V_2^2} = -\frac{r_3}{a_3}, \tag{32}$$

the cubic stiffness contributions cancel and the net linear term can be scaled.

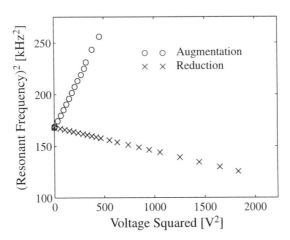

FIGURE 27. Graph of experimental data for augmentation and reduction actuators. The resonant frequency and the linear stiffness coefficient vary with the tuning voltage.

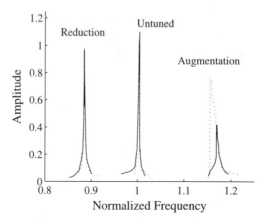

FIGURE 28. Experimentally determined amplitude vs. frequency responses. Solid lines indicate positive frequency sweeps and dotted lines indicate negative sweeps.

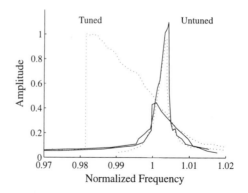

FIGURE 29. Cubic term tuning. Solid lines indicate positive frequency sweeps and dotted lines indicate negative seeps.

Electromechanical tuning of the resonant frequency or effective mechanical stiffness of a device is possible using a combination of electrostatic, fringing field actuators. Both the linear and nonlinear stiffness coefficients of a micromechanical device can be independently tuned. This approach requires actuators that contribute positive or negative linear, quadratic, and cubic stiffnesses. The fringing field actuators provide this capability. By shifting the alignment of the fingers, the actuator generates a positive linear stiffness, a negative linear stiffness, or even a quadratic stiffness. Through simple modifications of the finger spacing or relative widths, the cubic stiffness contribution can be designed to be positive or negative. The device described can be used to tune the cubic stiffness of an oscillator from 0.3×10^{11} to 4.0×10^{11} N/m^3 without affecting the linear stiffness.

The linear and nonlinear electromechanical tuning devices offer the potential for producing linear and nonlinear coupling of microstructures, for improving the linearity of MEMS, and for producing voltage programmable nonlinear MEMS. Feedback voltage operation can be used to control the frequency, amplitude and damping[56,69,70] responses of MEMS such as filters; and to generate real time dynamic gain control, calibration and nonlinear dynamics including bi-stable switching and tunable chaotic responses.

Integrated controlled or tunable damping on the micro/nano-scale remains a challenge. New methods and processes are needed to integrate controlled and stable dampers onto microstructures. However, with electronic components and electrostatic feedback, voltage tunable electrostatic damping like electrically adjustable stiffness can be used for controlling the system damping response or system Q[70].

The dominant dissipation processes for suspended microstructures are usually squeezed-film damping in air, water and other fluids[13,70,71] and internal friction[55,72] in vacuum. Static and dynamic friction and associated wear[73] between two surfaces in contact are of great importance in determining the reliability and mechanical response of rotating[9,74,75], linear[76] and torsional[55] MEMS. Internal friction of materials is usually quantified by a quality factor or Q factor[55,64,72]. Both bulk and surface properties of microstructures affect the magnitude and the stability of Q for high Q resonant MEMS. Room temperature vacuum Q's for RIE etched single crystal silicon (SCS) are in the range of $10^4 - 10^5$; a Q of 10^6 has been measured for chemically etched SCS[55]. Q's in air[7,70] are structure dependent and range from $Q \approx 1 - 200$. The Q's for 200 nm wide SCS beams and structures coated with 100 nm of silicon nitride[31,66,67,68] are lower than those of SCS beams; Q's for the nitride coated beams range from 1500 to 4500 in vacuum.

The high Q of SCS is very important for resonant, scan probe cantilever-tip devices for the detection of small forces[78]. Surface films, surface damage and surface contamination layers can affect both the Q and the resonant frequency of micro/nano-structures. Losses originating from dry-etched surfaces[55] and electron and ion irradiated surfaces[79] have been observed. Maintaining a stable surface structure and surface composition over time is a challenge for the stable operation of high Q resonant devices.

Thermally grown silicon dioxide and CVD ceramics used to make micro/nano-structures usually exhibit different stability characteristics and lower Q's than SCS. Thermally grown and plasma-enhanced chemical vapor deposition (PECVD) silicon dioxide resonant micro-beams (no metal coating) can be resonated by selective electron beam irradiation of the structure[80]. These resonant beam experiments show that thermal silicon dioxide is more stable than

PECVD material during continuous, high amplitude oscillation of the resonating beams.

Frequency, stiffness and Q tuning with externally applied voltages or internal feedback loops offer the capability to modify and control the parameters of micro/nanometer-scale structures and allows the MEMS to correct for process nonuniformities and changes in environmental conditions. For nanometer-scale structures, slight variations in the symmetry of suspended structures can adversely affect the desired operation of the device[66,67,68]. For example, if the resonant frequency or the stiffness of a structure is incorrect, an internal, real time elec-tromechanical feedback or tuning system can be used to adjust the parameters to their correct values.

VIII. SINGLE CRYSTAL SILICON CANTILEVERS AND TIPS

A. Introduction to Cantilevers and Tips

Cantilevers with integrated tips and probes are used for scanned probe micros-copy[48] including STM and atomic force microscopy (AFM). The cantilevers are usually formed from thin film structures that are released from the substrate. The stiffness is controlled by both the thickness of the thin film and by the length of cantilever, and bending occurs along the length of the cantilever. Thin film structures are used to fabricate both bending[48,50] and torsional[81] cantilevers. The bending of linear cantilevers can be detected by monitoring light deflected from the cantilever with a remote detector; the bending cantilever alters the light collected by the detector. A second method used to detect the bending of thin film cantilever uses the change in resistance of a piezoelectric thin film or piezoelectric cantilever material to monitor the bending[48,50]. The second method allows integration of the piezoelectric material and eliminates the need of a remote detector for each cantilever.

Cantilever structures are also common in robotics where both very stiff and very flexible cantilever structures − such as 'arms', 'legs' and 'fingers' − are used to build robotic systems. Robotic arms with multiple cantilevers and with rota-tional 'joints' are used to perform many tasks[82]. Micro-robotics also benefits from the use of cantilever structures with a wide stiffness variation and with integrated micro-actuators to produce rotation about a 'joint'. Rotation about an axis is accomplished with the SCS torsion bars and the torsional micro-actuators presented in Section III.C.

Cantilevers with integrated probes, tips and grippers, and torsional and linear micro-actuators can be used for nanometer-scale manipulation.

B. Torsional Cantilever

An SCS cantilever with an integrated torsional micro-actuator is shown in Figure 30. The cantilever structure[6,83] can be made very stiff in the vertical plane (z) by making b large and can be made very stiff in the x-y plane by using a lattice structure. The rotational stiffness is determined by the width a, height b, and length L_T of the torsional bar [Eq. (18)]. Cantilevers that exhibit both rotation and bending can be made by reducing the thickness b or by attaching thin film cantilevers to the structure. Thus, the design space is much improved, and cantilevers that exhibit a wide range of structural stiffness can be made with the same process and material and can be easily integrated with other system components.

Sensing the rotation of the torsional cantilever can be accomplished by monitoring piezo-resistance, light reflection or capacitance change of an integrated torsional micro-actuator.

Figures 1(d) and 30 highlight the elements of cantilever structure or arm, and the parallel-plate capacitor, torsional actuator. The torsional cantilever with a lattice structure offers a wide stiffness range for both in-plane and out-of-plane bending, and allows for the integration of torsional and linear micro-actuators, capacitor sensors, and other microstructures on the cantilever, including transistors, tips and probes.

C. Integrated Tips, Probes, and Grippers

Manipulators that interface with micrometer-scale and nanometer-scale objects require integrated micro-actuators. So an integrated batch fabrication process is desired in which the tips, probes and grippers are produced with the micro-actuators and microstructures. It is always a challenge to achieve 'complete' integration and batch fabrication of micro-instruments, since the

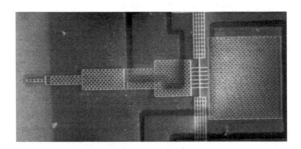

FIGURE 30. A scanning electron micrograph of a torsional cantilever (a=1.5 μm, b=12.3 μm, L_T=50 μm, L_C=950 μm).

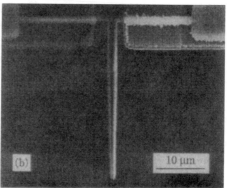

FIGURE 31. SEM images of the SCS tweezers (a) With no voltage applied, and (b) After a step function voltage is applied.

microstructures, electrical and thermal isolation structures, electrical conductors, and the required electronics must all be integrated on the chip with no assembly. The challenges include protecting what is made in previous process steps through the remainder of the process; providing a method and a structure to manipulate an external object or sense an external signal, and including a method to package the micro-instrument or micro-machine for the application.

An example of the integration challenge is the integration of a tip with micro-actuators such that the tip is positioned above the plane of the micro-actuators as demonstrated by the micro-STMs shown in Figure 1(d). The SCS tip on each STM is positioned 5 μm above the plane of the actuators and is mounted on an SCS tip-support structure. Since the tip is formed by high temperature (1100C) oxidation of silicon, the tip is made first and protected throughout the remainder of the process by a silicon dioxide[6] or silicon nitride cap[84]. Then the support pillar is made by an RIE step, so the 20 nm diameter tip is positioned above the plane of the actuators and suspended microstructures. After all the structures are made, the completed tips are silicided, e.g. nickel silicide, or coated with a metal or both. Very dense tip arrays can be fabricated using this approach, and using the SCREAM process tall SCS pillars can be made as shown in Figure 1(d) to position tips and grippers above the plane of the micro-actuators. Tips with integrated apertures to control the electric field at the tip have been fabricated using similar processing concepts [Figure 22].

Small grippers can be integrated with moving SCS microstructures. An example of a small SCS micro-tweezer[85] or gripper is shown in Figure 31.

Many larger grippers have been fabricated using micro-machining processes[8].

IX. ARRAY ARCHITECTURES AND NANOMETER-SCALE MANIPULATORS

A. Array Architectures

A theme of this chapter is array architectures for parallel operation of probes and tips for micromanipulation and nanometer-scale lithography. The array architecture incorporates large force micro-actuators to generate large displacements of a suspended backbone structure with an integrated array of tips. Each array element includes a tip supported on a small torsional micro-actuator that provides local, precision vertical (z) displacement of each tip. Figure 32 shows a *schematic* for a parallel architecture of 100 torsional micro-actuators, like those illustrated in Figure 16 (refer to Section III.C.2), that is supported on a moving backbone structure and connected to two interdigitated electrode micro-actuator arrays. If required, a small x-y micro-actuator such as that shown in Figure 8 can be integrated with each array element to correct for x-y misalignments.

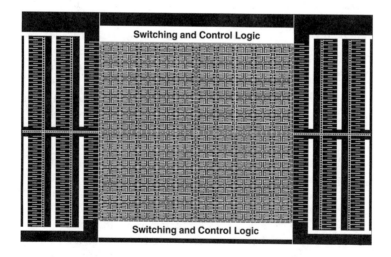

FIGURE 32. Array architecture for x and z tip displacements. A parallel array architecture of 10×10 torsional micro-actuators with integrated tips. The 100 tips are scanned horizontally (x) in parallel by the interdigitated electrode micro-actuators located at each end to the array [also refer to Figure 16].

The nested, micro-actuator array architecture like the 10×10 torsional micro-actuator tip-array shown in Figure 32 offers a high pixel or bit rate, because all the torsional micro-actuators (refer to Section VIII) are scanned in parallel in both x and y. Then, the pixel rate is set by the number of array elements and by the maximum frequency of operation of the z, torsional micro-actuator. We choose a frequency for the torsional micro-actuator of 5 MHz which is below the first resonant frequency. For a pixel area or bit area of 50 nm × 50 nm and a pixel or bit-pitch of 100 nm, each tip in the array addresses (±25 μm/100 Nm) or 500 pixels per line at 200 ns per pixel. For 100 line scans in parallel the array pixel-rate is 0.5 Gpixel/sec. or 0.5 Gbits/sec. The line scan frequency is 10 kHz which is below the first resonant frequency, f_{ox}, of the line scan micro-actuator support structure. This example is used to illustrate the use of the frequency scaling laws for MEMS and to highlight the substantial increase in speed, manipulations, pixels or bits per second produced by using simple but highly integrated array architectures for writing, indexing, probing or moving nanometer-scale objects.

The backbone structure acts as both a mechanical support structure and a support for routing wires to transistors integrated on the backbone structure. The use of 'on-the-backbone' electronics reduces the number of wires that must be routed off the structure via the springs and flexures that support the whole structure.

Partitioning of array components on two silicon chips offers many advantages. The arrays shown in Figures 11 and 32 incorporate x-axis scanning and z-axis motion of the tip. The y-axis can be scanned by the integration of a second set of micro-actuators perpendicular to x-axis actuators on the periphery to move the tip-array in the y direction. The single tip STMs shown in Figures 1(a,b) use this approach for x-y scanning. Another partitioning used to scan the y axis is to build the pair of y micro-actuators on a second 'silicon chip' which is placed over the 'chip' with the tip-array as shown in Figure 33(a). A third architecture partitions the tip-array with integrated torsional micro-actuators [Figure 16] onto one chip; the x-y micro-actuator and specimen stage or storage medium are integrated onto the second chip as shown in Figure 33(b). The correct partitioning-approach depends on the application: information storage, lithography, or nanometer-scale manipulation. Note that the partitioning of the tip-array and z micro-actuators on one 'chip' facilitates wiring and control of the individual tips and z axis torsional micro-actuators. Since the electronics and wires are on the non-moving substrate, only a few wires need to be routed onto each moving-tip microstructure. For many micro-robotic applications, the micro-machines and micro-robots with integrated electronics must be integrated on one 'silicon chip' which is then used to probe other objects.

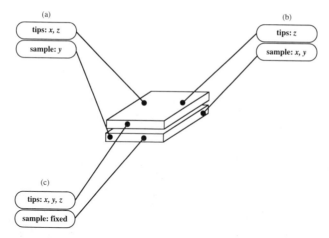

FIGURE 33. Examples illustrate partitioning of x, y and z actuators on two stacked silicon chips. In these examples the z actuators include the integrated tip. (a) x and z actuators on one chip and the y actuator on a second chip; (b) The x,y actuators are on one chip, and the array of z actuators are on the second chip, and (c) All the actuators are on one chip, and the second chip is the sample to be probed or patterned.

B. Nanometer-scale Manipulators and Micro-Robots

Arms, joints, grippers, actuators, sensors, communication and control are the 'stuff' of robots[82]. Micro-robots for nanometer-scale manipulation can be built by the integration of cantilevers, torsional micro-actuators, tips and probes, and linear actuators with wiring and external electronics. A few of the prototypical micro-robotic components are the microstructures, micro-actuators and micro-devices presented in this chapter.

As an example of a micro-robotic task for nanometer-scale manipulation, we look at the task of moving two tips over a substrate as shown in Figure 34. Each micro-robot arm consists of a lattice-structured cantilever. One cantilever supports an integrated tip. The two cantilevers are joined with two interdigitated electrode torsional micro-actuators which provide rotation about the axis of the torsion bar that connects and supports the two arms. The opposite end of the second arm is connected to interdigitated electrode torsional micro-actuators (Section III.C.2) that rotate the arm out-of-plane of the 'silicon chip'. The

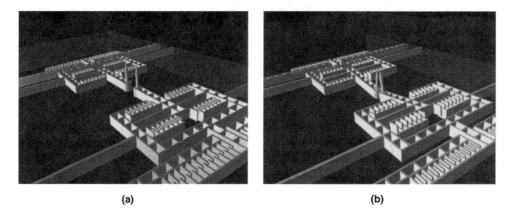

(a)   (b)

FIGURE 34. Two nanometer-scale probes integrated with two torsional actuators to rotate the probes out-of-plane of the silicon chip. Each torsional micro-actuator is displaced horizontally with integrated, interdigitated electrode micro-actuators located at each end of the structure. (a) No actuation, and (b) Voltages applied to both the torsional and linear actuator to position the high aspect-ratio probes.

second arm is also attached to a large force micro-actuator to move the tip in the x direction. A pair of these simple robots could position two tips, within a distance comparable to the tip radius.

A micro-robot architecture, using the simple components treated in this chapter, is illustrated by the stick-drawings shown in Figure 35. The building blocks are capacitor micro-actuators for x-y-z motion, and torsional micro-actuators for rotational motion at the joints. The connecting arms are the stiff, SCS lattice structures shown in Figure 34 with integrated torsional micro-actuators, flexures and springs.

Figure 36 shows a structure with two opposing tips within the plane of the silicon. The tips can be positioned with integrated micro-actuators. Additional degrees of freedom are possible by integrating small x-y and torsional capacitor micro-actuators attached to the tips.

The 'stick figure' micro-robots shown in Figure 35 illustrate how available SCS components can be integrated to perform robotic tasks. The key word is integration; a concept that has been and continues to be the strength of silicon-based processing and MEMS.

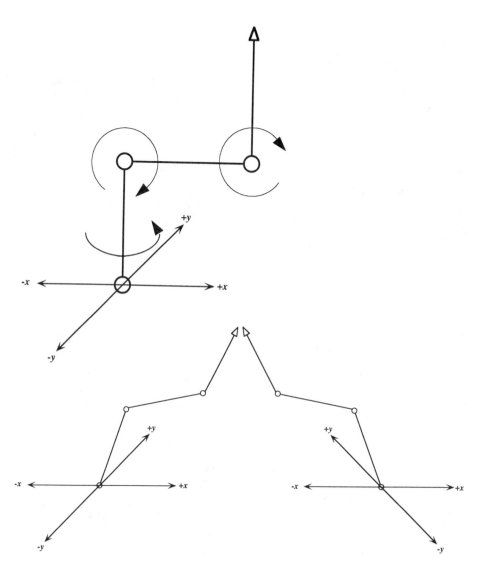

FIGURE 35. Stick figures showing that many degrees of freedom can be obtained using the linear and torsional micro-actuators mounted on a stiff backbone structure.

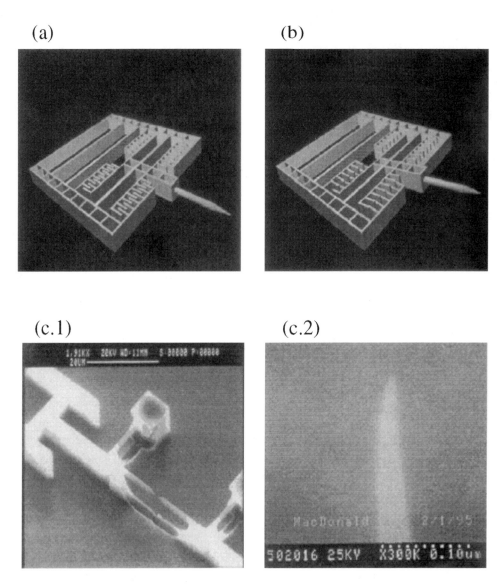

FIGURE 36. Lateral tip micro-robots showing how one or many tips can be positioned in-plane or out-of-plane of the silicon chip. (a) No voltage applied, (b) Rotation of tip out-of-plane, and (c) SCS lateral tips.

X. MICRO-INSTRUMENTS, MICRO-MACHINES AND NANOMETER-SCALING

Moving SCS structures with minimum features of a few hundred nanometers have been fabricated using electron beam lithography (EBL)[30,31,51,66-68] and characterized using scanning electron microscopy. A collage of these devices is shown in Figure 37. We can go much farther in scaling the processes and devices described in this chapter. Present EBL processes and equipment can be used to scale and build suspended structures with an MFS, a, less than 20 nm. The scaling laws show that there is much to be gained from reducing the MFS, since the force increases as $(1/a^3)$ for the parallel-plate [Eq. (8)] and as $(1/a^2)$ for the interdigitated electrode capacitor [Eq. (11)] micro-actuators. The angular rotation $\phi(rad)$ of a torsional actuator given by Eq. (18) scales as $(1/a^5)$. The performance of the small torsional actuator described by Eq. (19) improves by factors of 10^3 and 10^5 for a =250 nm and 50 nm respectively. So a few volts provides micrometers of z motion for the torsional actuator.

GaAs superlattice structures can also be scaled for MEMS, and the use of nanometer-scale quantum well structures is compatible with high aspect-ratio structures produced by chemically assisted ion beam etching[29].

Micro-machines and micro-instruments will have a major impact on nanometer-scale materials and structure characterization. Such small machines can produce >100 GPa of stress in sub-micrometer samples[59], and operate at high scan rates and low power. Furthermore the micro-machines can be used as components for larger machines. A smart tip can replace the tip in a larger STM or AFM; the smart tip with integrated micro-actuators can scan in three dimensions at much higher scan rates than the more massive tip attached to a piezo tube scanner or it can be used to scan a smaller area at higher resolution. A micro-loading or micro-fracture machine can be mounted on the stage of a scanning electron microscope (SEM), scanning Auger microscope (SAM), or a transmission electron microscope (TEM), and the specimens can be fractured[61], fatigued or buckled[59] in-situ with little or no modification of the electron microscopes.

Information on the 'torsional' fracture strength is necessary for the design of torsional micro-actuators and micro-robotic joints. An example of a micro-instrument[61] that is used to measure the fracture strength of a 4 μm long, 500 nm diameter SCS bar in torsion is shown in Figure 38. This small torsion rod turned 20 degrees before fracturing! The micro-instrument can be used to measure the fracture strength of SCS samples as small as 20 nm diameter. This is the first time such small samples have been fractured, and the preliminary measurements indicate an increase in fracture strength for these micro/nano-samples. Fracture strength, fatigue, yield strength can be determined on micro/nanometer-scale samples under extreme conditions of pressure, temperature and in harsh environments.

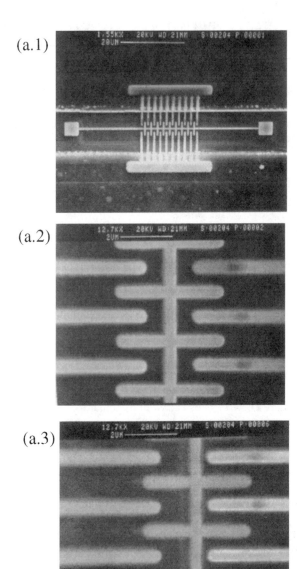

FIGURE 37. A collage of single crystal silicon devices with minimum features of 100–300 nm[67,68]. (a) Interdigitated electrode capacitor actuator with 250 nm SCS beams. (a.1) View of entire device, (a.2) V = 0, no displacement, and (a.3) V = 80 volts displacement = 400 nm.

(b.1)   (c.1)

(b.2)   (c.2)

(b.3)

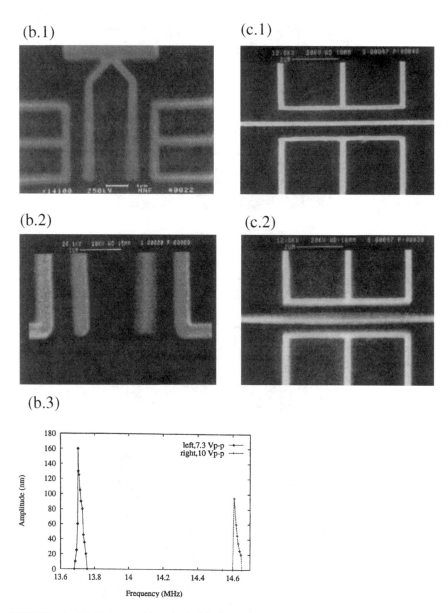

FIGURE 37. (b) Unsymmetrical tuning fork made from 200 nm SCS beams. (b.1) The entire device resonating; (b.2) Device showing only one time resonating, and (b.3) Amplitude-frequency response of tuning fork showing a different resonant frequency for each time. (c) 10 MHz SCS resonator made from 250 nm SCS beams (c.1) Device not resonating; and (c.2) Device resonating at 10 MHz.

152 N.C. MacDonald

(a)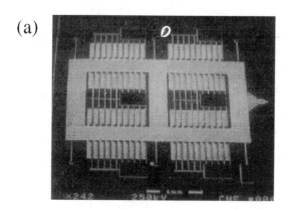

(b)

(c)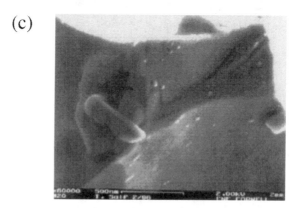

FIGURE 38. Torsional loading micro-machine used to fracture a single-crystal-silicon torsion rod 4 μm long and 500 nm in diameter. (a) The loading machine with capacitor actuator; (b) Torsional sample before fracture; (c) Fracture surface.

The fabrication of many unique micro-instruments for characterizing nanometer-scale specimens is possible using the MEMS technologies. MEMS-based micro-instrumentation will be used to perform detailed characterization of nanometer-scale samples and to determine their electrical, mechanical, chemical and thermal properties. So far internal friction[55], 'rubbing' friction and wear[86], material fracture[60], surface topography[6] and micro-beam buckling[59] have been measured using SCREAM-based micro-instruments.

Micro-robots for nanometer-scale manipulation should offer a suite of new experiments for nanometer-scale manipulation, lithography[90,91,92], writing, machining, fabrication and characterization[88,89] (refer to Chapter 11). The capability to make suspended, low capacitance tip-structures that are positioned in three dimensions with integrated micro-actuators opens many possibilities for nanometer-scale probing, electron tunneling, atom and molecule stacking, and spectroscopy experiments. The paradigm here is that these crude micro-machines will be used to build smaller, nanometer-scale structures and nanometer-scale electronic and electromechanical devices. A driving factor to build these micro-machines and micro-instruments is that the pieces can be integrated on a scale compatible with the experiment; the 'difficult components to integrate' such as electrical connections and isolation, micro-actuators and manipulators are built into the micro-instrument design methodology and the fabrication processes. Furthermore many devices and many samples can be fabricated or characterized during a single experiment under the same conditions.

Micro-instrumentation for micro/nanometer-scale fluidics[87] offers additional possibilities for characterizing chemical reactions and flow-based chemical mixing and synthesis. The integration of micro/nano-fluidics with microanalytical instrumentation offers possibilities to build microchemical cells, to analyze a few molecules in a nanometer-scale channel; and to characterize the mechanical and rheological properties of biological molecules. Also, nanometer-scale interactions such as the Casimir effect[93-96] can be investigated.

Nanolithography, nanometer-scale machining, nanometer-scale writing and reading, nanometer-scale fabrication and construction, and nanometer-scale manipulation are the products of micro-machines. Array architectures or the integration of many electromechanical devices working in parallel allows for massively parallel operation of micro-machines to perform the desired nanometer-scale operation with high throughput.

XI. SUMMARY AND CONCLUSIONS

Science, Engineering and technology continue to move forward and build on the advances and technology of new instruments, machines and processes. MEMS offers a new paradigm for these tools: micro-instruments, micro-machines, and micro-sensors, and the integration of all three. These new tools are applicable to probing the world of atoms, molecules, proteins, and cells; for measuring and characterizing nanometer-scale objects; and for writing nanometer-scale features and for nanometer-scale information storage. Many of the components and processes are available to build these micro-instruments and micro-machines. The future offers MEMS-based eyes, ears, noses, tongues and tactile sensing and includes micro-mass spectrometers, micro-optical systems, micro-interferometers, microanalytical instruments, micro-conveyors and much more. These micro-components and micro-instruments should improve and extend the performance of macroscopic instruments, sensors, vehicles and other electronic and electromechanical systems.

The two major themes for Chapter 3 are *array architectures* and *micro-robots* for nanometer-scale manipulation, with multiple degrees of freedom. We emphasized the importance of system level *integration* and *scaling* of the components: micro-actuators, support structures, electrical conductors, electrical and thermal isolations, tips, probes, joints and grippers. For the near future silicon, silicon-based structures, integrated circuits and processes were selected as 'todays' system approach for nanometer-scale manipulation. This approach uses the massive infrastructure that supports silicon-based technology including future generations of lithography and reactive ion etching equipment.

Another theme of Chapter 3 is that some of the pieces are in place to build MEMS-based micro-instrumentation for nanometer-scale manipulation, sensing and characterization. A few prototypical integrated micro-instruments have been made and characterized; and many additional 'pieces' still rely on the imagination and the inventiveness of the reader. There is a lot to be done, and the fun is to 'get-on-with-it' and build these micro-machines, micro-robots, micro-instruments and MEMS, so that the users of these micro-machines can move forward and explore, characterize, build, write and manipulate in the fascinating, complex and chaotic world of nanometer-scale systems.

"Now, you might say, 'Who should do this and why should they do it?' .. fun! "

Richard P. Feynman, "There's Plenty of Room at the Bottom", December 26, 1959[1].

ACKNOWLEDGEMENTS

The author acknowledges the editing, suggestions and preparation of Mr. Pete Hartwell, Mr. Scott Miller, Dr. Taher Saif and the assistance of Ms. Jessie Dimick in assembling the final manuscript. I also want to acknowledge the patience, tenacity and moral support of the editor, Dr. Greg Timp.

REFERENCES

1. Feynman, Richard P., *J MEMS* **1**, (1992); Transcript of a talk given by R.P. Feynman at the APS Meeting, California Institute of Technology, December 26, 1959.

2. MacDonald, G.A., *Sensors and Actuators* **A21-A23**, 303–307, (1990).

3. Nielsen, Neils J., *Hewlett-Packard Journal*, 4–10 (1985).

4. Baker, J.P., *et al.*, *Hewlett-Packard Journal* 6–15 (August 1988).

5. Hornbeck, L.J., IEEE International Electron Devices Meeting *Tech. Digest* 381–384 (1993).

6. Xu, Y., Miller, S.A., and MacDonald, N.C., *Appl. Phys. Lett.* **67**, 2305–2307 (1995).

7. Tang, W.C., Nguyen, T.H., and Howe, R.T., *Sensors and Actuators* **20**, 25–32 (1989).

8. Sze, S.M., *Semiconductor Sensors*, New York: John Wiley and Sons, Inc., 1994.

9. Mehregany, M., and Tai, Y.C., *J. Micromech. Microeng.* **1**, 73–85 (1991).

10. Linden, C., *et al.*, *J. Micromech. Microeng.* **2**, 122–132 (1992).

11. Markus, K.W., *et al.*, "MEMS Infrastructure: the Multi-User MEMS Processes (MUMPs), *Proceedings of the SPIE Micro-machining and Micro-fabrication Conference,* 1995, Vol. 2639–08, pp. 54–63.

12. Cole, B.E., *et al.*, "512×512 Infrared Scene Projector Array for Low-Background Simulations," *Tech Digest* from Solid-State Sensor and Actuator Workshop, Hilton Head, S.C., 1994.

13. Mastrangelo, C.H., and Hsu, C.H., *J. MEMS* **2**, 33–43 (1993); *J. MEMS,* **2**, 44–55 (1993).

14. Gardner, J.W., *Micro-sensors Principles and Applications*, Chichester, England: John Wiley and Sons Ltd., 1994.

15. Middelhoek, S., and Audet, S.A., *Silicon Sensors*, London: Harcourt Brace Jovanovich Publishing, 1989.

16. MacDonald, N.C., *et al.*, *Sensors and Actuators* **20**, 123–133 (1989).

17. Hoffman, W., *et al.*, *J. of Microelec. Eng.* 523–526 (1996).

18. Ehrfeld, W., *et al.*, "1988 LIGA Process: Sensor Construction Techniques via x-ray Lithography," *Tech. Digest* from IEEE Solid-State Sensor and Actuator Workshop, Hilton Head, SC, 1988.

19. Guckel, H., *et al.*, *J. Micromech. Microeng.* **1**, 135–138 (1991).

20. Ahm, C.H., and Allen, M.G., *J. MEMS* **2**, 15–22 (1993).

21. Shacham-Deamand, Y., *J. Micromech. Microeng.* **1**, 66–72 (1991).

22. Bean, K.E. *IEEE Trans. Electron Devices* **ED 25**, 1185–1192 (1978).

23. Kern, W. *RCA Reveiw* **39**, 278–307.

24. Danel, J.S., and Delapierre, G., *J. Micromech. Microeng.* **1**, 187–198 (1991).

25. Gianchandani, Y.B., and Najafi, K., *J. Microelectromech. Sys.* **1**, 77–85 (1992).

26. Tanghe, S.J., and Wise, K.D., *IEEE J. of Solid-State Circuits* **27**, 1819–1825 (1992).

27. Zhang, Z.L. and MacDonald, N.C., *J. Micromech. Microeng.* **2**, 31–38 (1992).

28. Shaw, K.A., Zhang, Z.L., and MacDonald, N.C., *Sensors and Actuators A* **40**, 63–70 (1994).

29. Zhang, Z.L., and MacDonald, N.C., *J. MEMS* **2**, 66–72 (1993).

30. Zhang, Z.L., and MacDonald, N.C., *J. Vac. Sci. Technol. B* **4**, 2538–2543 (1993).

31. Yao, J., Arney, S., and MacDonald, N.C., *J. of Micro-Electromechanical Systems* **1**, 14–22 (1992).

32. Jazairy, A., and MacDonald, N.C., *J. of Microelec. Eng.* 527–530 (1996).

33. Shaw, K.A., and MacDonald, N.C., "Integrating SCREAM Micro-Machined Devices With Integrated Circuits," *The Ninth Annual International Workshop on MEMS, (MEMS '96), IEEE Proc.* 1996, pp. 44–48.

34. Trimmer, W; Editor-in-Chief, *J. of Micro-Electromech. Sys.,* A Joint IEEE/ASME Publication on Microstructures, Micro-actuators, Micro-sensors, and Microsystems, **1,** March 1992.

35. Carr, W.N., and Guckel, H.; Editors-in-Chief, *J. of Micro-Mech. and Micro-Eng., Struc., Devices and Sys.,* Institute of Physics Publishing **1**No. 1, March 1991.

36. Middelhoek, S.; Editor-in-Chief, *Sensors and Actuators,* A Special Issue Devoted to Micromechanics **20**, (1989).

37. Reichl, H.; Editor, *Microsystem Technologies, Sensors Actuators System Integration,* Springer International Publishing **1**, (1994).

38. *IEEE Micro Electro Mechanical Systems (MEMS) Proceedings, An Investigation of Micro Structures, Sensors, Actuators, Machines and Systems,* Currently Eight Proceedings (Published in odd numbered years from 1981).

39. *Transducers International Conference on Solid-State Sensors and Actuators Digest of Technical Papers,* Currently Eight Conference Proceedings (Published in odd numbered years from 1981).

40. *Solid-State Sensor and Actuator Technical Digest,* Sponsored by the Transducers Research Foundation, Currently Six Proceedings (Published in even numbered years from 1984).

41. Jardine, A.P.; Editor, *SPIE Smart Structures and Materials 1995 Proceedings,* **2441**, San Diego, CA., February 27–28, 1995.

42. Motamedi, M.E., and Beiser, L.; eds., *SPIE Micro-Optics /Micromechanics and Laser Scanning and Shaping,* **2383**, San Jose, CA., February 7–9, 1995.

43. Hjort, K., Söderkvist, J., and Schweitz, J.A., *J. Micromech. Microeng.* **4**, 1–13 (1994).

44. Peterson, K.E., "Silicon as a mechanical material," in Proc. IEEE, **70**, 1982, pp. 420–457.

45. Feynman, R., *J. MEMS,* **2**, 4–14 (1993). (Manuscript from a talk given by Richard Feynman on February 23, 1983 at the Jet Propulsion Laboratory, Pasadena, CA.)

46. Gabriel, K.J., *Scientific American* **273**, 150–153 (1995).

47. Woodson, H.H., and Melcher, J.R., *Electromechanical Dynamics Part I: Discrete Systems; Electromechanical Dynamics Part II: Fields, Forces, and Motion; Electromechanical Dynamics Part III: Elastic and Fluid Media,* New York: John Wiley & Sons, 1968.

48. Wiesendanger, R., *Scanning Probe Microscopy and Spectroscopy, Methods and Applications,* Cambridge: Cambridge University Press, 1994, pp. 91–97.

49. Albrecht, T.R., *et al.,* *J. Vac. Sci. Technol. A* **8**, 317 (1990).

50. Minne, S.C., *et al.,* *J. Vac. Sci. Technol.,* B **13**, 1380–1385 (1995).

51. Yao, J.J., Arney, S.C., and MacDonald, N.C., *Sensors and Actuators A,* **40**, 74–84 (1994).

52. Ramo, S., Whinnery, J.R., and VanDuzer, T., *Fields and Waves in Communication Electronics,* Second Edition, New York: John Wiley & Sons, 1984.

53. Gilbert, J.R., Ananthasuresh, G.K., and Senturia, S.D., "3D Modeling of Contact Problems and Hysteresis in Coupled Electro-Mechanics," presented at The Ninth Annual International Workshop on MEMS, (MEMS '96), *IEEE Proc.,* San Diego, CA, February 11–15, 1996.

54. Timoshenko, S.P., and Goodier, J.N., *Theory of Elasticity* (1982).

55. Milhailovich, R.E., and MacDonald, N.C., (to be published in *Sensors and Actuators,* 1996.)

56. Tang, W.C., Lim, M.G., and Howe, R.T., *J. MEMS* **1**, 170–178 (1992).

57. Jazairy, A., and MacDonald, N.C., "Very High Aspect-Ratio Wafer-free Silicon Micromechanical Structures," SPIE's 1995 Symposium on Microlithography and Metrology in Micromachining, Postek, M.T., ed., *Proc. SPIE* **2640**, 1995, pp. 111-120.

58. Huang, X.T., Chen, L.-Y., and MacDonald, N.C., "A Low Temperature Process for Very High Aspect-Ratio Silicon Microstructures Using SOG Etch Mask," SPIE's 1995 Symposium on Microlithography and Metrology in Micromachining, Postek, M.T., ed., *Proc. SPIE* **2640**, 1995, pp. 178–183.

59. Saif, M.T.A., and MacDonald, N.C., be published in *Sensors and Actuators* 1996.)

60. Saif, M.T.A., and MacDonald, N.C., "Micro Mechanical Single Crystal Silicon Fracture Studies-Torsion and Bending," The Ninth Annual International Workshop on MEMS, (MEMS '96), *IEEE Proc.*, 1996, pp. 105–109.

61. Saif, M.T.A., and MacDonald, N.C., (to be published in *J. of Micro-Electromech. Sys.* 1996.)

62. Das, J.H., and MacDonald, N.C., *J. Vac. Sci. and Technol. B* **13**, 2432–2435 (1995).

63. MacDonald, N.C., "SCREAM Micro-Electro-Mechanical Systems," (Invited) Special Issue of *Journal of Microelectronic Engineering on Nanotechnology, (to be published July 1996).*

64. *Thomson, W.T., Theory of Vibration With Applications,* Third Edition, Englewood Cliffs, NJ: Prentice Hall, 1988.

65. Gorman, D.J., *Free Vibration Analysis of Beams and Shafts,* New York: John Wiley & Sons, 1975.

66. Yao, J.J., and MacDonald, N.C., *J. Micromech. Microeng.* **5**, 257–264 (1995).

67. McMillan, J.A., and MacDonald, N.C., "Nonlinear Vibrations of Submicron-Scaled Single Crystal Silicon Resonant Devices," presented at Proc. 184th Meeting of The Electrochemical Society, New Orleans, LS, 1993.

68. McMillan, J.A., "High Frequency Mechanical Resonant Devices," Ph.D. Dissertation, Cornell University, August 1993.

69. Adams, S.G., Bertsch, F., and MacDonald, N.C., "Independent Tuning of the Linear and Nonlinear Stiffness Coefficients of a Micromechanical Device," in *IEEE Proc.,* 1996, pp. 32–37.

70. Cho, Y-H., Pisano, A.P., and Howe, R.T., *J. MEMS* **3**, 81–87 (1994).

71. Nguyen, C.t.-C., and Howe, R.T., "Quality Factor Control for Micromechanical Resonators," in *IEEE International Electron Devices Meeting, Tech. Digest,* 1992, pp. 505–508.

72. Braginsky, V.B., Mitrofanov, V.P., and Panov, V.I., *Systems with Small Dissipation,* Chicago: The University of Chicago Press, 1985.

73. Beerschwinger, U., *et al.*, *J. Micromech. Microeng.* **4**, 95–105 (1994).

74. Tai, Y.C., and Muller, R.S., *Sensors and Actuators* **A21-A23**, 180–183 (1990).

75. Mehregany, M., Senturia, S.D., and Lang, J.H., "Friction and wear in micro-fabricated harmonic side-drive motors," in *Tech. Digest of 4th Int. Conf. on Solid-State Sensors and Actuators,* 1990, pp. 17–422.

76. Gabriel, K.J., *et al.*, *Sensors and Actuators* **A21-A23**, 184–188 (1990).

77. Prasad, R. and MacDonald, N.C., "Design, Fabrication and Measurements of Friction in SCREAM Micro-Devices," in *Transducers '95 - The 8th International Conference on Solid-State Sensors and Actuators,* **2**, 1995, pp. 52–55.

78. Smith, D.P.E., *Rev. Sci. Instrum.* **66**, (1995).

79. Mihailovich, R.E., and MacDonald, N.C., *J. Vac. Sci. and Technol. B* **13**, 2545–2549 (1995).

80. Ogo, I., and MacDonald, N.C., *J. Vac. Sci. and Technol. B* **12**, 3285–3288 (1994).

81. Bay, J., *et al.*, *J. Micromech. Microeng.* **5**, 161–165 (1995).

82. Mason, M.T., and Salisbury, Jr., J.K., *Robot Hands and the Mechanics of Manipulation,* Cambridge, MA: The Massachusetts Institute of Technology Press, 1985.

83. Miller, S.A., Xu, Y., and MacDonald, N.C., "Micro-Mechanical Cantilevers and Scanning Probe Microscopes," in *Proc. SPIE* **2640**, Postek, M.T., ed., 1995, pp. 45–52.

84. Spallas, J.P., and MacDonald, N.C., *J. Vac. Sci. Technol. B* **11** 437–440, (1993).

85. Yao, J.J., and MacDonald, N.C., *Scanning Microscopy* **6**, 939–942, (1992).

86. Prasad, R., and MacDonald, N.C., "Design, Fabrication and Measurements of Friction in SCREAM Micro-Devices," in *Transducers '95 - The 8th International Conference on Solid-State Sensors and Actuators,* **2**, 1995, pp. 52–455.

87. Carr, W.N., Editor-in-Chief, *Journal of Micro-mechanics and Micro-Engineering,* Special Issue on Micro-Fluidics, **4** Institute of Physics Publishing, (1994); Gravesen, P., Branebjerg, J., and Jensen, O.S., *Micro-mech. Microeng.* **3**, 168–182 (1993).

88. Niu, Q., Chang, M.C., and Shih, C.K., *Physical Review B* **51**, 5502–5505 (1995).

89. Avouris, P., *Acc. Chem. Res.* **28**, 95–102 (1995).

90. Minne, S.C., *et al.*, *Appl. Phys. Lett.* **66**, 703–705 (1995).

91. Shen, T.-C., *et al.*, *Science* **268**, 1590–1592 (1995).

92. Kramer, N., *et al.*, *J. Vac. Sci. Technol. B* **13**, 805–811 (1995).

93. Serry, F.M., Walliser, D., and Maclay, G.J., *J. MEMS* **4**, 193–205 (1995).

94. Casimir, H.B.G., "On the Attraction Between Two Perfectly Conducting Plates," in *Proc. Koninkl. Ned. Akad. Wetenschap,* **51**, 1948, pp. 793–795.

95. Brown, L.S., and Maclay, G.J., *Phys. Rev.,* **184**, 1272–1279 (1969).

96. Schwinger, J., DeRaad Jr., L.L., and Milton, K.A., *Ann. Phys.* **115**, 1–23 (1978).

Chapter 4

Limits of Conventional Lithography

D.M. Tennant

Bell Laboratories
Holmdel, NJ 07733

I. INTRODUCTION TO METHODS OF HIGH RESOLUTION LITHOGRAPHY

A. High Resolution Lithography: A Moving Target

Lithography is a key element in a cadre of planar processing methods used in advanced semiconductor manufacturing. Technology sectors, such as integrated circuits, flat panel displays, optoelectronic components, and advanced electronic packaging, all especially rely on it. Lithography as discussed in this chapter is used to pattern layered materials and is akin to the photographic process. In photography an imaging system is used to record an image in a silver-containing emulsion. In conventional lithography the emulsion is replaced by a thin radiation sensitive layer, usually a polymer, known as a resist. A pattern of radiation exposes the resist in order to alter its solubility in a chemical developer. The process relies on the highly non-linear response of the resist to produce well defined patterns after development.

While the type of exposing radiation, pattern delivery system, and the specific resist materials can take many forms, this basic lithographic process is an essential part of nearly all planar patterning processes. It currently accounts for over one-third of the cost of a silicon integrated circuit. Continuous improvement in lithographic systems and technology has therefore been at the heart of the rapid progress in the microelectronics industry. Twenty five years ago the resolution requirement for integrated circuits was 10 μm, while today it is approaching 0.25 μm and plans are being formulated for future systems beyond 0.1 μm. On another track, similar but more specialized methods and materials of advanced lithography have enabled countless researchers to explore and map out the limits of performance of electronic devices. As early as 1976 demonstrations of lithography in the 8–25 nm range had been achieved in the research lab[1] using electron beam lithography, albeit under very limiting conditions.

High resolution lithography has historically been an imprecise but extremely popular term used to describe the most advanced results to date – truly a moving target. Other terms which are currently trendy but are a bit more quantitative include: sub-micron (below 1 μm), deep sub-micron (below 0.35 μm), and nano-lithography (below 100 nm). High resolution lithography continues to be an important technological pursuit in research and development because it enables both high performance and unique device applications, and because of its economic importance in industry. As you will read about in other chapters in this volume, researchers have even engineered systems for modifying materials on the size scale of atomic distances.

This chapter, however, concentrates on conventional approaches to high resolution lithography, limits of these techniques, and several advanced structures made using these methods. We then go on to discuss the driving forces for fine line lithography, principally commercial production of electronic and photonic circuits and components. We examine the future requirements in these fields and discuss the shape future production lithography may take to meet these needs. We analyze the pros and cons of the leading candidates for achieving minimum linewidths of 0.1 µm and below in production. These candidates include: projection ion and e-beam systems, extreme ultraviolet (EUV) lithography, and X-ray proximity printing. We conclude with a discussion on the capabilities of future generation e-beam systems as flexible tools which continue to improve and therefore allow R&D in devices and circuits well in advance of available production tools. This chapter will help the reader to both distinguish the requirements of the R&D community from that of the manufacturing community while gaining an appreciation of the inexorable linkage between these two worlds.

B. Conventional vs. Niche and Experimental Techniques

As in the past, the future of high resolution lithography will not follow a single path. The method of choice for each task in the wide spectrum of lithography applications is typically a compromise among resolution, throughput, economics, process complexity and yield. Researchers pursuing nanofabricated structures at the limits of fabrication often do so with relaxed requirements in other areas such as patterning speed, pattern distortion, overlay accuracy, linewidth control, etc. Such a goal is often served by a laboratory fabrication method. Examples include modified imaging systems such as scanning – and scanning transmission – electron microscopes (SEM and STEM, respectively) modified for computer control[1-4], scanning tunneling and atomic force microscopes[5-7] or small scale X-ray printing systems[8].

On the other hand, large scale circuits which are derived from numerous sequential lithographic steps demand that the uniformity, reproducibility, linewidth control, and overlay accuracy adhere to strict industry standards. These standards are typically so stringent that the resolution capability lags several generations behind the laboratory fabrication systems listed above. Errors in linewidth, pattern placement, misalignments between levels, particulate contaminants, reticle defects, etc. can all result in disabling defects in VLSI circuits. Physically these can correspond to open or shorted circuits due to defects in metal

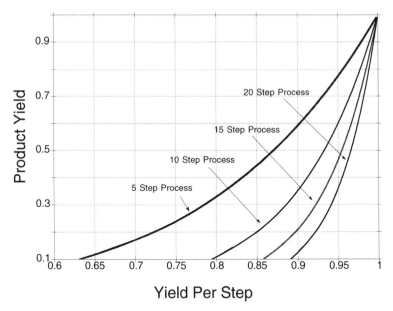

FIGURE 1. Plot of dependence of product yield on the average yield per lithography level shown for 5, 10, 15 and 20 step manufacturing processes.

layers or malformed layers resulting in underperforming circuits. Often, imperfections in relatively small fractional areas render whole chips unusable. As a simple estimate, the individual yields of processing levels can be multiplied to determine the total product yield. To illustrate this, Figure 1 plots the resulting product yield for varying individual yield values. A typical CMOS integrated circuit process might require fifteen or more lithography steps. From the plot you see that the individual step yield must be greater than 0.95 to produce a majority of working products. Yet, typical wafer labs have net product yields of 0.95! This implies an average yield per step of nearly 0.99.

Another portion of the lithographic spectrum is occupied by niche techniques in which significant printing speed is required of very restricted patterns. A good example of this is a grating produced by laser holographic printing. While not an appropriate method for general patterning, large areas of lines can be printed at rapid rates as is often done for photonic applications. These grating-type patterns have become key steps in the manufacture of components such as semiconductor lasers, filters, etc. for photonic integrated circuits (PICs) and optoelectronic integrated circuits (OEICs).

An over-simplified, but nonetheless useful, overview of lithography methods is shown in Figure 2[9]. Here for comparison are a wide range of lithographic accomplishments plotted to illustrate the resolution in Ångstroms and

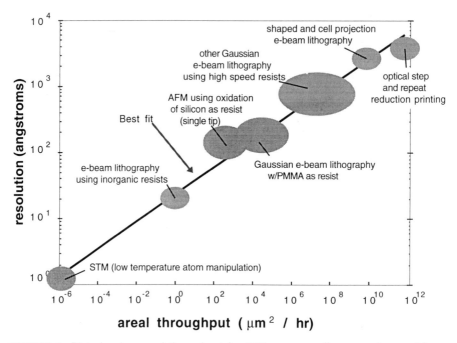

FIGURE 2. Plot showing areal throuput for 50% coverage (for example equal lines and spaces) at the plotted resolution for a wide range of lithographic methods which have been demonstrated to date. The solid line represents a phenomenological relationship between resolution and throughput given by the power law fit found in text.

corresponding areal writing speed in square micrometers per hour. The very finest resolution represents an extreme case of individual atom manipulation at cryogenic temperatures. The opposite extreme represents current silicon IC manufacturing capabilities, with many different research and development applications falling in between. It is rather remarkable that for over 18 orders of magnitude in throughput, these currently demonstrated techniques follow a simple power law:

$$resolution \approx 23 \ A_t^{0.2} \qquad (1)$$

where resolution is measured in Ångstroms and A_t is the areal throughput in μm^2/hr.

This empirical observation is more than just an amusement, it serves to emphasize the daunting challenge to be met by current laboratory methods before they seriously compete for a share of the manufacturing marketplace.

It is readily observed then that the limits of lithography are tempered by the accompanying requirements. To reflect this our discussion below will separate the ultra-high resolution methods used in the research community, primarily to explore the electronic and photonic properties of nanostructures at the limits of fabrication, from the high resolution lithographies being pursued because of their promise for meeting the future needs of near term development and manufacturing, primarily for the silicon integrated circuit industry.

II. LIMITS OF CONVENTIONAL HIGH RESOLUTION METHODS IN RESEARCH

A. Electron and Ion Focused Beam Lithography

1. Resolution Limits

In an ultra-violet (UV) optical exposure system the spatial distribution of the deposited energy is determined by the aerial image. It is usually assumed that resolution is limited in high quality optics by diffraction and not by photon statistics or resist limitations[10]. The Rayleigh criteria is often applied to estimate the diffraction-limited resolution:

$$resolution = k_1 \frac{\lambda}{NA} \tag{2}$$

where k_1 is a constant near 0.6, λ is the exposure wavelength, and NA is the numerical aperture of the lens. When this expression is used to estimate resolution in lithography applications, the value of k_1 can vary from that of the Rayleigh criterion, depending on resist contrast and processing conditions. The NA is defined as the sine of the convergence semi-angle at the focal point, which can be estimated for a simple lens by:

$$NA = \sin \left[\frac{D}{2\sqrt{(D^2/4)+f^2}} \right] \tag{3}$$

where D is the diameter of the lens and f is the focal length. For large field projection systems, the numerical aperture is usually limited to a value of the order 0.63 or lower. This means, as a rule of thumb, that for conventional optics the resolution is approximately equal to λ. This limitation prevents UV optical lithography from being useful for nanolithography and is the primary motivation for researchers in nanolithography to seek alternative methods of patterning.

Use of a much shorter wavelength, for example X-rays in the 1 to 1.5 nm range, is one way to improve on these diffraction limitations. However, the lack of refractive or reflective X-ray optics does not permit use of similar projection systems (although diffractive optics have been proposed[11]). Only contact and near-contact printing (discussed later in this chapter) have been used to realize improvement over the longer wavelengths. In the limit of perfect contact with a well collimated X-ray source (such as synchrotron), the resolution is approximated by:

$$resolution_{xray} = k_x (l\lambda_{xray} /2)^{1/2} \qquad (4)$$

where l is the resist thickness and k_x is a constant of the order of 1. Thus for $\lambda = 1$ nm, and a resist thickness of 0.5 μm the contribution of diffraction to resolution limit is about 15 nm. Since laboratory versions of such systems do not use synchrotrons but rather use point sources of finite size, there can be other contributions to edge blurring such as penumbral broadening (although for contact lithography this is very small, of the order 1 nm).

Since X-ray lithography requires mask making capability, only electron and ion beam lithography allow primary patterning directly from a design. Nanolithography as observed in the research lab is therefore more suited to using focused electron beam or ion beam systems. Among these, electron beam systems (e-beam) are the most common place. This owes, in part, to the rapid and widespread proliferation of scanning electron imaging microscopes. A rather modest priced scanning electron microscope can be readily adapted for lithography with commercially available modifications and control software. However, even the more sophisticated e-beam systems costing several million dollars are common at major research centers because of the versatility, established infrastructure of commercially available computer-aided design (CAD) tools, data-type conversion software, and resist materials. These attributes allow a wide spectrum of research and development projects to be supported by the same lithography system. It is worth noting that the application of the atomic force microscope to lithography is a rapidly growing research field which may someday also become a preferred inexpensive approach to nanofabrication. At present though the emphasis in this new research area is on developing the tools and methods of lithography and is not sufficiently turn-key.

The resolution limits of the various research lithography methods are determined by a combination of factors including the spatial distribution of the deposited energy, the granularity of the resist employed, the contrast of the resist developer system, and the statistics of the photons or particles deposited into individual pixels.

A typical modern e-beam system is shown in Figure 3. The electron optics consist of an electron emitter, or cathode, followed by a spot forming lens, a condenser lens, and an objective lens. The source is either a thermionic emitter such as W, a sharpened LaB_6 or CeB_6 tip, or a thermal field emission (TFE) source such as zirconiated tungsten. The lenses are typically magnetic although the probe forming lens in the TFE case is sometimes electrostatic. The electron source floats at the desired negative acceleration voltage, typically in the 5 to 100 kV range, with the anode, column and sample all at ground potential. Below or integrated within the objective lens is an electrostatic deflector which is responsible for the high speed point-to-point deflection of the focused beam. At the focal plane a positioning stage with laser interferometer feedback is provided to precisely locate the workpiece for patterning. Large area patterns are written pixel by pixel within a single scanning field. The sample is then translated to expose the next field. These fields are stitched together to form the whole pattern.

Focused ion beam systems (FIB) are similar in design to the e-beam system described but use a liquid metal ion source, typically Ga, and usually employ electrostatic lenses to better move the more massive ions.

Focused electron beam and ion beam systems are in a sense similar as described above but the differences in the physical phenomena that contribute to their limitations are substantial enough that in practice they are often optimized for different applications. Electrons are low in mass and have a large charge-mass ratio. The opposite is true for ions, especially massive ions like gallium. Electrons below about 250 kV do not displace atoms, therefore they do not do material mixing, they don't amorphize crystals like semiconductors, they don't act as dopants, and they do not sputter etch material. The electron stopping range in common substrate and resist materials for energies in the range of interest here (5–100 kV) is of the order 2–50 μm. Ions, on the other hand, interact more strongly with target materials causing damage, ion mixing, amorphizing crystals, changing optical properties, implanting dopants, sputter-etching and having stopping ranges as short as 10 nm!

The net result of these properties is that electrons tend to be more universal for patterning conventional resists. They penetrate through thick resists and do little damage to the underlying materials. The electron optical column is relatively clean and suffers minimal damage during extended use. The negative aspect is the large degree of elastic scattering that electrons undergo in the solid substrate beneath the resist. This results in a large interaction volume within the substrate. The electrons can emerge a large distance from the incident location causing a low dose exposure or fogging in proximity to the actual desired pattern. This backscattered electron dosage integrates in the resist as surrounding pixels are exposed. When the pattern is sparse or the backscatter yield of the substrate is very low, the non-linear resist response is somewhat tolerant. For dense patterns on typical bulk substrates, this dosage can be substantial and must be corrected. The correction is typically done in software designed to fracture the pattern

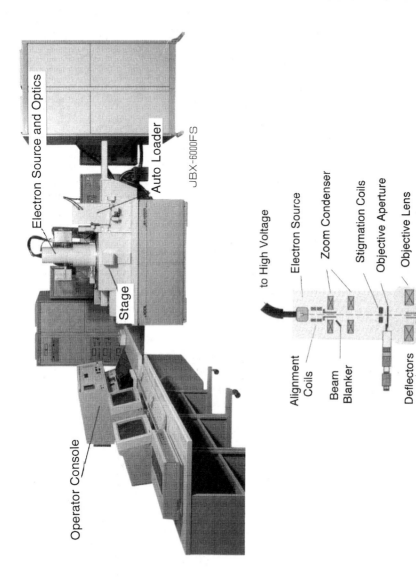

FIGURE 3. A representative high resolution electron beam exposure system. A model JBX 6000FS with a thermal field emission source from JEOL Ltd. is pictured with the major components labeled. The inset is a schematic view of a generic electron beam column (courtesy of JEOL).

and assign varying exposure adjustments[12] or by retaining a constant dose but altering the shape to compensate for the pattern distortions expected from the over/under exposures. This proximity effect correction can cause the pattern data to grow by factors of ten or even a hundred and consume considerable computer time.

Also worth mentioning is the low angle or "forward scattering " of electrons which broadens the incident beam as it passes through the resist before entering the substrate. This can produce a wider exposure at the substrate upon development and can even result in reentrant profiles in the resist. Forward scattering is negligible in higher energy exposures of 50 keV and above, but can be significant in low energy (e.g. 10 keV) exposures.

Ions undergo little long range redirection due to scattering. They interact strongly with the target and therefore tend to stop near the incident location. This eliminates the need for proximity correction as described for the electron case and decreases the dose requirement compared to electrons for exposing the same resists. For example the clearing dose for polymethyl methacrylate (PMMA), a high resolution but rather "slow" resist, requires a dose of 500 μC/cm^2 for 50 keV electrons but only 2 μC/cm^2 for ions of the same energy. This same effect, however, results in short stopping ranges even in polymers, requiring that high voltage ion systems (100 to 400 keV) be used in order to penetrate sufficiently thick layers of resist. This efficient transfer of energy by ions makes them very useful for micro-machining, however. Physical sputtering for removal and ion beam assisted deposition make these systems uniquely suited for repairing photomasks or prototype circuits by allowing computer controlled, highly localized direct repairs.

The resolution limits in practical focused beam systems are in large measure determined by the diameter of the focused spot used to expose the resist. The engineering of such optical columns[13,14] is a somewhat complicated optimization of some fundamental factors including: lens aberrations, spreads in the energy of the ion or electrons sources, uncertainties in the power supply voltages, Coulombic effects between the particles in the column, etc. There are also practical compromises which must be made. For example, a small spot obtained at a vanishingly small current is not useful. Similarly, if a smaller writing field is practical for the patterning task, resolution can be improved by design changes which would otherwise defocus and aberrate the beam at large off-axis scan positions. High quality systems of the e-beam variety have demonstrated beam diameters in the 1–10 nm range and 8–30 nm range for the ion beam systems.

Resists can also affect the minimum achievable resolution. In fact there are very few resists which are usable in the nanolithography regime. The most commonly used resist among nanofabricators is a positive tone resist, poly(methylmethacrylate) (PMMA). (In a positive acting resist, the irradiated regions become more soluble in a chemical developer, while in a negative acting resist, the irradiated region becomes less soluble.) The mean molecular weight for garden variety PMMA is in the range 10^5 to 10^6. Since these polymers tend to be balled-up when spin coated onto substrates, one would expect them to have a granularity approximately equal to the cube root of the molecular weight (this assumes an interatomic distance of about one Ångstrom), i.e., 3 to 10 nm. Fortunately, whole polymer molecules are not the quanta of exposure and development and the observed roughness in high resolution exposures is typically below 3 nm. While this seems small, when lithographic linewidths approach 10 nm, such graininess can be an intolerable uncertainty in the linewidth leading to a practical limit of the process.

PMMA is an example of a single component resist in which the incoming radiation completes the chemical change needed for exposure. There are also multi-component resists in which a process of chemical amplification is used to enhance the sensitivity. In chemically amplified resists the radiation causes a chemical release which upon a subsequent heating step completes the reaction needed for full exposure. If we treat each of these resist classes separately, in general, we find the lower the clearing dose (for positive resists), the lower the resolution of a resist since the reduced dose implies that a larger volume of resist is exposed per unit of energy.

Equally important for high resolution lithography is the contrast, γ, of the resist-developer system. A high contrast (positive) resist permits very little change in thickness in the under exposed regions and completely clears in regions dosed above a threshold value. This non-linear behavior can help to correct for imperfect energy deposition profiles due to leakage of energy in resolution-limited exposures and unwanted exposure due to electron scattering effects. Contrast is defined by,

$$\gamma = (\log_{10}(D_{l=0}/D_{l=1}))^{-1} \qquad (5)$$

where $D_{l=0}$ is the dose at which a large area exposure first clears and $D_{l=1}$ is the dose below which the full thickness of resist is retained. The conventional measure of resist contrast is obtained from a sensitivity curve which plots normalized resist thickness as a function of the log of exposure dose. The contrast is the absolute value of the measured slope at the 50% thickness point. High resolution resists such as PMMA can have contrasts in the range from 6 to 9 depending on process conditions such as developer strength, development time, and pre exposure bake time and temperature. A value below 2 is considered very low and would be a poor candidate for pattern transfer. Most resists fall in the range from 2 to 7 (although much higher values are possible). High contrast resists tend to produce steeper sidewalls in resist profiles. This enhances the fidelity of the pattern transfer process. A low contrast resist would both fare poorer for isolated fine lines and would rapidly degrade the remaining resist as the these sloping sidewalls are packed closer to one another.

Another important fundamental limitation on conventional electron beam lithography is the spatial extent of the exposure mechanism itself. As the primary electron beam passes through the resist film, so-called "fast secondary" electrons are generated which, in turn, break the bonds necessary to expose the resist. The spatial range of these electrons causes a cylindrical exposure volume to form concentrically about the incident beam. This process has been modeled by Kyser[15] and has been selectively plotted in Figure 4. The plots show the width of the cylinder for which the deposited energy is 1×10^{30} eV/C-cm^2 or higher for two different film thicknesses over the relevant incident energy range. If we assume that this deposited energy contour becomes the developed linewidth, one observes that narrower linewidths are obtained for thinner resist films and higher energies. These results indicate that sub-10 nm linewidths should be possible in polymeric resists.

While particle counting statistics can lead to another source of linewidth variation, it is usually not a major contributor for high resolution lithography unless very high-throughput is also required. Such particle counting may be more important for ions than electrons. Each ion so effectively transfers energy to the resist

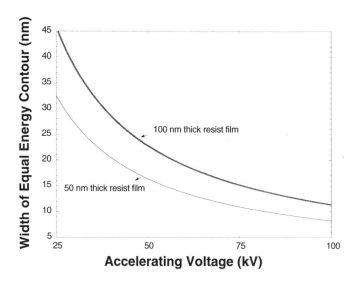

FIGURE 4. Dependence of equal deposited energy contours in 50 and 100 nm thick resist films on electron accelerating energy. Plots are fits to points calculated by Kyser[15] for an electron beam of zero width incident on the film with no supporting substrate to backscatter the electrons. The broadening of the exposed line is therefore due only to secondary electrons generated by the incident beam as it passes through the resist film. These plots assume absorbed energy contours of 1×10^{30} eV/C-cm^2.

that far fewer ions are required than electrons for exposure. As studied by Smith[10], however, the particle count estimates even in the high-throughput case depend in detail on the pixel size, permissible uncertainty in the linewidth, the resist contrast, and the physical broadening effects (especially associated with electron exposure). Using estimates in these categories for a specific grating pattern exposure, Smith concludes that at the maximum pixel transfer rate, 900 electrons is the minimum number per pixel while only 29 ions are required. For this comparison the significant difference arises between the electron and ion case because the linewidth variations are dominated by different effects. The electron case is limited primarily by the forward scattering distance which grades the exposure intensity in the resist. The ion case is primarily dominated by standard deviation in the number of incident ions, σ, per pixel which, due to the uncorrelated nature of the ion source emission, can be estimated by $\sqrt{N}$, where N is the average number of particles incident per pixel. Thus the dose variation which results causes linewidth variations even with minimal slope in the exposure profile.

2. Throughput Limits of Serial Tools

The simplest and most general lithography tools expose a single pixel (picture element) at a time. Scanning focused beam (eg. electron and ion) systems [described in the previous section] are of this type and offer the distinct advantage that arbitrary patterns can be written and modified using CAD (computer aided design) software. This eliminates the need for a physical mask to be fabricated prior to patterning of a resist layer. This, in fact, has made e-beam lithography the method of choice for mask making. This flexibility comes at a price, however. The number of pixels which need to be addressed increases linearly with the writing area and as the inverse square of the pixel to pixel spacing. The situation is, therefore, exacerbated for serial exposure tools when the highest resolution is desired because the required finer beam diameters have correspondingly lower beam flux. As an example, consider a generic focused electron beam system consisting of an electron source, followed by a condenser lens, a selectable aperture and an objective lens. When the optical column is operating such that the focused beam is not aberration limited (at larger beam sizes), the beam current varies as the inverse square of the diameter (i.e., for a fixed convergence angle the current density remains constant). Under these conditions, the exposure time dependence can be simply summarized as,

$$t = 4(D_0 A)/\pi d_0^2 \ J \qquad (6)$$

where t is the exposure time (in seconds), D_0 is the areal resist sensitivity (in C/cm^2), A is the writing area (in cm^2), d_0 is the beam diameter at focus (in cm), and J is the beam current density (in A/cm^2). A more fundamental relation can be obtained by substituting for the current density at the specimen in the above expression:

$$J = \pi \beta \alpha_0^2 \qquad (7)$$

where β is the source brightness (in A/cm^2-steradian) at the specimen and α_0 is the convergence semi-angle (in radians), yielding,

$$t = 4(D_0 A)/\pi^2 d_0^2 \beta \alpha_0^2 \qquad (8)$$

High resolution patterns can require exceedingly long direct write times even when written only in selected areas on a wafer. In part this is due to the desire to use both a conservative beam diameter[16] and a high resolution, high contrast resist (with the accompanying poor sensitivity) in order to improve process latitude. In a similar vein, uniformity improves with good depth of focus and low beam distortion. These requirements generally imply using a small convergence angle or low *NA* electron optical column configuration which results in a low current density beam. As an example, a system which employs a LaB_6 thermionic cathode with a source brightness of $1.0 \times 10^6 A/cm^2$-steradian when operated at 50 kV can correspond to current density of only 25 A/cm^2 (depending on the working distance and aperture size.) Under typical conditions, areal writing rates for PMMA resist[17] are about 0.42 mm^2/hr. If a modest writing area of 30 mm^2 is assumed, this translates into three days of writing!

Improvements which can be implemented to alleviate this throughput problem

TABLE 1. Dependence of electron beam writing time for a sample grating on resist sensitivity and electron source type.

System	Source	Resist	Sensitivity (nC / cm)	Current (nA)	Grating Rate (mm^2/hr)
JBX5DII	CeB_6	PMMA	4.5	0.25	0.42
JBX5DII	CeB_6	ZEP320	0.7	0.25	2.75
JBX6000FS	Zr/W TFE	PMMA	4.5	10.0	16.8
JBX6000FS	Zr/W TFE	ZEP320	0.7	10.0	110

are illustrated in Table 1. Switching from PMMA to a more sensitive resist is an important first step. If we choose ZEP320, due to its demonstrated high resolution, good contrast ($\gamma = 7$), reasonable etch resistance, and simple development process, as seen in Table 1, the immediate improvement in throughput from the resist can be about a factor of 6.5, reducing the same write to about 11 hours. A second major potential improvement can be realized with a higher brightness electron source, such as a Zr/W thermal field emitter (TFE). This source can provide substantially higher current densities (1000–2000 A/cm^2) which can result in an additional factor of 40 in writing speed (if the maximum deflection speed of the e-beam system is not reached). For the example in Table 1, a minimum shot dwell time of 0.167 μs (6 MHz) is assumed. Note that the combination of high brightness and sensitive resist in the bottom row of the table is nearly deflector speed limited but does allow a full factor of 40 improvement over the speed with the thermionic source. The three day write is reduced to about twenty minutes. In practice, there is some overhead associated with these writing times due to stage movement, system calibration, etc. However, these numbers serve to illustrate the major concerns. The limitations of single pixel exposure tools is evident and has caused decades of research and development of parallel or multi-pixel exposure systems. The progress in the area of tool development will be discussed in sections V and VI below.

B. Pattern Transfer Limitations

While we concentrate in this chapter on the limiting factors for producing high resolution resist images, this is almost never the final goal. The patterned resist usually serves as a mask for subsequent processing to transfer the pattern into other layers by means such as dry etching in a reactive plasma, wet chemical etching, ion implantation for electrical doping, or deposition of thin film layers. Among these, the dry etching step has proven to be the most widely used method of high resolution pattern transfer. Dry etching encompasses a number of different but related techniques including: reactive ion etching (RIE) (also sometimes called reactive sputter etching), plasma etching, reactive ion beam etching (RIBE), and chemically assisted ion beam etching (CAIBE). While these names sound much alike, each has become the calling card for a specific type of etching system. RIE and plasma etching both introduce a reactive gas into an evacuated process chamber and use an RF induced plasma to create reactive ion species (for a discussion of the physics of this process, the reader is referred to Chapman[18]). These reactive ions selectively etch the desired material by energetically combining with the elements of the sample to form volatile reaction products which are pumped away. The term RIE designates that the gas pressure in the chamber is quite low, usually in the 5 to 100 mTorr range in contrast to that of plasma etching in which the pressures are typically higher, in the few Torr range. This seemingly minor difference dramatically changes the mean free path of the ions and therefore can affect the anisotropy of the etching process. In RIBE and CAIBE, the ion species are created within a Kaufman-type ion gun then extracted into an evacuated chamber, striking the wafer. In the case of RIBE the ion species is a reactive species for the elements being etched. In CAIBE, a process gas such as chlorine is introduced onto the sample and an inert ion such as argon is extracted from the ion gun. The argon then imparts the energy required to induce a reaction between the chlorine and the sample. All of these methods are directional compared with wet chemical etching and therefore are used in high resolution pattern transfer where the intent is to accurately replicate the lithographically patterned resist mask.

Limitations in pattern transfer include: edge roughness, edge slope, trenching, lag and loading effects. Edge roughness is usually initiated by graininess in the resist mask which is then propagated in the material as the etch proceeds, and leads to intolerable variations in linewidth, especially as CDs (critical dimensions) shrink. Edge slope is also a form of linewidth variation in which the top and bottom of the etched structure vary in width. The sources of this include: trajectory distributions in the ion etching process; the etching activity of neutral but reactive species in the etch process which causes an undercutting of the resist mask; replication of an existing slope in the resist layer; and erosion of the resist mask at the edges thus baring more of the substrate to the reactive species as the etch proceeds.

Another limitation that can degrade pattern transfer quality is trenching. Trenching is caused by ions striking the sidewall of a pattern feature and being redirected like a specular reflection to the base of the feature. This causes an accelerated etch rate there and forms a trench around the perimeter of patterned features. A fourth limitation, RIE lag, is the slower etch rate observed in narrow, high aspect ratio trenches when compared to widely spaced features. To overcome lag, over-etching is required to etch completely through the layer. This leads to poor size control in different sized features. Finally, loading and microloading are etch rate changes caused by varying reactant to etch product ratios. Patterns with large etching areas produce higher concentrations of etch products than sparsely etched regions of a wafer. The lack of reactant species or the slowed removal of the products can rate limit the chemical reactions responsible for etching. This can change the etch rates from run to run or worse, from area to area on a single wafer, thus degrading the size control within complicated patterns. All of these limitations contribute to the margin of error in the pattern transfer process and must be controlled to continue to scale linewidth at the same pace as the imaging lithography permits.

III. EXAMPLES OF NANOFABRICATED STRUCTURES

Laboratory lithography (and especially e-beam lithography) provides a tool for exploring the performance of electronic, photonic, and mechanical devices on a deep submicron scale. In this section we present six examples taken from a cornucopia of evidence that dramatically enhanced performance can be achieved via miniaturization to the deep submicron scale. The six examples are: FETs; surface gated quantum devices; quantum dots and wires; gratings; zone plates; and mask making. Examples like these show why miniaturization is the primary means for improving performance.

The first example we cite below, the transistor, is the archetype. Shrinking the size transistors in an integrated circuit improves performance because the resulting increase in density and shorter carrier transit times and faster circuits result in the incorporation of more functionality within the same chip area for the same cost (about 4 dollars per cm^2 of silicon substrate.) Thus, functionality and the economics of higher performance and higher density chips on a wafer propels miniaturization.

A. Deep submicron Field Effect Transistors

Various chapters throughout this book illustrate how performance can be improved through miniaturization. For example, Chapter 2 chronicles the development of conventional bipolar transistors and Metal-Oxide-Semiconductor Field Effect Transistors (MOSFET) toward the 40 nm scale, and clearly explains the rationale: i.e. shrinking the active region of the device results in improvements in both the transconductance, g_m, and in the bandwidth, f_T, without compromising noise performance or cost. In particular, Figure 5 shows a cross-section through the active region of a 100 nm channel length MOSFET. The MOSFET incorporates two junction diodes and an MOS capacitor. The source and drain are represented by the two n^+ contacts in the p-type Si substrate. The gate electrode, usually heavily doped polycrystalline silicon, is the "metal" electrode of the MOS capacitor, and is separated from the substrate by a 4 nm thick SiO_2 dielectric. A positive gate bias induces a positive potential in the channel of the MOSFET, lowering the potential barrier between the source and drain. With a positive gate bias, the source acts like an electron emitter, while the drain behaves as an electron collector.

The 100 nm MOSFET is an example of the phenomenal performance that can be achieved in the deep submicron regime with $g_m = 570$ μS/μm; $f_T = 110$ GHz; and with good subthreshold characteristics at room temperature[19]. The high performance develops because the transconductance in saturation is inversely proportional to the channel length, L, and the oxide thickness, t_{ox}, at a constant drain current; while f_T is proportional to $t_{ox}^{-1/2} L^{-3/2}$ for a constant drain current. It is noteworthy that channel length in the device of Figure 5 is so small that high resolution electron microscopy reveals individual atoms in the active region of the device. Even this size and level of performance are expected to be surpassed by still smaller devices. Recently, MOSFETs with gate lengths as short as 40 nm and oxides thinner than 3 nm have been introduced. However, the DC characteristics of these devices only show g_m = 466 μS/μm with good subthreshold characteristics at room temperature[20,21].

The deterioration in the transconductance found in the 40 nm MOSFETs relative to 100 nm, may be due in part to the inability to scale the oxide thickness appropriately due to tunneling. Ultimately, as the size of the electronic devices shrinks to be comparable to the electron wavelength, parasitic quantum mechanical effects such as tunneling and quantization are supposed to be manifested in the transport characteristics. Consequently, it is expected that the ultimate performance of extreme sub-micron devices will be limited by tunneling through the thin oxide and through the p-n junctions, and by the minimum junction depth associated with the source and drain contacts to the quantized inversion layer.

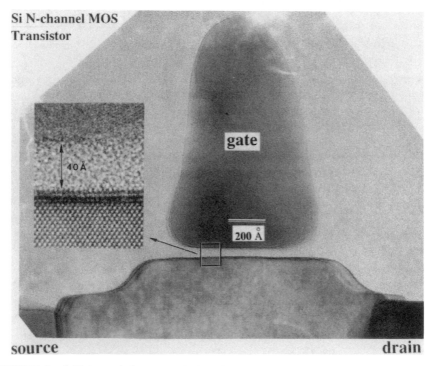

FIGURE 5. A high resolution transmission electron micrograph of a cross-section of an n-channel MOSFET. The inset shows a lattice image of the channel region. Each dot in the inset represents a column of pairs of silicon atoms [micrograph courtesy of Y.O. Kim].

B. Surface Gated Quantum Devices

Recognizing the quantum mechanical limitations of conventional transistors, the investigation of novel electronic devices that exaggerate the quantum mechanical aspects of the transport and attempt to make use of them has been vigorously pursued using laboratory nanolithography (with the tacit assumption that cheap manufacturing tools would eventually become available.) Chapters 13, 14 and 15 of this book provide a comprehensive treatment of the investigations into quantum wires and quantum dots in which electrons are dynamically confined on the scale of the electron wavelength to one and zero dimensions respectively.

A common feature of such devices is the degenerate two-dimensional electron gas (2DEG) present at the *GaAs/AlGaAs* interface in a heterostructure or at a *Si/SiO$_2$* interface. The electron gas is dynamically two-dimensional because of the vertical confinement to within an electron wavelength that occurs near the hetero-interface due to the repulsive potential barrier associated with the conduction band offset and the attractive potential associated with positively charged ionized donors. In addition to the vertical confinement, the electron gas is also usually constricted laterally to the size of an electron wavelength using the

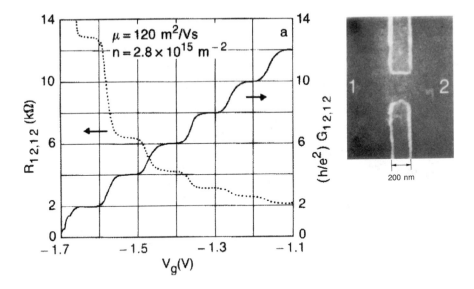

FIGURE 6. The quantized conductance $G_{12,12}$ (solid line) and resistance $R_{12,12}$ (dashed line) measured at 280mK through a ballistic quantum wire formed by using the split-gate structure shown in the inset. The gate voltage, V_g, applied to the split-gate electrodes controls the width of the quantum wire.

electrostatic potential provided by split-gate electrodes. For example, the inset to Figure 6 shows the top view of a 200 nm long split-gate structure that was fabricated using e-beam lithography. The split-gate structure is used to deplete the underlying 2DEG in a *GaAs/AlGaAs* heterostructure and form a quantum wire connecting two regions of 2DEG labelled 1 and 2 in the figure. An increasingly negative gate voltage, V_g continuously reduces the width of the quantum wire. It has been shown that channel conductance through such a device is quantized in steps of $2e^2/h$ as a function of gate voltage at 300mK as shown in Figure 6. The conductance quantization is an unequivocal signature of the quantization of the electron energy spectrum as a function of the wire width and of the ballistic nature of the transport within the constriction. Although the voltage gain (<1) and operating temperature (T<4K) of this particular device configuration is still impractically low (e.g. see Chapter 15), the promise of increased functionality has already been realized by Eugster *et al.*, who used this device as a novel 2-bit analog-to-digital converter[22].

It follows from Eugster's work that a 4-bit A:D converter could be implemented using 8 split-gate transistors with 16 conductance plateaus each, instead of the usual 16 comparators comprised of 3–5 conventional transistors.

C. Quantum Dots and Wires

Although the primary incentive for shrinking the scale of lithography so far has been the improvement of functionality and/or cost of an integrated electronic circuit, the performance gains found in miniaturized photonic and electro-mechanical devices (e.g. see Chapter 3) have provided added impetus. Exploiting the reduction of the active volume and the concomitant energy quantization,

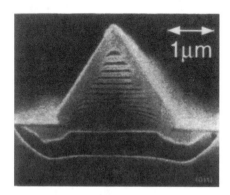

FIGURE 7. An electron micrograph of an array of *GaAs/AlGaAs* quantum wires formed by selective growth is shown on the right. A cross-section through the multiple layers that comprise each wire is shown on the left [reprinted from reference 28].

quantum wire lasers are supposed to show improved performance over quantum well lasers including[23]: an ultralow current threshold for lasing[24], an enchancement of the optical nonlinearities, an increase in the modulation bandwidth, and an increase in the differential gain and a narrowing of the optical gain spectra[25]. There are several growth techniques suited to producing quantum wire lasers (see Chapter 5)[26,27] that have already been reviewed elsewhere; one attempt used lithographic definition combined with etching and regrowth[28]. Figure 7 is a micrograph of such an array of quantum wires only 20 nm in width that were fabricated in *GaAs/AlGaAs* heterostructures using e-beam lithography.

D. Gratings

The simple periodic image produced by interfering laser beams can be used to fabricate captivatingly precise structures. It is such a spatially precise metric that it has recently been proposed as an on-wafer fiducial mark to allow spatial-phase locking for electron beam lithography in order to improve the long range spatial coherence (and therefore field stitching) in such systems[29]. Very fine period grating structures in quartz have been fabricated by laser holographic patterning followed by using a frequency doubling method to obtain grating pitches down to 99 nm[30]. As will be discussed below, the lightwave communication industry has perhaps the most commercially relevant interest in producing high quality grating structures. If lithographic alternatives to laser holographic methods of producing periodic structures are to play a role in this industry they must offer some advantage and have demonstrated adequate precision to serve as a stand-in for the holographic method. The challenge then becomes not one of resolution but rather one of accuracy and precision in the spatial frequency and coherence. Some specific examples of the state of this technology are discussed in the next section.

E. Zone Plates for X-Ray Microscopy

Zone plates were invented over 100 years ago as diffractive focusing elements for the visible portion of the spectrum. Today they are the most common optics used for high resolution scanning imaging in the soft X-ray range. In 1974 Kirz[31] proposed that the complex index of refraction would allow use of phase shifting zone plates in the EUV/ Soft X-ray portion of the spectrum (typically from 0.1–1 keV) in a manner similar to that known for visible light, thus allowing substantial improvement in X-ray collection efficiency.

Zone plates are structures comprised of concentric circular zones with edges of radii given by:

$$r_n^2 = mn\lambda f + m^2 n^2 \lambda^2 / 4,$$ (9)

where n is the zone number, f the focal length, λ the X-ray wavelength, and m is the diffractive order. Typically the radiation focused into the $m = +1$ diffractive order is sought for applications such as X-ray microscopy. All other orders are discarded. When an absorbing material is used to form an "amplitude" zone plate on an otherwise transparent supporting layer, it can be shown that only about 10% can be focused into the $m = +1$ order. However, 40% is possible if a π phase-shifting material with no absorption can be found to form a "phase" zone plate[31].

Scanning images obtained from such focused spots are limited by the Rayleigh resolution, d_1, which can be shown to correspond to the width of the outermost zone, $w_{n_{max}}$, as

$$d_1 = 1.22 w_{n_{max}}$$ (10)

FIGURE 8. Scanning electron micrograph of a phase zone plate made from germanium supported on a silicon nitride membrane. This zone plate was designed for use in a scanning transmission X-ray microscope operating in the wavelength range of 2.5 to 4.0 nm. The expected resolution is 30 nm.

High resolution and highly efficient phase zone plates represent a significant nanofabrication challenge since they have combinatorial requirements of high resolution; high aspect ratio metal structures, especially in the outer zones; accuracy in shape, placement, and duty cycle.

Sub-0.1 micron phase zone plates have been most commonly produced by e-beam lithography in a number of materials in various laboratories worldwide[32-36]. Anderson *et al.*[32] used electroplating of *Ni* to produce zone plates with outer zone widths of 30 nm. Highly efficient *Ge* phase zone plates were produced using reactive ion etching[34] in which a multilayer resist process served as sacrificial masking layer during the pattern transfer. A germanium phase zone plate with a 30 nm outermost zone produced by the method described in reference [35] is shown in Figure 8. Even higher resolution versions but with lower efficiency have also been recently reported using similar methods[36] with resolution down to 20 nm.

Clearly the challenge in this research area is to produce ever improving resolution while maintaining the highest possible diffraction efficiency. This requires good control during the lithography and methods of producing ever higher aspect ratio structures to maintain the high efficiency π-phase shifting zone plates.

F. Prototype Mask Making

As the IC industry considers its options for future high-throughput printing tools, the research and development teams producing these technologies have, in each case, issued a major challenge to the mask makers. In the optical lithography community, the attributes of the once passive role of the mask (or reticle) have changed dramatically. Enhancements to the optical step and repeat lithography method have necessitated new considerations such as optical proximity correction and phase-shifting in the mask in order to extend the practical limits of the UV and DUV reduction camera. Even more difficult challenges have been placed on mask making for the alternative lithographic technologies being proposed for manufacture of 0.18 μm CD IC's and below (e.g. 1GB DRAM). The production of prototype masks has therefore been a fertile research and development topic for the microfabrication community and represents an important class of best-in-kind of nanostructures. While more of the context for the various lithographic methods is provided in the next section, some of the highlights of the fabricated mask results are presented here.

1. Membrane Supported Transmission Masks

Two of the mask types, that of X-ray (XRPP) and electron projection (SCAL-PEL), require transmission masks which must be constructed on very thin membrane supporting layers to allow sufficient transmission in "light" field regions. These inherently require careful engineering of the strength and stress of the materials of the membrane to form a robust platform with a low stress "dark" field patterned region. The detailed requirements differ somewhat for these two mask types, however.

X-ray proximity masks have benefited from the most mature development effort to date among the advanced lithography methods listed above. These masks are characteristically formed on a 2 μm thick membrane about 1 inch in diameter supported by a thick (usually pyrex) ring. Since no bulk materials are transparent to X-rays in the wavelength range of 1 to 1.5 nm, the membrane is needed to support a patterned absorber (typically *Au* or *W*) while remaining highly transmissive in the unpatterned regions. Several areas of difficulty arise in fabricating these masks to meet industry requirements.

Transmission properties of materials typically used in the production of XRPP masks are tabulated below in Table 2, where T is the material thickness required (in nanometers) to allow 90%, 10%, and 1% intensity transmission at 1.0 nm, respectively. One readily sees that to avoid excessive absorption in the supporting membrane low Z, low density materials such as silicon and silicon nitride in thicknesses of the order 1 μm are needed and that comparable thickness of high density materials such as gold or tungsten are needed to provide high absorption and therefore good contrast in the patterned regions.

From the mask makers point of view, much of the difficulty arises from the inherent 1× nature of this printing method since it requires an enormous leap in resolution and placement accuracy for e-beam mask making tools. As shown in Figure 9, a typical IC generation undergoes about a 30% linewidth decrease in order to maintain the expected density improvement each generation. For a 1 GB DRAM using a 5× reduction system, the minimum feature size on the mask is 0.9 μm, while for a 1× XRPP mask this is reduced to 0.18 μm. Most of the progress toward this goal has been made in R&D labs and not yet made it into the commercial mask making offerings. At IBM, variable shaped e-beam systems, EL III+ and EL IV, have been commissioned for producing prototype masks at 0.25 μm CD level[38,39]. XRPP masks comprising smaller dimensions ≤ 0.1 μm

TABLE 2. Transmission Properties of Selected Materials at $\lambda = 1.0$ nm[37]

Material	$T_{90\%}$	$T_{10\%}$	$T_{1\%}$
Silicon	509	11162	22327
Silicon Nitride	241	5276	10553
Gold	18	400	798
Tungsten	21	463	926

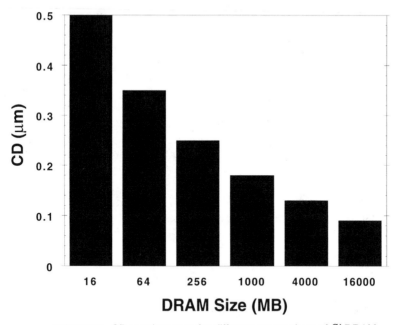

FIGURE 9. CD requirements for different generations of *Si* DRAM.

have been produced to research the limits of the process and to explore circuit performance at these size scales[40,41]. The finest dimensions in X-ray mask making, approximately 17.5 nm printed linewidths, were reported by Flanders[42], not for proximity printing but for contact X-ray printing using a corrugated polyimide membrane with edge-deposited *W* absorbers and illuminated with carbon *K* X-rays.

Another mask with similar attributes to the X-ray mask is that being proposed for use in SCALPEL (Scattering Angular Limitation Projection Electron Lithography). In this instance the mask comprises a very thin silicon or silicon nitride membrane (typically 0.1 μm) supported by a silicon grillage[43] with a patterned electron scatterer (as opposed to absorber) typically only 20–50 nm in thickness. While resolution is somewhat relaxed since this printing method employs reduction imaging (from 2 to 4) at the wafer plane, the challenges will be to make it low distortion and thermally ducted over relatively large areas (greater than 500 mm^2 for 1GB DRAM). The grillage segments the mask; thus reducing in one direction the support spacing in order to meet these requirements. Fabrication of SCALPEL masks is a rather interesting mix of selective crystallographic etching to form the grillage, stress controlled deposition of the scatterer and e-beam pattern definition. This process, for a single span, is illustrated in Figure 10.

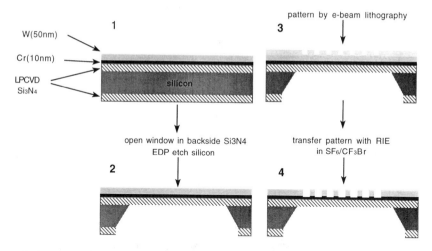

FIGURE 10. Process Steps used to prepare SCALPEL transmission masks[43].

2. Free Standing Transmission Masks

Free standing structures are required when masks, zone plates, or other transmissive components are required but extreme absorption prevents use of even a thin supporting membrane. This is the case for ion beam lithography[44], extreme ultraviolet lithography[45], and neutral atom optics[45]. For lithographic

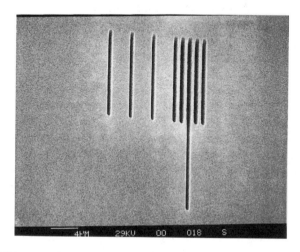

FIGURE 11. Ion beam lithography free standing stencil mask[44].

mask applications, heating due the absorption of energetic particles can lead to pattern distortion. Therefore very different requirements for the membrane mechanical properties are indicated when compared to the membrane masks discussed above. In this instance, a lower membrane stress is more tolerable than for that of say the 1× XRPP masks. Figure 11 is a scanning electron micrograph of an IPL free standing stencil mask fabricated for a 5× reduction system[44]. The narrowest features are about 0.5 μm lines and spaces.

Although in practical systems EUV masks are envisioned to be reflection masks, early experiments in the field used transmission masks with both patterned absorbers on supporting membranes and free standing stencil masks. Figure 12 shows free standing grating features as small as 0.1 μm, with 6:1 aspect ratio fabricated as a resolution test pattern for an EUV 1× camera operated by Bell Laboratories at Brookhaven National Laboratories[45,46].

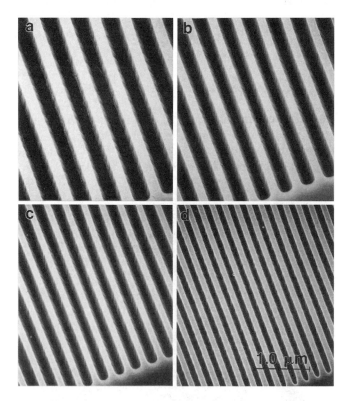

FIGURE 12. Extreme ultraviolet lithography transmission freestanding stencil mask used in a 1× experimental exposure system[45]. The finest lines are 0.1 μm .

IV. LITHOGRAPHIC REQUIREMENTS
FOR DEVELOPMENT AND PRODUCTION

A. Trends in VLSI and PIC's

Clearly, the major driving force for the development of high-throughput lithographic tools is the rapid growth and concomitant progress in the silicon integrated circuit industry. Only the high volume production of silicon VLSI integrated circuits (ICs) has had the economic breadth to sustain the expensive development costs of new lithographic tools and processes. Other semiconductor industries such as ASICs, *GaAs* ICs, and Optoelectronics benefit symbiotically from the rapid progress in tool development, often by operating with previous generation lithographic equipment or low throughput e-beam systems where higher resolution is needed. While the progress and future predictions of the silicon VLSI IC industry are discussed elsewhere in this volume, industrial groups such as the Semiconductor Industry Association (SIA) has assembled consensus predictions of future requirements and likely technology choices[47] in its publication of "The National Technology Roadmap for Semiconductors". The roadmaps of technology needs outline the critical level requirements in four major lithographic categories:

- lithography requirements which correspond to the level of integration and relevant size scales for integrated circuits looking forward through the year 2010;

- resist requirements corresponding to the various lithographic generations;

- mask or reticle requirements for the various lithography generations;

- wafer metrology to keep pace with the quality control and yield requirements.

Similarly, the document outlines likely technological pathways (with several options where appropriate) as they can best be determined with the present state of research and development. Certain general observations can be made from such a consensus document. First, that the silicon IC industry is actively planning for manufacture of circuits down to the 0.07 μm CD level. In DRAM terms, according to industry projections, this corresponds to 64 Gb on a single chip! Second, that optical lithographic options will be exhausted before new methods of manufacture are embraced.

There are exciting technological challenges involved in the development of the various alternate or follow-on lithography methods discussed above. However, the process of narrowing the field of candidates at appropriate generations is left to the market place and to the research funding mechanisms, and is therefore likely to be a politically as well as technologically charged debate. In any event, in order to maintain the overall progress needed to fulfill the promise of these predictions, each of the targeted areas listed above must be advanced together, and the infrastructure nurtured in order allow these emerging methods to be commercially viable.

While a much smaller component industry, another potential customer for advanced lithography is broad band lightwave communications. Advances in optoelectronic integrated circuits (OEICs) and photonic integrated circuits (PICs) are fueled by current trends which indicate that transmission capacity improvements will be achieved by migrating from time division multiplexing (TDM) to system designs which employ wavelength division multiplexing (WDM) architecture.

This system level change shifts the emphasis from ever higher speed components such as laser driver circuits and optical modulators to more highly parallel system designs. Compact, highly integrated versions of such systems will likely require semiconductor laser arrays to provide multi-wavelength light sources integrated with other optical components such as combiners, modulators, etc. Practical processing methods are therefore needed to produce corrugated-waveguide gratings which are the principle high resolution lithographic elements for many of the optical devices such as filters and distributed feedback (DFB) or distributed Bragg reflector (DBR) lasers that are expected to play an important role in PICs. The challenge for this rather narrow class of high resolution features as outlined in the previous section, is to produce an array of gratings closely spaced with precise features (in an economical way). Conventional two-beam UV laser holographic methods are not well suited to the manufacture of laser arrays where pitch variations in adjacent channels and abrupt shifts within the gratings are needed. Depending on the volume demand for these structures they may be produced using e-beam or ion beam writing, or, when possible, using a parallel printing technique to avoid the slow serial writing process. Direct write e-beam has been demonstrated to exhibit the needed control in applications to semiconductor lasers[48,49]. Frequent concern is voiced over the abrupt spatial errors which can occur where e-beam writing fields are "stitched" together to form a large area grating structure. This can be managed by either using a large writing field which accommodates the entire grating[50] or by using numerous small errors to minimize the effect[51].

Other issues such as e-beam throughput limitations can be addressed by producing grating masks which can in turn be used to economically print gratings. A good example of a niche printing method is near-field holographic (NFH) printing[50,52] using e-beam generated phase grating masks. It is an alternative method of printing grating structures using a modified UV contact aligner in which interference of diffracted light forms a holographic standing wave pattern near the grating mask. Precisely controlled grating periods in the e-beam written masks allow for the finely spaced frequency standards required for WDM lightwave communication systems. Figure 13 shows the grating periods measured for a 1×16 fused silica phase grating mask with a mean period of 243.63 nm

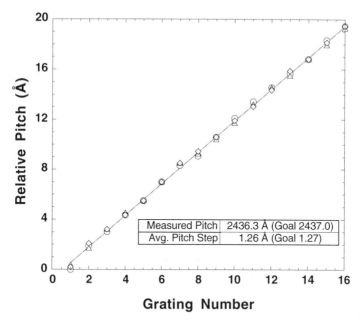

FIGURE 13. Measured periods for a fused silica phase grating mask comprising 16 grating pitches with an average pitch of 243.7 nm, spaced by only 0.127 nm[52].

and a design period change of 0.127 nm per channel. The lithographic challenge is the manufacture of the grating mask which requires 100 nm CD but with daunting precision requirements on grating periods and controlled period increments in the 0.1 − 0.2 nm range. The leverage here is that once produced, the mask can be printed repeatedly with well characterized results. It is intriguing that the spatial periods needed for many of the gratings required by WDM communications systems are within the resolution capability of the future *Si*-VLSI lithographic production tools discussed below.

B. Extensions and Limits of Optical Printing Techniques

While there is considerable debate over which generation of IC will see optical lithography give way to other methods of printing, the UV optical methods are generally considered the incumbent technology. But these technologies become more limited in resolution and depth of focus as linewidths shrink, making it likely that there will be follow-on technologies. Enhancements to the optical step-and-repeat lithography method have been employed to demonstrate further extension of the resolution capability of mercury i-line UV (365 nm) and

TABLE 3. Estimated Performance of Optical Stepper Lithography Systems

Wavelength	Resolution	Depth of Focus
365 nm	0.5 μm	1.5 μm
365 nm w/ enhancements	0.35 μm	1.5 μm
248 nm	0.35 μm	1.0 μm
248 nm w/ enhancements	0.25 – 0.18 μm	1.0 μm
193 nm w/ enhancements	0.18 – 0.13 μm	0.75 μm

DUV (248 nm and 193 nm) reduction lithographies.

These enhancements include: optical proximity correction, and phase-shifting masks, as well as non-mask related changes such as innovative off-axis illumination designs in order to extend the practical limits of the UV and DUV reduction camera[53]. The lithography community is divided on how the various wavelengths and enhancements will perform in production. Table 3 is therefore only an estimate of the expected performance of the various optical projection systems.

C. Need for New Tools

The need for new tools will likely result from a combination of factors, only some of which are related to the pure resolution capabilities of each method. For example, even within the optical lithography path, high-throughput which is characteristic of optical printing is difficult to maintain. Beyond "i-line" (365 nm) very different illumination sources are required: e.g. excimer lasers will replace mercury arc lamps. New condensers with more restrictive designs are used to "buy back" depth of focus, but also reduce the power delivered to the wafer. Issues such as these, and such as the cost of new resists make the "cost-of-ownership" and "cost-per-level" estimates more uncertain. Already, stepper manufacturers are marketing a "two-tool" approach for the IC industry, in which a high resolution tool with reduced throughput is used for critical levels and a high-throughput companion tool is used for relaxed CD levels. In this way the "average" throughput meets the industry requirements. Others are working on new optical designs which can take advantage of the enhancements while maintaining the required throughput.

Mask or reticle manufacture, inspection, repair costs, and availability also become important considerations. Optical phase-masks comprise the fine featured regions on an IC pattern. Using the index of refraction and the mask thickness, phase masks introduce a π-phase shift in the light to enhance resolution in the printed image. Local repair of defects in these areas is extremely difficult and may not be practical. With these technological and economic issues being aired, and the increasing importance of lithographic tools for future IC generations, the question becomes, "Will it be more fruitful to undergo a paradigm shift?" That is, do e-beams, ion beams, X-rays, or EUV optics schemes represent a better way?

V. FUTURE HIGH THROUGHPUT LITHOGRAPHIC ALTERNATIVES

While any departure from the optical printing methods currently employed for manufacturing will meet with some resistance, several leading edge choices are presented here. We note that each combines a different set of advantages, extendibility, and technological challenges, all of which must be considered before a system can become commercially viable.

A. Role of E-Beam Systems

The shaped beam EBL system, while lacking the highest resolution available in Gaussian beam systems, has been an important first step toward a parallel printing method using projected electrons. By illuminating a square aperture and deflecting the beam over shaping apertures, arbitrary rectangles are de-magnified and projected onto the mask or wafer. This allows multi-pixel exposures which can significantly reduce the writing time and relax the system requirements for high speed deflection. Such systems have traditionally been useful for prototyping or limited production of circuits and for mask making for optical stepper reticles. Recently several efforts have been reported in which the shaped aperture has been replaced by "character"[54] and "cell"[55] apertures which are "dial in" libraries of common shapes that appear in a given mask or circuit level. This is an attempt to dramatically improve the throughput of direct-write-on-wafer EBL to compete with optical step and repeat cameras and their sequel technologies. While the limited cell printing field (of order 5 μm on a side) and the currently envisioned resist materials prevent the full throughput of an optical stepper, the writing rates represent an impressive increase over serial exposure rates. The drawback to such systems is that they become less general in nature since the shaped apertures are unique to a narrow set of patterning tasks.

As the pixel count continues to increase by a factor of four per DRAM generation, it is probable that the writing speed enhancements, including character projection schemes which have been developed for their direct write prospects, may make an important contribution to the mask making community as well.

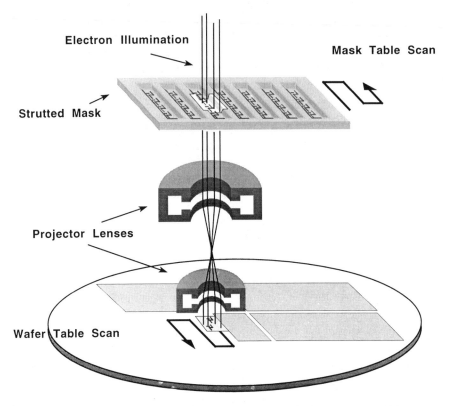

FIGURE 14. Schematic view of SCALPEL e-beam projection system which uses a scattering mask and a back focal plane aperture to derive image contrast.

In an effort to take advantage of the resolution that e-beam systems have demonstrated, while bridging the gap in throughput, several groups have advanced the notion of electron projection systems. While several full-field electron projection electron beam systems have been proposed in the past[56-58] as alternative methods to supplant optics, lingering problems have prevented their full development as manufacturing systems. These problems include the use of absorber masks, the need for very large field low-distortion electron optics, and electron-electron interactions[59].

One more recently proposed electron projection method that avoids some previous problems is called SCALPEL. A schematic view is depicted in Figure 14. The SCALPEL concept derives aerial image contrast from the use of a scattering mask rather than an absorptive one. The mask is illuminated with electrons over an area of the order of 1 mm^2 which pass through the clear areas in the thin membrane and are scattered by a thin metallic layer (typically 50 nm of tungsten) in the patterned areas of the mask. The scattered electrons are sorted and absorbed by an aperture placed at the back focal plane of the magnetic lens, forming a high contrast image. This image can then be further de-magnified by additional electron optical elements in the column. This contrast mechanism was first suggested by Koops[60] and has been incorporated in the proposed step-and-scan SCALPEL instrument being built at AT&T[59]. The step-and-scan approach complicates the mechanical design requiring controlled scanning of both the mask and wafer during exposure (at different rates) but reduces the requirement for large electron optical elements. Expected column designs which include 4× reduction optics, will allow currents of up to about 50 μA before space charge effects limits the resolution.

B. X-ray Proximity Printing (XRPP)

Perhaps the simplest method of advanced lithography, in principle, is X-ray proximity printing (XRPP). It has therefore enjoyed the most extensive development period of all the methods discussed in this section. In this method, collimated X-rays are incident on a membrane mask with a patterned absorber. As discussed above, the absorbers are typically high aspect ratio structures on a supporting membrane. Fabrication of these masks can have residual stress in the patterned metal which distorts the pattern by straining the membrane. Controlling this stress is especially important since the placement errors are not reduced by the usual factor of 4 or 5 inherent in other printing methods. Therefore mask material research has been a very active topic for some time[61,62].

The fundamental limit of resolution in XRPP is quite good when the mask is brought into contact with a resist coated wafer. In a practical manufacturing system where a fully inspected and repaired mask is expected to be rather costly, it is recognized that there needs to be a gap between the mask and wafer, however. The size of the gap, g, needed to prevent mask damage is a controversial topic, and a critical one since the diffraction limited resolution is given by:

$$resolution = k(\lambda g)^{1/2} \qquad (11)$$

where λ is the X-ray wavelength and k is a constant of order 1[63]. Since the various gap values quoted range from 5 μm to 40 μm for production environments, we calculate a corresponding range in resolution of from .07 to 0.2 μm for $\lambda = 1$ nm. The answer to this technical issue is very important for the extension of XRPP to future generations of IC manufacture. In the extreme case where contact X-ray printing is performed, the gap equivalent is one half the resist thickness. As illustrated in Figure 15, under contact conditions, very high resolution can be

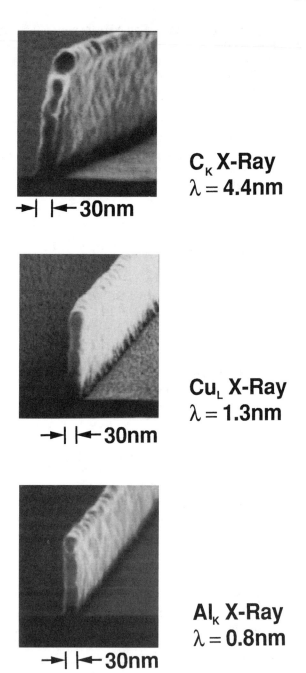

C$_K$ X-Ray
$\lambda = 4.4$nm

→| |←30nm

Cu$_L$ X-Ray
$\lambda = 1.3$nm

→| |←30nm

Al$_K$ X-Ray
$\lambda = 0.8$nm

→| |←30nm

FIGURE 15. 0.03 µm features in PMMA printed using contact X-ray lithography with 4.5 (Carbon $K\alpha$), 1.32 (Copper $L\alpha$), and 0.834 (Aluminum $K\alpha$) nm X-rays[64].

achieved with a wide range of X-ray wavelengths[64].

Another issue is the appropriate X-ray source for XRPP. Synchrotron sources at research labs have been used as high flux X-ray sources for exploratory work in this field. A major investment in XRPP technology at IBM resulted in Helios, the first commercial compact synchrotron, optimized for this application[65]. The concern at smaller IC companies is that more economical sources be available which are capable of similar performance. A laser plasma generated X-ray source has been built and studied for this use[66] and is potentially more economical for smaller volume semiconductor companies.

C. Ion Projection Lithography

Ion projection lithography (IPL) has been developed primarily by Stengl *et al.* in Austria since the late 1970's[67]. Current development of a full field projection system is being pursued by the Advanced Lithography Group (ALG), a consortium of industry, government, and universities. The ALG-1000 is several generations removed from the early demonstration system originally built. This latest tool is designed to print 20 mm × 20 mm fields at 3× reduction imaging using 150 keV hydrogen ions and that can be extendable to 0.1 μm IC design rules[60]. The use of high voltage and light ions allows reasonable depth penetration into resists and better manages shot noise compared with the early predecessors.

IPL is a controversial candidate for several reasons. Since the ions are so highly absorbed, the mask is required to be a free standing stencil type mask. This requires that a ring feature on a mask would have its center fall out during fabrication. Therefore either sub-resolution supports or a complementary mask system must be used. In addition, full field projection can suffer from distortions within the field and from space charge effects in which the charged particles repel each other at cross over points in the ion optics resulting in additional pattern dependent distortions. There are, however, engineering solutions to these potential problems which are being pursued[68].

One of the attractive features of ion lithography is the lack of long range scattering that is present in e-beam lithography. Where electron projection systems must anticipate dose variations in the resist for various sized and spaced pattern areas, ion beams tend to stay near their initial trajectory and require no dose adjustment.

D. Extreme Ultraviolet Lithography (EUVL)

A relative newcomer to the spectrum of advanced lithographic options is Extreme Ultraviolet Lithography or EUVL (also sometimes referred to as soft X-ray projection lithography, SXPL). EUVL is most similar in concept to optical step and scan lithography, but several major differences exist due to the significant changes in the optical properties of materials for wavelength decreases near 10 nm. First, reflective optical systems are required since all materials are too highly absorptive to fashion refractive lenses. Also since reflectivities are quite low for single coatings, the optical reflective coatings are comprised of multilayer reflectors which act as Bragg-type reflectors. Typical reflectors in EUVL, for use near 13 nm, consist of 40 layer pairs of molybdenum and silicon (each layer pair about 7 nm thick). Multilayer reflector in this regime have produced normal incidence reflectances of about 65%.

Figure 16 shows the design space for optical systems employing wavelengths near 13 nm[69]. If we require a depth of focus (DOF) of ± 1 μm and a resolution of 0.1 μm using *Mo/Si* coated optics at λ=13 nm, the optical system would have an *NA* between 0.05 and 0.1; quite modest compared with today's step-and-repeat cameras.

Laboratory-scale systems of varying reduction factors have been built and used to investigate the imaging properties and resist characteristics in the EUV regime[70-72]. Figure 17 shows a small area of PMMA resist exposed with an early 20× reduction Schwarzschild experimental system after development and

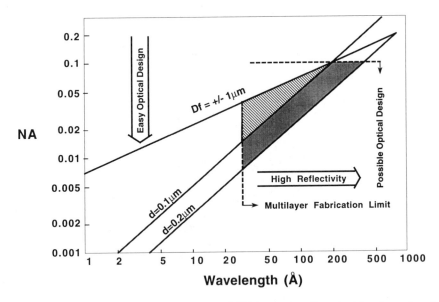

FIGURE 16. Sets of numerical aperture and EUV wavelengths capable of meeting the diffraction limited resolution and depth of focus requirements of a projection camera design. The shaded regions correspond to a bound of ± 1 μm of DOF and two different resolutions, *d* =0.2 and *d* =0.1 μm[70].

FIGURE 17. A 50 nm line and space pattern in PMMA resist, patterned by reduction EUV lithography[71].

gold coating to enhance the SEM image. The features represent 50 nm lines and spaces.

The membrane mask structures required to implement SCALPEL, IPL, and XRPP discussed above tend, in general, to be fragile, prone to distortion, and sensitive to heating effects although these issues are being aggressively addressed within the various development programs. EUVL differs from these other lithographies in that a thick reflective multilayer mask is also a viable alternative. As a result, a variety of multilayer coated reflective mask technologies have been reported[73,74] which can provide both high resolution features and good reflectance contrast between the patterned and unpatterned regions. Typical masks comprise a patterned absorber or composite absorber atop a multilayer reflector. Masks well suited for research and development of this new printing method are readily fabricated, but recent simulation and experiments on the effects of subsurface defects[75] warn that near zero defect multilayer mask blanks will be needed for this mask type to be commercially viable.

VI. TRENDS IN CONVENTIONAL LITHOGRAPHY
FOR RESEARCH

A. Vector Scan E-beam Systems

Researchers continuously strive to access a window on the future of new devices, circuits and technology. Toward this end, research e-beam tools are often sought to gain entry into new lithographic regimes. These can include better resolution to permit fabrication of smaller feature sizes, higher current densities to investigate new processing methods, and more pattern precision in order to allow critical alignment or meet requirements for spatial coherence. Trends in advanced e-beam system development has been toward: higher beam voltages (100 kV and beyond)[76-78]; higher brightness sources with optical column designs suited to nanolithography[76,77,79]; and compatibility with process gas introduction[79]. One remarkable result is shown in Figure 18, in which 5 nm lines on 20 nm centers were produced using an inorganic resist, not on a laboratory STEM adapted for EBL but rather on a full size, laser interferometer-staged (6 inch), computer-controlled, commercial e-beam lithography system.

While mechanical design and laser interferometers continue to improve, new

FIGURE 18. Developed pattern of 5 nm lines on 20 nm centers in silicon dioxide directly patterned by e-beam lithography using HF as a developer. This result would usually require current densities only found in laboratory STEM systems, but was produced here in a modified commercial EBL system[79].

ideas have also been demonstrated that promise to dramatically increase the spatial precision of e-beam systems. One idea employs a laser holographically produced grating pattern transferred to selective areas over an entire wafer to serve as a continuous fiducial mark that is scanned within each writing field. The signal is phase locked to the positioning feedback signal of the e-beam deflection system. This scheme was shown to reduce the field stitching errors down to about 2 nm in an EBL written grating pattern.

Another exciting prospect on the horizon is in the area of focused ion beams systems. FIB systems have become an important tool for repair of advanced masks. Sources for these systems have typically been either a liquid metal ion source (LMIS) like *Ga* or a cryogenic sourced light ion like hydrogen. *Ga* sources, popular because they are the easiest to make and enjoy a long lifetime (>1000 hrs.), are also efficient sources for micro-machining since they exhibit high sputter rates, while the hydrogen source is favored for lithography, to allow penetration of thicker resists. Gas ionization sources, pioneered by Kalbitzer[80], are being developed that would allow a variety of gases, such as argon, to be used for micro-machining operations such as reticle repair provided that these sources produce sufficiently high brightness. Currently, *Ga* sources used for this purpose cause staining due to the unwanted implantation of *Ga* atoms during the sputtering operation[81].

The continuing improvements in system specifications and capabilities in the research and development community for both electron and ion beam systems will also allow the larger industry to test its ability to meet the future manufacturing requirements for alignment tolerance, placement accuracy, resolution, linewidth control, repair, etc.

B. E-beam Micro-Column and Tip Array Systems

Scanned electron beam lithography is the only demonstrated approach which combines the needed resolution, low distortion, and overlay accuracy to accomplish complex circuit fabrication in the sub-100 nm regime. As described above, however, such systems are severely throughput limited. In response to this limitation, several groups have proposed multiple beam approaches to greatly increase the parallelism of such systems. Some of these approaches envision large arrays of scanning tunneling tips[82] or scanning force tips[83] both of which can be fabricated using large scale methods akin to silicon IC processing[84].

One potential problem with such an implementation is the strong interaction between the tip and the material surface. This has led a group at IBM to begin development on a micro-machined electron optical column called SAFE, for STM aligned field emission[85]. Although in the current phase the design is being studied for a single focusing, scanning micro-column, as shown in Figure 19, the

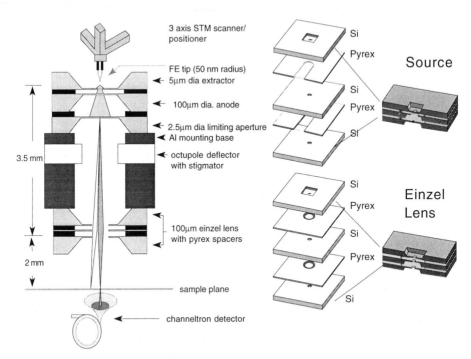

FIGURE 19. Schematic of the SAFE micro-column: a) functional diagram, b) assembly of silicon micro-machined column parts.

method is envisioned to comprise an array of 60 or so columns, allowing one column per chip. The most recent prototype SAFE micro-column has produced 10 nm resolution images, used as a scanning electron microscope[86].

Whether this method will be sufficiently mature in time for production is not clear. However, the notion of palmtop, high-throughput, nanolithography capability is a very exciting one indeed!

C. Arrays of AFM Tips for Parallel Writing

Another array type approach being proposed is one of massively parallel AFM lithography. AFM tips are conventionally produced using planar semiconductor processing methods. It is therefore plausible that large arrays of uniform high quality, pre-aligned tips can be manufactured on a single chip. An array 10^4 to

10^6 tips or more (as many as 10^9 could be required according to Figure 2) – if individually addressable – could provide yet another high-throughput approach to nanolithography in future manufacture. Since these methods are in their early proof-of-concept stages, issues such as reliability, uniformity, and placement accuracy need to be demonstrated. Early results, however, using more than one tip simultaneously are an exciting glimpse of future possibilities in lithography[83,87].

VII. Summary

We have attempted in this chapter to highlight the progress and limitations in the field of conventional lithographic technology and its role in the creation of a wide array of structures. The path is bifurcated by the separate needs of the research community which demands simple access to nanostructures down to 10 nm and beyond, and that of the semiconductor industry which is relentlessly marching toward manufacturing at the 100 nm size scale. The research community is pursuing a variety of methods outlined in this chapter, but is most broadly served by ever improving performance in electron beam lithography. The manufacturing community, with its host of concurrent requirements most notably, extremely high-throughput, cannot simply look upon high resolution as the goal for future systems.

Whether we see a paradigm shift from manufacture by optical lithography to X-rays or electron or ion beams, or whether the progression of optics continues toward shorter and shorter wavelengths, the next decade will be an exciting time in the field of lithography. It will reveal the fruits of the research and development efforts of the past two decades in lithographic technology.

REFERENCES

[1]   Broers, A.N., et al., Appl. Phys. Lett. **29**, 596 (1976).

[2]   Howard, R.E., et al., Appl. Phys. Lett. **36**, 598 (1980).

[3]   Beaumont, S.P., et al., Appl. Phys. Lett. **38**, 436 (1981).

[4]   Lee, K.L., and Ahmed, H., J. Vac. Sci. Technol. **19**, 946 (1981).

[5]   Dagata, J.A., et al., Appl. Phys. Lett. **56**, 2001 (1990).

[6]   Snow, E.S., Campbell, P.M., and McCarr, P.J., Appl. Phys. Lett. **63**, 749 (1993).

[7]   Snow, E.S., and Campbell, P.M., Appl. Phys. Lett. **64**, 1932 (1994).

[8]   Smith, H.I., *et al.*, *Microelectronics Engineering, "Special Issue on Nanotechnology"*, to be published.

[9]   A number of individual cases were calculated for the various lithography methods listed, then generous shaded figures were drawn to include those sets of points. The power law fit was performed using the individually plotted points.

[10] Smith, H.I., *J. Vac. Sci. Technol. B* **4**, 148 (1986).

[11] Hector, S.D., and Smith, H.I., *OSA Proceedings on SXPL,* Hawryluk, A.M., and Stuhlen, R.H., eds., **18**, 202 (1993).

[12] For an assessment of various electron proximity correction methods see: Owen, G., *J. Vac. Sci. Technol B* **8**, 1889 (1990).

[13] Goldstein, J.I., *et al.*, *Practical Scanning Electron Microscopy*, Goldstein, J.I., and Yakowitz, H., eds., New York: Plenum Press, 1977, pp. 21–47.

[14] Wilson R.G., and Brewer, G.R., *Ion Beams with Applications to Ion Implantation*, Huntington, NY: Krieger Publishing Company, (1979), pp. 129–260.

[15] Kyser, D.F. *J. Vac. Sci. Technol. B* **1**, 1391 (1983).

[16] Howard, R.E., *et al.*, *J. Vac. Sci. Technol. B* **1**, 1101 (1983).

[17] Tennant, D.M., *et al.*, *J. Vac. Sci. Technol. B* **12**, 3689 (1994).

[18] Chapman, B., *"Glow Discharge Processes"*, John Wiley & Sons, (1980), (especially Chapters 5 and 7).

[19] Yan, R.H., *et al.*, *IEEE Elec. Dev. Lett.,* **13**, 256–258 (1992).

[20] Fiegna, C., *et al.*, *IEEE Trans. Electron Dev.* **41**, 941–950 (1994).

[21] Ono, M., *et al.*, *J. Vac. Sci. Tech B* **13**, 1740 (1995).

[22] Eugster, C.C., Nuytkens, P.R., and del Alamo, J.A., *IEEE IEDM Tech. Digest* 495–499 (1992).

[23] Tiwari, S., and Woodall, J.M., *Appl. Phys. Lett.* **64**, 2211 (1994).

[24] Tiwari, S., *et al.*, *Tech. Dig. IEDM 1992* 859 (1992).

[25] Simhony, S., *et al.*, *Appl. Phys. Lett.* **59**, 2225 (1991).

[26] Tsukamoto, S., *et al. Appl. Phys. Lett.* **63**, 355 (1993).

[27] Tsuchiya, M., *et al. Phys. Rev. Lett.* **62**, 466 (1989).

[28] Weisbuch, C., and Vinter, B., *Quantum Semiconductor Structures*, San Diego: Academic Press (1991).

[29] Ferrera, J., *et al.*, *J. Vac. Sci. Technol. B* **11**, 2342 (1993).

[30]   Hawryluk, A.M., Smith, H.I., Ehrlich, D.J., *J. Vac. Sci. Technol. B* **1**, 1200 (1983).

[31]   Kirz, J., *J. Opt. Soc. Am.* **64**, 301–309 (1974).

[32]   Anderson, E.H., and Kern, D., in *X-Ray Microscopy III,* Michette, A., Morrison, G., and Buckley, C., eds., Berlin: Sringer-Verlag (1992).

[33]   Thieme, J., *et al.*, "X-Ray Optics and Microanalysis" *Inst. of Physics Conf. Ser.* **130**, Bristol, 1993, 527–530.

[34]   Tennant, D.M., *et al.*, *Optics Lett.* **16**, 621 (1991).

[35]   Tennant, D.M., *et al.*, *J. Vac. Sci. Technol. B* **6**, 1970 (1990).

[36]   Thieme, J., *et al.*, in *X-Ray Microscopy IV,* Erko, A.I., and Aristov, V.V., eds., Bogorodski Pechantnik: Chernogolvka (1995).

[37]   Windt, D.L., private communication.

[38]   Hartley, J., Groves, T., and Pfeiffer, H., *J. Vac. Sci. Technol. B* **9**, 3015 (1991).

[39]   Pfeiffer, H.C., *et al.*, *J. Vac. Sci. Technol. B* **11**, 2332 (1993).

[40]   Chou, S.Y., Smith, H.I., and Antoniadis, D.A., *J. Vac. Sci. Technol. B* **3**, 1587 (1985).

[41]   Chou, S.Y., Smith, H.I., and Antoniadis, D.A., *J. Vac. Sci. Technol. B* **4**, 253 (1986).

[42]   Flanders, D.C., *Appl. Phys. Lett.* **36**, 93 (1980).

[43]   Liddle, J.A., *J. Vac. Sci. Technol. B* **9**, 3003 (1991).

[44]   Mauger, P.E., *et al.*, *J. Vac. Sci. Technol. B* **10**, 2819 (1992).

[45]   Tennant, D.M., *et al.*, *J. Vac. Sci. Technol. B* **8**, 1975 (1990).

[46]   MacDowell, A.A., *et al.*, *J. Vac. Sci. Technol. B* **9**, 3193 (1991).

[47]   "The National Technology Roadmap for Semiconductors", San Jose: The Semiconductor Industry Association (1994).

[48]   Tiberio, R.C., *et al.*, *J. Vac. Sci. Technol B* **9**, 2842 (1991).

[49]   Zah, C.E., *et al.*, *Electron. Lett.* **25**, 650 (1989).

[50]   Tennant, D.M., *et al.*, *J. Vac. Sci. Technol. B* **10**, 2530 (1992).

[51]   Kjellberg, T., Schatz, R., *J. Lightwave Technol.* **10**, 1256 (1992).

[52]   Tennant, D.M., *et al.*, *J. Vac Sci Technol. B* **11**, 2509 (1993).

[53]   Levenson, M.D., *Japan. J. Appl. Phys.* **33**, 6765 (1994).

[54] Hattori, K., *et al.*, *J. Vac. Sci. Technol. B* **11**, 2346 (1993).

[55] Nakayama, Y., Okazaki, S., and Saitou, N., *J. Vac. Sci. Technol. B* **8**, 1836 (1990).

[56] Heritage, M.B., *J. Vac Sci Technol.* **12**, 1135 (1975).

[57] Frosien, J., Lischke, B., and Anger, K., *J. Vac Sci Technol.* **16**, 1827 (1979).

[58] Nakasuji, Suzuki and Shimizu, *Rev. Sci. Inst* **64**, 446 (1993).

[59] Berger, S., *et al.*, *Proc. SPIE* **2322**, 434 (1994).

[60] Koops, H.W.P. and Grob, J., *Springer Series in Optical Sciences: X-ray Microscopy* **43**, Berlin: Springer-Verlag, 1984.

[61] Heuberger, A., *J. Vac Sci Technol. B* **6**, 107 (1988).

[62] Warlaumont, J., *J. Vac Sci Technol. B* **7**, 1634 (1989).

[63] Fay, B., in *Microcircuit Engineering,* Ahmed, H., and Nixon, W.C., eds., Cambridge: Cambridge University Press , 1980, 323–353.

[64] Early, K., Schattenburg, M.L, and Smith, H.I., *Microelectron. Eng.* **11**, 317 (1990).

[65] Archie, C.N., *et al.*, *J. Vac. Sci. Technol. B* **10**, 3224 (1992).

[66] Frackoviak, J., *et al.*, *J. Vac Sci Technol. B* **9**, 3198 (1991).

[67] Stengl, G., *et al.*, *J. Vac Sci Technol.* **16**, 1883 (1979).

[68] Finkelstein, W., and Mondelli, A.A., *Semiconductor International* **55**, (1995).

[69] Brunger, W.H., *et al.*, *Microelectronic Eng.* **27**, 323–326 (1995).

[70] Kinoshita, H., *et al.*, *J. Vac Sci Technol. B* **7**, 1648 (1989).

[71] Bjorkholm, J.E., *et al.*, *J. Vac. Sci. Technol. B* **8**, 1509 (1990).

[72] Kubiak, G.D., *et al.*, *J. Vac. Sci. Technol. B* **12**, 3820 (1994).

[73] Hawryluk, A.M., *et al.*, *Proc. OSA* **12**, 45 (1991).

[74] Tennant, D.M., *et al.*, *Appl. Optics* **32**, 7007 (1993).

[75] Ngyugen, K.G., *et al.*, *J. Vac. Sci. Technol. B* **12**, 3833 (1994).

[76] Koek, B.H., *et al.*, *J. Vac. Sci. Technol. B* **12**, 3409 (1994).

[77] McCord, M.A., *et al.*, *J. Vac. Sci. Technol. B* **10**, 2764 (1992).

[78] Thoms, S., Beaumont, S.P., and Wilkinson, C.D.W., *J. Vac. Sci. Technol. B* **7**, 1823 (1989).

[79] Hiroshima, H., *et al.*, *J. Vac. Sci. Technol. B* **13**, 2514 (1995).

[80]   Wilbertz, C., *et al.*, *Nuclear Instrum. & Methods in Phys. Section B* **63**, Issue 1–2, 120 (1992).

[81]   Harriott, L.R., *J. Vac. Sci. Technol. B* **11**, 2200 (1993).

[82]   Binnig, B., *et al.*, *Phys. Rev. Lett.* **49**, 47 (1982).

[83]   Minne, S.C., Manalis, S.R., and Quate, C.F., *Appl. Phys. Lett.* **67**, 3918 (1995).

[84]   Zhang, Z.L., and MacDonald, N.C., *J. Vac. Sci. Technol. B* **11**, 2538 (1993).

[85]   Chang, T.H.P., *et al.*, *SPIE* **10**, 127 (1993).

[86]   Kratschmer, E., *et al.*, *J. Vac. Sci. Technol. B* **13**, 2498 (1995).

[87]   MacDonald, N.C., Chapter 3, this volume.

Chapter 5

Fabrication of Atomically Controlled Nanostructures and Their Device Application

H. Sakaki

University of Tokyo (RCAST)
4-6-1 Komaba, Meguro-ku, Tokyo, Japan

I. INTRODUCTION

During the last three decades, we have witnessed remarkable progress in the entire spectrum of semiconductor technology. In the area of epitaxy, for example, the emergence of molecular beam epitaxy (MBE) and organo-metallic vapor phase epitaxy (OMVPE) has allowed us to prepare quantum wells, tunneling barriers, and other ultra-thin layered structures, such as shown in Figures 1(a) and (b). In these structures, semiconductor films of specified thicknesses and compositions are deposited with the accuracy of one atomic layer (~0.3 nm). Artificial potential profiles $V(z)$ created in such structures are used to control the quantum-mechanical motion of electrons, providing a variety of electronic properties which are important both in solid-state physics and in advanced device applications[1–6]. Figure 2 illustrates several examples of such potentials. Indeed, these layered structures are now used as the core parts of high-performance devices, such as quantum well lasers, Stark modulators, quantum-well infrared detectors, high electron mobility transistors, and resonant tunneling diodes.

We have witnessed also remarkable advances in lithography and other patterning technology, by which very fine structures are laterally defined. Such lateral patternings have been mainly used to produce high-speed transistors, and very-large-scale integrated circuits with sub-micron feature sizes. Advanced lithography has been extended further to prepare a new class of laterally-defined structures with characteristic lengths of the order of 10–100 nm. Typical examples of such nanometer-scale structures are quantum wires (QWRs) and quantum boxes (QBs), in which electrons are quantum mechanically confined in two or three dimensions as shown in Figures 1(c) and 1(d). Because of the reduced degrees of freedom of motion, QWRs and QBs are predicted to and partly proven to possess a variety of new properties and device functions, which cannot otherwise be realized[7–10].

Although most of QWRs and QBs studied so far have feature sizes of about 100 nm, considerable effort has been recently made to develop novel epitaxial

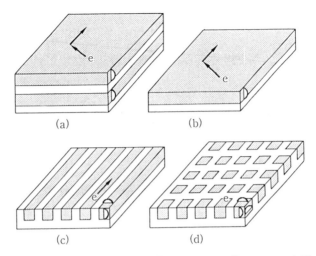

FIGURE 1. Schematic drawings of a multiple quantum well or a superlattice (SL) (a), a quantum well (QW) (b), a quantum wire (QWR) structure (c), and a quantum box (QB) or dot (QD) structure (d).

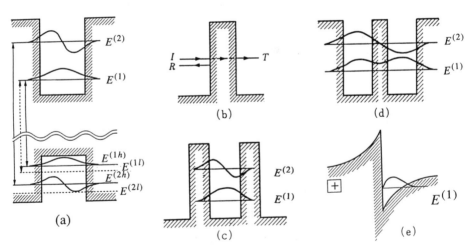

FIGURE 2. Potential profiles and eigenstates of electrons and holes in various quantum structures: (a) a quantum well (QW), (b) a single tunneling barrier, (c) a quantum well bound by double barriers, (d) a double (multiple) QW structure, (e) a heterojunction with an attractive field. Only in (a), the potential profile and eigenstates of heavy holes and light holes is shown.

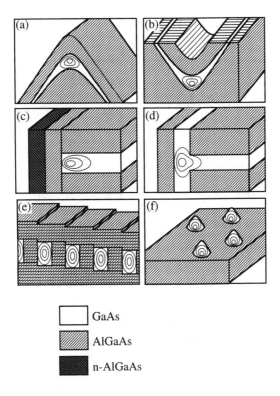

GaAs

AlGaAs

n-AlGaAs

FIGURE 3. Examples of epitaxially grown QWRs and QBs. A ridge QWR (a), a groove QWR (b), a (field-induced) edge QWR (c), a (T-shaped) edge QWR (d), a tilted superlattice (TSL) (e), and a self-assembled QB structure (f). See section VI.B for their details and references.

methods by which 10 nm-scale lateral structures are fabricated, as illustrated in Figure 3[11]. For example, the selective growth of *GaAs* onto very sharp *AlGaAs* ridge and/or groove structures is shown to be useful in fabricating 10–20 nm wide QWR structures [Figure 3(a)and (b)]. Similarly, the overgrowth of an $n-AlGaAs$ layer or an *AlGaAs/GaAs* QW layer onto the edge surface of a pre-grown QW structure is shown to yield 10 nm-scale edge QWR structures [Figure 3(c) and (d)]. Furthermore, it has been found that an alternate growth of a half monolayer of *GaAs* and *AlAs* on a vicinal plane with evenly spaced surface steps yields a laterally modulated in-plane superlattice (IP-SL) or a tilted SL structure [Figure 3(e)] with typical periods of 10–20 nm, while the lattice-mismatched growth of *InAs* onto a *GaAs* substrate just beyond its critical thickness results in the spontaneous formation of 10–20 nm-scale islands that function as QB or QD structures [Figure 3(f)].

In this article, we describe first the current state-of-the-art for fabricating ultra-thin layered structures and their structural characterizations. We then describe how electronic and photonic properties of nanometer-scale structures are influenced by interface roughness and other structural imperfections and show how important it is to control the thickness and other structural parameters in realizing advanced device performances. We discuss, in particular, effects of the interface roughness and film thickness variations on quantum-well lasers, modulators, and other photonic devices and also their influences on heterostructure FETs, resonant tunneling diodes, and other transport devices.

II. ATOMICALLY CONTROLLED FABRICATION OF DEVICE STRUCTURES

A. Roughness at Surfaces and Interfaces: Its Characterization by TEM, RHEED, and STM/AFM Studies

It is known from a number of studies that *GaAs/AlGaAs* quantum wells (QWs) and other hetero-structures prepared by molecular beam epitaxy (MBE) or by organo-metallic chemical vapor deposition (OMCVD) are usually of quite high, Δ, quality and yet their heterointerfaces always have some roughness, whose height is typically one or two atomic layers (see Figure 4)[12–32]. This interface morphology is determined mainly by the roughness of freshly-grown surfaces of *GaAs* (or *AlGaAs*) layers, on which an *AlGaAs* (or *GaAs*) layer is further deposited, as shown in Figure 4.

The presence of such roughness has been noticed in the photoluminescence (PL) studies of QWs, in which PL spectra are found to broaden substantially, when the well width L_z is reduced, as will be discussed in Section III.A[21–28]. Similarly, electron mobilities in narrow QWs are found to decrease rapidly, when L_z is set well below 10 nm, indicating the importance of interface roughness scattering[29–31]. The detail of this transport study will be described in Section IV.A. In this subsection, we review a series of studies to characterize interface structures by using the transmission electron microscope (TEM)[17–20], reflection high-energy electron diffraction (RHEED)[23–25,32], the scanning tunneling microscope (STM)[13–16], and the atomic-force microscope (AFM).

Cross-sectional transmission electron microscopy (TEM) appears at first sight to be the most straightforward way to characterize the morphology of *GaAs/AlGaAs* heterointerfaces[17–20]. One should note, however, that a standard way to take TEM images with a clear contrast for *GaAs* and *AlAs* is to launch the incident electron beam along the [100] axis of the specimen and make images by using only a (002) diffraction beam[20]. This approach gives a very

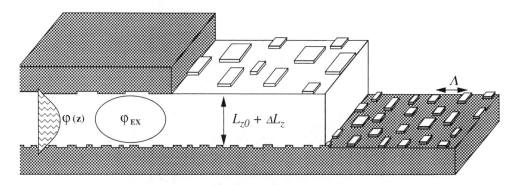

FIGURE 4. A schematic drawing of a *GaAs* quantum well with some interface roughness. Note that the morphology of a bottom interface is mainly determined by the roughness on the growth front of the bottom *AlAs* (or *AlGaAs*) layer, whereas the top interface is primarily determined by the smoothness of a *GaAs* layer during/after the growth. The parameter Λ is the spatial correlation length of the roughness.

good contrast due to the material sensitivity of (002) diffraction but it suffers from the loss of the spatial resolution. In contrast, lattice images with very high-resolution are usually made by using many diffraction beams, which tends to weaken the material-dependent contrast needed to assess precisely the heterointerfaces.

To resolve this problem, two novel TEM methods have been developed and representative images are shown in Figures 5(a) and (b). One of the methods is to slice the sample parallel to a (100) plane and launch the incident beam along the [100] axis of the sample and to make lattice images by using two pairs of (200) and (220) diffraction beams. This method is known as chemical imaging, since this approach gives rise to a simple 4-fold lattice image for *GaAs* and a more complex, centered 4-fold pattern as the lattice image of *AlAs*, which can be easily distinguished, as shown in Figure 5(b)[17,18]. The second method to achieve high resolution and high material contrast simultaneously is to set the incident beam along the [110] direction of a carefully sliced specimen and to form a lattice image by using a set of (111) diffracted beams[19]. It has been shown both theoretically and experimentally that when the thickness of a specimen is close to 14 nm and the defocusing is optimized, the amplitude of (111) diffraction beams becomes quite weak for *GaAs* but intense for *AlAs*, giving rise to a good contrast. To test this prediction, the TEM image of a superlattice consisting of 10 – monolayer – thick *GaAs* and one-monolayer-thick *AlAs* has been taken by

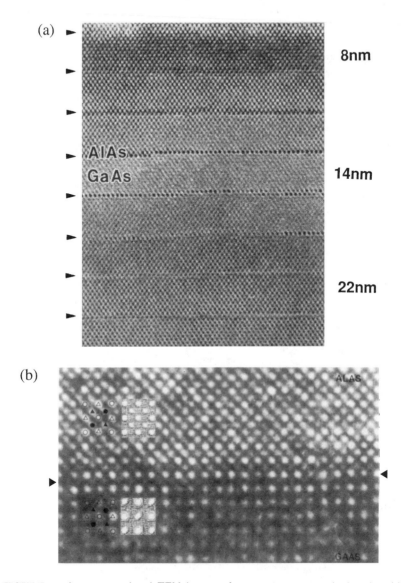

FIGURE 5. A cross-sectional TEM image of a *GaAs/AlAs* superlattice, in which one-monolayer thick *AlAs* and 10-monolayer thick *GaAs* were alternately deposited by MBE (after Ikarashi *et al.*[19]) (a). Another cross-sectional image of a *GaAs/AlGaAs* hetero-junction (after Ichinose *et al.*[18]) (b). The image (a) was taken with the incident beam along the [100] direction, whereas the image (b) was made with the incident beam along the [110] direction.

this method and the result is shown in Figure 5(a). Here, a cross-sectioned specimen is thinned to various thicknesses, ranging from 6 to 22 nm. Note that the maximum contrast is seen at the sample thickness of 14 nm, as expected. Here, a mono-layer thick *AlAs* embedded in *GaAs* is clearly seen with the interface roughness of, at most, one monolayer in height. This particular sample has been prepared with growth interruption (GI), indicating that the *GaAs* surface prepared with GI is quite smooth. This point will be further discussed later, since the surface morphology depends on a number of parameters. One should also be careful in the interpretation of TEM images, since these TEM images represent always an effective composition which is averaged over the thickness (10–20 nm) of a cross-sectioned specimen.

A convenient way to evaluate the surface smoothness during the MBE growth is to measure the intensity of a specularly-reflected electron beam in RHEED set up, since its intensity is known to decrease as the step density at the growth front increases[23–25,31–34]. Figure 6(a) is an example of such a measurement, in which the RHEED intensity is plotted as a function of time for *GaAs* and *AlAs* layers grown at 580C[24,25]. A clear intensity oscillation is seen, particularly in the initial stage of *GaAs* growth. This indicates that the morphology of the initial surfaces undergoes a quasi-periodic change, becoming rough at a sub-monolayer coverage and recovering its smoothness when the coverage reaches a full monolayer. This RHEED intensity oscillation is often used to measure precisely the number of deposited layers.

One notices also in Figure 6(a) that the intensity oscillation damps and approaches a dynamically stable state as the number of deposited layers increases, which indicates the surface of a freshly grown *GaAs* becomes increasingly rough. This roughening process was recently studied more directly using STM, which will be mentioned later. The nature of this dynamically stable surface will be revisited in Section III.B, concerning the linewidths of photoluminescence.

One can also see in Figure 6(a) that the RHEED intensity recovers and reaches a statically stable state when the growth of *GaAs* is suspended or interrupted for a minute or so. The RHEED intensity from a statically stable *GaAs* surface and that from a dynamically stable surface depend on substrate temperature T_s, as shown by the solid line and the chained line in Figure 6(b)[25]. Its significance will be discussed later. The recovery of RHEED seen in Figure 6(a) suggests that the *GaAs* surface is smoothed during the GI, as schematically illustrated in Figure 6(c). Indeed, a recent STM study has provided supporting evidence[16], as will be mentioned later.

The time evolution of the specular beam intensity during and after the growth of *AlAs* is somewhat different from that for *GaAs*, as shown in Figure 6(a). Note, in particular, that the intensity oscillation damps more distinctly and the weakened specular beam signal does not quickly recover. These findings indicate that the density of surface steps grows more rapidly on *AlAs* surfaces and those steps can not be removed as efficiently as those on *GaAs* surfaces. This is probably

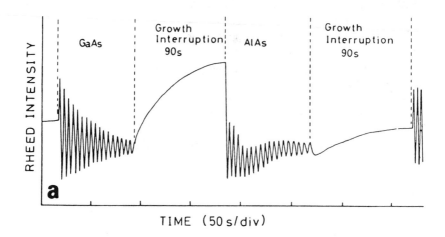

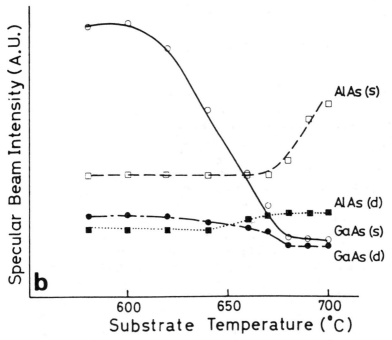

FIGURE 6. (a) The specular beam intensity of RHEED during the MBE growth of a *GaAs/AlAs* multi quantum well structure, consisting of 17 monolayers of each. (b) RHEED intensities from dynamically stable surfaces (d) during the growth, and those from statically stable surfaces (s) after the GI, are shown as functions of the substrate temperature, T_s, both for *GaAs* and *AlAs*.

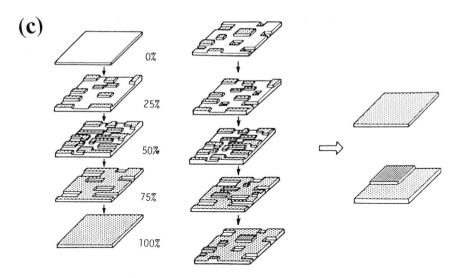

FIGURE 6. (c) The schematic illustration of the smoothing process during the GI (Tanaka *et al.*[25]).

because of stronger bonds of *AlAs* to the substrate and to the step sites. The nature of this *AlAs* surface will be discussed again in Sections III and IV, in connection with PL spectra and electron mobilities.

The results described so far are obtained in *GaAs/AlGaAs* structures which were grown at the substrate temperature of 580-600C with the (As_4/Ga) flux ratio of about 3.5. Surface morphologies of both *GaAs* and *AlAs* would change, if they were grown or annealed at different temperatures. Figure 6(b) shows intensities of RHEED from these materials measured during the growth (i.e., on dynamically stable surfaces) or after the GI (i.e., on static or annealed surfaces) as functions of the substrate temperature[25]. Note that the RHEED intensity from the static *AlAs* surface or after the GI increases as T_s is raised to 700C, while the intensity from the static *GaAs* surface tends to decrease. This suggests that as T_s increases, the density of steps decreases on static *AlAs* surfaces but it increases on static *GaAs* surfaces. Similarly, the surface morphology during the growth depends on T_s, and other growth-related parameters, such as the incoming fluxes, the material compositions, the orientations of the substrate, and the thickness of the grown layers. Hence, one must specify all of these conditions precisely to predict the morphology of interfaces.

The use of STM and AFM techniques to characterize the structure of heterointerfaces at the atomic scale appears at first sight to be straightforward but has become possible only recently[13–16]. There have been two main approaches:

one approach is to measure the morphology of freshly grown *GaAs* or *AlGaAs* surfaces by STM and/or AFM just before forming the heterointerface and estimate the structure of interfaces to be grown; the other scheme is to cleave a fully-grown sample and to characterize its cross-section by STM/AFM. In the following, we briefly describe features of these two approaches.

For the STM/AFM study of freshly grown *GaAs* and other III/V compounds, one must first stop the growth and rapidly cool down the sample in the right ambient to freeze the original surface morphology during the growth[16]. The sample is then transferred to a separate chamber for a subsequent STM/AFM study through an ultra-high vacuum (UHV) tunnel or through the air after covering the sample surface with a protective *As* layer. It has been found by such studies that the *GaAs* surfaces prepared by a long GI are usually quite smooth with typical step spacings of a few hundred nanometers or longer, consistent with the RHEED study. When a *GaAs* layer is newly deposited onto such a surface, the surface changes periodically between the smooth and rough state in correspondence with the RHEED intensity oscillation, as discussed earlier. As the growth proceeds, the number of islands and holes on the surface increases. By the time a few tens of monolayers are deposited, the synchronicity or the correlation of surface morphology at different locations is lost, as indicated by the damping of the RHEED oscillation. The roughening of the surface still continues until the film thickness reaches about 100 monolayers, at which a steady state morphology is established. On such a surface, four different levels of atomic planes are visible, though the major part of the surface is covered by the middle two levels. The spacing of steps distributes widely and is typically a few tens of nanometers in an in-plane direction and is somewhat longer along the other direction. These findings have provided supporting evidence on the surface morphology model of *GaAs* derived earlier from the RHEED study.

The cross-section of *GaAs/AlGaAs* heterostuctures exposed by the cleavage process has been also studied by STM. When the filled-state, valence-band related *As* sublattice is imaged by applying a negative bias (-2V) to the sample, some contrast is obtained between the two materials due to the different ionicities of *Ga* and *Al*, since they influence the charge transfer between the group III and *As* atoms. Similarly, by performing the I-V spectroscopy of tunneling current, the local band-alignment can be evaluated. The tunnel injection of electrons into *p*-type heterostructures is found to result in luminescence, from which the local bandgap information has been extracted. While a variety of information can be obtained by using cross-sectional STM, the precise evaluation of interface roughness on the atomic scale is not always possible, at present, because of the difficulty achieving high spatial resolution and high chemical resolution (or material identification) at the same time.

B. Effects of Interface Roughness and Size Fluctuations on Electronic Properties of Nanostructures

As discussed in II.A, heterointerfaces prepared by epitaxy have usually some kind of roughness, which makes the thickness or the width L_z of a *GaAs* quantum well to be position-dependent. Hence the well width can be expressed as

$$L_z(x,y) = L_{z0} + \Delta L_z(x,y), \tag{1}$$

where (x,y) is the two-dimensional co-ordinate along the heterointerface, L_{z0} is the average width of the quantum well and $\Delta L_z(x,y)$ represents the fluctuation of the well width. Since the i-th energy level E_z^i of electrons in the quantum well depends on the well width L_z, it becomes dependent on (x,y) and can be expressed as

$$E_z^i(x,y) = E_{z0}^i + \frac{dE_{z0}^i}{dL_{z0}} \Delta L_z(x,y), \tag{2}$$

in which the first term represents the average value of the i-th energy level and the second term indicates its (x,y) dependent component or fluctuation ΔE. In the limit of an infinite potential barrier, the energy fluctuation, ΔE, can be expressed as

$$\Delta E_z^i = \frac{d}{dL_z}[\frac{\hbar^2}{2m}(\frac{i\pi}{L_z})^2 \Delta L_z] = -\frac{\hbar^2}{m}\frac{i^2\pi^2}{L_z^3}\Delta L_z = -2E_z^i \times \frac{\Delta L_z}{L_z}. \tag{3}$$

Since this fluctuation of the quantum level becomes quite large when L_z is below 10 nm, the roughness affects both optical and transport properties of heterostructures. For example, it results in the broadening of both luminescence and absorption spectra of quantum wells (QWs) and affects the performances of QW lasers and modulators, and also infrared optical devices using intersubband transitions. These points will be discussed in Section III. The level fluctuation also influences the transport of 2D/1D electrons in ultra-thin films, wires, and related FET structures, since it gives rise to the scattering of electrons , as will be described in Section IV.

The level fluctuation affects also the energy dependence of transmission characteristics of electrons across the heterointerface in quantum-well and super-lattice structures, and influences the performance of resonant tunneling diodes and related devices. In these devices, the thickness variation ΔL_b of *AlGaAs* barrier layers also plays a very significant role, since the tunneling coefficient TT^* of electrons across a barrier is roughly proportional to $\exp(-2\beta L_b)$, where β is the decay constant of the evanescent wave in the barrier $(=\sqrt{2m^*(V-E)}/\hbar)$. The variation of the tunneling coefficient caused by the barrier thickness variation is, therefore, given by

$$\frac{\Delta TT^*}{TT^*} = -2\beta\Delta L_b. \tag{4}$$

Note that , when the thickness of an *AlAs* barrier (for which V=1eV) changes by one atomic layer, the tunneling coefficient will change by a factor of two. This issue will be discussed in Section V.

C. Macroscopic Variation of Film Thicknesses and Compositions

Semiconductor films grown epitaxially by MBE or MOCVD are known to have some macroscopic variations of film thicknesses and compositions when measured over the entire wafer. In the case of MBE, this variation is caused mainly by the non-uniformity of incoming fluxes of group III elements and dopant atoms, which respectively determine the film thickness and composition and the dopant concentration in epitaxial films. By using an MBE system with a large source-to-substrate distance and by optimizing the shape of effusion cells, it has been shown that the variation of thickness and alloy composition in MBE grown films can be reduced to less than a few percent of the intended values, even when they are tested over the substrate holder of 20 cm in diameter[35].

The macroscopic variation of film thicknesses and compositions leads, of course, to the spatial variation of various physical quantities and also key device parameters. For example, the variation of energy levels in a quantum well influences directly the gain and absorption spectra and results in the variation of operating wavelengths of lasers, and other QW photonic devices (see Section III). Similarly, when the thickness of the tunneling barrier varies from one sample to another, this will give rise to the variance of tunneling current (see Section V).

In the case of HEMTs or modulation-doped heterojunction FETs, the threshold voltage V_{th} is dependent on the thickness W_n of an $n-AlGaAs$ layer and on the concentration N_d of donors in the layer. Hence if W_n and/or N_d vary from one spot to another on a wafer, the threshold voltage V_{th} will vary from sample to sample. This issue will be discussed in Section IV more in detail.

III. CONTROL OF ENERGY LEVELS AND
PHOTONIC DEVICES

In this section we discuss first how the photoluminescence and other optical spectra of a quantum well broaden in the presence of interface roughness[21–28]. Then we examine how such spectral broadenings affect the performance of various photonic devices such as lasers, Stark modulators and intersubband detectors and lasers. Finally, we discuss the influence of size fluctuation on electronic properties of quantum wires and boxes.

A. Interface Roughness and Luminescence Spectra

As mentioned briefly in Section II, the presence of interface roughness leads to the spatial fluctuation of quantum well widths $L_z(x,y)$. This results in the position dependent energy levels of electrons, holes, and excitons, as suggested by Eq.(3). Consequently, characteristic energies of nearly all the optical transition processes become dependent on the coordinates (x,y) and give rise to inhomogeneous broadening of the optical spectra.

In the case of (PL) studies, photo-generated carriers form excitons and migrate over some distance L_d in the QW plane, until each exciton disappears by emitting a photon. The energy of such a photon represents the local (effective) energy gap $E_g^*(x,y)$ of the QW, or more precisely E_g^* minus the exciton binding energy E_b, where the local energy gap E_g^* is equal to the bulk energy gap plus the sum of quantized energies of electrons and holes.

It has been well established that the linewidth of PL spectra measured at high temperatures is dominated by the exciton-optical phonon interactions but the linewidth at low temperatures, less than 80K, is mainly determined by inhomogeneities of the local energy gap E_g^*, and therefore increases as the well width L_z is reduced, as suggested by Eq.(3). As an example, we show in Figure 7 the result of such measurements, where the PL spectra[23] and their linewidths at 77K are plotted as functions of L_z for two series of MBE grown *GaAs/AlAs* QWs[24,25]. Note that samples prepared without GI at the formation of the top heterointerfaces show broad luminescences whose linewidth is close to the energy-level broadening caused by the well width fluctuation of one mono-layer. In contrast, samples prepared with GI of 60s at the top heterointerfaces show much sharper PL spectra, which correspond to the effective well-width fluctuation of about a quarter of a mono-layer. These findings suggest that the roughness at the top interface dominate the PL broadening but can be significantly reduced by the GI. Somewhat puzzling is the fact that the PL linewidth is not much broadened by the bottom interface (or the *GaAs*-on-*AlAs* interface) irrespective of the way they are prepared, although the RHEED study has indicated the *AlAs* surface to be not smooth.

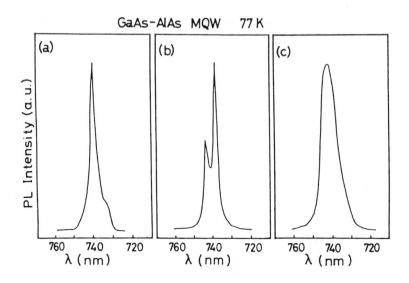

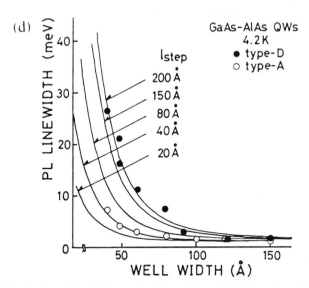

FIGURE 7. Spectra (a)–(c)[23] and the linewidths (d)[24,25] or the full width at half maximum (FWHM) of photoluminescence from two series of *GaAs/AlAs* QWs grown by MBE at 580C. Samples showing broader linewidths were prepared without GI, whereas those with sharper linewidths were formed with 60 second interruption at the top (or *AlAs*-on-*GaAs*) interface.

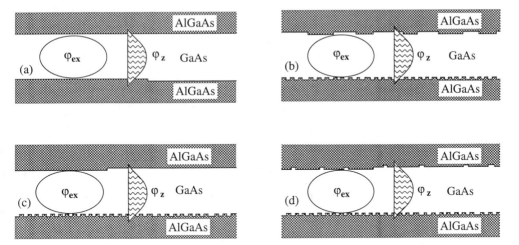

FIGURE 8. Models (a)-(d) of quantum wells having different types of interface morphologies. As described in the text, interfaces sensed by excitons are rough R, only when the lateral size Λ is comparable with the exciton diameter d. When Λ is far larger than or far smaller than d, interfaces are regarded to be truly smooth S or pseudo-smooth PS, respectively.

At this point, we remind the reader that the effective well-width fluctuation discussed above is only a phenomenological parameter, which expresses the magnitude of observed PL linewidths simply in the unit of $\dfrac{d[E_z^{(e1)}+E_z^{(h1)}]}{dL_z}$. For more quantitative study, one must define not only the amplitude Δ of roughness but its lateral size or its correlation length Λ as defined in Figure 4. In the following, we examine several representative cases, where Λ of the top and bottom interfaces are larger or smaller than the exciton diameter.

First we consider a case where Λ is far larger than the diameter d of excitons [See Figure 8(a)]. In such a case, the interface probed by each exciton is truly smooth, at least locally, and PL spectra should consist of one or a few sharp lines. A single sharp line appears when the width of the well is almost uniform over the area that is excited by a laser, while a multiple peak structure shows up when the probed region of a QW comprises a few regions whose widths differ from each other by one or two monolayers of *GaAs*. Incidentally, the diameter of a laser-excited area is typically 100 μm but can be less than 1 μm, when high-resolution microscope optics is used. When the excitation is done by a scanning near-field optical microscope (SNOM)[36,37] or by a focused electron beam in cathode luminescence (CL) set up, the area excited can be less than 0.1 μm. Even in such a case, the diameter of the probed area is still a few tenths of a micrometer or even larger because of the lateral diffusion of photogenerated (or electron beam induced) carriers.

As shown in Figure 8(b), when the correlation length Λ of roughness at one or two interfaces is of the same order of magnitude as the exciton diameter, the local width of the quantum well sensed by excitons becomes highly position-dependent. This gives rise to a broad PL peak, whose linewidth is comparable to the energy level fluctuation defined by Eq.(3). In fact, as shown by solid circles in Figure 7(b), the PL spectra of QWs prepared without GI are well explained by assuming the lateral size of roughness to be about $15-20$ nm.

If the correlation length Λ of interface roughness is far smaller than the exciton diameter, the potential energy sensed by an exciton will be a kind of potential averaged over the diameter of that exciton. Note that such an effective potential is less dependent on the position (x,y) and therefore does not broaden the PL spectrum much. Hence, this type of interface is often referred to as a pseudo-smooth interface. In fact, the *GaAs*-on-*AlAs* interface or the bottom interface of a *GaAs/AlAs* QW grown at 600C or less is found to have such a character by a number of studies. For example, *GaAs/AlAs* QWs prepared by smoothing the top interface with GI exhibit rather sharp PL spectra as shown by blank circles in Figure 7(b). The L_z dependence of PL linewidths can be well explained by assuming the top interface to be truly smooth but the bottom interface to be pseudo-smooth with the lateral size of roughness being around 4 nm, as illustrated in Figure 8(c). If grown under some other conditions, the interface structure could be somewhat different. For example, if grown at a higher temperature, some *GaAs* may leave step edges more easily or undergo an exchange reaction with *AlAs*. As a result, the interface may be modified as illustrated in Figure 8(d)[28].

B. Absorption Spectra and Stark/Wannier-Stark Modulators

Energy-level fluctuations caused by the interface roughness in QW structures influence not only their luminescence spectra but also their absorption spectra. They lead to the broadening and lowering of excitonic absorption peaks. Quantum-confined Stark modulators are electro-optic devices which make use of the electric-field induced deformation of QW eigenstates and the subsequent red-shift of the excitonic edge[39,40]. By this scheme, both the absorption coefficient $\alpha(\lambda)$ and the refractive index $n(\lambda)$ of QW structures are electrically modulated. Hence, in order to form efficient Stark modulators reproducibly, one must control the spectral shape of excitonic absorption peaks by controlling precisely the average well width as well as its spatial fluctuation especially in those devices with narrow QW structures.

Another important scheme of electro-optic modulation is the field-induced Wannier-Stark localization effect in a coupled multi-QW structure with a finite mini-band width. When a strong field is applied, the resonant coupling of quantum levels breaks down and the mini-band disappears. As a result, a part of the optical absorption resulting from the bottom part of the mini-band vanishes, leading to the blue shift of the absorption edge[41,42]. If the interface roughness or the well width fluctuation broadens the ground level energies of constituent QWs and the level broadening gets comparable with the mini-band width, then the electric-field induced change of absorption spectrum would be blurred or washed out. Hence, it is important to minimize the inhomogeneous broadening of quantum levels.

The use of QWRs and QB structures in place of QWs for the Stark modulator has been proposed[43]. In these lower dimensional electron systems the densities of states are more concentrated in a narrower range of energy, as shown in Figure 9, and therefore their optical absorption spectra will be concentrated in a narrower range of photon energy. The Stark shift of such absorption spectra will lead to a more efficient modulation of the absorption constant. Correspondingly, the field-induced electro-optical change of the retractive index would be greater in these systems. Moreover, in the energy region where the density of states is reduced

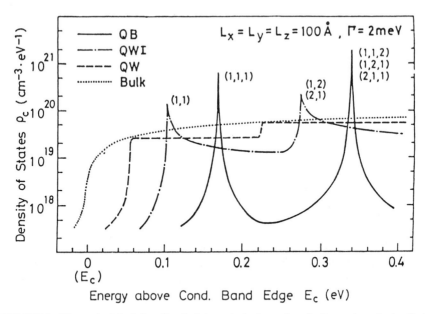

FIGURE 9. The calculated density of states of electrons in a bulk semiconductor (bulk), and those in a quantum well (QW), a QWR and a QB. Here the whole volume is filled with 10 nm-scale quantum structures and their level broadenings are set at 2 meV.

due to the the mini-gap formation, the absorption shows a unique spectrum, which is not a monotonic function of photon energy. These features can be used to create new types of electro-optic devices.

C. Laser Diodes and Injection Modulators

When electrons and/or holes are introduced into a semiconductor, its absorption constant $\alpha(h\nu)$ is reduced due to the filling of the conduction (valence) band states. The refractive index of the semiconductor also changes, as it is linked with α through the Kramers-Kronig relation. This carrier-induced bleaching and subsequent changes in refractive index take place mainly near the absorption edge, as the injected carriers occupy the low energy states of the respective bands, following the Fermi-Dirac distribution.

Because of the step-like density of states (DOS), the energy distribution of carriers introduced in QWs is usually narrower than that in bulk. This reduction of energy distribution goes even further in high-quality quantum wires and boxes, as their DOS is more concentrated in a narrower region of the energy. Hence, the carrier-induced bleaching and refractive-index change take place more efficiently in QWs, QWRs, and QBs than in bulk materials[44–46]. In addition to the concentrated DOS effect, the use of quantum structures is effective for reducing the number of injected carriers for bleaching, simply because of the small volume of optically active media in these devices. Note that the volume of active media in a quantum structure is on a 10 nm-scale which is much smaller than that of a waveguide, especially in a separate-confinement heterostructure (SCH) in which the lightwave is confined on a 100 nm-scale. These features of quantum structures make them attractive as materials for injection modulators. Indeed, QWs are used for optical modulators, where carriers in the injection electrodes or in the near-by reservoir are electrically driven into quantum wells.

To evaluate the bleaching effect quantitatively, we have calculated the change $\Delta\alpha$ of the absorption constant which is induced by injecting 10^{18} cm^{-3} electrons into *GaAs* QBs, QWRs, QWs, and bulk systems. The result is shown in Figure 10 as functions of energy $(E - E_g)$ both at 10K and 300K. Here the level broadening is set at 2 meV and the sizes of quantum structures are all set at 10 nm. Note that $\Delta\alpha$ of QBs is sharply peaked both at 10K and 300K, but such peaks broaden and lower significantly in bulk and in QWs, especially at 300K.

To be more exact in predicting the performance of injection modulators, one must take into account the exciton contribution to the optical spectra, since the excitonic peak increases both the absorption coefficient and the wavelength dependence of refractive index. The role of excitons may be even greater in quantum wire and box systems, since both the binding energy and the oscillator strength of excitons are known to be enhanced in these systems[47].

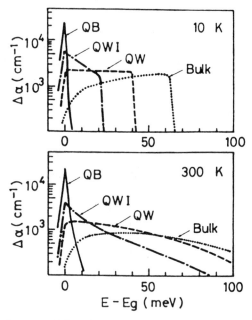

FIGURE 10. The changes of the absorption constant of *GaAs* QBs, QWRs, QWs and bulk calculated at 10K and 300K as functions of the energy $(E - E_g)$. Here the density of injected electrons is set at $10^{18}/cm^{-3}$ and the whole volume is assumed to be filled with 10 nm-scale QWs, QWRs or QB.

Nearly the same argument holds true for most semiconductor lasers, where the gain spectrum is determined mainly by the energy distribution of injected electrons and holes[48,49]. When the same number of carriers are injected, the gain spectrum would be higher and sharper in QBs, QWRs, and QWs than in bulk materials, which makes these quantum structures preferable over bulk materials as a laser medium. Indeed, QW layers are now widely used in making high-speed, low noise 1.55 μm lasers for optical communications, high-power lasers for optical pumping sources, and low-threshold lasers in surface-emitting geometry[50]. Recent experimental studies on QWR lasers and QB lasers will be again described in Section VI.

These advantages of QBs, QWRs, and QWs for injection lasers and modulators will be reduced, if the sharpness of DOS is weakened by the fluctuation in the quantum levels. Roughly speaking, the level broadening has to be less than 20 meV. Otherwise, the energy distribution of carriers would be comparable to the thermal energy kT at 300K, which is the energy spread of carriers in bulk materials. To be more precise, however, one must consider not only the size fluctuation but also some other factors. For example, in the case of QBs, one must take into account the influence of inter-box and intra-box Coulomb effects on the quantum levels and also the redistribution of injected carriers among different QB structures. Also one must consider the effect of tail states, which are known to be formed in the presence of high-density dopants and/or carriers. Since the role of these factors in quantum wires and boxes has not been completely clarified, the quantitative investigation of these effects will be an important subject of future study.

D. Intersubband Optical Devices

Electrons in the ground subband $E^{(1)}(k)$ of a QW structure (see Fig. 11(a) and (b)) can be excited to the second subband $E^{(2)}(k)$ by absorbing a mid-infrared photon with the energy $(E^{(2)} - E^{(1)})$. This intersubband transition results in an increase of electrical conductivity across the layers and this photoconductive response has been used to realize a new infrared detector, which covers the spectral range of $2 - 15$ μm[51 − 52].

In such devices, the response is proportional to the absorption constant, which has a delta-function like spectrum with its center at $h\nu = E^{(2)} - E^{(1)}$. The height of the absorption peak is inversely proportional to the peak width, ΔE, and the width $E^{(2)} - E^{(1)}$ is determined by such factors as the band nonparabolicity, the inter-well coupling, and the spatial fluctuations of the energy level spacing. Hence, the reduction of the size fluctuation of QWs is quite important to maximize the photo-response of the infrared detector, unless the spectral response is to be deliberately broadened. Note that the inhomogeneous broadening, ΔE, of the first excited level is four times as large as that of the ground level, when the barrier is high.

When electrons are injected electrically into an excited subband in a biased multi-QW structure, some fraction of electrons relax to a lower subband by emitting photons, although the majority of them relax by emitting phonons. It has recently been shown by Faist and Capasso[53,54] that this photon emission process can be used to realize a new infrared laser. In their devices (see Fig. 11(c) and (d)), electrons go through a cascade-type potential, in which injected electrons undergo the intersubband relaxation about 25 times before reaching the terminal electrode. Though the overall efficiency is low (less than a few percent), this "quantum cascade laser" is attractive for spectroscopic applications.

The peak gain of such lasers is inversely proportional to the spatial fluctuation of the level spacing $E^{(2)} - E^{(1)}$. Hence the reduction of inhomogeneous level broadening is essential to minimize the threshold pumping for the quantum cascade lasers. Potentially, the dramatic reduction of threshold current can be anticipated, if one suppresses this phonon emission rate. For example, it has been

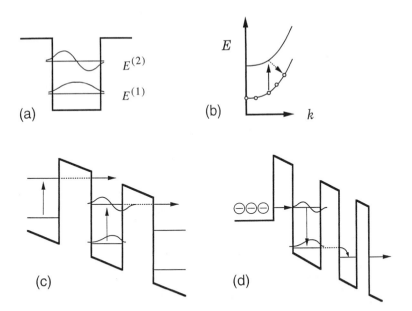

FIGURE 11. The intersubband transition in a QW (a) and its applications to infrared photodetectors[51,52] (b), and infrared quantum cascade lasers[53,54] (c) and (d).

pointed out that the optical phonon emission process may be suppressed in 10 nm − scale QB structures, when the energy spacing is far larger than the optical phonon energy of the constituent media. Even in such a case, it is important to minimize the size fluctuations of the box structures.

IV. FIELD EFFECT TRANSISTORS AND TRANSPORT OF 2D/1D ELECTRONS

Field effect transistors (FETs) are one of the most important devices in electronics. Modulation-doped heterostructure FETs, (MODFETs), known also as high electron mobility transistors (HEMTs), now play indispensable roles as a low-noise transistor in the area of microwave and satellite communication, while *Si* metal-oxide-semiconductor (MOS) FETs are the core device in LSI-based digital electronics[55–59]. Figure 12(a) and (b) show, respectively, the cross-sections of HEMTs fabricated on selectively-doped $n - AlGaAs/GaAs$ single- and double-heterojunction structures. Figure 12(c) illustrates their energy band diagrams.

In these FET devices, the channel length L_{ch} or the gate length L_g is constantly reduced, as it not only allows the low voltage operations of FETs but also

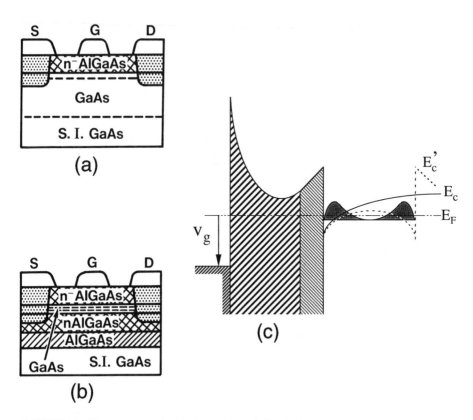

FIGURE 12. The cross-sectional views (a) and (b) of high-electron mobility transistors based on a selectively-doped (SD) single- and SD double-heterojunction[55–57]. Their energy band profiles are also shown (c).

determines the transit time ($\tau=L/v$) of carriers across the channel and, hence, the ultimate speed of FETs. There is also a trend to reduce the effective distance d^* between the gate and the channel, as it increases the specific gate capacitance $C_g(=\varepsilon^*/d^*)$ and allows an efficient modulation of carrier concentration N_s by gate voltage V_g. To reduce d^*, one needs to decrease the thickness d_{ch} of the channel as well as that d_{in} of the gate insulator. We discuss how FET characteristics are influenced by the reduction of d_{ch} and d_{in}.

Though the channel width W of practical FETs is usually longer than the channel length and far beyond 0.1 μm, there have been several attempts to squeeze the channel to the regime of the sub 0.1 μm, where the quantum size effect along the width direction plays an important role. In the last part of this section, we discuss unique properties of such quantum wires.

A. Reduction of Gate Length and Ultimate Speed of FETs

The transit time, τ, of carriers across the gate, which is equal to the gate length, L_g, divided by the drift velocity, v, of electrons, is an extremely important parameter as it determines the ultimate switching time, $\tau_s(=\tau)$, of FETs and the cut-off frequency, $f_t(=1/2\pi\tau)$, of FETs. Hence, the reduction of L_g is essential to fabricate high frequency FETs. Figure 13 shows the cut off frequency f_t measured on a series of *InGaAs*-channel HEMTs as a function of L_g[59]. It is clear that f_t increases almost in proportion to $1/L_g$, reaching 300 GHz for $L_g = 0.05\,\mu m$. As the slope of f_t versus $1/L_g$ characteristic is equal to $(v/2\pi)$, we find the drift velocity of electrons to be 1.5×10^7 cm/s for L_g greater than 0.2 μm. This value is close to the saturation velocity of electrons, which is reached when electrons are accelerated by high electric fields to a level where optical phonons are frequently emitted. We notice in Figure 13 that the increase in f_t tends to saturate when L_g is reduced below 0.15 μm. This saturation is caused mainly by the series resistance, R_s, in the ungated part of the channel, as the cut-off frequency is reduced from the intrinsic value by a factor $R_{ch}/(R_{ch}+R_s)$. Note that this correction becomes important when the channel resistance gets comparable with the series resistance R_s.

In case the transit time τ of electrons becomes comparable with or smaller than their energy relaxation time via optical phonon emission, the electron velocity is expected to overshoot beyond the usual saturation velocity, v_s, and may lead to a cut-off frequency higher than normally expected. This velocity

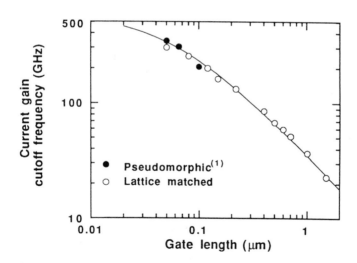

FIGURE 13. The cut-off frequency, f_t, of *InGaAs* HEMTs measured as a function of the gate length, L_g (after Enoki *et al.*[59]).

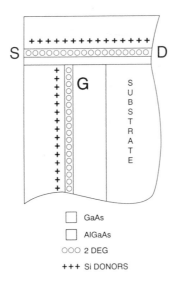

FIGURE 14. The schematic illustration of a novel FET with a 10 nm-long gate, prepared by the cleaved edge overgrowth (after Stormer *et al.*[60]).

overshoot effect, however, is usually overshadowed by the increasing role of series resistance and has not clearly been demonstrated yet.

Some attempts have been made to reduce the gate length L_g down to the scale of 10 nm. Figure 14 shows the schematic cross-section of one such FET structures, first developed by Stormer *et al.*[60]. Here the gate region consists of a 10 nm-thick conductive layer formed by growing selectively doped $n^+ - AlGaAs/GaAs$ layer in the middle of an undoped layer. The channel layer is formed in the second step by depositing a selectively doped $n - AlGaAs/GaAs$ two dimensional electron gas (2DEG) layer onto the cleaved edge plane of the pregrown layer, in which the 10 nm-thick gate layer is embedded. In such an FET, both the series resistance effect in the channel and the fringing field effect of the gate electrode play important roles. Nonetheless, electrons are efficiently accelerated even at a low drain voltage, resulting in a rather high transconductance.

B. Control of Thickness and Impurity Profiles of Gate Insulators in HEMT Structures

As discused earlier, the gate capacitance c_g per unit area and the transconductance g_m of FETs are nearly proportional to the inverse of the thickness of their gate insulators. Hence the *AlGaAs* layer in HEMTs and the SiO_2 layer in MOS-FETs are made quite thin as long as the gate leakage current remains low. In MOSFETs, the SiO_2 layer can be 5–10 nm in thickness because of the high barrier height. In contrast, the *AlGaAs* layer of HEMTs is usually made much thicker (30–60 nm) not only to keep the leakage current low, but also to manipulate the band bending $V(z)$ in the heterojunction and, hence, the threshold voltage V_{th}.

As shown in Figure 12, the potential $V(z)$ inside the *AlGaAs* layer decreases first linearly with the distance z from the heterojunction, for z inside of the undoped *AlGaAs* spacer layer (thickness W_{sp}). Once the position z is inside of the doped $n-AlGaAs$ layer (thickness W_a), the potential profile $V(z)$ becomes a parabolic function of z, whose curvature is determined by the donor concentration N_d. By simple algebra, one can show that the threshold voltage V_{th} at temperature $T=0$ can be expressed approximately as,

$$V_{th} = (\phi^M - \Delta E_c + E^{(1)}) - (eN_D/2\varepsilon) W_d^2 + F^{dep} (W_d + W_{sp})$$

Here, ϕ^M is the Schotty barrier height ($\approx 1 eV$) at the metal-*AlGaAs* contact, ΔE_c is the band discontinuity ($\approx 0.25 eV$) at the heterojunction, $E^{(1)}$ is the quantization energy ($\leq 0.05 eV$) in the space charge region, ε is the permittivity of *AlGaAs* ($\approx 11\varepsilon_0$), and F^{dep} is the electric field of the heterojunction ($\approx 10 kV$/cm), when the system is at the threshold condition. By using this equation, one can readily show that the threshold voltage is very sensitive to changes in W_d changes only slightly. For example, if W_d changes from 40 nm to 41 nm, V_{th} would decrease by 70 meV when $N_d = 10^{18} cm^{-3}$. This indicates the need for a very precise control of W_d, especially when a large number of HEMTs are incorporated in large scale integrated circuits. For the same reason, the spatial variation or sample-to-sample fluctuation of donor concentration N_d would result in a substantial change of threshold voltage and must be minimized.

C. Effects of Well Width and Roughness Scattering

As discussed earlier, to increase the transconductance of FETs, one must thin down, not only the gate insulator, but also the conduction channel. When electrons are confined in very thin quantum well (QW) channels, however, the role of potential fluctuations resulting from the interface roughness increases, which leads to the reduction of electron mobility[29,30]. Figure 15 shows the mobility of electrons in *GaAs/AlAl* QW structures measured at 4.2K as a function of well width L_z. This QW was prepared with GI at the top interface so that the interface roughness appears predominantly on the bottom (or *GaAs*-on-*AlAs*) interface. Note that the mobility drops almost in proportion to L_z^6, becoming as low as

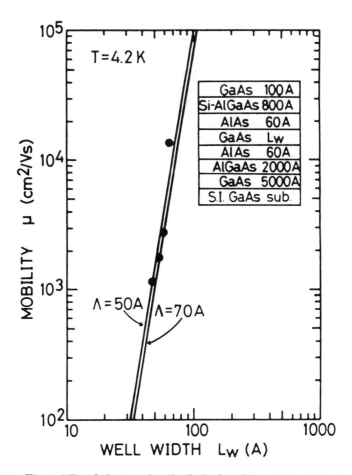

FIGURE 15. The mobility of electrons in selectively doped *GaAs/AlAs* quantum well structures measured at 4.2K as a function of the well width L_z. Two solid lines are
roughness-dominated mobilities calculated for the roughness of one mono-layer in height
and the lateral correlation length of 5 nm and 7 nm, respectively (Sakaki *et al.*)[29].

10^3 $cm^2/V \cdot s$ at $L_z = 5$ nm. This L_z^6 dependence is a characteristic feature of
roughness scattering, since the strength of scattering potential is inversely proportional to L_z^3, as indicated by Eq.(3). Note here that the roughness scattering plays
almost a negligible role, once the well width exceeds 20 nm.

Mobility data in Figure 15 provide quantitative information on the interface
roughness, when they are compared with the theory of roughness-dominated
mobilities μ_{ifr}, in detail. As discussed earlier, μ_{ifr} is proportional to the square of
the roughness-induced potential and can be expressed as

$$\mu_{ifr} = \frac{L_z^6}{\Delta^2} g(N_s, T, \Lambda),$$

where $g(\cdot)$ represents a factor that depends on the electron concentration (i.e. the Fermi energy), temperature, and the lateral correlation length Λ of interface roughness. Normally, the roughness $\Delta(r)$ is assumed to have an auto-correlation of the Gaussian form $<\Delta(r)\Delta(r')> = \Delta^2 \exp\cdot(r\cdot r')^2/\Lambda^2$. As mentioned earlier, data of Figure 15 can be well explained, if we assume the interface roughness to be one atomic layer in height Δ and its correlation length Λ to be 5–7 nm.

In principle, the correlation length Λ and Δ can be independently determined if we measure and analyze the mobility at different temperatures T or electron concentrations N_s[30]. For example, the N_s dependence of low-temperature

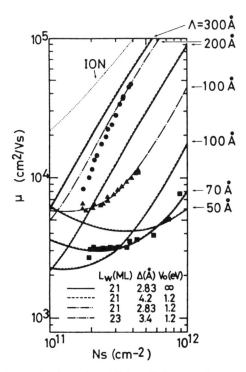

FIGURE 16. Roughness-dominated mobilities of electrons in *GaAs/AlAs* quantum wells as functions of the electron concentration Ns. Solid lines are theoretical values calculated for the roughness of one monolayer in height and the correlation length Λ of 5–30 nm (Noda *et al.*[30]).

mobility reflects the Fermi-energy (or the Fermi-wavelength) dependence of scattering rates, which is primarily determined by the lateral correlation length Λ of interface roughness. Solid lines in Figure 16 show examples of such analysis, where roughness-dominated mobilities in *GaAs/AlAs* QWs are plotted against N_s. Note that the N_s dependence of μ is independent of the amplitude of roughness but determined mainly by Λ. For example, when Λ is 20 nm or larger, the mobility increases rapidly with N_s, since the Fermi wavenumber gets much larger than the main Fourier component, $2\pi/\Lambda$, of the roughness. When Λ is in the range of 5–7 nm, the N_s dependence of mobility is more complicated. The minimum mobility as a function of N_s occurs where the Fermi wavelength is close to the correlation length of the roughness.

Electron mobilities of a *GaAs/AlAs* QW prepared by a standard MBE process with the GI at the top interface were measured at 4.2K and are shown by squares in Figure 16. The data fit well with a theoretical prediction for Λ of 7 nm. Mobilities of another sample grown by a modified MBE method to slightly smooth the bottom interface, are shown by triangles. This set of data are better explained by setting Λ to be 10 nm. These results indicate that one of the heterointerfaces has a roughness with a correlation length of about 7 nm.

Since the top interfaces of these QWs were prepared with the GI so as to enlarge the correlation length of the roughness, the substantial reduction of mobilities shown in Figures 15 and 16 should be ascribed mainly to the roughness at the bottom (or *GaAs*-on-*AlAs*) interface. This is in a marked contrast to the photoluminescence of quantum wells, where the 20 nm-scale roughness at the top interface plays a very important role in broadening the spectrum, unless such a roughness is removed by the GI process.

D. Reduction of Channel Width and Transport in Quantum Wires

While the trend to scale down the channel length of FETs is constantly driven to achieve higher switching speed and lower operation voltage, an effort to reduce the channel width W_{ch} has been made mainly for the enhancement of packing densities and for the reduction of currents and power consumption. This effort of width reduction, however, will not go too far below a 0.25 μm, since W_{ch} must be above some value so as to supply sufficient currents to drive capacitive loads and interconnects in actual circuits.

Apart from these trends in FET-based electronics there are other motivations for reducing the channel width down to the scale of 10–100 nm. One of them is the scientific curiosity to disclose features unique to a one-dimensional electron gas (1DEG), whose motion is quantized not only along the thickness (z) direction but the width (y) direction of the channel[7–10].

In order to achieve such lateral confinement, various methods have been developed. The most straightforward among them is an approach in which a wire pattern is defined on top of a selectively doped $n-AlGaAs/GaAs$ heterojunction by electron beam or other advanced lithographic method and this pattern is used to etch off the uncovered region so as to form a narrow mesa-type QWR structure. Since the carrier depletion layer of $0.1-0.2$ µm will be formed in the surface region of the wire, the electrically conductive part of the wire is always narrower than the metallurgical width of the wire by $0.2-0.4$ µm. Hence the confinement of electrons is achieved by an electrostatic field in the depletion layer, which tends to form a rather soft confinement potential.

Another method to define a QWR is a "split-gate" scheme, in which a pair of narrowly spaced metal gates are formed on top of a selectively doped $n-AlGaAs/GaAs$ heterojunction with a typical spacing of 0.3 µm[61]. By applying a negative gate voltage V_g with respect to the channel, electrons under the gate can all be depleted, resulting in the lateral confinement of a narrow channel region between the gate. The channel conductance G of such a split-gate device with a submicron gate length was studied as a function of V_g at low temperatures. It was found that G in this case is determined by the ballistic transport of one-dimensional electrons, which increases with V_g in a stepwise manner by $(2e^2/h)$, every time a higher lateral mode of electrons is newly introduced[61].

In most cases, the electron transport in QWRs is not ballistic but is dominated by scattering. For example, when electron waves are elastically scattered in multi-mode QWRs at low temperatures, electrons may undergo either constructive or destructive interference, since their phase information is maintained. This process modifies the transmission coefficient of electrons through the wire, and leads to either the increase or the decrease of wire conductance. If the phase of such electron waves is modulated by magnetic fields or by gate voltages, the conductance fluctuates with the typical amplitude of (e^2/h). This is known as universal conductance fluctuation (UCF) and has been studied as a characteristic phenomenon of multi-mode QWRs[62].

When the width of $GaAs$ QWRs is set at 30 nm or less, the spacing between the ground subband and the higher subbands gets sufficiently large so that nearly all the electrons are accomodated in the ground subband. In such single mode QWRs, the electronic motion is strictly one-dimensional (1D) and both the elastic and inelastic scattering processes will be modified[63]. For example, the scattering of 1D electrons by ionized impurities is predicted to be suppressed, especially when the concentration N_{1D} of 1D electrons is 10^6 cm^{-1} or higher. The suppression occurs because a 1D electron with the Fermi wavenumber k_f can be elastically scattered only to another state with $-k_f$ with a probability for scattering that is quite low since it is caused only by the high-frequency Fourier component $V_q(q=2k_f)$ of the impurity potential. Indeed, from the simple perturbational analysis, impurity-dominated mobility is found to exceed 10^7 cm^2/Vs, even when the impurity-to-wire distance d is 10 nm and N_{1D} is 10^6 cm^{-1}[63].

The process of optical phonon emission plays a key role in limiting the electron velocity at high electric fields. This process has been analyzed theoretically by Yamada *et al.*[64]. They have shown that the optical-phonon emission rate can be substantially reduced in a single-mode QWR when its Fermi energy is well above the optical phonon energy (36 meV). This reduction of electron-phonon interactions originates mainly from the peculiarity in the density of states and is predicted to enhance the saturation velocity by a factor of two or even more. If such increase of v_s is realized, and if these high-velocity electrons are used in the FET channel, the transit time of the electrons will correspondingly be reduced and the cut-off frequency will be substantially improved.

As discussed in Section IV.C, two-dimensional electrons in a thin quantum well are strongly scattered by interface roughness, even when the roughness is only one monolayer in height. Similar consequences are expected for one-dimensional electrons in narrow quantum wire structures. To quantify this effect, we have calculated the roughness dominated mobility of 1DEG confined in a 10 nm-wide edge quantum wire structure, where the amplitude Δ of interface roughness is one monolayer in height and and its correlation length Λ ranges from 5 to 20 nm. The result of the calculation is shown in Figure 17, in which mobilities are plotted as functions of the electron concentration N_{1D}. Note first that the

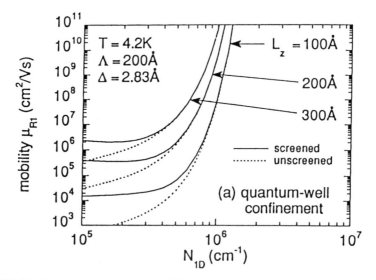

FIGURE 17. Roughness-dominated mobility of electrons in a quantum wire as functions of electron concentration. Solid lines are calculated with the screening taken into account, whereas broken lines are for the unscreened case. See the text for details (after Motohisa and Sakaki[65]).

result of Figure 17 is not sample-specific but contains a general message, since the roughness-dominated mobility is proportional to L_z^6/Δ^2. One can clearly see that mobility increases rapidly when N_{1D} exceeds 10^6 cm^{-1}, just as in the case of impurity scattering. This is due to the fact that electrons with a large Fermi momentum interact only with the roughness of high spatial frequency and can rarely be scattered, unless the correlation length Λ is very short. These results indicate that the optimum design of wire width is important to achieve high electron mobilities in quantum wires.

V. Resonant Tunneling in Double Barriers and Related Devices

In this section, we discuss the importance of controlling the thickness and interface roughness of constituent films in resonant tunneling (RT) diodes, super-lattices, and other devices[66–67].

A. Peak Current and Barrier Thickness in Resonant Tunneling Diodes

As discussed earlier, the peak current, I_p, of of an RT diode (RTD) depends critically on the barrier thickness L_b. This is because I_p is proportional to the width, ΔE, of the transmission peak, TT*(E), which is proportional to $\exp(-2\beta L_b)$, where β is the decay constant of the evanescent wave in the barrier, and is given by $\sqrt{2m(V-E)}/\hbar$[38,66,73,74]. In $GaAs$ RTDs with $AlAs$ barriers, where the effective mass, m^*, is 0.07 m and the effective barrier height, V, is about 1 eV, the peak current, I_p, and the transmission peak, ΔE, change almost by a factor of 2, if L_b is varied by only one atomic layer (=0.28 nm). Figure 18 (a) shows the resonant tunneling current measured in $GaAs/AlAs$ RTDs as a function of the barrier width, L_b[73]. For comparison, theoretical curves are also shown and found to be in reasonable agreement with the data. Another way to demonstrate this dependence is to measure the tunnel escape time, τ_t, of electrons from a QW via thin $AlAs$ barriers[74]. Figure 18(b) shows the result of such a measurement. Note that the escape time is inversely proportional to the width, ΔE, of the transmission peak and the exponential increase of τ_t proves the exponential decrease of the resonant tunneling current with L_b.

One should note that, to make use of RTDs in high-speed digital circuits and other practical applications, the device-to-device variation of peak currents must be minimized. Hence, it is essential to reduce the run-to-run variation of L_b, as well as the non-uniformity of L_b within a wafer, to a tolerable level.

At present, the variation $\Delta L_b/L_b$ of the average barrier thickness can be less than 3%. Hence, when L_b is below 3 nm, ΔL_b can be less than 0.1 nm. This means that the variation of I_p can be reduced to 25% or less even for RTDs with

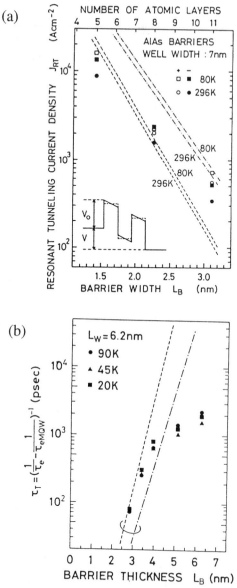

FIGURE 18. The peak and valley current in *GaAs/AlAs* resonant tunneling diodes as functions of barrier width (a)[73]. The tunnel-escape time, τ_t, of electrons from a *GaAs* quantum well through *AlAs* barriers with different barrier thicknesses (b) (Tsuchiya *et al.*[74]).

AlAs barriers. In reality, however, since the local thickness of barriers cannot be controlled continuously, but is quantized in the unit of one atomic layer (-0.28 nm), one must take into account effects of interface roughness on the RT process of electrons[75]. When the lateral size of such roughness is larger than the Fermi wavelength of electrons, electrons may tunnel preferentially through specific regions of a double barrier structure, where the barrier is locally thin.

B. Effects of Well Width Variations in RT Diodes and Superlattices

In RTDs, the well width plays an important role, since the voltage V_p at which the diode current is maximum is approximately equal to twice the ground quantum level $E^{(1)}/e$. Hence the fluctuation of the well width results in the variation in the quantum level $E^{(1)}$, and thereby affects the peak voltage V_p[68,75]. This shift of V_p, however, is not very critical in many cases, since the energy distribution of three-dimensional electrons in the emitter is normally broad to start with and the resonant condition is usually satisfied over a relatively broad range of voltage.

The well width variation plays a more critical role in coupled quantum wells and superlattices[66–69,71,72], where the resonant coupling of discrete quantum levels in neighboring wells determines their electronic structures. If this coupling is sufficiently strong, the mini-band width ΔE_{sl} of a superlattice or the level splitting ΔE_{sas} of a coupled well structure exceeds the inhomogeneous broadening ΔE^i of an otherwise discrete level. If the system inhomogeneity is too large to satisfy this condition, the coupling of neighboring wells takes place only locally and the simple notion of level splitting and the mini-band must be modified. For example, the electronic states in such an inhomogeneous superlattice tend to be localized and a negative conductance and other novel properties resulting from the negative-mass effect of the mini-band will be suppressed.

As pointed out in Section III.B, when a strong electric field F_z is applied to a superlattice having the period of d, the interwell coupling is suppressed and a series of Stark ladder states are formed with the energy spacing of $eF_z d$[41,42]. When an electron wavepacket is locally generated (for example, by a femtosecond laser pulse) as a superposition of a few ladder states, the center of the wavepacket oscillates back-and-forth in real space with the frequency of $eF_z d/h$. This "Bloch oscillation" phenomenon has been intensively studied[71,72]. Apparently, homogeneities of the system must be good enough to allow coherent oscillations of a wavepacket at least over a few superlattice periods. In reality, the oscillation damps even in high quality SLs due to the scattering by phonons and other carriers.

C. Interface Roughness and Transport in
Resonant Tunneling Diodes and Superlattices

As mentioned in Sections V.A and V.B, the deviations of average barrier thickness L_b and/or well width L_z in RTDs and SLs from the designed values lead to significant variations in current-voltage characteristics, mainly because of the thickness dependences of the tunneling probability and the quantum level energy. One must be aware also that, even when the average thicknesses of barriers and wells are precisely controlled, the interface roughness, if present, may affect microscopic processes of electron transport and characteristics of these devices.

In the idealized model of *RT* diodes, we assume both momentum and the energy conservation in calculating both the peak current and valley current. In reality, however, the presence of optical phonons, interface roughness, and other perturbations are known to open new channels, which would be otherwise forbidden. It has been theoretically shown that the main current of a RTD is not much modified by these factors but the valley current is substantially enhanced at room temperature mainly by phonon-assisted inelastic tunneling processes[66,76]. At low temperatures, the valley current may be enhanced by those tunneling processes, in which the momentum change between the initial state and the final state is provided by the interface roughness or by the phonon emission process.

The transport of electrons in layered superlattices is determined by the miniband relation $E_z(k_z)$. It means that, if the wavenumber k_z of a test electron is frequently changed by scattering (or deflection) processes, the uniqueness of k_z dependent transport, such as a negative conductance, will be averaged over various k_z and smeared out. Hence, it is important to minimize the roughness of heterointerfaces in layered superlattice structures or by the phonon emission process.

An alternative scheme to reduce the detrimental effect of electron scattering is to use a linear array of 10 nm-scale coupled quantum boxes. In such structures, the motion of electrons is allowed only along its length direction and interacts strongly with the periodic potential. In addition, the effect of optical phonon scattering can be substantially reduced by properly setting the structures' parameters. For example, the intra-miniband scattering can be eliminated when the mini-band width is set less than the optical phonon energy $\hbar\omega_{opt}$ (≈ 35 meV in *GaAs*), while the inter-miniband transitions can be suppressed by setting the mini-gap layer to the $\hbar\omega_{opt}$[77].

VI. SUMMARY AND PROSPECTS

A. Control of Film Thickness and Interface Smoothness

In this chapter, we have described the current state of epitaxial technology to show that both the film thickness and the roughness at the heterointerface in layered structures can be controlled with the accuracy of about one molecular layer. For further improvements, one must develop new techniques in which the roughness of the growth front is to be automatically removed and the thickness of a deposited layer is to be controlled precisely in the unit of a monolayer.

For the purpose of such precise thickness control, the atomic-layer epitaxy (ALE) technique has been developed[78,79]. ALE uses a chemically selective absorption process to deposit precisely one monolayer of source material on a surface which is then converted into a semiconductor film of one monolayer thickness. Though attractive, this ALE process is usually done at relatively low temperatures with quite a slow growth rate and is prone to impurity incorporation. In addition, ALE does not usually heal the roughness of the original surface, either.

To prepare completely smooth growth fronts and flat heterointerfaces, one must develop a newer epitaxy technique, by which materials are selectively deposited nowhere except along these edges of atomic steps. Though a few growth processes allow the preferential incorporation of materials along the edges, the site-selectivity of deposition processes is not quite complete. Hence, a practical method to get smooth interfaces is either to use the GI process or to prepare pseudo-smooth interfaces, both of which were discussed earlier.

B. Growth of 10 nm-Scale Quantum Wires and Dots

Although the majority of work on QWs and QDs was done on lithographically defined 100 nm-scale samples at low temperatures[7–10], further work is under way to prepare 10 nm-scale QWRs and QDs and to explore their new properties at higher temperatures and to disclose their potential for device applications. These structures are indispensable for such purposes, since their level spacings and charging energies must be comparable to or larger than the thermal energy, kT, and the Fermi energy, E_f, in order to maintain the quantum limit condition and/or single-electron effects, even at 300K. Although the fabrication of such structures was difficult, it has finally become feasible, thanks to new developments in epitaxial methods. In the following, we review the recent progress of these approaches.

1. Edge Quantum Wires

As described earlier, if one exposes the edge surface of a *GaAs/AlGaAs* QW structure and subsequently overgrows onto it an *n − AlGaAs* layer, electrons will be confined two-dimensionally in a modulation-doped QWR channel of Figure 2 (c)[80-85]. The cross-sectional size of such edge QWRs can be 10 nm, as it is determined by the thickness, L_z, of the *GaAs* QW in one direction and by the thickness, L_y, of a modulation-doped channel in the other direction. Transport properties of electrons in such edge QWRs have been studied for the well width, L_z, of 14–100 nm to show the formation of unique subband structures of one-dimensional (1D) electrons[81–85]. Very recently, the quantization of channel conductance has been observed in narrow and long samples, indicating that the interface roughness plays only a small role in edge QWR structures[83].

Another type of edge QWRs can be prepared, if one overgrows a second *GaAs* QW, B, onto the edge surface of a pregrown QW layer, A[86–92]. Electrons and holes can be confined along the T-shaped intersection zone of two QWs, A and B, as illustrated in Figure 3(d). By using this T-shaped edge QWR scheme, the formation of one-dimensional (1D) excitons has been demonstrated. For example, the three photoluminescence spectra (PL), shown in Figure 19(a), are taken from 5 nm-scale *AlGaAs/GaAs* T-shaped wires formed by depositing QW2 of slightly different thicknesses. Figure 19(b) shows PL spectra from three *AlAs/GaAs* T-shaped wires. Here the average thickness of QW2 is varied between 5.3 nm and 5.5 nm . Note that when the lateral confinement is tightened by intersecting 5 nm-scale QWs with *AlAs* barriers, a strong lateral confinement energy and a substantial enhancement of the exciton binding energy have been observed[92]. By placing T-shaped QWRs inside a waveguide cavity, lasing with unique properties has also been achieved[89].

The edge QWR approach is one of the most promising ways to fabricate 10 nm − scale QWRs with geometrical precision comparable to that for QWs. This approach, however, has a drawback, especially when edge QWRs are formed on a narrow cleavage plane, since subsequent processing on such a small area is quite tedious and inefficient. One way to relax this difficulty is to dispense with the cleavage process by making use of multi-QW structures grown on patterned (100) substrates. This selective growth results in a well-defined facet, consisting of a (100) top plane and two (111) side planes. As the edge surfaces of the QWs are automatically exposed at the (111) plane , one can overgrow a QW or *n − AlGaAs* layer onto this (111) plane and form edge QWRs without resorting to the cleavage process. The effectiveness of this approach has already been demonstrated[84,85].

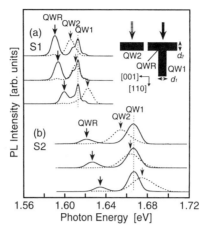

FIGURE 19. Photoluminescent spectra of two sets S1 and S2 of 5 nm-scale T-shaped quantum wires formed at the intersection of QW1 and QW2. See the text for details (after Someya *et al.*[92]).

2. *Groove and Ridge Quantum Wires*

The epitaxial growth of *GaAs* and *AlGaAs* on appropriately-patterned (100) *GaAs* substrates leads to the formation of very sharp groove or ridge structures, consisting of two (111) side facet planes and a very narrow (100) plane with the width as small as 20–10 nm[93–100]. Figure 20(a) shows an example of a *GaAs* ridge, prepared by the MBE growth on a mesa-shaped substrate. This results from the inter-facet material diffusion and the selectivity of growth rates on crystallographic planes. When a *GaAs* QW is deposited on such a sharp groove or ridge having the width of 20 nm or less, 2D electrons in the QW will be laterally confined along the top of the ridge or the bottom of the groove, since the QW gets thicker there than elsewhere, as shown schematically in Figure 3(a) and (b).

Cross-sectional shapes of such groove/ridge QWRs have been studied and found to depend on such growth parameters as the substrate temperature and the *As* pressure. Figures 20(b) and (c), for example, show that the MBE growth tends to heal geometrical irregularities and lead to a very sharp and uniform ridge on which a 10 nm-scale wire is formed. By adopting a medium-to-low substrate temperature ($\approx 520 - 580$C) and a high *As* pressure, very high-quality QWRs have been achieved, in which sharp luminescent spectra indicative of the 1D nature of

(a)

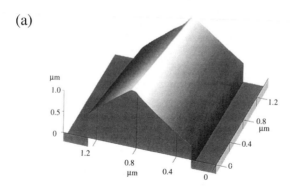

(b)

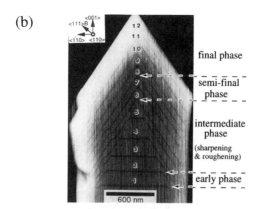

(c)

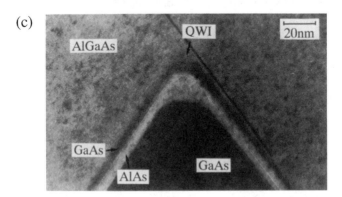

FIGURE 20. (a) An AFM image of an MBE grown *GaAs* ridge. (b)An SEM cross-sectional image of a *GaAs* ridge with thin *AlAs* marker layers. (c)A TEM image of a ridge quantum wire structure after Koshiba *et al.*[97]. (See color plate.)

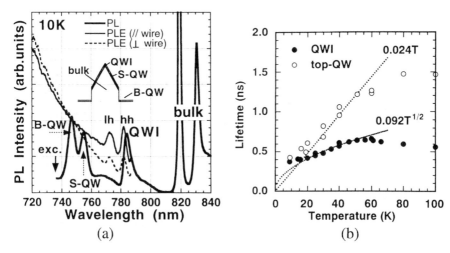

FIGURE 21. (a) Photoluminescence spectra and (b) the temperature dependence of a radiative lifetime of a ridge wire structures.

excitons have been observed, as shown in Figure 21. The square-root temperature dependence of lifetime shown in Figure 21(b) indicates that excitons have 1D freedom of motion. Kapon *et al.* vertically stacked groove QWRs in a waveguide structure, by MOVCD and achieved laser operation with a very low threshold current[94]. Similarly, selective growth on patterned substrates was successfully used to make QD structures[96,100].

3. Self-Assembled Dots and Related Structures

When an *InAs* layer of 1.7–2 monolayer in thickness is deposited on a *GaAs* substrate that has about 7% mismatch with the *InAs*, the layer grows in the Stranski-Krastanov mode and pyramidal islands of 15–20 nm in width and $3-7$ nm in height are automatically formed with the typical area concentration of about $5 \times 10^{10} \, cm^{-2}$ as shown in Figure 22[101–112]. Such islands, when embedded inside of a *GaAs* or similar matrix, can confine electrons and holes tightly, and form quantum dot (QD) states. Similarly, the formation of *InP* dots on a lattice mismatched *InGaP* substrate and other material combinations is found to be equally feasible. The zero-dimensional feature of quantum states in *InAs* pyramids has been demonstrated by various spectroscopic studies, such as capacitance-voltage analysis[104], PL and PL excitation studies[101–103], and far-infrared spectroscopy[104].

The width of the PL spectra from *InAs* QDs is typically 30 meV and is usually dominated by the inhomogeneous broadening associated with the size fluctuation

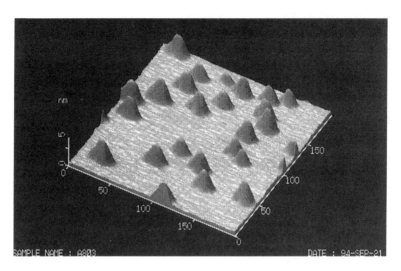

FIGURE 22. AFM image of *InAs* dots on *GaAs*[105]. (See color plate.)

of QDs as shown in Figure 23 (S1). This broadening lowers the peak value of the density of states and, hence, needs to be reduced to about 10 meV or below in order to make low-threshold QD lasers, as mentioned earlier. Recently, the broadening has been reduced significantly by adopting a novel coupled QD struc-ture in which an *InAs* QD layer and a thin *GaAs* spacer are alternately stacked in such a way that a second QD sits just on top of the previous one through the assis-tance of the strain effect[107,108]. Indeed, lasers using the multilayer *InAs* QDs have been fabricated and the reduction in the temperature dependence of the threshold current has been reported.

These dots may be used for novel electronic devices as well. For example, the use of *InAs* QDs as single-electron trap sites in novel heterostructure FETs with programmable threshold voltage has been demonstrated[110]. Figure 24(a) shows the device structure and Figure 24(b) shows the electron concentration N_s as functions of the gate voltage V_g. Note that the threshold voltage of this FET shifts upward, once each dot captures one electron.

While *InAs* dots are formed usually at random, site selective growth can be achieved by patterning substrates. For example, dots are shown to be preferen-tially formed on such sites as the bottom of a narrow groove, and the concave edge of step patterns or bunched steps[111,112]. In addition, the strained region of overgrown *GaAs* just above an underlying dot is found to work as a nucleation site for the second *InAs* layer[107,108]. Novel properties may be anticipated in these site-controlled dot structures, especially in connection with inter-dot

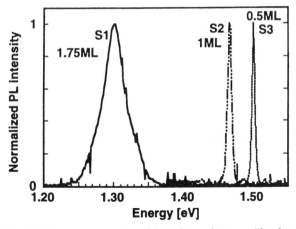

FIGURE 23. Photoluminescence spectra of *InAs* dot and *InAs* wetting layer[106].

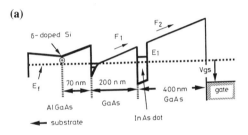

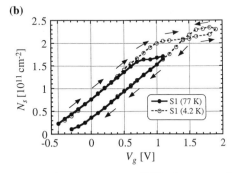

FIGURE 24. The schematic of a quantum trap FET, where *InAs* dots are embedded near the channel (a) and its characteristics (b). By raising the gate voltage above a critical value, each *InAs* dot captures one electron to reduce the 2D electron in the channel (after Yusa and Sakaki[110]).

tunneling processes.

4. Other Approaches and Their Prospects

There are several other methods for the fabrication of 10 nm-scale wires and dots. One well-known approach is the growth of a tilted superlattice (TSL) proposed by Petroff *et al.*, in which a half-monolayer (ML) of *GaAs* and another half ML of *AlAs* are alternately deposited onto a vicinal substrate which is misorientated from the (100) orientation by one or two degrees[113–119]. Since quasi-periodic step structures appear on such a vicinal plane, *GaAs* (or *AlAs*) deposited on the substrate may migrate and get incorporated preferentially at the edge of these steps. If this process is repeated alternately, the in-plane composition will be modulated leading, in principle, to a laterally modulated *GaAs/AlAs* TSL structure, which is schematically shown in Figure 3(e). In practice, however, TSLs actually grown have shown that the composition modulation in nominally *GaAs/AlAs* TSL is weak and gradual, modulating the *Al* content x only between 0.4 and 0.6. In addition, the straightness of steps and the periodicity of steps could be easily lost, unless the polishing of initial substrates and the growth conditions are carefully optimized.

The growth of TSL in other material combinations has been tried also. Better modulation of composition has been achieved in both *GaSb/AlSb* TSLs and *InAs/GaAs* TSLs[119]. The different surface chemistry and strains in these systems appear to play positive roles. In particular, the 7% lattice mismatch between *InAs* and *GaAs* seems to reduce the mixing in *InAs/GaAs* TSL when grown on *InP* substrates.

Under some growth conditions, mono-layer steps on vicinal planes tend to bunch and develop into macroscopic steps, which are 1–50 nm in height. For example, such bunched steps are found to be formed on such *GaAs* surfaces as (100) vicinal plane prepared by MOCVD[120], vicinal (110) planes formed by MO-MBE[121], and MBE-grown vicinal (111) B plane[122]. Quantum wire structures are shown to be formed by depositing thin *GaAs/AlGaAs* quantum wells on some of these corrugated surfaces, as *GaAs* tends to be incorporated preferentially in the vicinity of macro-steps. Similarly, laterally-modulated 2D electron systems can be formed by growing an $n-AlGaAs$ layer onto corrugated surfaces of *GaAs*[122]. The size and periodicity of these structures, however, are not yet uniform enough to satisfy stringent conditions required for practical device applications.

Several unique attempts have also been made to control the local deposition or reaction of materials by using focused beams of ions, electrons, or photons for nano-fabrication. In a case where focused ion beams are used, 10 nm structures are hard to make, since a very narrow beam with the diameter less than 100 nm is available only for high energy ions and these beams tend to spread inside of the

20nm→| |←

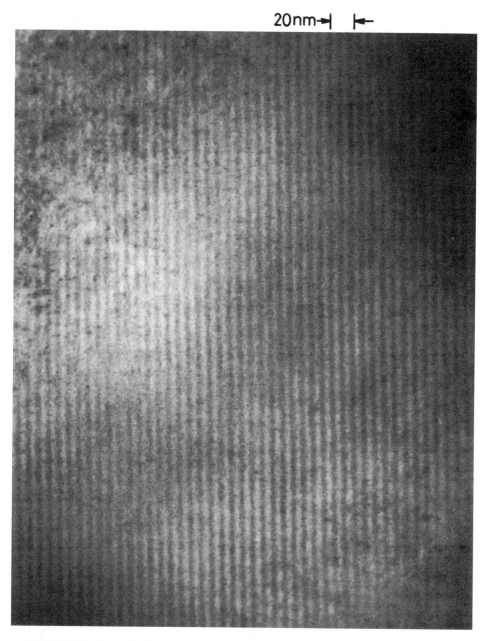

FIGURE 25. Transmission electron micrograph of an MOCVD-grown *AlAs/GaAs* tilted superlattice (after Fukui *et al.*[117]).

substrate and also create point defects. In the case of electrons beams, various chemical reactions can be locally initiated on substrate surfaces with the spatial resolution of about 10 nm. For example, the dissociation of hydro-carbon molecules or an oxide film can be induced by the electron beam to form 10 nm-scale stripes of deposited carbon or oxide holes, respectively. These patterns can then be transferred onto semiconductor crystals by the subsequent etching process.

Very fine electron beams with the width less than 10 nm can be supplied not only by the conventional electron optics but also by such new techniques as double-beam diffraction optics (i.e. electron holography arrangement) and a scanning tunneling tip. For all of these approaches, however, the temperature of a substrate is normally maintained at room temperature, at which the surface often degrades by contamination. The use of higher substrate temperatures reduces the contamination problem but severely degrades the spatial resolution.

In the case of photon beams, the minimum feature size of optical patterns is usually subject to the diffraction limit and is often well above 100 nm. McClelland and Prentiss (see Chapter 10) have developed a unique way to overcome this limit by steering an otherwise uniform beam of atoms with a spatially modulated standing wave of light. In this scheme, the force of light deflects atoms and makes them land onto a very narrow region of the substrate. With this approach, Timp *et al* has succeeded in producing 10–50 nm wide stripe patterns of *Na*. When this method is successfully extended to semiconductor materials this can be an attractive scheme of nanofabrication, because of its compatibility with MBE technique and its high throughput.

C. Concluding Remarks

Thanks to remarkable progress both in heterostructure epitaxy and *Si* MOS-FET technology, the formation and the subsequent use of 10 nm-scale semiconductor layered structures for various fields of electronics and condensed matter physics has now been well established. Qualitatively similar approaches have been successfully used in the fields of magnetic and other materials systems, proving the effectiveness of nanometer film technology.

Various lithographic techniques and selective epitaxial techniques, originally developed for micrometer-scale lateral structures, have also made impressive advances in recent years, and now allow the reproducible fabrication of 50–100 nm-scale structures. Some selected new techniques stated in Section VI.B are shown to permit the formation of even 10 nm-scale lateral structures. Although the role of impurities, surface states and defects becomes increasingly important in these structures, some of the fabrication techniques, especially those based on epitaxial growth with short or no GIs, are nearly free from these problems, and capable of yielding high-quality 10 nm-scale quantum wire and box (dot) structures.

Further developments in materials manipulation will permit us to explore more extensively into untrodden areas of lower dimensional systems and bring forth deeper understandings on these systems. Visionary attempts to design, create and explore newer structures will continually expand the frontier of our research as well as the forefront of modern technology.

REFERENCES

[1]   *Semiconductors and Semimetals,* Dingle, R., ed., **24**, Boston: Academic Press, 1987.

[2]   Weisbush, C., and Vinter, B., *Quantum Semiconductor Heterostructures,* Boston: Academic Press, 1991.

[3]   *Physics of nanostructures,* Davies, J.H., and Long, A.R., eds., Bristol: Institute of Physics Publ., 1992.

[4]   Cardona, M., and Pinczuk, A., *Solid State Commun.* **92**, No. 1-2 (1994).

[5]   See a series of related papers in *A Special Issue of IEEE J. Quantum Electronics,* **QE-22**, Chemla, D., and Pinczuk, A., eds., 1986.

[6]   For recent progress of MBE, see *Molecular beam epitaxy: Key Paper in Physics,* Cho, A.Y., ed., American Inst. of Phys., 1994. For OM VPE, see, for example, *J. Cryst. Growth* **124**, No. 1-4 Stringfellow, G.B., Coleman, J.J., eds., (1992).

[7]   *Nanostructured Systems, Semiconductors and Semimetals* **35**, Reed, M., ed., New York: Academic Press, 1992. On quantum wires, see the article by Timp, G., (p. 113). On quantum point contacts, see the article by Van Houten, H., Beenakker, C.W.J., and Van Wees, B.J., (p. 9).

[8]   Sakaki, H., *Surface Sci.* **267**, 623 (1992).

[9]   Meirav, U., and Foxman, E.B., *Semicond. Sci. Technol.* **10**, 255 (1995).

[10] *Single Charge Tunneling,* Grabert, H., and Devorter, M.H., eds., New York: Plenum, 1992; and Geerlings, L.J., Harmans, C.P.J.M., and Kouwenhoven, L.P., *Physica B* **189**, (1993).

[11] For a recent review, see: *NATO-ASI series E,* Eberl, K., Petroff, P.M., and Demester, P., eds., **298**, Kluwer, 1995.

[12] Sakaki, H., in *Semiconductor Interfaces at the Subnanometer Scale,* Salemink, H.W.M., and Pathley, M.D., eds., Kluwer, 1993.

[13] Feenstra, R.M., *et al.*, in *Semiconductor Interfaces at the Subnanometer Scale,* Salemink, H.W.M., and Pathley, M.D., eds., Kluwer, 1993, p. 127.

[14]   Johnson, M.B., *et al.*, ibid, 207, 1993 and references therein.

[15]   Salemink, H.W.M., *et al.*, ibid, 151, 1993 and references therein.

[16]   Orr, B.G., *et al.*, in *Nanostructures and Quantum Effects,* Sakaki, H., ed., Springer, 1994. Also, Johnson, M.D., *et al.*, *Appl. Phys. Lett.* **64**, 484 (1994).

[17]   Ourmazd, A., in *Semiconductor Interfaces at the Subnanometer Scale,* Salemink, H.W.M., and Pashley, M.D., eds., Kluwer, 1993, p.139; and also *Phys. Rev. Lett.* **62**, 933 (1989).

[18]   Ichinose, H., *et al.*, *J. Electron Microscopy* **36**, 82 (1987).

[19]   Ikarashi, N., *et al.*, *Jpn. J. Appl. Phys.* **32**, Part-1, 2824, (1993); and *Appl. Phys. Lett.* **60**, 1360 (1992).

[20]   Petroff, P.M., *J. Cryst. Growth* **44**, 5 (1978).

[21]   Weisbuch, C., *Solid State Commun.* **38**, 709 (1981).

[22]   Goldstein, L., *et al.*, *Jpn. J. Appl. Phys.* **22**, 1489 (1983).

[23]   Sakaki, H., Tanaka, M., and Yoshino, J., *Jpn. J. Appl. Phys.* **24**, L417 (1983).

[24]   Tanaka, M., Sakaki, H., and Yoshino, J., *Jpn. J. Appl. Phys.* **L155**, (1986). Tanaka, M., and Sakaki, H., *J. Cryst. Growth* **81**, 153 (1987).

[25]   Tanaka, M., and Sakaki, H., *Superlattices and Microstructures* **4**, 1988, p. 237; and *J. Appl. Phys.* **64**, 4503 (1988).

[26]   Fukunaga, T., Kobayashi, K.L., and Nakashima, H., *Jpn. J. Appl. Phys.* **24**, L510 (1985). Binberg, D., *et al.*, *J. Vac Sci. and Tech.* **B5**, 1141 (1987).

[27]   Tu, C.W., *et al.*, *J. Cryst. Growth* **81**, 159 (1987).

[28]   Warwick, C.A., and Kopf, R.F., *Appl. Phys. Lett.* **60**, 386 (1992); and Gammons, D., Shamabruiks, B.C., and Katzer, D.S., *Phys. Rev. Lett.* **67**, 1547 (1991).

[29]   Sakaki, H., *et al.*, *Appl. Phys. Lett.* **51**, 1934 (1987).

[30]   Noda, T., Tanaka, M., and Sakaki, H., *et al.*, *Appl. Phys. Lett.* **56**, 51 1990.

[31]   Tokura, Y., *et al.*, *Phys. Rev. B, Condens. Matter,* **46**, No. 23, 15558–61 (1992).

[32]   Neave, J.H., *et al.*, *Appl. Phys. A.* **31**, 1 (1983).

[33]   See, for example, Vvedensky, D.D., in *Semiconductor Interfaces at the subnanometer Scale*, Salemink, H.W.M., and Pashley, M.D., eds., Kluwer, 1983, p. 217.

[34]   Shitara, T., *et al.*, *Phys. Rev. B* **46**, 6815 (1992).

[35]   Saito, J., *et al.*, *J. Cryst. Growth* **81**, 188 (1987).

[36] Hess, H.F., *et al.*, *Science* **264**, 1720, (1994); and also Grober, R.D., *et al.*, *Appl. Phys. Lett.* **64**, 1421 (1994).

[37] Brunner, K., *et al.*, *Appl. Phys. Lett.* **64**, 3300, (1947); and also Enenner, A., *et al.*, *Phys. Rev. Lett.* **72**, 3382 (1994).

[38] Miller, D.A.B., Weiner, J.S., and Chemla, D.S., *IEEE J. Quantum Electronics* **QE-22**, 181h (1986).

[39] Chelma, D.S., in *Physics and Applications of Quantum Wells and Superlattices,* Mendez, E.E., and Von Klitzing, K., eds., NATO ASI series B Physics 170, Plenum Press, 1987, p. 423.

[40] Chelma, D.S., Miller, A.B., and Smith, P.W., *Semiconductor Semimetals,* Willardson, R.K., and Beer, A.C., eds., Academic Press, 1987, vol. 24, p. 279.

[41] Mendez, E.E., Agullo-Rueda, F., and Hong, J.M., *Appl. Phys. Lett.* **56**, 2545 (1990).

[42] Voisin, P., *et al.*, *Phys. Rev. Lett.* **61**, 1639 (1988).

Miller, D.A.B., *et al.*, *Appl. Phys. Lett.* (1988).

[44] Sakaki, H., Kato, K., and Yoshimura, H., *Appl. Phys. Lett.* **57**, 2800 (1990).

[45] Zucker, J., *et al.*, *Electronics Lett.* **28**, 2206 (1992).

[46] Tsang, W.T., in *Semiconductor Semimetals,* Willardson, R.K., and Beer, A.C., eds., Academic Press, vol. 24, 1987, p. 397; see also Arakawa, Y., and Yariv, A., *IEEE J. Quantum Electronics* **QF-22**, 1987 (1986).

[47] Ranyai, L., *et al.*, *Phys. Rev. B* **36**, 6099 (1987).

[48] Arakawa, Y., and Sakaki, H., *Appl. Phys. Lett.* **40**, 893 (1982).

[49] Asada, M., Miyamoto, Y., and Suematsu, Y., *IEEE J. Quantum Electronics* **22**, 1915 (1986).

[50] See, for example: Slusher, R.E., and Weisbuch, C., *Solid State Commun.* **92**, 149 (1994).

[51] Esaki, L., and Sakaki, H., *IBM Tech. Disclos. Bulletin* **20**, 2456 (1977).

[52] Levine, B., *J. Appl. Phys.* **74**, R-1 (1993).

[53] Faist, J., *et al.*, *Science* **264**, 553 (1994).

[54] Kazarinov, R.F., and Suris, R.A., *Sov. Phys. Semicond.* **5**, 207 (1971).

[55] Sakaki, H., *IEEE J. Quantum Electronics* **QE-22**, 1845 (1986).

[56] For review on HEMTs, see the following authors in *Semiconductor and Semimetals,* Willardson, R.K., and Beer, A.C., eds., Academic Press, vol. 24, 1987: Morkoc, H., and Unlu, H., (p. 105); Linh, N.T., (p. 203); Abe, M., *et al.*, p. 249.

[57]   Mimura, T. *et al.*, *Jpn. J. Appl. Phys.* **19**, L225 (1980).

[58]   Dingle, R., *et al.*, *Appl. Phys. Lett.* **33**, 665 (1978).

[59]   Enoki, J., *et al.*, *Jpn. J. Appl. Phys.* **33**, 798 (1994).

[60]   Stormer, H.L., *et al.*, *Appl. Phys. Lett.* **59**, 1111 (1991).

[61]   van Wees, B.J., *et al.*, *Phys. Rev. Lett.* **60**, 842 (1988).

[62]   Lee, P.A., and Stone, A.D., *Phys. Rev. Lett.* **55**, 1622 (1985).

[63]   Sakaki, H., *Jpn. J. Appl. Phys.* **19**, L735 (1980).

[64]   Yamada, T., and Sone, J., *Phys. Rev. B* **40**, 6265 (1989).

[65]   Motohisa, J., and Sakaki, H., *Appl. Phys. Lett.* **60**, 1315 (1992).

[66]   *Resonant Tunneling in Semiconductors: Physics and Applications* Chang, L.L., and Mendez, E., eds., NATO ASI series, Kluwer, (1991).

[67]   Capasso, F., Mohommed, K., and Cho, Y., *IEEE J. Quantum Electronics* **QE-22**, 1953 (1986).

[68]   Sakaki, H., Matsusue, T., and Tsuchiya, M., *IEEE J., Quantum Electronics* **QE-25**, 2498 (1989).

[69]   Esaki, L., and Tsu, R., *IBM J. Res. Develop.* **14**, 65 (1970).

[70]   Chang, L.L., Esaki, L., and Tsu, R., *Appl. Phys. Lett.* **24**, 593 (1974).

[71]   Wascke, C., *et al.*, *Phys. Rev. Lett.* **70**, 3319 (1993).

[72]   Roskos, H.G., *et al.*, *Phys. Rev. Lett.* **68**, 2216 (1992).

[73]   Tsuchiya, M., and Sakaki, H., *Jpn. J. Appl. Phys.* **25**, L185 (1986).

[74]   Tsuchiya, M., Matsusue, T., and Sakaki, H., *Phys. Rev. Lett.* **59**, 2356 (1987).

[75]   Tsuchiya, M., and Sakaki, H., *Appl. Phys. Lett.* **49**, 88 (1986).

[76]   Tsuchiya, M., and Sakaki, H., *Jpn. J. Appl. Phys.* **30**, 1164 (1991).

[77]   Sakaki, H., and Noguchi, H., *Jpn. J. Appl. Phys.* **28**, L314 (1989); and Noguchi, H., Leberton, J.P., and Sakaki, H., *Phys. Rev. B* **47**, 15593 (1993).

[78]   Nishizawa, J., Abe, H., and Kurabayashi, T., *J. Electrochem Soc.* **132**, 1194 (1985).

[79]   Ushi, A., and Sunakawa, H., *Jpn. J. Appl. Phys.* **25**, L212 (1981); and Ozeki, M., *et al.*, *J. Vac. Sci. Tech. B* **5**, 1184 (1986).

[80]   Sakaki, H., *Jpn. J. Appl. Phys.* **19**, L735 (1980); and *J. Vac. Sci. and Technol. B*, (1981).

[81] Stormer, H.L., *et al.*, *Appl. Phys. Lett.* **58**, 726 (1991).

[82] Ohno, Y., *et al.*, *Phys. Rev. B* **52**, R11619 (1995).

[83] Yacoby, A., Stormer, H.L., *et al.*, *Phys. Rev. Lett.* **77**, 4612 (1996).

[84] Fukui, T., and Ando, S., *Electronics Lett.* **25**, 410 (1989).

[85] Nakumaura, Y., *et al.*, *Appl. Phys. Lett.* **64**, 2552 (1994).

[86] Chang, Y.C., Chang, L.L., and Esaki, L., *Appl. Phys. Lett.* **47**, 1324 (1985).

[87] Pfeiffer, L., *et al.*, *J. Cryst. Growth* **127**, 849 (1993).

[88] Goni, A.R., *et al.*, *Appl. Phys. Lett.* **61**, 1956 (1992).

[89] Wegscheider, W., *et al.*, *Phys. Rev. Lett.* **71**, 4071 (1993).

[90] Pfeiffer, L.N., *et al.*, *Low-Dimensional Structures Prepared by Epitaxial Growth or Regrowth on Patterned Substrates,* Eberl, K. *et al.*, eds., NATO, ASI series, E. 298 Kluwer, 93, 1995.

[91] Someya, T., Akiyama, H., and Sakaki, H., *Phys. Rev. Lett.* **74**, 3664 (1995); and *J. Appl. Phys.* **79**, 2522 (1996).

[92] Someya, T., Akiyama, H., and Sakaki, H., *Phys. Rev. Lett.* **76**, 2965 (1996); and also *Appl. Phys. Lett.* **66**, 3672 (1995).

[93] Kapon, E., Hwang, D.M., and Bhat, R.B., *Phys. Rev. Lett.* **63**, 430 (1989).

[94] Kapon, E., *et al.*, *Surf. Sci.* **267**, 593 (1992).

[95] Tsukamoto, S., *et al.*, *J. Appl. Phys.* **71**, 533 (1992).

[96] Arakawa, Y., *Solid State Electronics* **37**, 523 (1994).

[97] Koshiba, S., *et al.*, *Appl. Phys. Lett.* **64**, 363 (1994).

[98] Koshiba, S., *et al.*, *J. Cryst. Growth* **150**, 332–326 (1995).

[99] Akiyama, H., *et al.*, *Phys. Rev. Lett.* **72**, 924 (1994).

[100] Fukui, T., *et al.*, *Appl. Phys. Lett.* **58**, 2018 (1991).

[101] Goldstein, L., *Appl. Phys. Lett.* **47**, 1099 (1985).

[102] Leonard, D., *et al.*, *Appl. Phys. Lett.* **63**, 3203 (1993); and Ahopelto, J., *et al.*, *Jpn. J. Apl. Phys.* **32**, L32 (1993).

[103] Schmidt, O.G., *et al.*, *Electronics Lett.* **32**, 1302 (1996); and Shoji, H., *et al.*, *IEEE Photon Tech. Lett.* **7**, 1985 (1990).

[104] Drexler, H., *et al.*, *Phys. Rev. Lett.* **73**, 2252 (1994).

[105] For a recent review, R. Nötzel *Semicond. Sci. and Technol.* **11**, 1365 (1996).

[106] Sakaki, H., *et al.*, *Appl. Phys. Lett.* **67**, 3433 (1995).

[107] Xie, Q., *et al.*, *Phys. Rev. Lett.* **75**, 2542 (1995).

[108] Solomon, G.S., *et al.*, *Phys. Rev. Lett.* **76**, 952 (1996).

[109] Sugiyama, Y., *et al.*, *Jpn. J. Appl. Phys.* **35**, 1320 (1996).

[110] Yusa, G., and Sakaki, H., *Electronics Lett.* **32**, 491 (1996).

[111] Mui, D.S.L., *et al.*, *Appl. Phys. Lett.* **66**, 1620 (1995); Sseifert, *et al.*, ibid **68**, 1684 (1996).

[112] Kitamura, M., *et al.*, *Appl. Phys. Lett.* **66**, 3663 (1995).

[113] Petroff, P.M., Gossard, A.C., and Wiegmann, W., *Appl. Phys. Lett.* **45**, 620 (1984). Petroff, P.M., *et al.*, *J. Cryst. Growth* **111**, 360 (1991).

[115] Miller, M.S., *et al.*, *J. Cryst. Growth* **111**, 323 (1991).

[116] Bloch, J., Bockelmann, V., and Laruelle, F., *Solid State Electronics* **37**, 527 (1994); Tanaka, M., and Sakaki, H., *Appl. Phys. Lett.* **54**, 1326 (1989).

[117] Fukui, T., and Saito, H., *Jpn. J. Appl. Phys.* **29**, L731 (1990).

[118] Kanbe, H., *et al.*, *Appl. Phys. Lett.* **58**, 2969 (1991).

[119] Nakata, Y., *et al.*, *Proc. 9th Int. Conf. on MBE* , 1996; (*J. Cryst. Growth* 1997 to be published.)

[120] Kasu, M., and Fukui, T., *Jpn. J. Appl. Phys.* **31**, 964 (1992); and Ishizaki, J., Ohkuri, K., and Fukui, T., *Jpn. J. Appl. Phys.* **35** (Part-1) 1280 (1996).

[121] Inoue, K., *et al.*, *J. Cryst. Growth* **127**, 1041 (1993).

[122] Nakamura, Y., Koshiba, S., Sakaki, H., *Proc. of 9th Int. Conf. on MBE* 1996, (*J. Cryst. Growth* 1997 to be published.)

Chapter 6

Chemical Approaches
to Semiconductor Nanocrystals
and Nanocrystal Materials

Louis Brus

Columbia University, Chemistry Department
New York, N.Y., 10027

I. INTRODUCTION

A. Nanoscience

At present "nanotechnology" is a vision rather than a reality. We do not have practical, manufacturable methods to make complex materials, machines, and electrical circuits on the 1-100 nm scale. However, there is a flourishing "nanoscience" research effort involving ideas and methods drawn from chemistry, physics, and engineering science. "Nanoscience" presently is in a discovery stage, uncovering new physical processes and effects , and learning how to use these processes in new devices and designed materials. The key technological issue is control of natural processes to make assemblies of nanometer components in useful ways.

The 1-100 nm "nanoscale" length is intermediate between the traditional realms of synthetic chemistry, and VLSI lithographic processing as employed in electronics. The two main approaches to nanoscience, colloquially "bottom-up" and "top-down", represent extensions of these methods, respectively, into the intermediate territory. The "top-down" approach uses lithography (e-beam, x-ray) or scan probe methods (tunneling, force, and near-field optical microscopies) to create and explore nanometer structures. In this present AIP monograph, many chapters describe such elegant "top-down" science.

The "bottom-up" approach develops chemical synthetic "self-assembly" methods to create and explore such structures. One might suspect that, as objects and devices decrease in size to tens and hundreds of atoms in diameter, chemical ideas and methods must become useful and efficient. In this chapter, I describe semiconductor nanocrystals made by chemical synthesis. In favorable cases, high quality nanocrystals with controlled surfaces can be made in gram amounts, and can be used as building blocks for new materials and devices. Nanocrystals are also used to explore the size dependence of electronic, optical, and structural properties, as I will describe.

Our original motivation in 1983 was to understand the evolution, with decreasing crystallite size, of molecular properties from bulk properties. This area of science will affect device design, as component size in integrated circuits approaches nanometer scale. At present in 1995 additional goals are possible, because of the continuing increase in synthetic crystallite quality and monodispersity, and chemical control of surface structure. For example, one might now speculate on ultimate chemical self-assembly of nanometer circuits, perhaps even as a replacement for lithographic circuit technology, several decades into the next century. I return to this question in **Final Remarks.**

B. Chemical Synthesis:

Chemical synthesis remains something of an art form, in the sense that theory does not quantitatively predict how to make new materials and specific molecules. Nevertheless, synthesis is extraordinarily powerful, when optimized for specific reactions. Proteins of nanometer size, and organic molecules of almost arbitrary complexity and design, can now be made from simple reagents. These synthetic abilities underlie the biotechnology, drug, and polymer industries; we hope to use and extend this body of knowledge in the development of nanotechnology.

One of the largest inorganic cluster molecules directly synthesized (by Fenske and coworkers in 1993[1]), and crystallized from liquid solution for x-ray structural determination, is $Cu_{146}Se_{73}(PPh_3)_{30}$ in Figure 1. This molecule is 2 nm by 4 nm in size, and can be considered to be a capped nanocrystal of the layered semimetal Cu_2Se, with surface Cu atoms bonded to triphenyl phosphine groups. The nanocrystal is essentially monodisperse, although the x-ray structure does show some surface disorder. The dense surface capping layer of organic groups insulates the inorganic core from the environment. In fact a series of such specific cluster molecules, corresponding to especially stable cluster molecules, has been made. Such chemical synthesis is a type of "natural self-assembly", in that first, the capped cluster molecule forms spontaneously from molecular reagents in solution, and second, the cluster molecules crystallize out of solution. However, we presently only partially understand how to control and modify the self-assembly process. This family of reactions is not yet as useful as MBE synthesis of $GaAs:AlGaAs$ superlattices, for example, where layers of arbitrary thickness with near perfect interfaces can be made. Nevertheless, this example shows the potential power of chemical synthesis in nanoscience.

The semiconductor nanocrystal synthesis that is most highly developed, and closest to being a useful methodology, is the organometallic synthesis of CdS and $CdSe$, both II-VI semiconductors with sp^3 tetrahedral bonding and direct band gaps in the visible. In 1988 my colleagues and I first made organically capped $CdSe$ nanocrystals of controlled size, as pure and stable powder samples[2]. The reactions initially used inverse micelle water pools as microscopic reactors. More

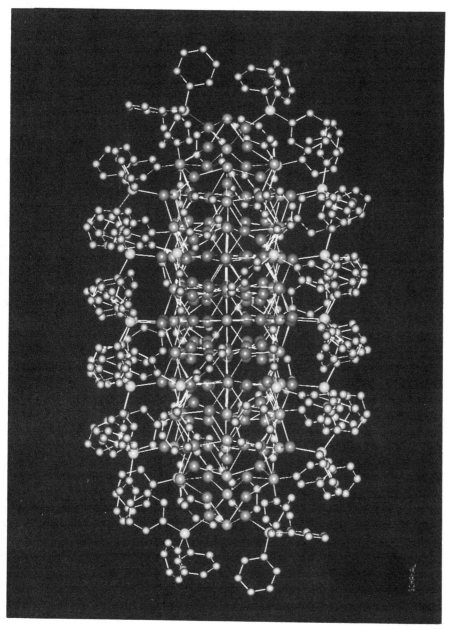

FIGURE 1. Structure of the compound $Cu_{146}Se_{73}(PPh_3)_{30}$ as determined by x-ray crystallography in reference [1]. Hexagons are phenyl groups, Ph. Color of atoms: red, Se; large green, P; small green, C; blue, Cu. (See color plate.)

recently improved reactions have been carried out in near 300C liquids composed of capping molecules such as trioctyl phosphine oxide[3]. By separating the initial nucleation event from the subsequent growth stage, narrow size distribution (~3 %), highly crystalline samples can be made in arbitrary diameters up to 10 nm in size. In a subsequent section, I will return to the question of materials made from such nanocrytals.

C. Physical Size Regimes

Figure 2 shows a schematic diagram of the size regimes that occur as a cluster of a few atoms grows into a bulk semiconductor[4,5,6]. The smallest clusters do not exhibit the bulk unit cell and are molecular. This happens because the rather open sp^3 diamond lattice is only stable in long polymerization lengths, and small fragments rehybridize into other structures to eliminate broken surface bonds. For example, Si_{10} adopts a metallic-like close packing structure[7]. Si_{45} shows strong distortion from the diamond lattice[8]. In the unique case of C where sp^2 (graphite) and sp^3 (diamond) phases are equally stable, fullerenes and nanotubes form.

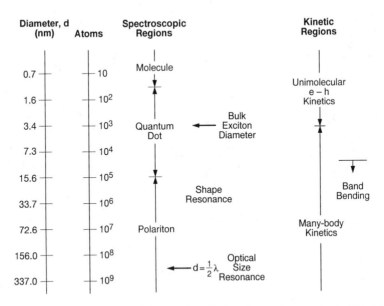

FIGURE 2. Schematic diagram of the semiconductor size regimes, adapted from reference [6]. Note logarithmic vertical scale. λ is the internal optical wavelength.

As size increases the structure eventually shows the bulk unit cell. This is the quantum nanocrystal or quantum dot regime. This transformation occurs at smaller size if surface capping groups are present that preserve sp^3 local surface bonding, as occurs with Si clusters capped with H atoms, for example. In nanocrystals, excited electronic states are discrete and molecule-like, due to three dimensional quantum confinement. Tight confinement occurs for diameters less than the bulk Bohr exciton diameter: here the electron and hole are individually confined with little correlation. Weak confinement refers to diameters a few times larger than the exciton diameter. Here the exciton center-of-mass is confined in the nanocrystal: one consequence is a giant exciton oscillator strength at low temperature. Bulk band structure is attained when the diameter is large with respect to the Bohr exciton diameter. In following sections I discuss the physical properties of $CdSe$ and Si nanocrytals in the tight confinement limit.

Both molecular and nanocrystal regimes exhibit weak coupling to the radiation field. However, as a nanocrystal increases in size it begins to spatially modify the radiation field. In this polariton regime, the crystallite's optical spectra show shape and size Mie resonances. In the lowest order, volume size resonance, the particle acts as an optical cavity, with a diameter equal to one half an internal optical wavelength. Finally, larger (several micrometer sized) particles exhibit surface polariton modes, as utilized, for example, in the "whispering gallery" microdisk laser[9].

These three size regimes – molecular, quantum nanocrystal, and polariton – refer principally to spectroscopic properties. There are also regimes relating to the size dependence of carrier kinetics. As the crystallite optical excitation probability generally scales with volume, larger crystallites are more likely to contain multiple electron-hole pairs than smaller nanocrystals, under steady state conditions. In small nanocrystals the decay kinetics is principally that of a single pair in the tight confinement limit. This is essentially molecular decay kinetics. However, in larger crystallites, multiple pairs interact and can decay efficiently by many body processes such as Auger recombination. The Auger size dependence is especially important in Si nanocrystal luminescence[6].

Recombination in bulk semiconductors is often controlled by rare traps or impurities at the *ppm* or lower level. In this case, "impurity size exclusion" occurs in nanocrystals, when size decreases to the point where there is less than one impurity per nanocrystal. One trap can catalyze recombination in just one nanocrystal; this may be a far smaller volume than in the bulk if mobilities are high. For example, in indirect gap $AgBr$ as used in photographic film, the luminescence quantum yield increases substantially in nanocrystals when I atom impurities and other bulk traps are excluded[10,11].

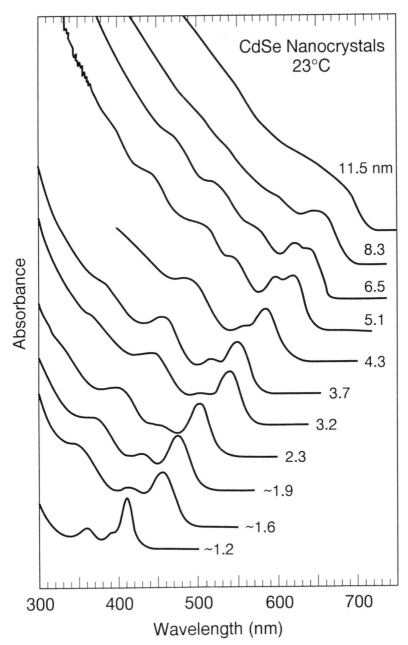

FIGURE 3. Colloidal *CdSe* optical spectra as a function of diameter at a temperature of 23C, adapted from reference [3].

II. ELECTRONIC AND OPTICAL PROPERTIES

A. Direct Gap *CdSe*

1. Volume Confined States

Figure 3 shows the room temperature optical spectra of a colloidal solution of *CdSe* nanocrystals, capped with surface organic groups, as a function of size. 11.5 nm crystallites have almost bulk like spectra, with continuous optical absorption beginning at the bulk band gap near 7000Å. Smaller nanocrystals show a blue-shifted band gap, and discrete, molecule-like excited states. In the smallest ~1.2 nm nanocrystal, the band gap has shifted to higher energy by about 1.5 eV. Careful structural characterization shows that the particles are essentially wurtzite single crystals, slightly elliptical in shape. This figure shows the development of bulk optical spectra with increasing size from tight confinement to weak confinement; in *CdSe* the exciton Bohr diameter is about 6 nm.

Figure 4 is a photograph of colloidal solutions of 1.2 nm, 4.3 nm, and ~10 nm *CdSe* nanocrystals. In all three cases, optical absorption dominates Raleigh scattering as the particle size is small with respect to a visible wavelength. The color changes from black in 10 nm nanocrystals, red in 4.5 nm nanocrystals, and finally to light yellow in 1.2 nm nanocrystals, as the band gap increases.

A simple model of 3 dimensional, volume confinement of a bulk-like electron and hole explains these spectra semiquantitatively . In the Bloch wavefunction $\Psi = \Phi_k e^{i(k \cdot r)}$, Φ_k is a strong function of k and develops sp^3 character as k increases across the Brillouin zone. Averaged over the zone, Φ_k is sp^3-like; this directed valence character imparts chemical stability to the lattice. But near Γ, the conduction band Φ_k is s-like, while the valence band Φ_k is p-like (as is also the case in *GaAs* and *InP*). In the Luttinger Hamiltonian for the electron-hole pair near Γ the electron kinetic energy is characterized by a scalar, isotropic effective mass. The hole kinetic energy is a 6 fold tensor due to the p atomic spatial degeneracy; this tensor includes the ~0.4 eV spin orbit coupling which creates the split-off valence band[12-14].

If this Hamiltonian is solved for confinement in a sphere with infinite surface barrier, the calculated band gap increases somewhat faster with decreasing size than the measured values. However, the pattern of various discrete states in Figure 3 can be assigned as the dipole allowed transitions of the volume confined states in Figure 5[15]. The pattern is different than the light and heavy hole pattern of superlattices, due to the high symmetry spherical confinement in nanocrystals. All but one of the observed transitions originate on various discrete hole levels, and terminate on the lowest, totally symmetric, 1S electron level. In the model, these states all have nodes on the nanocrystal surface. Discrete states resulting from 3D confinement are in all respects the same as molecular orbitals, as this term is used in molecular electronic structure theory. In the tight confinement limit, when the individual confinement energies of the electron and hole are large with respect to the Coulomb interaction, the bandgap size dependence ($1S_e \rightarrow 1S_h$ energy) on radius, R, is approximated by[16,17]:

$$E(R) = E_g + \frac{\hbar^2 \pi^2}{2R^2}\left[\frac{1}{m_e} + \frac{1}{m_h}\right] - \frac{1.8e^2}{\varepsilon R} + \cdots \qquad (1)$$

This formula depends only upon known bulk parameters: the two effective masses, m_e and m_h, and the dielectric constant at optical frequencies, ε.

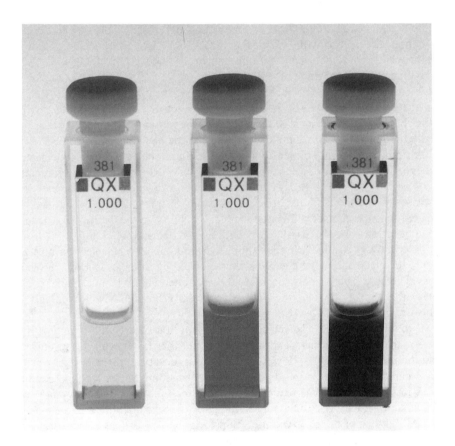

FIGURE 4. Photograph of colloidal *CdSe* nanocrystals dissolved in 4-ethylpyridine. Left: 1.2 nm diameter and 3.0 eV bandgap. Center: 3.5 nm diameter and 2.3 eV bandgap. Right: Bulk particles, 1.8 eV bandgap. (See color plate.)

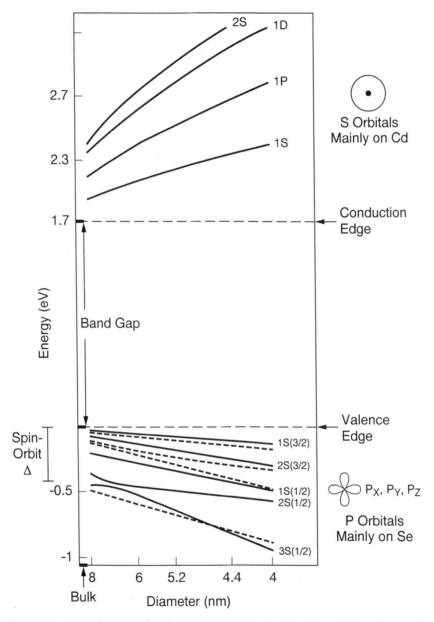

FIGURE 5. *CdSe* volume confined molecular orbitals as a function of diameter, adapted from reference [14].

2. *Surface States*

The volume confinement model ignores surface states. In a 3 nm nanocrystal with perhaps 800 atoms, about one third are on the surface. In principle there is one surface state for each broken surface chemical bond. The energies of these surface states depend upon local surface bond reconstruction, and possible bonding to capping molecules or solvent. Typically, with little reconstruction or capping, surface states will exist inside the nanocrystal band gap, acting as deep traps and quenching luminescence. Bare *Si* nanocrystals appear to belong to this case. Correspondingly, if a nanocrystal luminescences with high yield, then it is unlikely there are surface states deep in the gap.

In bulk wurtzite *CdSe*, reconstruction on nonpolar cleavage planes in vacuum removes surface states from the gap, and puts these surface states into resonance with the volume band states[18]. Surface *Se* atoms have a lone pair surface state, energetically near the top of the valence band, which can act as a isoenergetic trap for photogenerated holes. In *CdSe* nanocrystals, it has been proposed that the photogenerated hole localizes on the surface on a 100 femtosecond time scale, while the electron remains in a 1S internal orbital. The evidence for this comes from detailed studies of luminescence kinetics and band gap lineshape[19,20]. This remains an uncertain subject: the volume confined states are quite polarizable, and it is not clear if localization reflects environmental perturbation, surface irregularity, or intrinsic nanocrystal properties. Certainly there is wavefunction dynamical evolution between absorption and luminescence. The lineshape also reflects coupling to the lowest confined acoustical phonon[21,22], and has an electronic fine structure due to exchange splitting and shape effects. The Frohlich coupling to the LO phonon is also size dependent[23].

B. Onion Shell Nanocrystals

Broadly speaking, in nanocrystals the optical spectrum can be tuned via the size and choice of material. The surface (including possible capping molecules) determines the recombination kinetics, the solubility in various media, and the affinity for physical adsorption and charge transfer across the surface, all without much effect on the optical spectra.

Complete surface passivation (of both anionic and cationic surface atoms in compound semiconductors) creates an ideal quantum dot (zero dimensional exciton) without surface traps. In *CdSe*, this could occur by encapsulation in a higher band gap semiconductor or insulator. In 1982 it was observed that colloidal *CdS* particles show increased room temperature luminescence when Zn^{++} ions were adsorbed, to give a surface coating of *ZnS*[24]. Band gap emission is significantly enhanced when a $Cd(OH)_2$ layer is adsorbed on *CdS* nanocrystals at negative *pH*[25]. *ZnS* layers have been grown on *CdSe* seeds, and vice versa, by sequential

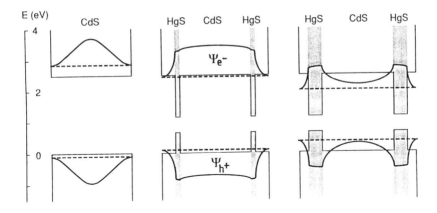

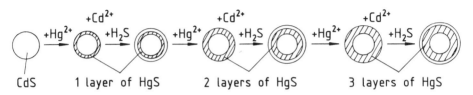

CdS   1 layer of HgS   2 layers of HgS   3 layers of HgS

FIGURE 6. Bottom: schematic diagram of sequential aqueous precipitation scheme for synthesis of shell nanocrystals. Top: electron and hole wavefunctions across the nano-crystal diameter. Left is a pure *CdS* particle with both electron and hole in 1S spatial wavefunctions. Center is an onion particle with a 0.3 nm *HgS* shell (one monolayer). Right is an onion particle with a 0.9 nm *HgS* shell. Adapted from reference [27(b)].

precipitation in the inverse micelle microreactors mention earlier[26]. While both *ZnS* and *CdSe* are zinc-blende type structure, the lattice mismatch is large (13%) and epitaxial growth was not observed.

CdS with band gap 2.5 eV is lattice matched to *HgS* with band gap 0.5 eV; both materials are direct gap. In a remarkable experiment, three layer onion nano-crystals *CdS*|*HgS*|*CdS* were grown in 1994 by sequential adsorption and dis-placement aqueous reactions, as shown in Figure 6[27]. Layered nanocrystal elec-tronic structure can be modeled by incorporating band offsets in the radial poten-tial energy function, in analogy with the envelope approximation in superlattice modeling[28,29]. Calculation shows that electron and hole are both local-ized in a 1 nm thick *HgS* shell, if grown on a 3.5 nm *CdS* seed and covered with a 1 nm *CdS* outer shell. These particles luminesce with high quantum yield, apparently from the intermediate *HgS* shell. This example shows the potential in design of

shell semiconductor nanoparticles; future progress depends critically upon optimizing synthesis to create sharp interfaces and precisely defined dimensions.

C. Indirect Gap Silicon

How does the indirect gap nature of diamond lattice *Si* develop with increasing size? Does it develop more quickly or slowly than the band gap itself? This
is potentially a technological, as well as a basic scientific question, because optical data interconnections may be required on some future generations of VLSI *Si*
chips. Partially direct gap like nanocrystalline *Si* layers might be used for electro-
luminescent optical interconnects. The efficient visible photoluminescence and
quantum nature of porous *Si* thin films at 23C, discovered only four years ago,
supports the idea that nanocrystalline *Si*, with a larger band gap due to 3D
confinement, might be a useful optical material[30,31].

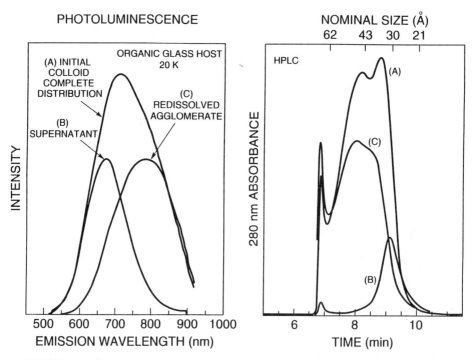

FIGURE 7. Left hand side: *Si* nanocrystal luminescence spectra of an initial colloid *A*,
and of two fractions *B* and *C* derived from *A*. Right hand side: corresponding high pressure liquid chromatograms. Adapted from reference [33(b)].

Because of the covalent nature and high bond energy of diamond lattice *Si*, organometallic liquid phase synthesis of *Si* nanocrystals has been reported only under both high pressure and temperature conditions[32]. However, luminescent colloids of crystalline *Si* with a surface 0.8 nm thick oxide shell, can be made by a high temperature aerosol process followed by nanocrystal scrubbing from the aerosol into ethylene glycol[33]. Narrowed size fractions can be made by size exclusion chromatography and size selective precipitation. Figure 7 shows visible emission spectra and size distributions for two fractions derived from a broader distribution. The smaller fraction emits near 6500Å, with a measured quantum yield of 5.6% at 23C, increasing to near 100% below 50K. In spite of the fact that the quantum yield is high, the lifetime remains long: near 50 microseconds at 23C increasing to approximately 2 milliseconds near 15K.

The luminescence of this smaller fraction is broadened by the residual size distribution. Laser excitation on the low energy side of the distribution, near 7100Å, excites only the larger particles in this distribution, and yields structured and yet further narrowed emission, as shown in Figure 8. This luminescence near 7400Å shows the TO phonon steps that earmark indirect gap, vibronically induced emission. Similar, better resolved structure is present in laser excited luminescence of porous-silicon[34,35].

These lifetime and spectral data show that *Si* essentially remains an indirect gap semiconductor, even in the 1.2–1.5 nm *Si* core diameter nanocrystals that emit near 6500Å. The luminescence is vibronically induced, and the electron-hole pair radiative lifetime is very long – on the order of 10^{-4} sec. In *Si* the indirect nature of the band gap forms more quickly with increasing size than does the band gap numerical value. The *Si* conduction band minimum traverses almost completely across the Brillouin zone from the valence band maximum at Γ; the electron phase changes almost completely from one unit cell to the next, with respect to the hole phase. Thus, intra-unit-cell optical transition dipoles will destructively interfere in nanocrystals made of just several unit cells. The numerical value of the gap is not related to phase difference, but instead to the effective masses and the dielectric constant as in equation (1).

In *Si* nanocrystals, room temperature luminescence quantum yields increase not because coupling to the radiation field is stronger in confined systems, but because radiationless processes which dominate bulk *Si* emission are significantly weaker in nanocrystal *Si*[36]. This weakening of nonradiative processes is caused by loss of carrier mobility over macroscopic distances. Auger nonradiative decay is slower as electron-hole pairs in separate nanocrystals are electrically isolated. Similarly, deactivation at rare deep traps and impurities is decreased, as a defect can only quench emission from the one crystallite containing it.

It is challenging to physically characterize such small surface oxidized, Si core nanocrystals. Near edge *Si* x-ray absorption data show three *Si* "phases" present: crystalline *Si*, SiO_2, and interfacial SiO_x, in the nanocrystals emitting near 6500Å[37]. The SiO_2 shell is 0.8 nm thick, while the interface region is

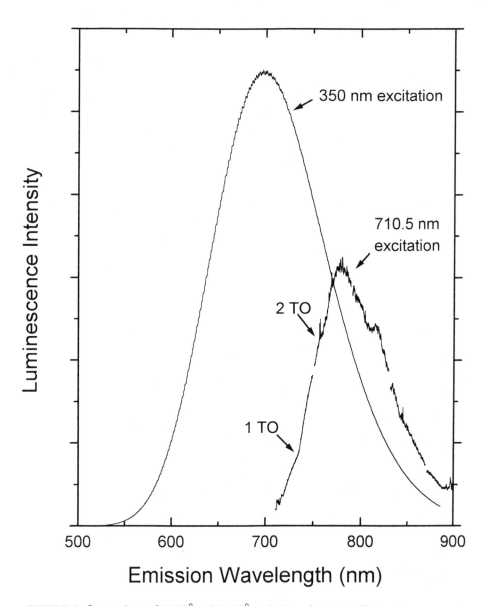

FIGURE 8. Comparison of 3500Å and 7105Å excitation of *Si* crystallite luminescence, at 15K in an organic glass host.   Adapted from reference [6]. 3500Å excitation excites all particles in the size distribution, while 7105Å excitation excites only the larger particles in the distribution.

about one monolayer. This thin surface oxide is a remarkably efficient surface passivation, as shown by the high luminescence quantum yield. Yet it is thin enough that charge can tunnel across the oxide on slow time scales; this tunneling allows a common Fermi level to be set in dense films of nanocrystals, as in porous *Si* as we shall describe.

III. STRUCTURAL PHASE TRANSITIONS

A. Melting

As nanocrystal size decreases, surface energy becomes an increasing factor in structural stability: for very small sizes the diamond lattice is not thermodynamically stable with respect to isomerization into molecular structures in which the bulk unit cell is not present. This process is driven by rehybridization of surface bonds. For a given size, the surface energy can be varied somewhat by the choice of surface capping species. For example, *H* atom termination stabilizes diamond-lattice-like structures in *Si* cluster molecules.

For a given size, surface energy also influences the structural phase diagram in the temperature-pressure plane. *CdS* nanocrystals (as well as other materials such as metals and molecular crystals) show a large, reversible, and congruent melting point depression[38]. A 3 nm diameter *CdS* nanocrystal in vacuum melts at about 700K; the bulk melting point is 1678K. Melting is observed by imaging and electron diffraction in a high resolution transmission electron microscope (TEM). Simple thermodynamic models, which predict a lowering of the melting point if the surface energy of the solid is higher than that of the liquid, describe the data semiquantitatively.

In the liquid phase synthesis carried out near 300C, melting point depression may enhance annealing into crystalline rather than amorphous nanocrystals. Additionally, the ability of Lewis base solvents such as triakylphosphine oxide to bond to *Cd* atoms facilitates surface atom mobility and annealing.

B. Solid-solid Transformation

Under pressure, II-VI semiconductors transform from four-coordinate wurtzite to six-coordinate rocksalt structure. In bulk *CdSe* this transformation occurs at 2 GPa. In 4.4 nm diameter *CdSe* nanocrystals, a wurtzite single nanocrystal transforms into a rocksalt single nanocrystal at about 6 GPa, as observed by Bragg x-ray diffraction in Figure 9[39]. In the data, size can be determined by Debye-Scherrer broadening of the diffraction lines.

The fact that the transition pressure is elevated as size decreases is hard to understand thermodynamically. It may be that the transformation is kinetically

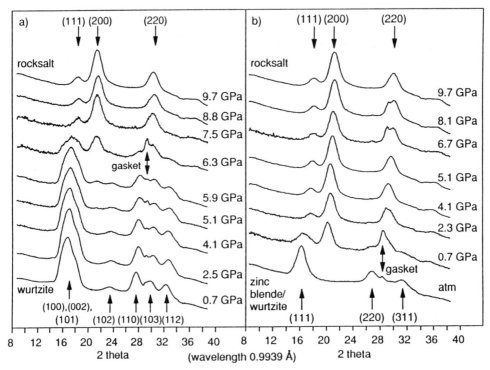

FIGURE 9. Bragg x-ray powder patterns of 4 nm *CdSe* nanocrystals as a function of pressure, adapted from reference [39]. a) increasing pressure; b) decreasing pressure.

controlled. If a nearly round wurtzite single nanocrystal coherently transforms ("isomerizes") into a rocksalt particle, then structural models show that the rock salt nanocrystal will have high index surface planes exposed. The rocksalt nanocrystal made this way at 23C will not be the thermodynamically stable shape; this higher energy rocksalt shape may explain why the transition pressure is elevated. (As shape has not been observed in the high pressure rocksalt phase, this point remains speculative.) In both the bulk and nanocrystal phase transitions, there is a large hysteresis in the phase transition with pressure, indicating the presence of a large kinetic barrier at 23C. With increasing pressure the rock salt structure appears near 6 GPa, while with subsequent decreasing pressure the wurtzite structure reappears only below 1 GPa.

This discussion suggests the future possibility that high pressure bulk phases might be kinetically stabilized in nanocrystals at room temperature and pressure, using some combination of hysteretic pressure cycling and surface capping chosen to mimic the local bonding of the high pressure phase. Actually, certain yeasts synthesize nearly monodisperse 2 nm diameter, essentially amorphous

CdS particles capped with polypeptides[40]. The quite broad Bragg diffraction pattern is better fit by rocksalt than by wurtzite local structure. This result appears to reflect the local surface bonding on these quite small particles. The biological function of *CdS* synthesis in yeasts is defense again heavy metal Cd^{++} poisoning in the local environment.

IV. MATERIALS AND DEVICES

A. "Wet" Nanocrystal Films, and Porous Silicon

ZnO is a 3.34 eV direct gap wurtzite semiconductor, with a Bohr exciton diameter of 4 nm. Dense yet randomly packed nanocrystal *ZnO* films on transparent conductive substrates can be simply made by spin coating with concentrated aqueous *ZnO* colloids, followed by brief heating at a few hundred degrees C. Particles 2-6 nm in diameter show the expected quantum size effects in their optical spectra, both for isolated nanocrystals in the colloid, and in the dense film. The nanocrystals are thought to have a surface passivating monolayer of $Zn(OH)_2$.

If a "wet", several micrometers thick, *ZnO* film is used as an electrode in an electrochemical cell, with a 1M electrolyte penetrating into the interstitial spaces between nanocrystals, then the optical properties are reversibly voltage dependent as shown in Figure 10[41]. The quantum confined, band edge absorption bleaches as the film is charged with electrons; simultaneously the photoluminescence changes from trap-like to band edge as deep traps are filled. This is a remarkable result: the film optical properties are quantum confined, yet electrical transport occurs in the film apparently by electron hopping from nanocrystal to nanocrystal across thin passivating layers.

TiO_2 is a 3.4 eV ultraviolet band gap semiconductor. A similar "wet" TiO_2 film of partially sintered 10 nm particles makes an efficient liquid junction solar cell, if the particles are surface derivatized with a dye molecule that adsorbs visible light[42]. After optical excitation of the dye, an electron is injected into the TiO_2 particle. (Alternately, small *PbS* and *CdS* nanocrystals adsorbed on the TiO_2 particle can be used as photosensitizer in place of the dye[42(c)].) The hole in the dye molecule is transferred to an oxidizable solute molecule in the interstitial electrolyte. With voltage applied across the device, the electron moves from particle to particle, while the oxidized molecule diffuses through the interstitial electrolyte across the film in the opposite direction. While charge collection across the film is slow (milliseconds), separation efficiencies of 80% have been reported across TiO_2 films 10 μm thick. In these films, an interfacial mechanism of charge separation is at work. The high efficiency occurs because the electron moves in the solid phase, while the hole moves in the liquid phase. Large area TiO_2 particulate thin films are environmentally benign and inexpensive; TiO_2

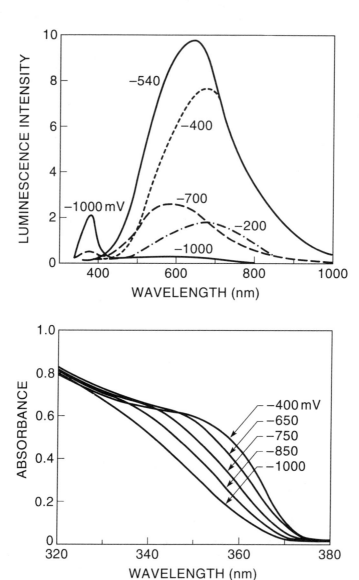

FIGURE 10. Top: photoluminescence of a wet, dense *ZnO* nanocrystal film electrode in an electrochemical cell, as a function of applied voltage. The broad feature at 600 nm is deep trap emission, and the sharper feature at 380 nm is band gap emission. Bottom: Corresponding bleach of the band edge adsorption as a function of voltage. Adapted from reference [41(a)].

powders are made in huge volume for the paint and ceramics industries. The possibility of significant cost reduction drives the development of these TiO_2 solar cells.

Porous Si is a yellowish film that grows epitaxially on wafer Si, under anodic electrochemical etching in HF. The film is a rather open, branched network of crystalline Si wires of undulating diameter. These wires might be considered lines of partially fused nanocrystals of variable diameter. The etching process appears to be self-limiting, in that the reduction in wire size accompanying etching stops at ~2 nm size, when the injection energy of a bulk Si hole (the etching agent) into a wire becomes large due to quantum confinement in the wire[31]. The films show visible photoluminescence similar to the Si nanocrystal emission

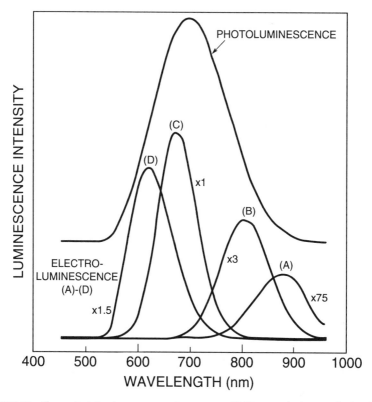

FIGURE 11. Top: photoluminescence of a porous Si film anode in an electrochemical cell. Bottom: electroluminescence as a function of applied voltage. Adapted from reference [43(b)].

described previously[30].

"Wet" porous-Si also shows efficient electroluminescence, in electrochemical diodes where electrons are injected across the Si wafer: film epitaxial interface, and holes are injected from molecules on the interstitial interfaces between the electrolyte and the wires. In a sense, this is the reverse process of the dye sensitized, TiO_2 particle solar cell. The electroluminescence spectrum is narrower than the photoluminescence spectrum, and can be voltage tuned across the photoluminescence, as shown in Figure 11[43].

The photoluminescence is broad as it represents emission from a wide nanocrystal size distribution. In electroluminescence, as voltage is increased and the film accepts electrons, the larger crystallites are charged at lower potential than the smaller crystallites, and eventually are multiply charged. The narrowed, tunable electroluminescence appears to come from crystallites that have just one electron when a hole is injected; if there is more than one electron, then Auger nonradiative recombination dominates radiative recombination. Tunability is possible because the optical and charging properties are size dependent, yet a common Fermi level can be established in the film, on a slow time scale, and holes can be injected, if it is "wet". This situation is partially analogous to the ZnO experiments.

B. Nanocrystal Light Emitting Diodes

The porous-Si electroluminescence suggests the possibility of using nanocrystals as luminescent centers. Color, and band offsets for charge injection, can be size tuned; also, nanocrystals might be more stable than organics in current carrying devices. The practical problem is to design contacts, or a conductive host, that efficiently inject equal numbers of electrons and holes into the nanocrystals, without nonradiative recombination or progressive degradation.

If one simply embeds nearly monodisperse $CdSe$ nanocrystals in an amorphous polymer film, made from both a hole conductor and an electron conductor, then $CdSe$ electroluminescence occurs at low voltage bias[44]. At higher bias, the host polymer emits as well. An alternate design involves a (macroscopic) junction between a dense $CdSe$ nanocrystal layer, and a conductive polymer layer. If the $CdSe$ layer carries electrons, and the polymer layer carries holes, then similar (bias dependent) electroluminescence is observed[45]. These experiments are simply demonstrations of principle and are not optimized; yet they do show the potential usefulness of nanocrystals as device building blocks. In these diodes, as well as in the "wet" nanocrystal layers described above, the fundamental transport physics under applied field is not well understood.

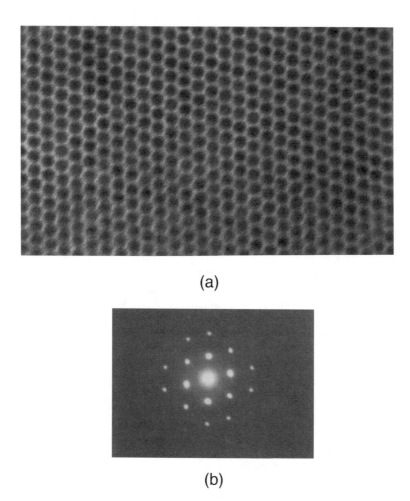

(a)

(b)

FIGURE 12. (a) A TEM micrograph of a thin section of 3D supercrystal made from 4.8 nm *CdSe* nanocrystals. This (101) projection shows hexagonal packing as evident from the diffraction pattern in (b).(C.B. Murray and M.G. Bawendi,private communication.)

C. Self-Assembly of Nanocrystals

Gem opals are naturally occurring, ordered 3D close packed crystals of nearly monodisperse, spherical, several hundred nanometer diameter silica particles[46]. Opals diffract light efficiently, and are near in structure to the proposed photonic band gap materials[47]. Loosely speaking, opals are "colloidal crystals, or super-crystals" in which polarizable objects are held together by van der Waals forces. Synthetic opals are formed when nearly (<5%) monodisperse colloidal silica particles aggregate slowly under thermodynamically reversible conditions. In the limit of small, identical particles, eg. identical large molecules, this "self-assembly" process is simply normal crystallization. Crystals of several capped ~1 nm CdS cluster molecules have been reported[48], in addition to the Cu_2Se cluster crystal described in the introduction. Macroscopic 3D crystals of 6.9 nm Fe_2O_3 particles[49], and (just recently) trioctyl phosphine oxide capped $CdSe$ nanocrystals of various sizes, have been similarly made[50]. Figure 12 shows a TEM image and electron diffraction pattern of a hexagonal closest-packed, 3D crystal made from 5 nm $CdSe$ nanocrystals.

In terms of electrical properties, nanocrystal supercrystals are similar to crystals of C_{60}: the individual object(C_{60} or nanocrystal) has discrete states, and, from this quantum mechanical basis, a macroscopic band structure develops in the supercrystal. In C_{60} crystal, the valence and conduction bands are about 0.5 eV wide, and a wide variety of electrical properties can be achieved by interstitial doping. The narrow band widths and the strong electron-phonon coupling on one C_{60} sphere lend themselves to superconductivity.

The band structures of the 3D supercrystals of capped $CdSe$ nanocrystals have not been measured. With trioctyl phosphine oxide surface capping, the organic spacer layer is about 1 nm thick, and thus charge transfer should be quite slow. (Such supercrystals show efficient excited state energy transfer, however, via the near fields of the electronic transition dipoles.) One can imagine, however, use of short conducting organic capping molecules, or even partial loss of capping molecules with subsequent point bonding of nanocrystals in the supercrystal. (This latter structure is somewhat similar to porous-Si) By methods of this sort, one might tune the electronic band widths for a given nanocrystal size in the supercrystal. Perhaps characteristic charge transport times can be varied from milliseconds, as in the "wet" dense particle films, to picoseconds.

Remarkable phenomena occur when supercrystals are formed from dense colloids containing mixtures of large A and small B spheres, of radii ratio ~0.58[51]. In specific ranges of volume fractions, ordered superlattice macroscopic crystals form of the type AB_2 (AlB_2 structure type) and AB_{13} ($NaZn_{13}$ structure type). These two superlattice structural types are commonly seen in intermetallic compounds. These structures have been observed to spontaneously crystallize at 23C from concentrated organic solutions containing large A and small B $PMMA$ polymer particles, even if long range van der Waals forces between particles are eliminated by use of an index matched solvent. Freezing of the superlattice solid from the dense colloid thus appears to be driven by the entropy.(editor's word choice.)

Both superlattice structures in fact were first observed in a gem opal, where they were presumably formed by natural hydrothermal processing of colloidal

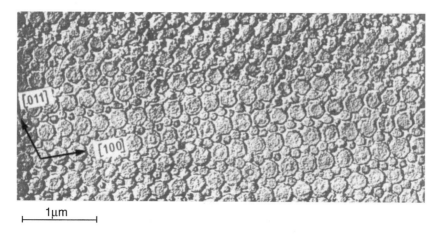

FIGURE 13. A scanning electron micrograph of an AB_2 superlattice structure found in a naturally occurring gem opal composed of two different size silica spheres. The larger sphere is 180 nm in diameter. Adapted from reference [51(a)].

silica, which somehow happened to be present in two different sizes of narrow distribution[51(a)]. In the AB_2 structure, close-packed planes of large A spheres are spaced by planes of B spheres as shown in Figure 13. In a hypothetical AB_2 superlattice made from large $CdSe$ A nanocrystals and small silica B particles, the silica spheres would act as insulators between planes of close packed $CdSe$ nanocrytals. The electrical properties would be anisotropic. In the parent super-crystal made from closest packing of $CdSe$ A nanocrystals only, electrical properties would be essentially isotropic. This is a rich potential area of materials science, when one considers opal-like structures made from nanocrystals of different materials, sizes, and surface chemistries. There may be a new inorganic structural chemistry here, with all the various nanocrytals serving as a new periodic table of "atoms".

V. FINAL REMARKS

In the context of nanotechnology, semiconductor nanocrystals are building blocks. Size and choice of material determine the optical and electrical properties, while surface chemistry determines charge transfer and electron-hole recombination kinetics. The quantum mechanics of three dimensional confinement is now moderately well understood. The $CdSe$ family shows the potential of chemical synthesis to create gram amounts of high quality nanocrystals, with chosen and variable surface chemistry. Within ten years all major semiconductors should be available at this level of precision and quality.

The "wet" ZnO, TiO_2, and porous Si nanocrystal films are not purposefully designed, yet they offer electronic possibilities not present in bulk crystalline semiconductors. The existence of an interstitial electrolyte enables the efficient separation of a photoexcited electron and hole. It also allows a film to be electrically doped by choice of a dissolved redox couple. Molecules or smaller nanocrystals can be chemically bound at the nanocrystal: electrolyte interface. The film optical properties show the quantum confinement of the individual nanocrystals, yet slow electron transport is possible by hopping. All this happens at 23C in a technology that is inexpensive on a per unit area basis. However, "wet" nanocrystal films do not have the high mobilities of bulk single crystals, and it remains to be seen if they have adequate stability in long term use.

In the 3D $CdSe$ supercrystal, and especially in the superlattice AB_2 and AB_{13} opal structures, we see the possibility of more purposefully designed nanocrystal films. Assembly can be controlled by efficient packing related to the ratio of sphere diameters, by specific chemical bridging molecules chosen to bind one type of surface to another, and by the use of different materials as well as different sizes. This will be a very rich area of materials science.

It seems likely that fundamentally new aspects of transport physics will be uncovered as the "wet" nanocrystal films, the nanocrystal/conducting organic light emitting diodes, and the opal-like nanocrystal materials are further developed.

At present we have no concept of how to assemble nanocrystals into a complex, designed circuit, with anything like the generality of present micrometer scale lithographic silicon technology. In fact we need to convince ourselves that this is a practical and worthwhile goal on the nanometer scale. Over long time scales, unexpected discoveries can change human perspective on the importance and relative priority of such heroic tasks. Just last year it was first realized that "quantum computation" offers a major scaling advantage in parallel processing, when compared with macroscopic computer algorithms[52]. "Quantum Computation" conceptually operates by coherent wavefunction propagation in a microscopic system. Loosely speaking, nature herself does the computation. Perhaps some new idea such as this will provide the momentum necessary for a major assault on creation of a practical nanotechnology. Somehow, chemistry and electronic manufacturing will ultimately merge on the nanometer scale.

REFERENCES

[1]   Krautscheid, H., Fenske, D., Baum, G., and Semmelman, M., *Angew. Chem. Int. Ed. Engl.* **32**, 1303 (1993).

[2] Steigerwald, M. L., Alivisatos, A. P., Gibson, J. M., Harris, T. D., Kortan, R., Muller, A. J., Thayer, A. M., Duncan, T. M., Douglass, D. C., and Brus, L. E., *J. Am. Chem. Soc.* **110**, 3046 (1988).

[3] Murray, C. B., Norris, D. J., Bawendi, M. G., *J. Am. Chem. Soc.* **115**, 8076 (1993).

[4] Efros, Al. L., Efros, A. L., *Sov. Phys. Semicond.* **16**, 1209 (1982).

[5] Brus, L., *J. Phys. Chem.* **90**, 2555 (1986).

[6] Brus, L., Szajowski, P., Wilson, W., Harris, T., Schuppler, S., and Citrin, P., *J. Am. Chem. Soc.* **117**, 2915 (1995).

[7] (a) Raghavachari, K., *Phase Transitions* **24–26**, 61 (1990); (b) Rothlisberger, U., Andreoni, W., and Giannozzi, P., *J. Chem. Phys.* **96**, 1248 (1992).

[8] Rothlisberger, U., Andreoni, W., and Parrinello, M., *Phys. Rev. Lett.* **72**, 665 (1994).

[9] McCall, S., Levi, A., Slusher, R., Pearson, S., Logan, R., *Appl. Phys. Lett.* **60**, 289 (1992).

[10] (a) Johansson, K. P., McLendon, G. P.; and Marchetti, A. P., *Chem. Phys. Lett.* **179**, 321 (1991); (b) Johansson, K. P., Marchetti, A. P., and McLendon, G. P., *J. Phys. Chem.* **96**, 2873 (1992); (c) Marchetti, A. P.; Johansson, K. P., and McLendon, G.P., *Phys. Rev. B* **47**, 4268 (1993); (d) Chen, W.; McLendon, G., Marchetti, A., Rehm, J. M., Freedhoff, M., and Myers, C., *J. Am. Chem. Soc.* **116**, 1585 (1994).

[11] Kanzaki, H., Tadakuma, Y., *Solid State Commun.* **80**, 33 (1991).

[12] Chestnoy, N., Hull, R., Brus, L., *J. Chem. Phys.* **85**, 2237 –2242 (1986).

[13] Xia, J. B., *Phys. Rev. B* **40**, 8500–8506 (1989).

[14] Ekimov, A. I., Hache, F., Schanne-Klein, Richard, D., Flytzanis, C., Kudryatsev, I. A., Yazeva, T. V., Rodina, A. F., Efros, Al. L. *J. Opt. Soc. Am. B* **10**, 100–107 (1993).

[15] Norris, D. J., Sacra, A., Murray, C. B., Bawendi, M. G., *Phys. Rev. Lett.* **72**, 2612–2615 (1994).

[16] (a) Rossetti, R., Nakahara, S., Brus, L. *J. Chem. Phys.* **79**, 1086 (1983); (b) Brus, L., *J. Chem. Phys.* **80**, 4403 (1984).

[17] Kayanuma, Y., *Solid State Commun.* **59**, 405 (1986).

[18] Wang, Y., and Duke, C., *Phys. Rev. B* **37**, 6417 (1988).

[19] Bawendi, M. G., Wilson, W. L., Rothberg, L., Carroll, P. J., Jedju, T.M., Steigerwald, M. L., Brus, L.E., *Phys. Rev. Lett.* **65**, 1623–1626 (1990).

[20] Bawendi, M.G., Carroll, P.J., Wilson, W. L., Brus, L.E., *J. Chem. Phys.* **96**, 946–1004 (1990).

[21] Alivisatos, A., Harris, T., Carroll, P., Steigerwald, M., Brus, L., *J. Chem. Phys.* **90**, 3463 (1989).

[22] Schoenlein, R. W., Mittleman, D. W., Shiang, J. J., Alivisatos, A. P., Shank, C. V., *Phys. Rev. Lett.* **70**, 1014–1017 (1993).

[23] Shiang, J., Goldstein, A., Alivisatos, A., *J. Chem. Phys.* **92**, 3232 (1990).

[24] Rossetti, R., Brus, L., *J. Phys. Chem.* **86**, 4470 (1982).

[25] Spanhe, L., Hasse, H., Weller, H., Henglein, A., *J. Am. Chem. Soc.* **109**, 5649 (1987).

[26] Kortan, A., Hull, R., Oplia, R., Bawendi, M., Steigerwald, M., Carroll, P., and Brus, L., *J. Am. Chem. Soc.* **112**, 1327 (1990).

[27] (a) Eychmuller, A., Mews, A., and Weller, H., *Chem. Phys. Lett.* **208**, 59 (1993); (b) Mews, A., Eychmuller, A., Giersig, M., Schooss, D., and Weller, H., *J. Phys. Chem.* **98**, 934 (1994).

[28] Schooss, D., Mews, A., Eychmuller, A., and Weller, H., *Phys. Rev. B* **49**, 17072 (1994).

[29] Haus, J., Zhou, H., Honma, I., and Komiyama, H., *Phys. Rev. B* **47** , (1993).

[30] Canham, L.T., *Appl. Phys. Lett.* **57**, 1046 (1990).

[31] Lehmann, V.; and Gzsele, U., *Appl Phys. Lett.* **58**, 856 (1991).

[32] Heath, J., *Science* **258**, 1131 (1992).

[33] (a) Littau, K.A., Szajowski, P.F., Muller, A.J., Kortan, R.F., and Brus, L.E., *J. Phys. Chem.* **97**, 1224 (1993); (b) Wilson, W., Szajowski, P., Brus, L., *Science* **262**, 1242 (1993).

[34] (a) Calcott, P.D.J., Nash, K. J., Canham, L. T., Kane, M. J., Brumhead, D., *J. Phys. Condens. Matter* **5**, L91 (1993).; (b) Calcott, P.D.J., Nash, K.J., Canham, L.T., Kane, M.J., Brumhead, D., *J. Lumin.* **57**, 257 (1993).

[35] (a) Suemoto, T., Tanaka, K., Nakajima, A., and Itakura, T., *Phys. Rev. Lett.* **70**, 3659 (1993); (b) Suemoto, T., Tanaka, K., and Nakajima, A., *J. Phys. Soc. Jpn. (Supplement B)* **63**, 190 (1994).

[36] Brus, L., *J. Phys. Chem.* **98**, 3575 (1994).

[37] Schuppler, S., Friedman, S. L., Marcus, M. A., Adler, D. L., Xie, Y.-H., Ross, F. M., Harris, T. D., Brown, W. L., Chabal, Y. J., Brus, L. E., and Citrin, P. H., *Phys. Rev. Lett.* **72**, 2648 (1994); and *Phys. Rev. B* (to be published).

[38]   Goldstein, A., Echer, C., Alivisatos, A., *Science* **256**, 1425 (1992).

[39]   Tolbert, S., and Alivisatos, A., *Science* **265**, 373 (1994).

[40]   Dameron, C., Reese, R., Mehra, R., Kortan. A., Carroll, P., Steigerwald, M., Brus, L., Winge, D., *Nature* **338**, 596 (1989).

[41]   (a) Hoyer, P., Eichberger, R., Weller, H., *Ber. Bunsenges. Physik. Chem.* **97**, 630 (1993).; (b) Hoyer. P., and Weller, H., *Chem. Phys. Lett.* **221**, 379 (1994).

[42]   (a) O'Regan, B., Gratzel, M., *Nature* **353**, 737 (1991); (b) Nazeeruddin, M., Kay, A., Rodicio, I., Humphry-Baker, R., Muller, E., Liska, P., Vlachopoulos, N., and Gratzel, M., *J. Am. Chem. Soc.* **115**, 6382 (1993); (c) Vogel, R., Pohl, K., and Weller, H., *Chem. Phys. Lett.* **174**, 241 (1990).

[43]   (a) Bsiesy, A., Muller, F., Ligeon, M., Gaspard, F., Herino, R., Romestain, R., and Vial, J., *Phys. Rev. Lett.* **71**, 637 (1993); (b) Bsiesy, A., Muller, F., Ligeon, M., Gaspard, F., Herino, R., Romenstain, R., and Vial, J., *Appl. Phys. Lett.* **65**, 3371 (1994).

[44]   Dabbousi, B., Bawendi, M., Onitsuka, O., and Rubner, M., *Appl. Phys. Lett.* **66**, 1316 (1995).

[45]   Colvin, V., Schlamp. M., and Alivisatos, A., *Nature* **370**, 6488 (1994).

[46]   (a) Sanders, *J. Acta. Cryst. A* **24**, 427 (1968); (b) Pieranski, P., *Contemp. Phys.* **24**, 25 (1983).

[47]   Ho, K. M., Chan, C. T., and Soukoulis, C. M., *Phys. Rev. Lett.* **65**, 3152 (1990).

[48]   Herron, N., Calabrese, J., Farneth, W., Wang, Y., *Science* **259**, 1426 (1993).

[49]   Bentzon, M., van Wonterghem, J., Morup, S., Tholen, A., and Koch, C., *Phil. Mag. B* **60**, 169 (1989).

[50]   C. B. Murray and M. Bawendi, private communication 1994.

[51]   (a) Sanders, J.,*Phil. Mag. A* **42**, 705 (1980); (b) Bartlett, P., Ottewill, R., and Pusey, P., *Phys. Rev. Lett.* **68**, 3801 (1992); (c) Bartlett, P., and Pusey, P., *Physica A* **194**, 415 (1993).

[52]   Shor, P. W., *Proceedings of 35th Annual Symposium on Fundamental Computer Science,* IEEE Computer Science (Nov. 1994), p. 124.

Chapter 7

Nanotechnology in Carbon Materials

M.S. Dresselhaus

*Department of Electrical Engineering and Computer Science
and Department of Physics
Massachusetts Institute of Technology
Cambridge, Massachusetts 02139, USA*

G. Dresselhaus

*Francis Bitter National Magnet Laboratory
Massachusetts Institute of Technology
Cambridge, Massachusetts, 02139, USA*

R. Saito

*Department of Electronics Engineering
University of Electro-Communications
Chofugaoka, Chofu, 182 Tokyo, Japan*

I. INTRODUCTION

This article reviews carbon materials from the standpoint of nanotechnology. The unusual features of carbon materials stem from the ability of carbon to form materials with vastly different structures and hence with vastly different properties. The major focus of this review is on the ability of carbon to form zero dimensional quantum dots of subnanometer dimensions in the form of fullerenes, and one-dimensional quantum wires in the form of carbon nanotubes. The structure and properties of fullerenes and carbon nanotubes are reviewed in this article in the context of nanotechnology.

Carbon is unique as an electronic material. It can be a good metal in the form of graphite, a wide gap superhard semiconductor in the form of diamond, a superconductor when intercalated with appropriate guest species, or a flexible polymer when reacted with hydrogen and other species. Furthermore, carbon-based electronic materials provide examples of materials showing the entire range of dimensionalities from fullerenes which are 0D quantum dots, to carbon nanotubes which are 1D quantum wires, to graphite which is a 2D layered anisotropic material, and finally to diamond, a 3D wide gap semiconductor. Within this family of electronic materials, fullerenes and carbon nanotubes are the newest additions and are the subject of this review. Whereas graphite is the ground state for a system containing a huge number of carbon atoms, there is a large energy per carbon atom associated with the edge sites for a small graphene sheet (defined as a single isolated layer of the graphite lattice). Thus, to avoid the occurrence of edge sites, small numbers of carbon atoms form closed shell configurations such as fullerenes and carbon nanotubes [see Figure 1].

Although fullerenes and nanotubes share many common features, their differences regarding structure and properties are sufficiently large to merit separate discussion in this review.

Fullerene solids differ from conventional electronic materials because, in common with most polymeric materials, the fullerene molecule is the fundamental building block of the crystalline phase. Unique features about fullerenes include the structural perfection and reproducibility of these subnanometer building blocks. Since all C_{60} fullerene molecules (0.7 nm diameter) are identical to one another, they perhaps represent the most reproducible currently available nanostructure. At the present cost of approximately \$100/g and a purity of 99+%, C_{60} is already attractive for use as a monodisperse, reproducible, self-assembled nanostructure. Present trends of rapidly decreasing costs and increasing purity through better and cheaper fabrication technology are making C_{60} even more attractive relative to other nanostructures for specific applications. The unique

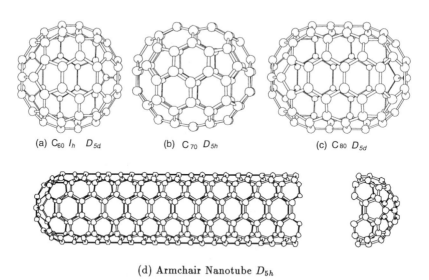

(a) C_{60} I_h D_{5d} (b) C_{70} D_{5h} (c) C_{80} D_{5d}

(d) Armchair Nanotube D_{5h}

FIGURE 1. Examples of closed shell fullerene configurations which avoid dangling bonds: (a) C_{60}, (b) C_{70}, (c) C_{80} and (d) an armchair carbon nanotube. Also indicated on the figure are the point group symmetries of the various structures. We note that D_{5d} is a subgroup of the icosahedral group I_h, exhibiting inversion symmetry.

structural features of the building blocks are reflected in their unique electronic structures and properties which are summarized in this review article. Another attractive feature of fullerenes is their ready accessibility to doping, charge transfer, and the resulting control of the electronic properties of the nanostructures. As a result of their unique structures and properties, potential applications for fullerene-based electronic materials are suggested. Following the discussion on the quantum dot fullerenes, the corresponding discussion on 1D carbon nanotubes is presented.

In early gas phase work, a molecule with 60 carbon atoms was established experimentally by mass spectrographic analysis[1] as a relatively stable form relative to other carbon clusters and it was conjectured that C_{60} is a closed cage molecule with icosahedral symmetry. The name of "fullerene" was given to this family of gas phase molecules by Kroto and Smalley[1] because of their resemblance to the geodesic domes designed and built by R. Buckminster Fuller[2]. The name "buckminsterfullerene" or simply "buckyball" was given specifically to the C_{60} molecule. In the early gas phase work, the fullerene molecules were produced by the laser vaporization of carbon from a graphite target in a pulsed jet of helium[1,3].

Definitive verification that the C_{60} molecule was indeed the shape of a regular truncated icosahedron came somewhat later, from NMR studies showing a single NMR line associated with a ^{13}C nucleus[4,5]. Since every carbon atom of the C_{60} molecule is in an identical site location, only one NMR line is expected for the 60 carbon atoms located at the vertices of a regular truncated icosahedron. The identification of C_{60} as a molecule with icosahedral symmetry was further strengthened by observation of the characteristic four-line infrared absorption spectrum[6,7], from such a highly symmetrical molecule having many (174) vibrational degrees of freedom.

In the fall of 1990 a new type of condensed matter, based on C_{60}, was synthesized for the first time by Krätschmer, Huffman and co-workers[8], who found a simple carbon arc method for preparing gram quantities of C_{60}, which had previously only been available in trace quantities in the gas phase[1,9]. The availability of much larger quantities of fullerenes provided a great stimulus to this research field. It was soon found[10] that the intercalation of alkali metals into C_{60} to a stoichiometry M_3C_{60} (where $M = K$, Rb, Cs) greatly modifies the electronic properties of the host fullerenes and yields a conducting material with a relatively high superconducting transition temperature[11]. The first reported superconductor in the fullerene family was K_3C_{60} with $T_c \sim$ 19K[11], and subsequent work has revealed superconductivity with $T_c \sim$ 40K in Cs_3C_{60} under a pressure of 12 kbar[12]. The discovery of superconductivity[12,13] in these compounds further spurred research activity in the field of C_{60}-related materials. The unusual structure of molecular fullerenes leads to unusual optical transport and magnetic properties, with possible applications as photoconductors, diode rectifiers, optical limiters, photorefractive materials, bonding agents, STM tip coatings, etc.[14].

II. FULLERENES AS NANOSTRUCTURES

A. Structural Properties

In the preparation of fullerenes by any of the conventional synthesis methods (by carbon arc discharge, laser pyrolysis, or combustion flames) the fullerene species of greatest abundance by far is C_{60}, the most stable of the fullerenes and the fullerene with the greatest symmetry. Every C_{60} molecule is identical, except for the possible presence of the ^{13}C isotope, with a 1.1% natural abundance which can substitute randomly for ^{12}C in the caged molecule. C_{60} molecules thus form a very small (0.71 nm diameter) monodisperse nanostructure of high (icosahedral I_h) symmetry. Because of the simplicity of C_{60} relative to other fullerenes and its relatively high abundance, most of the discussion of fullerenes in this review is for C_{60}.

Since the structure and properties of fullerene solids are strongly dependent on the structure and properties of the constituent fullerene molecules, we first review the structure of the molecular building blocks, which is followed by a review of the structure of the corresponding molecular solids, representing self-assembled ordered arrays of these nanostructures.

1. Structure of C_{60}

To a good approximation, the 60 carbon atoms in C_{60} are located at the vertices of a regular truncated icosahedron. As mentioned above, every carbon site on the C_{60} molecule is equivalent to every other site [see Figure 1(a)], consistent with a single sharp line in the NMR spectrum[4,5]. The average nearest neighbor carbon-carbon ($C-C$) distance a_{C-C} in C_{60} is very small (0.144 nm) and is almost identical to that in graphite (0.142 nm). Each carbon atom in C_{60} (and also in graphite) is trigonally bonded to three other carbon atoms and 20 of the 32 faces on the regular truncated icosahedron are hexagons, the remaining 12 being pentagons. Thus, we may consider the C_{60} molecule as a "rolled-up" graphene sheet (a single layer of crystalline graphite) which forms a closed shell, in keeping with Euler's theorem, which states that a closed surface consisting of hexagons and pentagons has exactly 12 pentagons and an arbitrary number of hexagons[15]. The introduction of pentagons gives rise to curvature in forming a closed surface. To minimize local curvature, the pentagons become separated from each other in the self-assembly process, giving rise to the isolated pentagon rule, an important rule for stabilizing C_{60} clusters. The high reproducibility of C_{60} in the self-assembly process relates to the fact that the smallest cluster to obey the isolated pentagon rule is C_{60} and that there is only one way to assemble 60 carbon atoms in a closed cage configuration which obeys the isolated pentagon rule. That structure is shown in Figure 1(a).

The symmetry operations of the icosahedron consists of the identity operation, 6 five-fold axes through the centers of the pentagonal faces giving rise to 24 independent symmetry operations, 10 three-fold axes through the centers of the hexagonal faces resulting in 20 independent symmetry operations, and 15 two-fold axes through centers of the edges joining two hexagons. Each of the 60 rotational symmetry operations is then compounded with the inversion operation, resulting in 120 symmetry operations in the icosahedral point group I_h[16] [see Figure 1(a)]. I_h is the highest possible point group symmetry for any molecule. The diameter of the C_{60} molecule is 0.710 nm (see Table 1), treating the carbon atoms as points[1,17,18].

From Euler's theorem it follows that the smallest possible fullerene is C_{20} which would form a regular dodecahedron with 12 pentagonal faces[14], but this structure is energetically unfavorable in accordance with the isolated pentagon rule, because of its high local curvature and high strain. Since the addition of a single hexagon adds two carbon atoms, all fullerenes C_{n_C} must have an even number of carbon atoms n_C, in agreement with the observed mass spectra for fullerenes[3]. An estimate for the diameter of a fullerene can be found from the relation for an icosahedral fullerene $d_i = a_{C-C}\sqrt{15n_C/2\pi}$ where $a_{C-C} = 0.144$ nm is the average nearest-neighbor carbon–carbon distance. Each fullerene C_{n_C} can thus be considered as a nanostructure with diameters less than 3 nm for $n_C < 10^3$.

Although each carbon atom in C_{60} is equivalent to every other carbon atom, the three bonds emanating from each atom are not equivalent [see Figure 1(a)]. Each of the four valence electrons of each carbon atom is engaged in covalent bonds, so that two of the three bonds on the pentagon perimeter are electron-poor single bonds, and one between two hexagons is an electron-rich double bond. Consistent with the x-ray diffraction evidence, the structure of C_{60} is further stabilized by introducing a small distortion of the bond lengths to form the Kekulé structure of alternating single and double bonds around the hexagonal face. The single bonds that define the pentagonal faces are increased from the average bond length of 0.144 nm to a length $a_5 = 0.146$ nm, while the double bonds between adjacent hexagons are decreased in length to $a_6 = 0.140$ nm[30,50]. We note that the icosahedral I_h symmetry is preserved under these distortions. Since each carbon atom has its valence requirements fully satisfied, a solid composed of C_{60} molecules is expected to form a van der Waals-bonded crystal which is nonconducting (an insulator or a semiconductor).

TABLE 1. Physical constants for C_{60} molecules and for crystalline C_{60}

Quantity	Value	Reference
Average $C-C$ distance	0.144 nm	[19]
$C-C$ bond length on a pentagon	0.146 nm	[5,20]
$C-C$ bond length on a hexagon	0.140 nm	[5,20]
C_{60} mean molecule diameter	0.710 nm	[18]
Moment of inertia I	1.0×10^{-43} kg m^2	[21]
Volume per C_{60}	1.87×10^{-22}/cm^3	–
Number of distinct C sites	1	–
Number of distinct $C-C$ bonds	2	–
Binding energy per atom[a]	7.40 eV	[22]
Heat of formation (per g C atom)	10.16 kcal	[23]
Electron affinity	2.65 ± 0.05 eV	[24]
Cohesive energy per C atom	1.4 eV/atom	[25]
Spin-orbit splitting of $C(2p)$	0.00022 eV	[26]
Ionization potential	7.58 eV	[27]
Optical gap[b]	1.9 eV	[22]
fcc lattice constant	1.417 nm	[28]
$C_{60}-C_{60}$ distance	1.002 nm	[17]
$C_{60}-C_{60}$ cohesive energy	1.6 eV	[29]
Mass density	1.72 g/cm^3	[17]
Molecular density	1.44×10^{21}/cm^3	[17]
Transition temperature (T_{01})	261K	[34]
Vol. coeff. of thermal expansion	$6.1 \times 10^{-5} K^{-1}$	[37]
Band gap (HOMO-LUMO)	1.7 eV	[38]
Work function	4.7 ± 0.1 eV	[39]
Thermal conductivity (300K)	0.4 W/mK	[43]
Electrical conductivity (300K)	1.7×10^{-7} S/cm	[44]
Static dielectric constant	$4.0 - 4.5$	[45,46]
Melting temperature	1180°C	[47]
Sublimation temperature	434°C	[48]
Heat of sublimation	40.1 kcal/mol	[48]
Latent heat	1.65 eV/C_{60}	[49]

[a] The binding energy for C_{60} is believed to be ~0.7 eV/C atom less than for graphite, although literature values for both are given as 7.4 eV/C atom. The reason for the apparent inconsistency is attributed to differences in calculational techniques.

[b] Calculated value for the optical band gap for the free C_{60} molecule.

2. Structure of C_{70} and Higher Fullerenes

In the synthesis of C_{60}, larger molecular weight fullerenes C_{n_C} ($n_C > 60$) are also formed, by far the most abundant being C_{70}. However, significant quantities of C_{76}, C_{78}, C_{84}, and higher mass fullerenes have also been isolated and studied in some detail[14].

C_{70} has been found to exhibit a rugby-ball shape[51], and its form can be envisioned either by adding a ring of 10 carbon atoms or a belt of 5 hexagons around the equatorial plane of the C_{60} molecule normal to one of the five-fold axes [see Figure 1(b)]. In contrast to the C_{60} molecule with I_h symmetry, the C_{70} molecule has the lower symmetry D_{5h} which is a subgroup of I (lacking inversion symmetry). Fullerenes often form isomers, since a given number n_C of carbon atoms can correspond to closed cage molecules C_{n_C} with different geometrical structures[51,52], each distinct structure referring to a different isomer. For example, C_{80} might be formed in the shape of an elongated rugby ball prepared by adding two rows of 5 hexagons normal to a five-fold axis of C_{60} at the equator [see Figure 1(c)]; an icosahedral form of C_{80} can also be constructed as another of the 7 distinct isomers of C_{80}, obeying the isolated pentagon rule and having symmetries D_{5d}, D_2, C_{2v}, D_3, D_{5h}, and I_h[53]. The three fullerenes obeying the isolated pentagon rule, and having just a single isomer are C_{60} (I_h), C_{70} (D_{5h}), and C_{76} (D_2). Isomers belonging to group I (rather than I_h), and not having inversion symmetry, are expected to give rise to nanostructures exhibiting right- and left-handed optical activity.

3. Structure of Metallofullerenes

In addition to the nanostructures formed purely from carbon-containing fullerenes, as described above, there are metallofullerenes which contain a metal ion within the fullerene cage[54] which is discussed in this subsection, and metal-coated fullerenes containing layer(s) of metal species external to the fullerene cage[55,56] (discussed in Section II.A.4).

Regarding metallofullerenes, many species can be inserted into the interior hollow core of the C_{60} molecule to form an endohedrally doped molecular unit, and the insertion of one, two, or three metal species inside a single fullerene cage is common[57]. The endohedral fullerene configuration has, for example, been denoted by $La@C_{60}$ for one endohedral lanthanum in C_{60}, or $Y_2@C_{82}$ for two Y atoms inside a C_{82} fullerene[58].

Slow progress with the isolation and purification of sufficient quantities of metallofullerene materials has, however, delayed study of the structure and properties of metallofullerenes in the solid state. Upon preparation of endohedrally doped fullerenes, the $M@C_{60}$ species is thought to be the most common in the soot, but after sublimation of the soot, the most common form is $M@C_{82}$, with lesser amounts of $M@C_{74}$ and $M@C_{80}$ also present[59]. Studies of $La@C_{82}$ confirm the relatively high stability of this material[60].

From an historical perspective, soon after it was realized that C_{60} formed a closed-cage molecule, efforts were made to insert guest ions within the fullerene cage[1,61]. Mass spectra suggesting endohedral doping of C_{60} with a single La ion were soon observed, and this early observation played an important role in supporting the hypothesis that C_{60} was a closed-cage molecule[62,63]. In this pioneering work it was found that laser photodissociation could be used to remove C_2 units from the endohedral shell, thereby yielding smaller endoful-lerenes in a process which "shrink wrapped" the carbon cage around the endohedral core, eventually leading to $La@C_{44}$, the smallest member of the lan-thanum metallofullerene series[63]. Shrink wrapping has also produced $U@C_{28}$ with a cage that is smaller than that reported for the smallest stable empty ful-lerene (C_{32}). Furthermore, evidence has also been given for endohedral ful-lerenes with other column IV-B ions, such as $Ti@C_{28}$, $Zr@C_{28}$, and $Hf@C_{28}$. Infrared and Raman spectra for $La@C_{82}$ indicate that the cage structure is not much altered by La introduction[64,65].

The stability of an endohedral fullerene $M@C_{n_C}$ depends upon the dopant species M, the number of carbon atoms in the fullerene cage n_C and the shape of the fullerene C_{n_C} isomer. For example, C_{82} has nine isolated-pentagon iso-mers[66], four of which (with symmetries C_2, C_{2v}, C_{3v}, and C_2) have been identified by liquid chromatography[60,67], and two metallofullerene isomers of $La@C_{82}$ (with C_2 and C_{3v} symmetries) have been identified in the electron paramagnetic resonance (EPR) spectra[68,69].

It is widely believed that the dopants are inside the cage, although not neces-sarily at the central position[70] [see Figure 2]. Model calculations for minimiz-ing the energy of $La@C_{82}$ show the La to be significantly displaced (by 1.49 nm) from the center of the nearly spherical carbon cage [see Figure 2(b,c)], with the smallest $La-C$ distances (of the six nearest neighbors to La) at 2.53–2.56 nm[64]. Charge transfer between the endohedral La and the fullerene shell results in the formation of the La^{3+} ion surrounded by a C_{60}^{3+} anion with a strong electrostatic $La-C$ attraction between them. This electrostatic interaction gives rise to the above-mentioned displacement of the La^{3+} ion from a central position of the fullerene cage as the La^{3+} ion bonds with a single carbon atom (or a few carbon atoms) on the fullerene cage. The charge transfer also leads to the forma-tion of a dipole moment. Because of the relatively large size of the fullerene shell, the dipole moment of endofullerenes tends to be large (2–4 Debye)[64] and may strongly affect the solubility of $La@C_{82}$ in various solvents such as CS_2 and toluene[64]. Differences in solubility and in dipole moment could be exploited in isolating and purifying endofullerenes.

For the case of three La ions in C_{82}, model calculations show that the three La ions form an equilateral triangle, as shown schematically in Figure 2(d)[71,72].

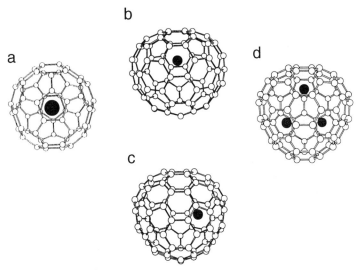

FIGURE 2. Structural models for various endofullerenes. (a) One possible structural model for $M@C_{60}$, with M at the center of the C_{60} cage[73]. (b, c) Two different structural models for $La@C_{82}$, with the La at two different off-center positions within the C_{82} cage[74]. (d) A structural model for $Sc_3@C_{82}$ (assuming C_{3v} symmetry of the C_{82} cage), where black balls represent the three equivalent Sc^{3+} ions, which rapidly reorient within the C_{82} cage as an equilateral triangle[64].

An interesting recent discovery is the endohedral doping of C_{60} with the rare gas atoms of *He* and *Ne*[75], though only low concentrations ($\sim 10^{-6}$) of the C_{60} cages are doped with these rare gases.

Thus far, only very small quantities of endohedrally doped fullerenes have been prepared, and most studies of metallofullerenes have been limited to the gas-phase. However, isolation of sufficient quantities of $Sc@C_{74}$ and $Sc_2@C_{74}$ has allowed vacuum deposition of a monolayer of these species onto a clean *Si* (100) surface, where scanning tunneling microscopy techniques could be used to obtain a diameter of ~0.95 nm for these endohedral fullerenes, indicating the presence of some charge delocalization over the fullerene cage[76]. As synthesis techniques improve and larger samples of isolated and purified metallofullerenes become available, it is expected that both the structure and properties of metallofullerenes in the solid state will be clarified.

4. Metal-Coated Fullerenes

It has also been demonstrated that ordered metal-coated fullerene clusters can be synthesized[55,56] using a vapor synthesis method. Dopants, which have been successfully used to coat fullerenes, satisfy size constraints and are exohedral dopants for fullerenes in the crystalline phase. For alkali metal-coated fullerenes, only monolayer coatings of metal atoms have thus far been demonstrated, such as for the case of $Li_{12}C_{60}$ where a lithium is believed to reside over the centers of each pentagonal face of C_{60}[77,78]. Work is in progress using other alkali metals for coating fullerenes[56].

A remarkable multilayer metal structure can also be grown over the C_{60} surface using small alkaline earth atoms such as Ca, Sr, and Ba. This is accomplished when the temperature and pressure conditions are appropriately regulated. Photoionization time-of-flight mass spectrometry measurements show well-defined peaks in the singly charged and doubly charged spectra, corresponding to $Ba_{32}C_{60}$, consistent with placing one Ba over each of the 12 pentagonal faces and over each of the 20 hexagonal faces, as shown schematically in Figure 3.

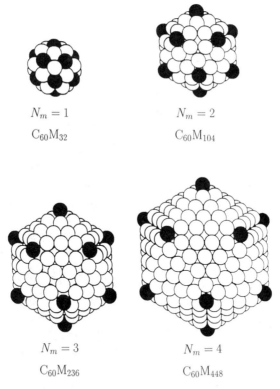

$N_m = 1$

$C_{60}M_{32}$

$N_m = 2$

$C_{60}M_{104}$

$N_m = 3$

$C_{60}M_{236}$

$N_m = 4$

$C_{60}M_{448}$

FIGURE 3. Proposed arrangements in multilayer metal-covered fullerenes of the atoms $M = Ca$ in the first four layers surrounding a C_{60} molecule. The M atoms over the icosahedral vertices of C_{60} are shaded and N_m denotes the number of metal layers[55].

The Ba atoms have just the right size (radius is 0.198 nm) to cover the C_{60} shell[55,56]. In support of this model is the observation of metal-coated $Ba_{37}C_{70}$ clusters based on C_{70}, consistent with five additional hexagons in C_{70} relative to C_{60}. Even more remarkable is the formation of ordered structures with multiple Ca shells, which are implied by mass spectra, where well-resolved peaks for $Ca_x C_{60}$ are observed for $x = 32,104,236$, and 448, corresponding to $N_m = 1,2,3$, and 4 ordered layers of Ca atoms emanating from the C_{60} core, and shown schematically in Figure 3. In this figure the black balls denote the Ca atoms lying over the pentagonal faces, and white balls denote other Ca atoms. For the $M_{32}C_{60}$ cluster, each white ball is over a hexagonal face. As the next layer is deposited, three white balls are closely packed relative to the lower layer, giving rise to 60 additional white balls and 12 additional black balls over the pentagonal faces, and corresponding to the $M_{104}C_{60}$ stoichiometry ($32 + 12 + 3 \times 20 = 104$). Likewise, the buildup of the white balls in the third and fourth layers is 6 and 10 per hexagonal face, respectively, thereby accounting for the peaks observed in mass spectra at $M_{236}C_{60}$ and $M_{448}C_{60}$, respectively. Although little is known about the physical properties of metal-coated fullerene nanoparticles, they could be of interest for nanotechnology, either as isolated well-ordered clusters or in the form of ordered arrays of clusters.

B. Synthesis

Synthesis methods for fullerenes are rapidly improving, thereby increasing yields and lowering costs. Likewise, separation and purification methods are also improving rapidly making available C_{60} samples with purities well beyond 99%, and sufficient quantities of purified C_{70} for many physical measurements. Substantial progress has also been made with the separation of still higher mass fullerenes, and of metallofullerenes so that physical measurements of the properties of these nanoclusters is now beginning. Through the use of scanning tunneling microscopy and electron energy loss spectroscopy, physical measurement on periodic arrays of higher mass fullerenes and of metallofullerenes, as occur in crystal lattices, is now starting in a number of laboratories.

Fullerenes are usually synthesized using an arc discharge between graphite electrodes (20 V, 60 A) in approximately 200 torr of He gas. The heat generated at the contact point between the electrodes evaporates carbon to form soot and fullerenes, which condense on the water-cooled walls of the reactor. This discharge produces a carbon soot which can contain up to ~15% fullerenes: C_{60} (~13%) and C_{70} (~2%). The fullerenes are next separated from the soot according to their mass (which is proportional to the number of carbon atoms in the fullerene molecule) using liquid chromatography and a solvent such as toluene for the chromatography column. Extraction and purification steps follow the separation to yield powder samples of specific fullerenes.

Metallofullerenes are prepared by endohedral doping of guest species such as rare earth, alkaline earth or alkali metal ions into the interior of the fullerene molecule. The synthesis is carried out by impregnating the positive electrode with graphite powder mixed with the desired dopant. During the arc discharge process, the dopant species is released into the plasma gas and becomes entrained within the fullerenes. Sophisticated liquid chromatography techniques are used for concentrating the minute amounts of a given metallofullerene that are prepared in the synthesis process, which is followed by further separation and purification steps[54]. Thus far, only small quantities of endohedrally doped fullerenes have been prepared and only limited investigations of the physical properties of endohedrally doped materials have been reported[54]. Metal-coated fullerenes are prepared in the vapor phase.

Property measurements of fullerenes are made either on powder samples, films or single crystals. C_{60} powder is obtained by vacuum evaporation of the solvent from the solution. Single crystals and polycrystalline films are then prepared from these purified powders. Fullerene films are prepared by vacuum sublimation of the fullerenes on substrates selected for the specific use of the films. Single crystals are best grown by the vacuum sublimation technique[79], though growth from solution yields crystals of lower chemical purity because of the incorporation of solvent during crystallization.

C. Forming Arrays of Fullerene-related Nanostructures

In this section we discuss the formation of arrays of fullerenes and the degree of ordering for fullerene-related nanostructures that is provided by single crystal growth, film growth on a substrate, and intercalation. From the standpoint of nanotechnology, the deposition of C_{60} on a substrate will in general result in a disordered array of fullerene molecules, but if the substrate is chosen for lattice matching [e.g., *GeS* (001)] and the growth conditions are appropriately controlled, epitaxial fullerene growth can occur[80]. Single crystal growth provides another method for producing ordered arrays of fullerene nanostructures.

In all cases, whether single crystal growth, film growth or intercalation is involved, the fullerene building blocks retain their molecular structural integrity in forming these nanocrystalline arrays. However, in the case of intercalation or the deposition of fullerenes on selected substrates, there is charge transfer between the fullerenes and their local environment, as discussed below.

1. Crystalline C_{60}

In the crystalline phase, the C_{60} molecules crystallize into a cubic structure with a lattice constant of 1.417 nm, a nearest neighbor $C_{60} - C_{60}$ distance of

1.002 nm[17] and a density of 1.72 g/cm^3 (corresponding to 1.44 x 10^{21} C_{60} molecules/cm^3). At room temperature, the molecules are rotating rapidly with full rotational freedom, and the centers of the molecules are arranged on a face centered cubic (fcc) lattice with one C_{60} molecule per primitive fcc unit cell [see Figure 4(a)], or 4 molecules per simple cubic unit cell [see Figure 4(b)]. The pertinent space group is O_h^5 or $Fm\overline{3}m$. This structure is established directly by x-ray and neutron diffraction[37,81,82]. Because of their rapid rotation at room temperature, all the C_{60} molecules in the array are equivalent, so that an ordered phase for the fullerenes is achieved upon crystallization.

Below about T_{01} = 260K, the C_{60} molecules lose most of their rotational degrees of freedom, so that the residual rotational motion in the low temperature phase below T_{01} occurs only along the four $\langle 111 \rangle$ axes[37,82,84], each molecule rotating about a different $\langle 111 \rangle$ axis [see Figure 4]. The structure of solid C_{60} below ~260K thus becomes simple cubic (space group T_h^6 or $Pa\overline{3}$), with the four distinct molecules shown in Figure 4(b) being inequivalent in the low temperature phase [see also Figure 5(a)][37,82,85], thereby resulting in a lowering of the symmetry. As the temperature is lowered further, additional ordering of the C_{60} molecules occurs as the rotations about the $\langle 111 \rangle$ axes become hindered, whereby adjacent C_{60} molecules develop correlated orientations.

In the low temperature phase, the [100] cubic axes pass through three mutually orthogonal 2-fold molecular axes on the icosahedron (the centers of the electron-rich hexagon-hexagon edges), and four $\langle 111 \rangle$ cubic axes pass through centers of the hexagonal molecular faces. Since the fullerene molecule has no four-fold axis, the molecules are ordered on the cubic lattice relative to their twofold axis. There are, however, two standard orientations of the C_{60} molecule

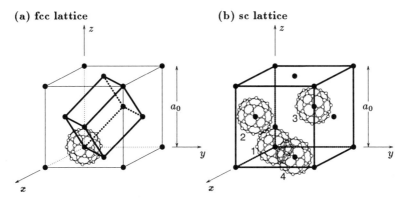

(a) fcc lattice   **(b) sc lattice**

FIGURE 4. Crystal structures for (a) the high temperature fcc phase of C_{60} and (b) the low temperature sc phase of C_{60}.

about its two-fold axes, which can be obtained from one another by a rotation about a three-fold axis. Thus when the C_{60} molecules crystallize in the simple cubic structure for temperatures below T_{01}, they will have equal probability for orienting their two-fold axes along either of the standard orientations, therefore introducing a type of disorder to the system which is called *merohedral disorder*. Although intermolecular interactions result in preferred orientations between adjacent fullerenes and some partial molecular alignment occurs, orientational (merohedral) disorder persists down to the lowest temperatures because of the basic incompatibility of the icosahedral symmetry and cubic symmetry with regard to the four-fold axes.

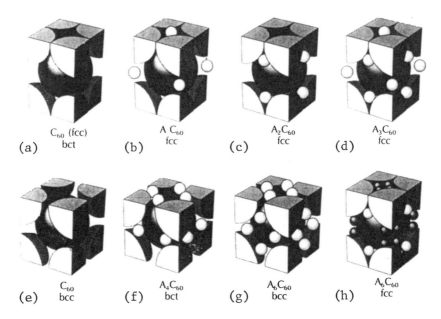

(a) C_{60} (fcc) bct   (b) $A\,C_{60}$ fcc   (c) A_2C_{60} fcc   (d) A_3C_{60} fcc

(e) C_{60} bcc   (f) A_4C_{60} bct   (g) A_6C_{60} bcc   (h) A_6C_{60} fcc

FIGURE 5. Crystal structures for the alkali metal fullerides (a) undoped fcc C_{60}, (b) MC_{60}, (c) M_2C_{60}, (d) M_3C_{60}, (e) bcc C_{60}, (f) M_4C_{60}, and two structures for M_6C_{60}, (g) M_6C_{60} (bcc) for $M=K,Rb,Cs$, and (h) M_6C_{60} (fcc), which is appropriate for $M=Na$ [83]. The large balls denote C_{60} molecules and the small balls are alkali metal ions. For fcc M_3C_{60}, which has four C_{60} molecules per cubic unit cell, the M atoms can be on either octahedral or tetrahedral sites. Undoped solid C_{60} also exhibits the fcc crystal structure at room temperature, but in this case all tetrahedral and octahedral sites are unoccupied. For (f) bct M_4C_{60} and (g) bcc M_6C_{60}, all the M atoms are on distorted tetrahedral sites. For (h) we see that four Na ions can occupy an octahedral site of this fcc lattice. The M_1C_{60} compounds (for $M=Na,K,Rb,Cs$) crystallize in the rock-salt structure.

2. Crystalline C_{70}

Because of the rugby-ball shape of the C_{70} molecule [see Figure 1(b)], the structure for crystalline C_{70} (which can be considered as an array of C_{70} nanostructures) is more complex than that for C_{60}[86–88], evolving through several distinct crystal structures as a function of temperature. At high temperature ($T \gg 340K$), the fcc phase ($a = 1.501$ nm) with freely rotating molecules is most stable, but since the ideal hexagonal close packed (hcp) phase with $c/a = 1.63$ is almost equally stable, fcc crystals of C_{70} tend to be severely twinned and show many stacking faults. As the temperature is lowered, the fcc structure is continuously transformed by deformation into a rhombohedral structure with the long diagonal 3-fold axis aligned parallel to the $\langle 111 \rangle$ direction of the fcc structure[87]. At lower temperatures, a hexagonal phase (space group $P6_3/mmc$) with C_{70} molecules rotating with full rotational symmetry is stabilized. This C_{70} crystal phase has lattice constants $a = b = 1.056$ nm and $c = 1.718$ nm and a nearly ideal c/a ratio of 1.63. A transition to another hcp occurs at ~337K, but with $a = b = 1.011$ nm and a larger c/a ratio of 1.82 for the lower temperature phase. This larger c/a ratio is associated with the orientation of the C_{70} molecules along their long axis, as the free molecular rotation (full rotational symmetry) that is prevalent in the higher temperature phase freezes into a rotation about the 5-fold axis of the C_{70} molecule[87]. As the temperature is further lowered to ~270K, the free rotation about the c-axis also becomes frozen, resulting in a monoclinic structure with the unique axis along the c-axis of the hcp structure, and the monoclinic angle β is close to 120°. In the low temperature phases ($T < 337K$), the main 5-fold axes for the C_{70} molecules are aligned, and the two-fold axes assume some degree of correlation. But because of the incompatibility of the D_{5h} point group operations with those of the space group, orientational misalignment with respect to the three-fold and two-fold axes persists to the lowest temperatures.

3. Crystalline Phases for Higher Mass Fullerenes

Because of the small sample sizes currently available for the higher mass fullerenes C_{n_C} ($n_C > 70$), structural studies of arrays of these nanostructures by conventional methods (such as x-ray diffraction and neutron scattering) are difficult. Therefore, only a few single crystal structural reports have, thus far, been published regarding the higher mass fullerenes. The most direct structural measurements have come from scanning tunneling microscopy (STM) studies[89,90] and selected area electron diffraction studies using electron energy loss spectroscopy (EELS)[91].

STM measurements have been done on C_{76}, C_{80}, C_{82}, and C_{84} fullerenes adsorbed on *Si* (100) 2×1 and *GaAs* (110) surfaces[89,90], providing detailed information on the local environment of the higher-mass fullerenes in the crystal lattice. These STM studies show that all of these higher fullerenes crystallize in an fcc structure with $C_{n_C} - C_{n_C}$ nearest-neighbor distances of 1.13 nm, 1.10 nm, 1.174 nm, 1.21 nm for C_{76}, C_{78}, C_{82}, and C_{84}, respectively, and fcc lattice constants are obtained from these distances by multiplication by $\sqrt{2}$. The more massive fullerenes are less mobile rotationally, so that the orientational phase transition at T_{01}, discussed above for crystalline C_{60} and C_{70}, is driven to higher temperatures (above room temperature) in the higher mass fullerenes, if the phase transition occurs at all. Since higher mass fullerenes generally have multiple isomers, the fullerenes on the fcc lattice sites are not expected to be identical, and therefore disorder of the higher-mass fullerene nanostructures on the lattice arrays is observed, due to both the presence of multiple isomers and orientational misalignment (as discussed above for crystalline C_{60} and C_{70} for their low temperature phases). Similar arguments can be made for preparing crystalline arrays of metallofullerenes, although less is currently known about crystalline forms of metallofullerenes, or for that matter metal-coated fullerenes.

Electron diffraction measurements have provided detailed results on lattice constants and grain sizes of very small crystalline samples of the higher fullerenes. Such measurements were made on thin films of higher mass fullerenes prepared by sublimation of a purified powder of a given C_{n_C} fullerene species onto an *NaCl* substrate. The *NaCl* substrate was then dissolved to yield a freestanding fullerene film with grains of ~100 nm in size[91]. By indexing the electron diffraction peaks observed for C_{60}, C_{70}, C_{76}, and C_{84} to an fcc lattice, the lattice constants a_{fcc} shown in Figure 6 are obtained as a function of $\sqrt{n_C}$, where n_C is the number of carbon atoms in C_{n_C}. Results for the lattice constant are in good agreement with the corresponding STM results for the $C_{n_C} - C_{n_C}$ nearest neighbor distances.

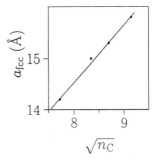

FIGURE 6. The lattice constant for the higher-mass fullerenes C_{n_C} as a function of $\sqrt{n_C}$, where $60 \leq n_C \leq 84$, as obtained by electron diffraction measurements[91].

4. Fullerenes on Surfaces

Crystalline substrates provide another method for forming ordered arrays of fullerene nanostructures. The bonding of fullerenes to some substrates is strong while to others it is weak. Furthermore, metals tend to have a lower work function than semiconductors. Because of the band gap between the occupied valence band and the empty conduction band in semiconductors, more energy is usually required to remove an electron from its highest-lying occupied state to the vacuum level, as compared to a metal. Thus it is expected that charge is more easily transferred between a fullerene and the substrate when the substrate is a metal. Consequently, fullerenes should bind more strongly to metal surfaces than to semiconductor or insulating surfaces. This simple argument is, in general, applicable to real C_{60} interfaces with substrates, except in cases where the semiconductor has a high density of dangling bonds at the surface, such as for the $Si(111)$ and $Si(100)$ surfaces. The binding of C_{60} to oxides such as SiO_2 and sapphire ($\alpha-Al_2O_3$) is weak, while the binding to metals such as Al, $Au(111)$, $Ag(111)$, and $Cu(111)$ is strong. In cases where the bonding of C_{60} to a surface is strong compared to $C_{60}-C_{60}$ bonding, it is easy to form a dense single monolayer of C_{60} on the surface. Such a C_{60} monolayer could, in principle, bond two dissimilar metals together. The effect of the substrate is important for the first two fullerene monolayers, after which the fullerene films begin to grow crystalline phases similar to those found in bulk C_{60}.

Low-temperatures tend to favor the growth of commensurate epilayers of small grain size in cases where the bonding is strong and approximate lattice matching can occur. Such growth, however, is often accompanied by a great deal of strain that is relieved by the formation of a large number of defects in the epilayer. Because of enhanced surface diffusion, higher temperatures favor growth that exhibits much larger structural perfection on incommensurate epilayers. An example of incommensurate growth is the formation at higher temperatures (470K) of nearly incommensurate defect-free $C_{60}(111)$ regions of ~100 nm in size on $GaAs(110)$[92]. An example of commensurate overlayer formation is the $c(2\times4)$ overlayer of C_{60} on $GaAs(110)$[93]. Since the sublimation of C_{60} molecules from a surface becomes appreciable in the temperature range 300–500C, or lower from some surfaces, film growth of fullerene nanostructures is normally carried out at temperatures below which sublimation is pronounced.

A wide variety of substrates have been used for C_{60} deposition, including $Si(111)$, $GaAs(110)$, $GeS(001)$, sapphire, $Au(111)$, $Au(100)$, $Cu(111)$, $CaF_2(111)$, mica, H-terminated $Si(111)$, cleaved MoS_2, $NaCl$, and other alkali halides[94]. Suitable substrates for C_{60} require a large cohesive C_{60}-substrate so that intercalation does not occur and that the integrity of the metal surface is preserved upon fullerene adsorption. Furthermore, studies of the fullerene-substrate interface interaction have been carried out for a variety of fullerenes, including C_{60}, C_{70}, C_{76}, C_{78}, C_{82}, and C_{84}[94,95], showing that the reduction of symmetry tends to result in a preferred alignment of the fullerenes during film growth, as imposed by steric considerations.

5. Intercalated Fullerene Arrays

Intercalation provides a method for forming arrays of fullerene anions, which in some cases can be conducting or even superconducting. Almost all studies of intercalated fullerene arrays have employed alkali metal guest species as exohedral dopants [see Figure 5], though some studies have also been carried out on alkaline earth dopants, and a very few on rare earth dopants[14]. In the intercalation process, the guest species is introduced into the interstitial positions between adjacent molecules (exohedral locations) [see Figure 5], showing a strong resemblance to the intercalation of donors into graphite intercalation compounds[96], except that the guest atoms in Figure 5 are relatively small (relative to the fullerene anions) and fit into the interstitial spaces with only small increases in lattice constant for the doped compound. Charge transfer can take place between the M atoms and the fullerene molecules, so that the M atoms become positively charged ions and the molecules become negatively charged anions with predominately delocalized electrons. With exohedral doping, the conductivity of fullerene solids can be increased by many orders of magnitude[96].

Among the alkali metals, Li, Na, K, Rb, and Cs and their alloys have been used as exohedral dopants for C_{60}[97,98]. The doping of the C_{60} with alkali metals can be achieved in a two temperature oven, similar to the apparatus used to prepare alkali-metal graphite intercalation compounds[96,99]. Some success has also been achieved with the intercalation of alkaline earth dopants, such as Ca, Sr, and Ba[28,100,101]. In the case of the alkaline earth dopants, two electrons per metal atom M are transferred to the C_{60} molecules for low concentrations of metal atoms, while for high concentrations of metal atom the charge transfer is less than two electrons per alkaline earth ion. In general, the alkaline earth ions are smaller than the corresponding alkali metals in the same row of the periodic table. For this reason, the crystal structures formed with alkaline earth doping are often different from those for the alkali metal dopants. Doping fullerenes with acceptors has been more difficult than with donors because of the high electron affinity of C_{60} (see Table 1). Although a very few stable compounds with acceptor-type dopants have been synthesized[102,103], acceptor doping has not been important for the synthesis of electronic materials based on fullerenes.

The chemical modification of fullerene nanoparticles has become a very active research field, largely because of the uniqueness of the C_{60} molecule and the variety of chemical reactions that appear to be possible[104,105]. Many new fullerene-derivatives have already been synthesized and characterized chemically. Fullerene chemistry thus offers a means for the preparation of a large variety of new nanostructures with properties that can be specifically tailored. A few crystal structures have already been determined, but almost no studies of their electronic properties have been reported until now.

D. Electronic Properties

On the basis of the Raman, infrared, and optical studies, it is concluded that fullerenes form highly molecular solids. Thus, their electronic structures are expected to be closely related to the electronic levels of the isolated fullerene molecules.

The most extensive calculations of the electronic structure of fullerenes so far have been done for C_{60}. Representative results for the free C_{60} molecule are shown in Figure 7(a) where it is seen that because of the molecular nature of solid C_{60}, the electronic structure for the solid phase would be expected to be closely related to that of the free molecule[107]. It has further been shown that a molecular approach can explain a large body of experimental observations pertinent to the optical properties of C_{60}, which are difficult to explain on the basis of a one-electron energy band picture.

According to the molecular approach, many-electron-orbitals for the "valence" and "conduction" electron states are constructed for the π-electrons on the C_{60} molecule, since the lower-lying σ orbitals are filled. The number of π-electrons on the C_{60} molecule is 60 (i.e., one state per carbon atom). In relating the levels of icosahedral C_{60} to those with full rotational symmetry, we note that 50 π-electrons fully occupy the angular momentum states through $l = 4$, and the remaining 10 electrons are available to start filling the $l = 5$ state.

In full spherical symmetry, the $l = 5$ state can accommodate 22 electrons, and the splitting of the $l = 5$ state in icosahedral symmetry is into the $H_u + F_{1u} + F_{2u}$ irreducible representations. The level of lowest energy by Hund's rule is the 5-fold H_u level which is completely filled by the 10 available electrons. Neglecting any thermal excitation, the two 3-fold F_{1u} and F_{2u} (some authors refer to these levels as T_{1u} and T_{2u}) levels in icosahedral symmetry are empty. Whether C_{60} is considered in terms of a spherical approximation or its icosahedral symmetry, the ground state is non-degenerate with a total angular momentum of $J = 0$. This special circumstance has been suggested as being in part responsible for the high stability of the C_{60} molecule[108]. Since the H_u highest occupied molecular orbital (HOMO) for C_{60} is fully occupied and the F_{1u} lowest unoccupied molecular orbital (LUMO) is completely empty, C_{60} is a semiconductor.

Transport measurements show that C_{60} is a semiconductor with a very high resistivity ($> 10^8\ \Omega$-cm). However, by doping C_{60} with alkali metals, alkaline earths, or other donor species, the resulting charge transfer greatly reduces the resistivity and in some cases can yield metallic conduction. Because of the similar parity for the HOMO and LUMO levels, electric dipole transitions across the band gap are symmetry forbidden. Thus the optical absorption at the absorption

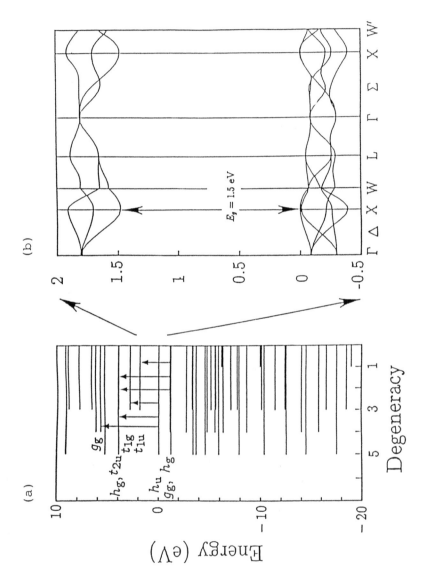

FIGURE 7. Calculated electronic structure of (a) an isolated C_{60} molecule and (b) FCC solid C_{60} where the direct band gap at the X-point is 1.5 eV[106].

edge is weak in intensity and involves vibronic excitations to excitonic states both for the free molecules and in the solid state.

Assuming one electron to be transferred to the C_{60} molecule per alkali-metal atom dopant, the LUMO levels are expected to become half-occupied at the alkali metal stoichiometry $M_3 C_{60}$ and totally occupied at $M_6 C_{60}$, leading to a filled-shell configuration for the $F_{1u}(t_{1u})$ band [see Figure 7]. Thus $M_6 C_{60}$ would be expected to be semiconducting with a band gap between the F_{1u} and F_{2g} (T_{1u} and T_{2g}) levels [see Figure 7], while $M_3 C_{60}$ should be metallic, provided that no bandgap is introduced at the Fermi level by a Peierls distortion. A large increase in optical absorptivity in $M_x C_{60}$ with alkali metal doping is observed as $x \rightarrow 3$ and this observation supports the metallic nature of $M_3 C_{60}$, in agreement with the large increase in electrical conductivity observed at $M_3 C_{60}$ for $M = K$, Rb. Below ~0.5 eV, the free electron contribution to the optical properties is dominant, and above ~0.5 eV, interband transitions from the partially occupied F_{1u} level to the higher lying F_{1g} level take place. It should be emphasized that the level-filling arguments for $M_x C_{60}$ pertain to the free molecule as well as to the levels in the solid state.

Fermi surface calculations for $K_3 C_{60}$ indicate a hole Fermi surface around the Γ-point of the Brillouin zone, and this hole surface contributes about 12% to the density of states at the Fermi surface[109]. These calculations also indicate a larger and more complicated Fermi surface consisting of both electron and hole orbits which contribute the remaining 88% to the density of states. Calculated values for m^* based on a band model yield $1.3 m_e$ (where m_e is the free electron mass) for the conduction band of C_{60} and $1.5\ m_e$ and $3.4 m_e$ for the valence band carriers.

Because of the weak van der Waals interaction of the fullerene molecules with each other and with the alkali metal dopants, solid $M_3 C_{60}$ is close to being a molecular solid, having energy levels with little dispersion, and thus giving rise to a very high density of states near the Fermi level. This property is important in understanding superconductivity in the $K_3 C_{60}$, $Rb_3 C_{60}$, and related compounds.

III. CARBON TUBULES AS NANOSTRUCTURES

In addition to quantum dot 0D fullerene nanostructures discussed in Section II, it is also possible to synthesize tubular fullerenes which can be classified as 1D quantum wires. The field of carbon nanotube research was greatly stimulated by the initial report of the experimental observation of carbon tubules[110] and the subsequent report of conditions for the synthesis of large quantities of nanotubes[111,112]. Various experiments carried out thus far (high resolution TEM, STM, resistivity, Raman scattering and susceptibility) are consistent with identifying the carbon nanotubes with rolled up cylinders of graphene sheets of sp^2 bonded carbon atoms. Because of their very small diameters (down to ~0.7 nm), carbon nanotubes are prototype hollow cylindrical quantum wires. They can be prepared both as monolayer and multilayer nested concentric fullerenes, and theoretical calculations have been carried out for both cases.

Formally, carbon nanotubes and fullerenes have a number of common features despite their different dimensionalities. The theoretical literature focusses mainly on single-walled tubules, cylindrical in shape with caps at each end, such that the two caps can be joined together to form a fullerene [see Figure 1(d)]. Formally, the cylindrical portions of the tubules consist of a single graphene sheet, rolled up to form the cylinder. In this section we discuss the structure and properties of carbon nanotubes and their remarkable electronic properties.

A. Observation of Carbon Nanotubes

The earliest observations of carbon tubules with very small (nanometer) diameters[110,113,114] were based on high resolution transmission electron microscopy (TEM) measurements, providing evidence for μm-long tubules, with cross-sections showing several concentric coaxial tubes and a hollow core. In Figure 8, the first published observations of carbon nanotubes are shown[110]. Here we see only multi-layer carbon nanotubes, but one tubule has only two coaxial carbon cylinders [Figure 8(b)], and another has an inner diameter of only 2.3 nm [Figure 8(c)][110]. These carbon nanotubes were prepared by a carbon arc process (typical dc current of 50–100 A and voltage of 20–25 V) where carbon nanotubes in the form of tubule bundles are found on the negative electrode, as the positive electrode is consumed in the arc discharge[116]. Typical lengths of the arc-grown tubules are ~1 μm, giving rise to an aspect ratio (length to diameter ratio) of 10^2 to 10^3. Because of their small diameter, involving only a small number of carbon atoms, and because of their large aspect ratio, carbon nanotubes are classified as 1D carbon systems. Most of the theoretical work on carbon nanotubes emphasizes their 1D properties. In the multilayered carbon nanotubes, the measured (by high resolution TEM) interlayer distance is 0.34 nm[110], in good agreement with the value of 0.339 nm for the average equilibrium interlayer separation, obtained from self-consistent electronic structure calculations[117,118].

Although very small diameter (less than 10 nm) carbon filaments, were observed many years earlier on vapor grown carbon fibers[119,120], no detailed systematic studies of such very thin filaments were reported in the 1970's and 80's. A direct stimulus to the systematic study of carbon filaments of very small diameters came from the discovery of fullerenes by Kroto, Smalley, and coworkers[1], and subsequent developments resulting in the synthesis of gram quantities of fullerenes by Krätschmer, Huffman and coworkers[8]. These recent developments heralded the entry of many scientists into the field, together with many ideas for new carbon materials, and bringing new importance to carbon systems of nanometer dimensions. Independently, Russian workers also reported discovery of carbon tubules and nanotube bundles, but generally having much smaller aspect ratios, and hence they called their tubules "barrelenes"[121,122]. These barrelenes have similarities to fullerenes reported by Wang and Buseck[123] with length to diameter ratios or 10 or less.

A major advance was made when it was found that single-wall nanotubes could be formed in an arc discharge chamber using a catalyst, such as *Fe, Co* and other transition metals, during the synthesis process[124,125]. Single-wall tubules are remarkably flexible, and bend into curved arcs with radii of curvature

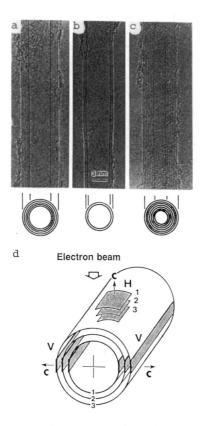

FIGURE 8. The observation of N concentric carbon tubules with various inner diameters d_i and outer diameters d_o reported by Iijima using TEM. (a) $N = 5$, $d_o = 6.7$ nm, (b) $N = 2$, $d_o = 5.5$ nm, and (c) $N = 7$, $d_i = 2.3$ nm, $d_o = 6.5$ nm. Each cylinder is described by its diameter and chiral angle. The sketch (d) indicates how the interference pattern for the parallel planes labeled H are used to determine the chiral angle θ, which is the angle between the tubule axis and the nearest zigzag axis defined in Figure 10[110].

as small as 20 nm. This flexibility suggests excellent mechanical properties, consistent with the high tensile strength and bulk modulus of commercial and research-grade vapor grown carbon fibers[126]. Lengths of 700 nm have been reported for a 0.9 nm diameter tubule, yielding a large length to diameter ratio (aspect ratio) of ~800, though not as large as the aspect ratio (~10^4) found in the larger (> 100 nm diameter) vapor grown carbon fibers[127]. The single-wall nanotubes, just like the multi-wall nanotubes (and also conventional vapor grown carbon fibers), have hollow cores along the axis of the tubule.

The diameter distribution of single-wall carbon nanotubes is of great interest for both theoretical and experimental reasons, since theoretical studies indicate that the physical properties of carbon nanotubes are strongly dependent on tubule diameter. Early results for the diameter distribution of Fe-catalyzed single-wall nanotubes [Figure 9] show a diameter range between 0.7 nm and 1.6 nm, with the largest peak in the distribution at 1.05 nm, and with a smaller peak at 0.85 nm[124]. The smallest reported diameter for the single-wall carbon nanotubes is 0.7 nm[124], the same as the smallest diameter expected theoretically (0.71 nm) for a tubule based on C_{60}[115]. Qualitatively similar results, but differing in detail, were obtained for the Co-catalyzed nanotubes, with a peak in the distribution at 1.3 nm[125]. These experimental results indicate that the diameter distribution of the single-wall nanotube involves predominantly small diameter tubules.

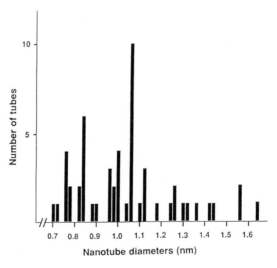

FIGURE 9. Histogram of the single-wall nanotube diameter distribution for Fe-catalyzed nanotubes[124].

B. Structure of Carbon Nanotubes

The structure of carbon nanotubes has been explored by high resolution TEM and STM, yielding direct confirmation that the nanotubes are cylinders derived from the honeycomb lattice (graphene sheet). Evidence that the tubules are cylinders and are not scrolls comes from observation of the same numbers of walls on the left and right hand sides of thousands of TEM images of nanotubes [see e.g., Figure 8]. The chirality of a carbon nanotube is conveniently explained in terms of its 1D unit cell as shown in Figure 10(a). Measurements of the tubule diameter d_t are conveniently made by using STM and TEM techniques. Measurements of the chiral angle θ have been made by high resolution TEM[124].

While the ability to measure the diameter d_t and chiral angle θ of individual single-wall tubules has been demonstrated, it remains a major challenge to determine d_t and θ for specific tubules used for an actual property measurement, such as electrical conductivity, magnetoresistance or Raman scattering.

The circumference of any carbon nanotube is expressed in terms of the chiral vector $\vec{C}_h = n\hat{a}_1 + m\hat{a}_2$ which connects two crystallographically equivalent sites on a 2D graphene sheet [see Figure 10(a)][115]. The construction in Figure 10 shows the chiral angle θ between $\vec{C}_h$ and the zigzag direction ($\theta = 0$) and the unit vectors $\hat{a}_1$ and $\hat{a}_2$ of the hexagonal honeycomb lattice. Whereas the zigzag tubule with its tubule axis along a three-fold axis of the fullerene has $\theta = 0°$, the armchair tubule corresponds to a chiral angle of $\theta = 30°$ [Figures 10(a) and 11]. An ensemble of chiral vectors specified by pairs of integers (n,m) denoting the vector $\vec{C}_h = n\hat{a}_1 + m\hat{a}_2$ is given in Figure 10(b)[128]. The intersection of OB with the first lattice point determines the fundamental 1D translation vector $\vec{T}$ and hence the unit cell of the 1D lattice [Figure 10(a)].

The cylinder connecting the two hemispherical caps is formed by superimposing the two ends of the vector $\vec{C}_h$ and the cylinder joint is made along the two lines OB and AB' in Figure 10(a). These two lines are perpendicular to the vector $\vec{C}_h$ at each end[115]. The chiral tubule, thus generated has no distortion of bond angles other than distortions caused by the cylindrical curvature of the tubule. Differences in chiral angle θ and in the tubule diameter d give rise to differences in the properties of the various graphene tubules. In the (n,m) notation for $\vec{C}_h = n\hat{a}_1 + m\hat{a}_2$, the vectors $(n,0)$ or $(0,m)$ denote zigzag tubules and the vectors (n,n) denote armchair tubules. All other vectors (n,m) correspond to chiral tubules[128]. In terms of the integers (n,m), the tubule diameter d_t is given by

$$d_t = \sqrt{3}\, a_{C-C}(m^2 + mn + n^2)^{1/2}/\pi \qquad (1)$$

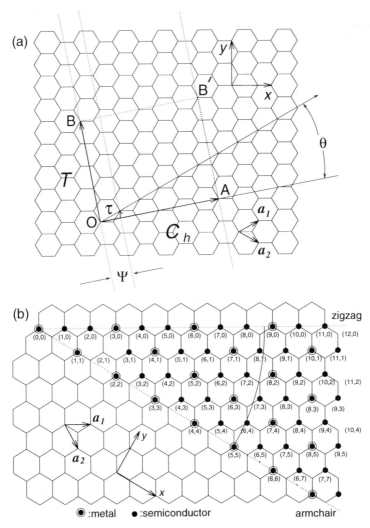

FIGURE 10. (a) The chiral vector $\overrightarrow{OA}$ or $\overrightarrow{C}_h = n\hat{a}_1 + m\hat{a}_2$ is defined on the honeycomb lattice of carbon atoms by unit vectors $\hat{a}_1$ and $\hat{a}_2$ and the chiral angle θ with respect to the zigzag axis. Along the zigzag axis $\theta = 0°$. Also shown are the lattice vector $\overrightarrow{OB} = \overrightarrow{T}$ of the 1D tubule unit cell and the rotation angle ψ and the translation τ which constitute the basic symmetry operation $R = (\psi|\tau)$ for the carbon nanotube. The diagram is constructed for $(n,m) = (4,2)$. (b) Possible vectors specified by the pairs of integers (n,m) for general carbon tubules, including zigzag, armchair, and chiral tubules. The encircled dots denote metallic tubules while the small dots are for semiconducting tubules[115].

and the chiral angle θ is given by

$$\theta = tan^{-1}(\sqrt{3}\,n/(2m+n)). \qquad (2)$$

The number of hexagons, N, per unit cell of a chiral tubule specified by integers (n,m) is given by

$$N = \frac{2(m^2+n^2+nm)}{d_R} \qquad (3)$$

where d_R is given by

$$d_R = \begin{cases} d & \text{if } n-m \text{ is not a multiple of } 3d \\ 3d & \text{if } n-m \text{ is a multiple of } 3d. \end{cases} \qquad (4)$$

and we note that each hexagon contains two carbon atoms. As an example, application of Eq.(3) to the (5,5) and (9,0) tubules yields values of 10 and 18, respectively, for N. These unit cells of the 1D tubule contain, respectively, five and nine

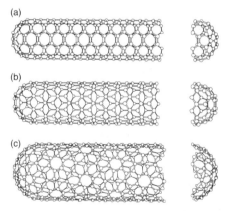

FIGURE 11. By rolling a graphene sheet (a single layer from a 3D graphite crystal) as a cylinder and then capping each end of the cylinder with half of a C_{60} molecule, a "C_{60} tubule" one layer in thickness is formed. Shown here is a schematic model for a tubule arising from (a) an armchair cap with a 5-fold axis, (b) a zigzag cap with a 3-fold axis, and (c) a chiral (10,5) nanotube with caps associated with C_{140}[115].

unit cells of the 2D graphene lattice, each 2D unit cell containing two hexagons of the honeycomb lattice. These 1D unit cells are used in the application of zone-folding techniques to obtain the electronic dispersion relations of carbon nanotubes, as discussed below.

Because of the special atomic arrangement of the carbon atoms in a C_{60} tubule, substitutional impurities are inhibited by the small size of the carbon atoms. Furthermore, the screw axis dislocation, the most common defect found in bulk graphite, is inhibited by the monolayer structure of the C_{60} tubule. For these reasons, we expect relatively few substitutional or structural impurities in carbon nanotubes.

C. Electronic Structure of Carbon Nanotubes

From the standpoint of nanostructures, fullerene tubules are interesting as examples of a one-dimensional periodic structure along the tubule axis. Confinement of the structure in the radial direction is provided by the monolayer thickness of the tubule. In the circumferential direction, periodic boundary conditions apply to the enlarged unit cell that is formed in real space and the subsequent zone folding that occurs in reciprocal space. We can then expect to observe 1D dispersion relations for electrons and phonons in C_{60} −derived tubules.

A number of methods have been used to calculate the 1D electronic energy bands for fullerene tubules[129–133] and all relate to the 2D graphene honeycomb sheet used to form the tubule. These calculated results for the 1D electronic structure show that for small diameter graphene tubules, about 1/3 of the tubules are metallic and 2/3 are semiconducting, depending on the tubule diameter d_t and chiral angle θ. Metallic conduction in a fullerene tubule is achieved when

$$2n + m = 3q \tag{5}$$

where n and m are integers specifying the tubule diameter and chiral angle and where q is an integer. Tubules satisfying Eq. (5) are indicated in Figure 10(b) as large circles and these are the metallic tubules. The small circles in this figure correspond to semiconducting tubules.

Dispersion relations are shown for metallic tubules $(m,n) = (5,5)$ and $(9,0)$ in Figures 12(a) and (b)[128,134], respectively, and for a semiconducting tubule $(m,n) = (10,0)$ in Figure 12(c)[134]. Figure 10(b) shows that all armchair tubules are metallic, but only 1/3 of the possible zigzag tubules are metallic[128]. It is surprising that the calculated electronic structure can be either metallic or semiconducting depending on the choice of (n,m), although there is no difference in the local chemical bonding between the carbon atoms in the tubules, and no doping impurities are present[128]. These surprising results can be understood on the

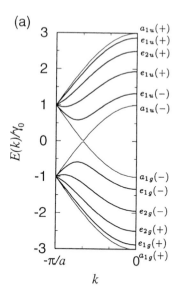

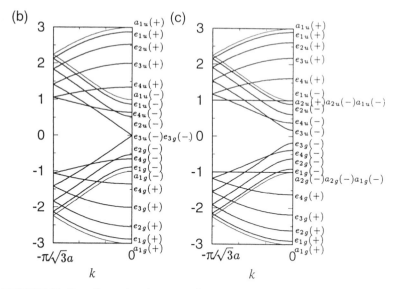

FIGURE 12. One-dimensional energy dispersion relations for (a) armchair (5,5) tubules, (b) zigzag (9,0) tubules, and (c) zigzag (10,0) tubules labeled by the irreducible representations of the point group $D_{(2n+1)d}$ at $k=0$. The A-bands are non-degenerate and the E-bands are doubly degenerate at a general k-point[128,134,135].

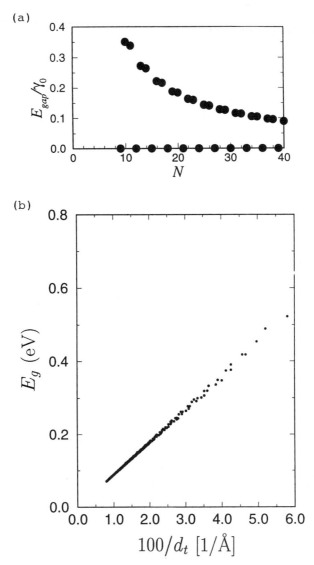

FIGURE 13. The dependence of the energy gap (normalized to the nearest neighbor overlap energy) on tubule diameter. (a) The dependence of the energy gap on the number of carbon atoms N along the circumference of zig-zag tubules[137]. (b) The energy gap E_g for a general chiral carbon nanotube as a function of $100 \text{ Å} / d_t$, where d_t is the tubule diameter in Å.

basis of the electronic structure of 2D graphite which is a zero gap semiconductor[136] with bonding and antibonding π bands, degenerate at the K-point (zone corner) of the hexagonal 2D Brillouin zone. The periodic boundary conditions for the 1D tubules permit only a few wave vectors to exist in the circumferential direction. Metallic conduction occurs when one of these wave vectors passes through the K-point of the 2D Brillouin zone, where the valence and conduction bands are degenerate because of the symmetry of the 2D graphene lattice.

Metallic 1D energy bands are generally unstable under a Peierls distortion. However, the Peierls energy gap obtained for the metallic tubules is found to be greatly suppressed by increasing the tubule diameter, so that the Peierls gap quickly approaches the zero-energy gap of 2D graphite[130,135]. Thus if we consider finite temperatures or fluctuation effects, such a small Peierls gap can be neglected. As the tubule diameter increases, more wave vectors become allowed for the circumferential direction, the tubules become more two-dimensional and the semiconducting band gap disappears, as is illustrated in Figure 13(a) which shows the diameter dependence of the semiconducting band gap, and its $\sim 1/d_t$ diameter dependence [Figure 13(b)]. At a tubule diameter of $d_t \sim 3$ nm [Figure 13(b)], the bandgap becomes comparable to thermal energies at room temperature. Calculation of the electronic structure for two concentric tubules shows that pairs of concentric metal-semiconductor or semiconductor-metal tubules are stable[137].

From these results one could imagine designing an electronic device consisting of two concentric graphene tubules with a smaller diameter metallic inner tubule surrounded by a larger diameter semiconducting (or insulating) outer tubule. This concept could be extended to the design of tubular metal-semiconductor all-carbon devices without introducing any doping impurities[128].

D. Tunneling Conductance of Nanotube Junctions

The tunneling conductance between two carbon nanotubes of different diameters has recently been considered[138], and it was shown that the joint between two carbon nanotubes can be uniquely specified in terms of their individual chiral vectors and the pentagon and heptagon that must be introduced in the junction region between them [see insets to Figure 14]. The conductance [see Figure 14] between two metallic nanotubes is found to be ballistic with some reflection effects occurring in the junction region, while a metal-semiconductor nanotube junction shows tunneling behavior across the junction. Oscillatory phenomena, perhaps related to the observed universal conductance fluctuations[139], are also found.

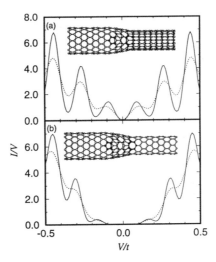

FIGURE 14. Calculated Conductance I/V for (a) (12,0)–(9,0) and (b) (12,0)–(8,0) zigzag carbon nanotubes (see insets), as a function of voltage V in units of the nearest-neighbor overlap integral γ_0. Two different Gaussian broadening values, $\Delta E/\gamma_0 = 0.33$ (solid line) and $\Delta E/\gamma_0 = 0.50$ (dotted line), are used for each of the nanotube junctions. The estimated energy gap for the (8,0) semiconductor nanotube is $0.55\gamma_0$[138].

E. Experimental Studies of Electronic Structure of Carbon Nanotubes

Experimental measurements to test the remarkable theoretical predictions of the electronic structure of carbon nanotubes are difficult to carry out because of the strong dependence of the predicted properties on tubule diameter and chirality. The experimental difficulties arise from the great experimental challenges in making electronic or optical measurements on individual single-wall nanotubes, and further challenges arise in making such demanding measurements on individual nanotubes that have been characterized with regard to diameter and chiral angle (d_t and θ). Despite these difficulties, pioneering work has already been reported on experimental observations relevant to the electronic structure of individual nanotubes or on bundles of nanotubes, as summarized in this section.

The most promising present technique for carrying out sensitive measurements of the electronic properties of individual nanotubes is scanning tunneling spectroscopy (STS) because of the ability of the tunneling tip to probe most sensitively the electronic density of states of either a single-wall nanotube[140], or the outermost cylinder of a multi-wall tubule. With this technique, it is further

possible to carry out both STS and scanning tunneling microscopy (STM) measurements at the same location on the same tubule and therefore to measure the
tubule diameter concurrently with the STS spectrum. It has also been demonstrated that the chiral angle θ of a carbon tubule can be determined using the STM
technique[141] or high-resolution TEM[110,124,142–144].

Several groups have thus far attempted STS studies of individual nanotubes[144,145]. Although still preliminary, the study which provides the most
detailed test of the theory for the electronic properties of the 1D carbon nanotubes, thus far, is the combined STM/STS study by Olk and Heremans[145]. In
this STM/STS study, more than nine individual multilayer tubules with diameters
ranging from 1.7 to 9.5 nm were examined. STM measurements were taken to
verify the exponential relation between the tunneling current and the tip-to-tubule
distance in order to confirm that the tunneling measurements pertain to the tubule
and not to contamination on the tubule surface. Barrier heights were measured to
establish the valid tunneling range for the tunneling tip. Topographic STM measurements were made to obtain the maximum height of the tubule relative to the
gold substrate, thus determining the diameter of an individual tubule[145]. Then
switching to the STS mode of operation, current–voltage (I–V) plots were made
on the same region of the same tubule as was characterized for its diameter by the

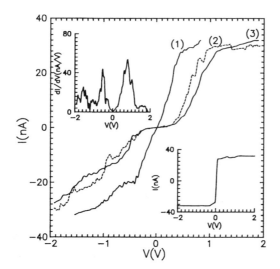

FIGURE 15. Current–voltage I vs. V traces taken with scanning tunneling spectroscopy (STS) on individual nanotubes of various diameters: (1) $d_t = 8.7$ nm,
(2) $d_t = 4.0$ nm, and (3) $d_t = 1.7$ nm. The top inset shows the conductance vs. voltage
plot for data taken on the 1.7 nm nanotube. The bottom inset shows an I–V trace taken
on a gold surface under the same conditions[145].

STM measurement. The I–V plots for three typical tubules are shown in Figure 15.

The results provide evidence for one metallic tubule with $d_t = 8.7$ nm [trace (1)] showing ohmic behavior, and two semiconducting tubules [trace (2) for a tubule with $d_t = 4.0$ nm and trace (3) for a tubule with $d_t = 1.7$ nm] showing plateaus at zero current and passing through $V = 0$. The dI/dV plot in the upper inset provides a crude attempt to measure the density of states, the peaks in the dI/dV plot being attributed to $(E_0 - E)^{-1/2}$ dependent singularities in the 1D density of states. Finally, the results for all the semiconducting tubules measured in this study by Olk and Heremans[145] showed a linear dependence of their energy gaps on $1/d_t$, the reciprocal tubule diameter, consistent with the predicted functional form [see Figure 13(b)].

Measurements of the temperature-dependent resistance and magnetoresistance have been recently reported for a single carbon nanotube 20 nm in diameter using 4 attached electrical contacts[139]. The results show negative magnetoresistance, evidence for weak localization, and evidence for universal quantum fluctuations. All observed phenomena appear to relate to 2D behavior. It is not currently understood whether the 2D behavior is due to the relatively large diameter of the tubule or to the interaction between carbon atoms on adjacent tubules.

IV. CARBON ONIONS

Closely related to the multi-wall carbon nanotubes and fullerenes are hollow carbon spherules [see Figure 16(b)] In commercial products, structures consisting of concentric spherical graphitic shells and ranging from 10 nm to 1 μm outer diameter are commonly called carbon blacks[146,147,148]. Ugarte has recently reported[149] the formation of hollow concentric carbon spheres upon intense electron beam irradiation of sooty carbon particles [see Figure 16]. Of particular interest in the recent work is the observation of stable carbon spherules or "onions" with an inner diameter of 0.71 nm, corresponding to the diameter of the C_{60} molecule. It is found that if enough energy is provided to form concentric spherical shell structures, their formation is favored over the coaxial carbon nanotube structures[149]. In this work, spherical shells up to diameter dimensions of 10 nm have been synthesized, similar to the smallest dimensions reported for the spherical shells of carbon blacks. Though containing a large amount of strain energy, the spherical shells contain no dangling bonds, making the spheres more stable than graphite sheets under some circumstances[151].

For structures involving few carbon atoms, the 3D planar graphite structure is not the most stable, because of the large concentration of dangling bonds. In this regime the small closed cage molecules gain stability, because of the absence of dangling bonds, despite the high penalty of the curved surfaces and the concomitant strain energy. The stabilization of a concentric stacking of double-layered

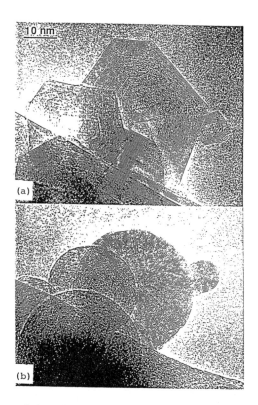

FIGURE 16. High-resolution electron micrographs of graphitic particles: (a) as obtained from an electric arc deposit, the particles display a well-defined faceted structure and a large inner hollow space, and (b) the same particles after being subjected to intense electron irradiation. The particles in (b) are commonly called carbon "onions" and show a spherical shape of concentric carbon spheres[150].

$C_{60}@C_{240}$ has been calculated[152], showing an optimized interlayer separation of 0.352 nm and a stabilization energy of 14 meV per carbon atom.

For the case of carbon "onions", the growth of carbon layers is believed to begin at the surface and progress toward the center[146,153]. The onion-like carbon particles are stabilized by the energy gain from the weak van der Waals interaction between adjacent carbon shells[154–156]. *Ab initio* calculations show that for C_{240}, a spherical fullerene is slightly more stable (−7.07 eV/carbon atom) than a polyhedral fullerene (−7.00 eV/carbon atom)[157] and that concentric spherical shells of $C_{240}@C_{540}$ are more stable than their concentric icosahedral counterparts[150]. The size (number of carbon atoms) at which the total energy of a closed surface particle becomes equal to a planar graphite sheet remains unanswered both theoretically and experimentally. The synthesis and purification of macroscopic quantities of quasispherical onion-like particles with a small size distribution remain a major challenge to studying the properties of carbon onions.

V. SELECTED POTENTIAL APPLICATIONS

Since all C_{60} molecules are identical and since C_{60} can be synthesized to high purity and in large quantities (gram quantities) at relatively low cost (~$100/g), the C_{60} molecule is attractive for a variety of nanostructure applications. Since C_{60} molecules are strongly bonded internally, chemically inert, nearly spherical, and nonpolar due to symmetry, and weakly bonded to other C_{60} molecules by van der Waals forces, these molecules can be considered to be nanostructures of subnanometer size (diameter of 0.7 nm) which can be manipulated using STM probe tips[158], just as is done for noble gas atoms. Moreover, some other manipulation methods may be possible for C_{60} that are not possible for any other species. For instance, it has been proposed[159], that a C_{60} molecule can be "rolled" (diffused) along the surface of a suitable ionic substrate by a rotating external electric field, utilizing the large size and polarizability of C_{60} as illustrated in Figure 17.

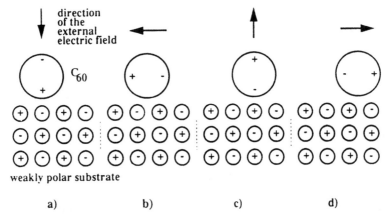

FIGURE 17. The principle of moving C_{60} molecules by a rotating external electric field. An isolated molecule is assumed to be adsorbed on an ideally flat, weakly polar substrate. A rotating external electric field induces a dipole moment that exceeds the effect of the substrate dipoles on the molecule; the successive directions of the field are shown in (a) through (d). The large diameter of the molecule should allow coupling to successive poles, thereby generating lateral movement[159].

A recent tribological study[160] suggested that nanocrystalline C_{60} islands grown on certain types of substrates [e.g., $NaCl(011)$] could be used as transport devices for fabrication processes of nanometer-sized machines. These 0.5–3 monolayer C_{60} islands might play the role of a tiny "nanosled" transporting larger molecules (e.g., biomolecules) to a desired location[160].

The C_{60} molecular nanostructures can be arranged in ordered arrays either by the growth of single crystals (3D) or by the epitaxial growth of C_{60} monolayers on selected substrates (2D)[14]. Since fullerenes are all-carbon molecules of interest for biotechnology applications, fullerene-based nanodevices may be capable of biological interfacing. A single layer of fullerenes could be of interest

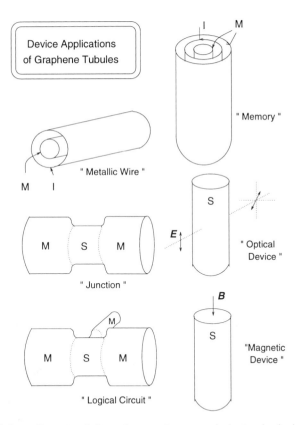

FIGURE 18. Schematic presentation of several proposed electronic device applications for carbon nanotubes[163].

for bonding two silicon surfaces together by tying up the dangling bonds on the *Si* surfaces. Fullerenes could also be used to enhance the resolution of STM tips, utilizing the uniform structure of the C_{60} molecules[14].

To interconnect nanoscale devices, it may someday be possible to utilize carbon nanotubes for nanowire applications. By varying the chirality and diameter of carbon nanotubes, it should be possible to alter the properties of the nanotubes from metallic to semiconducting. Such nanotubes could, in their own right, lead to nanoscopic electronic devices based on concentric semiconducting and metallic carbon tubules as shown in Figure 18. The significance of carbon nanotubes as electronic materials is the demonstration of quasi-1D cylindrical wires with a large aspect (length-to-diameter) ratio. There is also the possibility of the application of such filled nanotubes with regard to low-dimensional transport, magnetism, and superconductivity. The heterostructures shown in Figure 18 could function as an insulation-clad metallic wire, a tunnel junction, a capacitor in memory devices, or a transistor in switching circuitry[161,162]. Because the nanotubes themselves do not require any doping by impurities, as conventional semiconductors do, but acquire their electronic properties from their geometry, the resulting devices should be highly thermally stable and have high intrinsic mobility. Schematics for several devices have been suggested [see Figure 18].

While the field of nanotechnology based on carbon nanotubes is still about halfway between science and science fiction, it is a very active fast-growing field whose promise of nanoscale molecular devices and orders of magnitude increase in the integration density of electronic components is too appealing to be ignored. The discovery of fullerenes has given a significant boost to the field of nanotechnology by providing an abundance of stable, highly symmetric, nonreactive, and relatively large molecules that can, in principle, be manipulated one at a time. Fullerenes at semiconductor interfaces can be utilized to modify electronic device behavior on nanometer length scales. In such applications, C_{60} anions alone represent an almost ideal spherical capacitor that can store charge. Together with other carbon structures, such as tubules (serving as nanowires), a new carbon-based nanotechnology can be envisioned.

ACKNOWLEDGMENTS

We gratefully acknowledge the helpful discussions with Professors M. Endo and P.C. Eklund. We are also in debt to many other colleagues for assistance. The research at MIT is funded by NSF grant DMR-95-10093. One of the authors (RS) acknowledges the Japan Society for the Promotion of Science for supporting his visit to MIT by the US-Japan co-operative research program, and the MIT authors (GD and MSD) acknowledge corresponding support by NSF INT 94-90144 from the US-Japan program. Part of the work by RS is supported by a

Grant-in Aid for Scientific Research in Priority Area "Carbon Cluster" (Area No. 234/05233214) from the Ministry of Education, Science and Culture, of Japan.

REFERENCES

[1]   Kroto, H.W., *et al. Nature* (London) **318**, 162–163 (1985).

[2]   Fuller, R. Buckminster, *The Artifacts of R. Buckminster Fuller: A Comprehensive Collection of His Designs and Drawings*, Marlin,W., ed., New York: Garland Publishing, 1984.

[3]   Rohlfing, E.A., Cox, D.M., and Kaldor, A., *J. Chem. Phys.* **81**, 3322 (1984).

[4]   Taylor, R., *et al.*, *J. Chem. Soc. Chem. Commun.* **20**, 1423–1425 (1990).

[5]   Johnson, R.D., Meijer, G., and Bethune, D.S., *J. Am. Chem. Soc.* **112**, 8983–8984 (1990).

[6]   Krätschmer, W., Fostiropoulos, K., and Huffman, D.R., in "Dusty Objects in the Universe," *Proceedings of the 4th Inter. Workshop of the Astronomical Observatory of Capodimonte*, Bussolettii, E., and Vittone, A.A., eds., Dordrecht: Kluwer Academic Publishers, 1989.

[7]   Krätschmer, K., Fostiropoulos, K., and Huffman, D.R., *Chem. Phys. Lett.* **170**, 167–170 (1990).

[8]   Krätschmer, W., *et al. Nature* (London) **347**, 354–358 (1990).

[9]   Curl, R.F., and Smalley, R.E., *Science* **242**, 1017 (1988).

[10] Haddon, R.C., *et al.*, *Nature* (London) **350**, 320 (1991).

[11] Hebard, A.F., *et al.*, *Nature* (London) **350**, 600 (1991).

[12] Palstra, T.T.M., *et al.*, *Solid State Commun.* **93**, 327 (1995).

[13] Hebard, A.F., *Physics Today* **45**, 26 (1992).

[14] Dresselhaus, M.S., Dresselhaus, G., and Eklund, P.C., *Science of Fullerenes and Carbon Nanotubes*, New York: Academic Press, 1995, (in press).

[15] Dresselhaus, M.S., Dresselhaus, G., and Eklund, P.C., *J. Mater. Res.* **8**, 2054 (1993).

[16] Dresselhaus, G., Dresselhaus, M.S., and Eklund, P.C., *Phys. Rev. B* **45**, 6923 (1992).

[17] Stephens, P.W., *et al. Nature* (London) **351**, 632 (1991).

[18] Johnson, R.D., Bethune, D.S., and Yannoni, C.S., *Accounts Chem. Res.* **25**, 169 (1992).

[19] David, W.I.F., *et al.*, *Nature* (London) **353**, 147 (1991).

[20] Johnson, R.D., *et al.*, in *Videotape of Late News Session on Buckyballs, MRS, Boston*, Fleming, R.M., *et al*, eds., Pittsburgh, PA: Materials Research Society Press, 1990.

[21] Christides, C., *et al.*, *Phys. Rev. B* **49**, 2897 (1994).

[22] Saito, S., and Oshiyama, A., *Phys. Rev. B* **44**, 11532 (1991).

[23] Beckhaus, H.D., *et al.*, *Angew. Chem. Int. Edn. Engl.* **31**, 63–64 (1992).

[24] Lichtenberger, D.L., *et al.*, *Chem. Phys. Lett.* **176**, 203 (1991).

[25] Tokmakoff, A., Haynes, D.R., and George, S.M., *Chem. Phys. Lett.* **186**, 450 (1991).

[26] Dresselhaus, G., Dresselhaus, M.S., and Mavroides, J.G., *Carbon* **4**, 433 (1966).

[27] de Vries, J., *et al.*, *Chem. Phys. Lett.* **188**, 159 (1992).

[28] Kortan, A.R., *et al.*, *Nature* (London) **355**, 529 (1992).

[29] Jost, M.B., *Phys. Rev. B* **44**, 1966 (1991).

[30] Fischer, J.E., *et al.*, *Science* **252**, 1288 (1991).

[31] Lundin, A., and Sundqvist, B., *Europhys. Lett.* **27**, 463 (1994).

[32] Lundin, A., *et al.*, *Solid State Commun.* **84**, 879–883 (1992).

[33] Shi, X.D., *et al.*, *Phys. Rev. Lett.* **68**, 827 (1992).

[34] Fischer, J.E., *Mater. Sci. Eng.* **B19**, 90–99 (1993).

[35] Samara, G.A., *et al.*, *Phys. Rev. Lett.* **67**, 3136 (1991).

[36] Kriza, G., *et al.*, *J. Phys. I, France* **1**, 1361 (1991).

[37] Heiney, P.A. *et al.*, *Phys. Rev. B* **45**, 4544–4547 (1992).

[38] Rao, A.M., *et al.*, *Science* **259**, 955–957 (1993).

[39] Gensterblum, G., *et al.*, *Phys. Rev. B* **50**, 11981 (1994).

[40] Manfredini, M., Bottani, C.E., and Milani, P., *Chem. Phys. Lett.* **226**, 600 (1994).

[41] Grivei, E., *et al.*, *Solid State Commun.* **85**, 73 (1993).

[42] Grivei, E., *et al.*, *Phys. Rev. B* **48**, 8514 (1993).

[43] Yu, R.C., *et al.*, *Phys. Rev. Lett.* **68**, 2050–2053 (1992).

[44] Wen, C., *et al.*, *Appl. Phys. Lett.* **61**, 2162–2163 (1992).

[45] Ren, S.L., *et al.*, *Appl. Phys. Lett.* **59**, 2678 (1991).

[46] Wang, Y., *et al.*, *Phys. Rev. B* **45**, 14396–14399 (1992).

[47] Fischer, J.E., and Heiney, P.A., *J. Phys. Chem. Solids* **54**, 1725 (1993).

[48] Pan, C., *et al.*, *J. Phys. Chem.* **95**, 2944 (1991).

[49] Kim, E., Lee, Y.H., and Lee, J.Y., *Phys. Rev. B* **48**, 18230 (1993).

[50] Fischer, J.E., Heiney, P.A., and Smith III, A.B., *Accounts Chem. Res.* **25**, 112 (1992).

[51] Diederich, F., and Whetten, R.L., *Accounts Chem. Res.* **25**, 119 (1992).

[52] Kikuchi, K., *et al.*, *Nature (London)* **357**, 142 (1992).

[53] Manolopoulos, D.E., and Fowler, P.W., *Chem. Phys. Lett.* **187**, 1 (1991).

[54] Bethune, D.S., *et al.*, *Nature* (London) **366**, 123 (1993).

[55] Zimmermann, U., *et al.*, *Phys. Rev. Lett.* **72**, 3542 (1994).

[56] Zimmermann, U., *et al.*, *Carbon* **33**, 995 (1995).

[57] Smalley, R.E., *Mater. Sci. Eng.* **B19**, 1–7 (1993).

[58] Smalley, R.E., *Accounts Chem. Res.* **25**, 98 (1992).

[59] Yannoni, C.S., *et al.*, *Metals* **59**, 279 (1993).

[60] Kikuchi, K., *et al.*, *Chem. Phys. Lett.* **216**, 67 (1993).

[61] Heath, J.R., *et al.*, *J. Am. Chem. Soc.* **107**, 7779–7780 (1985).

[62] O'Brien, S.C., *et al.*, *J. Chem. Phys.* **88**, 220–230 (1988).

[63] Weiss, F.D., *et al.*, *J. Am. Chem. Soc.* **110**, 4464 (1988).

[64] Poirier, D.M., *et al.*, *Phys. Rev. B* **49**, 17403 (1994).

[65] Nagase, S., *et al.*, *Chem. Phys. Lett.* **201**, 475–479 (1993).

[66] Fowler, P.W., and Manolopoulos, D.E., *Nature* (London) **355**, 428 (1992).

[67] Kikuchi, K., *et al.*, *Nature* (London) **357**, 142 (1993).

[68] Suzuki, S., *et al.*, *J. Phys. Chem.* **96**, 7159 (1992).

[69] Hoinkis, M., *et al.*, *Phys. Chem. Lett.* **198**, 461 (1992).

[70] Wang, X.-D., *et al.*, *Jpn. J. Appl. Phys.* **B32**, L866 (1993).

[71] Shinohara, H., *et al.*, *Mater. Sci. Eng.* **B19**, 25–30 (1993).

[72]   van Loosdrecht, P.H.M., *et al.*, *Phys. Rev. Lett.* **73**, 3415–3418 (1994).

[73]   Saito, S., Sawada, S., and Hamada, N., *Phys. Rev. B* **45**, 13845 (1992).

[74]   Nagase, S., and Kobayashi, K., *Chem. Phys. Lett.* **228**, 106–110 (1994).

[75]   Saunders, M., *et al.*, *Science* **259**, 1428 (1993).

[76]   Wang, X.-D., *et al.*, *J. Appl. Phys.* **32**, L147 (1993).

[77]   Martin, T.P., *et al.*, *J. Chem. Phys.* **99**, 4210 (1993).

[78]   Kohanoff, J., Andreoni, W., and Parrinello, M., *Chem. Phys. Lett.* **198**, 472 (1992).

[79]   Meng, R.L., *et al.*, *Appl. Phys. Lett.* **59**, 3402 (1991).

[80]   Gensterblum, G., *et al.*, *Appl. Phys. A* **56**, 175 (1993).

[81]   Fleming, R.M., *et al.*, *Nature (London)* **352**, 787 (1991).

[82]   Heiney, P.A., *et al.*, *Phys. Rev. Lett.* **67**, 1468 (1991).

[83]   Fleming, R.M., *et al.*, in "Clusters and Cluster-Assembled Materials," *MRS Symposia Proceedings,* Averback, R.S., Bernholc, J., and Nelson, D.L., eds., Pittsburgh, PA: Materials Research Society Press, 1991, pp. 691–696.

[84]   David, W.I.F., *et al.*, *Europhys. Lett.* **18**, 219 (1992).

[85]   Fleming, R.M., *et al.*, *Nature (London)* **352**, 701 (1991).

[86]   Vaughan, G.B., *et al.*, *Science* **254**, 1350 (1991).

[87]   Verheijen, M.A., *et al.*, *Chem. Phys.* **166**, 287 (1992).

[88]   VanTendeloo, G., *et al.*, *Phys. Rev. Lett.* **69**, 1065 (1992).

[89]   Wang, X.-D., *et al.*, *Phys. Rev. B* **47**, 15923 (1993).

[90]   Li, Y.Z., *et al.*, *Phys. Rev. B* **47**, 10867 (1993).

[91]   Armbruster, J.F., *et al.*, *Z. Phys. B* **95**, 469–474 (1994).

[92]   Li, Y.Z., *et al.*, *Science* **253**, 429 (1991).

[93]   Li, L.Z., *et al.*, *Science* **252**, 547 (1991).

[94]   Weaver, J.H., and Poirier, D.M., in *Solid State Physics*, Ehrenreich, H., and Spaepen, F., eds., New York: Academic Press, 1994, ch.1, p.1.

[95]   Poirier, D.M., *et al.*, *Physik D: Atoms, Molecules and Clusters* **26**, 79 (1993).

[96]   Dresselhaus, M.S., and Dresselhaus, G., *Advances in Phys.* **30**, 139–326 (1981).

[97]   Rosseinsky, M.J., *et al.*, *Nature* (London) **356**, 416 (1992).

[98]   Tanigaki, K., *et al.*, *Nature* (London) **352**, 222 (1991).

[99]   Hérold, A., in *Physics and Chemistry of Materials with Layered Structures*, Lévy, F., ed., New York: Dordrecht Reidel, 1979, p.323.

[100]   Kortan, A.R., *et al.*, *Chem. Phys. Lett.* **223**, 501 (1994).

[101]   Chen, Y., *et al.*, *Phys. Rev. B* **45**, 8845–8848 (1992).

[102]   Datars, W.R., *et al.*, *Solid. State Commun.* **86**, 579–582 (1993).

[103]   Datars, W.R., *et al.*, *Phys. Rev. B* **50**, 4937 (1994).

[104]   Taylor, R., and Walton, D.R.M., *Nature* (London) **363**, 685 (1993).

[105]   Olah, G.A., *et al.*, *Carbon* **30**, 1203–1211 (1992).

[106]   Saito, S., and Oshiyama, A., *Phys. Rev. Lett.* **66**, 2637 (1991).

[107]   Negri, F., Orlandi, G., and Zerbetto, F., *J. Am. Chem. Soc.* **114**, 2910 (1992).

[108]   Saito, R., Dresselhaus, G., and Dresselhaus, M.S., *Phys. Rev. B* **46**, 9906 (1992).

[109]   Erwin, S.C., and Pederson, M.R., *Phys. Rev. Lett.* **67**, 1610 (1991).

[110]   Iijima, S., *Nature* (London) **354**, 56 (1991).

[111]   Ebbesen, T.W., and Ajayan, P.M., *Nature* (London) **358**, 220 (1992).

[112]   Ebbesen, T.W., *et al.*, *Chem. Phys. Lett.* **209**, 83–90 (1993).

[113]   Endo, M., Fujiwara, H., and Fukunaga, E., *Meeting of Japanese Carbon Society* 1991, pp. 34–35 (unpublished).

[114]   Endo, M., Fujiwara, H., and Fukunaga, E., *Second C_{60} Symposium in Japan* 1992, pp. 101–104 (unpublished).

[115]   Dresselhaus, M.S., Dresselhaus, G., and Saito, R., *Phys. Rev. B* **45**, 6234 (1992).

[116]   Ebbesen, T.W., *Annu. Rev. Mater. Sci.* **24**, 235–264 (1994).

[117]   Charlier, J.C., and Michenaud, J.P., *Phys. Rev. Lett.* **70**, 1858–1861 (1993).

[118]   Charlier, J.C., *Carbon Nanotubes and Fullerenes,* PhD thesis, Catholic University of Louvain, May 1994, Department of Physics of Materials.

[119]   Endo, M., PhD thesis, University of Orleans, Orleans, France, 1975, (in French).

[120]   Endo, M., PhD thesis, Nagoya University, Japan, 1978, (in Japanese).

[121]   Kosakovskaya, Z.Ya., Chernozatonskii, L.A., and Fedorov, E.A., *JETP Lett. (Pis'ma Zh. Eksp. Teor.)* **56**, 26 (1992).

[122]   Gal'pern, E.G., *et al.*, *JETP Lett. (Pis'ma Zh. Eksp. Teor.)* **55**, 483 (1992).

[123] Wang, S., and Buseck, P.R., *Chem. Phys. Lett.* **182**, 1 (1991).

[124] Iijima, S., and Ichihashi, T., *Nature* (London) **363**, 603 (1993).

[125] Bethune, D.S. *et al.*, *Nature* (London) **363**, 605 (1993).

[126] Dresselhaus, M.S., *et al.*, "Graphite Fibers and Filaments," *Springer Series in Materials Science,* Berlin: Springer-Verlag, Vol. **5**.

[127] Endo, M., *Chemtech* **18**, 568 (1988), September issue.

[128] Saito, R., *et al.*, *Appl. Phys. Lett.* **60**, 2204 (1992).

[129] Saito, R., Dresselhaus, G., and Dresselhaus, M.S., *Chem. Phys. Lett.* **195**, 537 (1992).

[130] Mintmire, J.W., Dunlap, B.I., and White, C.T., *Phys. Rev. Lett.* **68**, 631 (1992).

[131] Hamada, N., Sawada, S.I., and Oshiyama, A., *Phys. Rev. Lett.* **68**, 1579 (1992).

[132] Harigaya, K., *Chem. Phys. Lett.* **189**, 79 (1992).

[133] Tanaka, K., *et al.*, *Chem. Phys. Lett.* **191**, 469 (1992).

[134] Saito, R., *et al.*, "Electrical, Optical and Magnetic Properties of Organic Solid State Materials," *MRS Symposia Proceedings, Boston,* Chiang, L.Y., Garito, A.F., and Sandman, D.J., eds., Pittsburg, PA: Materials Research Society Press, 1992, p.333.

[135] Saito, R., *et al.*, *Phys. Rev. B* **46**, 1804–6242 (1992).

[136] Painter, G.S., and Ellis, D.E., *Phys. Rev. B* **1**, 4747 (1970).

[137] Saito, R., Dresselhaus, G., and Dresselhaus, M.S., *J. Appl. Phys.* **73**, 494 (1993).

[138] Saito, R., Dresselhaus, G., and Dresselhaus, M.S., *Phys. Rev. B* (1995) submitted August 4, 1995: MS BH5687.

[139] Langer, L., *et al.*, *Extended Abstracts of the Carbon Conference* p. 348 (1995).

[140] Wang, S., and Zhou, D., *Chem. Phys. Lett.* **225**, 165 (1994).

[141] Ge, M., and Sattler, K., *Science* **260**, 515 (1993).

[142] Dravid, V.P., *et al.*, *Science* **259**, 1601 (1993).

[143] Amelinckx, S., *et al.*, *Science* **267**, 1334 (1995).

[144] Zhang, Z., and Lieber, C.M., *Appl. Phys. Lett.* **62**, 2792–2794 (1993).

[145] Olk, C.H., and Heremans, J.P., *J. Mater. Res.* **9**, 259 (1994).

[146] Speck, J.S., *J. Appl. Phys.* **67**, 495 (1990).

[147] Kmetko, E.A., in *Proceedings of the First and Second Conference on Carbon,* Mrozowski, S., and Phillips, L.W., eds., Buffalo, New York: Waverly Press, 1956, p.21.

[148] Heidenreich, R.D., Hess, W.M., and Ban, L.L., *J. Appl. Crystallogr.* **1**, 1 (1968).

[149] Ugarte, D., *Nature* (London) **359**, 707 (1992), (see also: ibid Kroto, H.W., p.670.)

[150] Ugarte, D., *Carbon* **33**, 989 (1995).

[151] Kroto, H., *Nature* (London) **359**, 670 (1992).

[152] Yosida, Y., *Fullerene Sci. Tech.* **1**, 55 (1993).

[153] Ugarte, D., *Chem. Phys. Lett.* **207**, 473 (1993).

[154] Ugarte, D., *Europhys. Lett.* **22**, 45 (1993).

[155] Maiti, A., Bravbec, J., and Bernholc, J., *Phys. Rev. Lett.* **70**, 3023 (1993).

[156] Tománek, D., Zhong, W., and Krastev, E., *Phys. Rev. B* **48**, 15461 (1993).

[157] York, D., Lu, J.P., and Yang, W., *Phys. Rev. B* **49**, 8526 (1994).

[158] Stroscio, J.A., and Eigler, D.M., *Science* **254**, 1319 (1991).

[159] Viitanen, J., *J. Vacuum Sci. Tech. B,* **11**, 115–116 (1993).

[160] Lüthi, R., *et al.*, *Science* **266**, 1979–1981 (1994).

[161] Saito, S., *et al.*, *Mater. Sci. Eng.* **B19**, 105 (1993).

[162] Saito, R., *et al.*, *Materials Science and Engineering* **B19**, 185–191 (1993).

[163] Saito, R., (1994), private communication.

[164] Schmid, H., and Fink, H.W., *Nanotechnology* **5**, 26–32 (1994).

[165] Jin, C.M., *et al.*, in *Proceedings of the First Italian Workshop on Fullerenes: Status and Perspectives,* Taliani, C., Ruani, G., and Zamboni, R., eds., Singapore: World Scientific, 1992, pp.21–29.

[166] Smalley, R.E., and Norlander, P., "Open Tube Growth," private communication 1993.

Chapter 8

Self-Assembly and Self-Assembled Monolayers in Micro- and Nanofabrication

James L. Wilbur and George M. Whitesides

Department of Chemistry
Harvard University
Cambridge, Massachusetts 02138

I. INTRODUCTION

Despite the extraordinary success of current techniques for microfabrication, new techniques are needed. One reason is scale: optically based lithography is reaching the lower limits for the size of features it can produce (~100 nm). Another is efficiency: methods such as electron beam lithography are presently linear processes and will require significant development if they are to be used for large scale, high volume processing. Other considerations such as capital and processing costs, waste management, environmental concerns, and the degree of perfection of the final structures may also force the development of new methods for microfabrication.

This chapter discusses self-assembly– its principles, achievements and potential– as a new approach to micro- and nanofabrication. We will focus primarily on self-assembled monolayers (SAMs) as a model system that illustrates what can be accomplished in fabrication by self-assembly. We describe in turn: i) characteristics of self-assembly that make it promising as a technique for micro- and nanofabrication; ii) structural characteristics of SAMs that are relevant to problems in fabrication; iii) methods of patterning SAMs in the plane of the monolayer; iv) examples of applications that have used SAMs for fabrication, and v) issues that must be addressed in the future as SAMs become more widely used for applications in micro- and nanofabrication.

II. SELF-ASSEMBLY

Molecular self-assembly is the spontaneous organization of molecules into stable, structurally well-defined aggregates[1,2]. Principles drawn from biology[1-5] have helped in understanding the fundamentals of self-assembly, and have stimulated the development of new strategies and new applications. It is now possible to use molecular self-assembly to achieve molecular-scale resolution in the fabrication of certain types of structures.

The basic principles of molecular self-assembly are found throughout biology[1-5]: protein folding and aggregation[6], and pairing of base pairs in DNA are two well-known examples[7]. These principles are: i) molecular self-assembly uses multiple weak, reversible interactions– hydrogen bonds, ionic bonds and van der Waals interactions– to assemble individual molecular subunits into stable aggregates; ii) the structure of the final product represents a thermo-dynamic minimum that results from equilibration of these interactions; iii) the minimum energy structure is predetermined by the characteristics of the initial subunits: information required for assembly is coded in the shape and properties (especially surface properties, but in principle, others as well) of these subunits; iv) the number of types of subunits in a particular system is usually small (four base pairs in DNA, for example) to simplify the amount and type of information required to determine the final structure of the aggregate, and v) the individual subunits can be relatively small and easily synthesized; these characteristics may allow for molecular level control in the placement of individual atoms.

How do these principles of molecular self-assembly apply to problems in fabrication? Why use self-assembly for microfabrication? Three answers stand out. First, molecular self-assembly is a strategy for fabrication at thermodynamic minima[5,8]: it thus rejects defects and is a route to structures having a high degree of perfection[2]. Second, it is a route to well-defined structures having dimensions in a range (1 nm to 1000 μm) that is difficult to achieve conveniently by other methods: structures of this scale are too large for atom-by-atom assembly (via covalent synthetic chemistry) and too small for current methods in microlithography. Third, it is efficient relative to other methods of microfabrication since molecular-scale order is achieved spontaneously (at room temperature).

The application of molecular self-assembly to the formation of non-biological systems has many precedents: these include hydrogen-bonded aggregates[9], templated crystals[10,11], colloids[12], micelles[13-15], liquid crystals[16], emulsions[14], and artificial peptide tubules[17-20]. Other topics related to self-assembly and relevant to its application to fabrication include liquid-metal mirrors[21-23], microlens arrays prepared by melting micromachined drops[24], bubble rafts[25] and phase-separated polymers[26].

III. SELF-ASSEMBLED MONOLAYERS

Self-assembled monolayers (SAMs) are the most widely studied and best developed of the non-biological self-assembled systems[27]. SAMs were preceded historically by Langmuir-Blodgett (LB) monolayers[28], which have been studied extensively and are useful for many applications (fabrication of optical

TABLE 1. Monolayers and multilayers formed by self-assembly on inorganic substrates.

MONOLAYERS

SURFACE	LIGAND	BINDING	REFERENCES
Au	RSH, ArSH (thiols)	RS–Au	27,30,36–42
Au	RSSR′ (disufides)	RS–Au	41,43
Au	RSR′ (sulfides)	RS–Au	44
SiO_2, glass	$RSiCl_3, RSiOR_3$	siloxane network	31,32,45–48
Si	$[RCOO]_2$ (neat)	R–Si	49
Si	$RCH{=}CH_2$, $[RCOO]_2$	$R{-}CH_2CH_2{-}Si$	50
GaAs	RSH	RS–GaAs	33–35
Ag	RSH, ArSH	RS–Ag	51–57
Cu	RSH, ArSH	RS–Cu	53,58–60
metal oxides	RCOOH	$RCO_2^-\cdots MO_n$	61–73
metal oxides	RCONHOH	$RCONHOH \cdots MO_n$ $RCONHO^-\cdots MO_n$	74,75
Pt	RSH, ArSH	RS–Pt	76–80
Pt	RNC	RNC–Pt	76

MULTILAYERS

SURFACE	LIGAND	BINDING	REFERENCES
SiO_x, Au, Ge	RPO_3^-, Zr^+	ionic multilayers	81–89
Si/SiO_2, Au	polyelectrolytes, silicates	ionic multilayers	88,90
SiO_2	$HO{-}R{-}SiCl_3$	siloxane network	46,91,92
Au, Cu, Ag	RCOOH, $HS(CH_2)_n\,COOH$	metal–ligand multilayers	60

coatings or multilayers, for example). LB films, however, are neither convenient to prepare nor sufficiently robust for most applications. SAMs, in contrast, are robust, simple to generate and can be formed from a wide variety of ligands and supports (Table 1). The most widely-studied SAMs are alkanethiolates on gold[29,30] and alkylsiloxanes on silicon dioxide[31,32]. In this chapter we will focus on the former. SAMs of alkanethiolates on gold (RS-Au) are a model system that illustrates what can be accomplished with self-assembled monolayers. Furthermore, the ease of preparation of the RS-Au system makes it ideal for prototyping further developments. For some important applications in microfabrication, however, SAMs of alkanethiolates on gold do not have the required characteristics: gold is not a suitable material for silicon processing in microelectronics, for example, in part due to the low barrier for diffusion of gold into silicon. For this reason, we will also discuss (although in less detail) two systems that are alternatives to the RS-Au system: i) alkylsiloxanes on Si/SiO_2 because of the technological relevance of a monolayer system that allows processing directly on Si/SiO_2, and ii) alkanethiolates on $GaAs$[33-35]. All systems based on n-alkyl groups have intrinsic limitations for use in oxidizing and high-temperature environments; SAMs to address these issues are in only the early stages of design.

A. General Structure

SAMs of alkanethiolates on gold[29,30] form by spontaneous adsorption of alkanethiols ($X(CH_2)_n SH$) [27,30,36-42] and dialkyldisulfides ($X(CH_2)_n S - S$ $(CH_2)_m Y$)[41,43] (from the liquid or vapor phase) onto a clean gold surface according to:

$$X\text{-}R\text{-}SH + Au \rightarrow X\text{-}R\text{-}S^- Au(I) \cdot Au(0)_n + \tfrac{1}{2}H_2 \qquad (1)$$
$$\tfrac{1}{2}(X\text{-}R\text{-}S)_2 + Au \rightarrow X\text{-}R\text{-}S^- Au(I) \cdot Au(0)_n.$$

Related but less stable systems form from dialkylsulfides ($X(CH_2)_n S(CH_2)_m Y$), as well[44]. Alkanethiols (RSH) are most often used because: i) they are more soluble than disulfides (RSSR) or sulfides (RSR); ii) they are compatible with a wide variety of organic functional groups; iii) the rate of formation of SAMs on gold is ~1000-fold faster for thiols than for disulfides[41,93]; and iv) many thiols are commercially available. Mixed monolayers that contain more than one functional group can be prepared by simultaneous coadsorption of two different thiols.

The details of the mechanism[40] of the reaction summarized by Eq. (1) are not known[94]. Hydrogen gas (H_2) is believed to be the product of the reaction, but it has not been unambiguously established. Although it is generally thought that the sulfur is present on the gold surface in the form of a thiolate[95], the possibility that the sulfur exists as a disulfide has also been suggested[96].

Gold films are typically prepared by electron beam (e-beam) or thermal evaporation of gold onto a flat solid support such as a polished silicon wafer (with a native oxide), a glass slide or mica; titanium or chromium is often used to improve adhesion. The thickness of the evaporated gold films is typically between 5 nm and 300 nm: the thickness can be used to control the hysteresis in wetting, the conductivity, and the opacity of the substrate[97]. The gold in the resulting films is polycrystalline, and the domain size and surface roughness increases with the thickness of the gold film. Alkanethiolate SAMs also form on colloidal gold[12,98,99].

TABLE 2. Methods for the characterization of self-assembled monolayers of alkanethiolates on gold.

METHOD	INFORMATION	REFERENCES
wettability	interfacial free energy of surface	39,41,69,100–104
ellipsometry	thickness of adsorbed layer(s)	37,40
x-ray photoelectron spectroscopy (XPS)	chemical composition	30,41,53,105,106
surface raman scattering	structure of monolayer	77,107–115
transmission (high energy) electron diffraction	structure of monolayer	116
low energy helium diffraction	structure of monolayer	51,55,117–119
x-ray diffraction	structure of monolayer	51,55,71,96,120
infrared spectroscopy	structure of monolayer	37,52,53,105,112,121,122
electrochemical methods	degree of perfection of monolayer, electrical properties of monolayer	37,94,102,107,123–130
scanning tunneling microscopy (STM)	structure and composition of monolayer	119,129–141

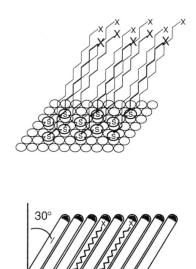

FIGURE 1. Schematic illustration of the molecular-level structure of a self-assembled monolayer of n-alkanethiolates on gold. Figure is not drawn to scale.

Table 2 summarizes the methods that have been used to characterize the structure and composition of SAMs. SAMs of alkanethiolates on gold [Figure 1] have a highly-ordered quasi-crystalline structure. The sulfur atoms [shown in Figure 1 as thiolates and not disulfides] adsorb to the surface of the gold in the three-fold hollow sites between gold atoms. The driving force for the formation of the SAM are the sulfur-gold bond (~44 kcal/mol) and the van der Waals interactions between the alkyl chains (~1.5–2 kcal/mol per CH_2)[95]. Electron diffraction studies (both high and low energy) and scanning tunneling microscope (STM) imaging reveal that the sulfur atoms adopt a $(\sqrt{3} \times \sqrt{3})R30°$ overlayer structure on $Au[111]$; in this configuration, the distance between adjacent sulfur atoms at ~0.5 nm. The alkyl chains extend out (~ 2.2 nm for a SAM of $CH_3(CH_2)_{15}SH$ on Au) from the plane of the surface of the gold; the terminal groups (X in Eq.(1)) are at the interface between the monolayer and the surrounding environment. Infrared spectroscopy, Raman spectroscopy, and X-ray diffraction have shown that, in general, the chains are fully extended, tilted with respect to the surface normal (usually around 30° for n-alkylthiolates on gold with eight or more CH_2 groups), and in a nearly all *trans* configuration.

The structure of the monolayer in the direction perpendicular to the plane of the gold surface can be controlled by varying the length of the alkyl chains: well-established techniques for organic synthesis make it possible to obtain resolution of 0.1 nm in the best circumstances[95]. Organic and inorganic functional groups can be incorporated into and at the end of the alkyl chains: hydrocarbons, fluorocarbons, acids, esters, alcohols, nitriles and amines are a few of the many possible functional groups.

B. Structure at the Atomic Scale

Figure 2(a) shows a high-resolution STM image (from Poirier and Tarlov[142(a)]) of a SAM prepared from decanethiol ($CH_3(CH_2)_9SH, C_{10}$) on gold. The darkest features in the image are single atom-deep pits (terraces or steps) in the Au surface. These pits are not defects in the monolayer: the surface of the Au inside the pits is covered with a SAM[99]. The dark lines in the image are boundaries between domains (differing in the angle of the twist about the axis of the alkyl chains) in the SAM lattice. The inset in Figure 2(a) shows a molecular resolution image of one of these domains.

The atomic-scale structure of SAMs depend on the temperature of the SAM. A high-resolution STM image (taken at 25C)[119] of a C_{10} SAM on gold that was vacuum annealed at 75C shows three important changes in the SAM with heating [Figure 2(b)]. First, several striped domains are evident: these domains correspond to regions of the surface where the SAM has desorbed during annealing[99,119,142(a)]. Second, the size of the domains has increased, indicating that the thiolates (or a thiolate/gold complex[99]) are mobile. Third, the pits in the gold surface are not present in the annealed sample. Several mechanisms for the healing of the pits have been suggested[138,142(b),143].

The structure of SAMs also changes with the length of the alkyl chains. SAMs with longer alkyl chains ($> C_6$) are polycrystalline[142(a)]; SAMs from alkanethiols shorter than C_6 have significant disorder. Figure 2(c) shows an STM image of a SAM of butanethiol (C_4) on Au (from Poirier and co-workers[144]). In addition to the gold pits (darkest features), there are large domains where the SAM is a disordered 2-D liquid. These disordered domains were not observed for either C_8 or C_{10} SAMs[142(a)]. The striped domains in the image correspond to regions where the SAM has partially desorbed in the vacuum of the STM (without heating) and crystallized into another phase [heating was required to create similar features in the C_{10} SAM in Figure 2(b)].

C. Defects in SAMs

Defects in SAMs are directly relevant to the suitability of SAM as materials for micro- and nanofabrication. Estimates of the minimum number or density of defects over larger areas (several cm^2) range from two to several thousand

FIGURE 2. (a) STM image of a SAM formed from the adsorption of decanethiol (C_{10}) on Au. Dark features are single-atom depressions in the surface of the gold. Dark lines are domain boundaries in the lattice of the SAM. The inset shows a molecular resolution image of the SAM. (b) STM image of a C_{10} SAM following annealing at 75C. See text for details. (c) STM image of a C_4 alkanethiolate SAM on Au. Disordered liquid-phase regions are evident; striped regions correspond to a phase induced by partial desorption of the SAM. These images were provided by Poirier and co-workers. See text for citations.

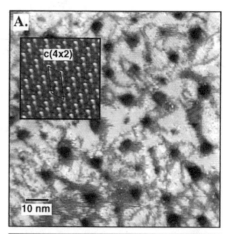

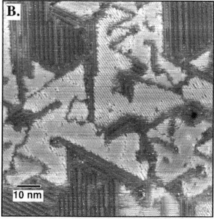

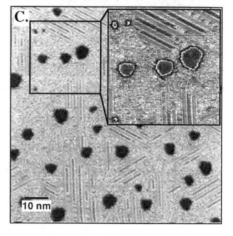

pinholes/cm^2; the higher number is probably correct[145-147]. Electrochemical experiments[37], which determine and average density of defects over a large area, have found that the fractional area of defects for a monolayer formed from octadecanethol was $<6\times10^{-6}$. A stringent test of the density of pinhole defects in SAMs can be performed by exposing the SAM/Au substrate to a wet-chemical etchant that selectively etches the gold (forming holes) in regions where the SAM is defective or missing and amplifying these holes in the gold by etching the underlying silicon. Under the best conditions identified in these studies[147], the lowest density of etchable defects obtained was 90 defects/mm^2.

IV. PATTERNING SAMS IN THE PLANE OF THE MONOLAYER

Patterning SAMs in the plane of the monolayer is one key to their use in microstructure fabrication. Here we focus on the development of simple, low-cost techniques that use self-assembly. These techniques provide alternatives to the methods of fabrication used historically. At present, their level of development is not sufficient to evaluate whether they will ultimately compete with better-established methods of lithography for small-scale (sub-micrometer) fabrication. They are, however, exceptionally convenient for fabrication at the micrometer scale and larger, and may, even at this early stage, compete effectively with other printing techniques such as silk-screening.

Table 3 lists several methods for the fabrication of patterned SAMs. Of these methods, microcontact printing (μCP)[146,148,149] represents a significant technological departure from photolithographic or electron beam lithography

TABLE 3. Techniques for the fabrication of SAMs patterned in the plane of the mono-layer.

METHOD	SCALE OF FEATURES	REFERENCES
microcontact printing (μCP)	100 nm – cm's	146,148,149
micromachining	100 nm – μm's	93,150,151
photolithography/lift-off	> 1 μm	152
photochemical patterning	> 1 μm	153,154
photo-oxidation	> 1 μm	155 – 158
focused ion beam writing	~ μm's	159
electron beam writing	25 – 100 nm	35,160–168
STM writing	15 – 50 nm	161,165,169 – 171
microwriting with pen	~10 – 100 μm	172,173

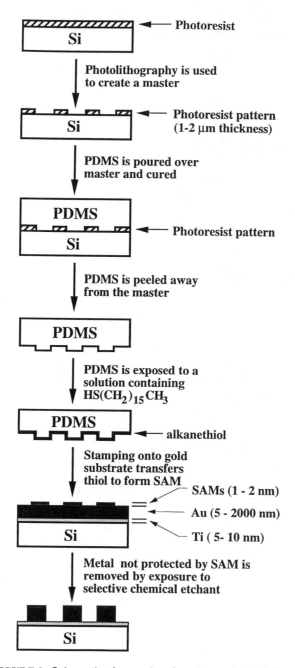

FIGURE 3. Schematic of procedure for microcontact printing.

methods. For this reason, we will focus our discussion predominantly on µCP.

A. Microcontact Printing

Microcontact printing uses a patterned elastomeric "stamp" to transfer a SAM-forming "ink" to an appropriate surface: patterned SAMs form in regions where the elastomeric stamps contact the surface [Figure 3]. The stamp is fabricated by casting a polydimethylsiloxane (PDMS) elastomer on a master having the desired pattern in 3-dimensional relief. Masters can be prepared using standard photolithographic techniques, electron beam lithography, micromachining, or from materials having microscale surface relief features.

The best developed system for µCP uses alkanethiol "ink" to form a patterned alkanethiolate SAM on a gold surface. In a typical process[146,148,149], a photolithographically produced master is placed in a plastic petri dish. A 10:1 ratio (weight:weight or volume:volume) mixture of SYLGARD silicone elastomer 184 and SYLGARD silicone elastomer 184 curing agent (Dow-Corning Corporation) is poured over the master, degassed at reduced pressure (~ 1 torr), cured for 1–2 hours at 60C, and gently peeled from the master. The stamp is "inked" with alkanethiol, dried in a stream of N_2 gas, and applied to the surface of the gold; very light pressure is applied to assure complete contact between the stamp and the surface. The stamp is then gently peeled from the surface. Following removal of the stamp, the patterned gold surface can be exposed to a chemical etchant that selectively removes the gold in regions not derivatized by µCP, and if desired, the underlying support (see below). Alternatively, further derivatization of the surface is possible, either by µCP again with a second stamp, or by washing the surface with a different alkanethiol.

As an elastomer, PDMS is sufficiently deformable to allow good conformal contact between stamp and the surface, even for surfaces with significant (nanometer to several micrometers)[174] relief or roughness; this contact is essential for efficient transfer of the alkanethiol "ink" to the gold film. Although deformable, PDMS is still sufficiently rigid to retain its shape: we have successfully generated patterns with lines as small as 200 nm in width by µCP[175]. The surface of PDMS has a low interfacial free energy[176] ($\gamma = 22.1$ dynes cm^{-1}) and the stamp does not adhere to the gold film. The fact that PDMS is an elastomer also plays a critical role in the inking procedure, by enabling the stamp to absorb the alkanethiol ink by swelling. A rigid stamp would also be difficult to remove from the master and the substrate.

Microcontact printing on gold surfaces can be conducted with a variety of alkanethiol "inks". Alkanethiols that do not undergo reactive spreading[177-180] (after application to the gold film) are required for formation of small features with high resolution. For stamping in air, we use autophobic alkanethiols such as hexadecanethiol. Microcontact printing of other non-autophobic alkanethiols, for example, $HS(CH_2)_{15}COOH$, can be conducted by stamping under water[181].

Figure 4 shows a scanning electron microscope (SEM) image of a complex pattern produced by standard lithographic techniques that was used as a master for µCP, and an SEM image of the patterned SAM formed by µCP with a stamp cast from this master[146]. The complexity and scale of these features is typical of patterned SAMs that µCP can produce routinely.

The primary advantage of µCP as a technique for fabrication is simplicity. It is also inherently parallel: large (many cm^2) areas can be patterned with the application of a single stamp. A number of stamps can be fabricated from a single

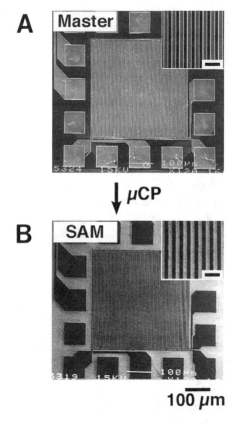

FIGURE 4. (A) Scanning electron micrograph of a master that was used to cast stamps for microcontact printing. The master had regions with 3-dimensional relief: the difference in height between these regions, and differences in the materials on the surfaces of these regions gave rise to the contrast in the SEM images. (B) Patterned SAMs formed by µCP with a stamp cast from the master in (A). Dark regions correspond to SAMs formed by µCP; light regions correspond to underivatized gold. The scale bars in the insets correspond to 10 µm.

master, and individual stamps can be used hundreds of times without degradation in performance. Issues that need to be addressed for the further development of μCP include: i) assessing and minimizing the density of defects in structures produced by μCP; ii) developing methods for achieving smaller scale (< 100 nm) patterns, and iii) developing techniques for controlling the registration of patterns over large scales and during multiple printing steps.

Some progress has been made in extending the basic μCP process to achieve smaller scales. Patterned SAMs with features as small as 100 nm have been fabricated by μCP using mechanical compression of the stamp[182], controlled reactive spreading of the thiol during stamping[183], and stamps cast from masters prepared by anisotropic etching of silicon[175].

Microcontact printing has been extended to systems other than SAMs of alkanethiolates on gold. Microcontact printing of alkyltrichlorosilanes on Si/SiO_2 and glass[184], alkanethiols on copper[185], and alkanethiols on silver[185] has been demonstrated. Microcontact printing on Si/SiO_2 and glass is desirable for several reasons: i) it is promising for microelectronics and microdevice fabrication since it allows for fabrication on Si/SiO_2 surfaces without intervening metal layers; ii) alkylsiloxane monolayers have higher thermal stability (~150C) than SAMs of alkanethiolate on gold, and iii) patterned SAMs on glass allow for applications where optically transparent supports are required.

B. Electron beam and STM Writing

Electron beam and STM writing offer the possibility of small-scale (< 50 nm), high-resolution lithography. High-resolution resists for lithography are difficult to obtain, in part, because they need to be sufficiently thin to reduce shadowing of the sample by the resist (which can reduce the resolution of the features). Because SAMs are thin (~2 nm), and yet still offer excellent corrosion resistance to several wet-chemical etches (see below), they are being explored as resists for electron beam and STM lithography. Several monolayer/substrate systems have been investigated, including alkanethiolates on gold[160,161,167,169-171], alkanethiols on $GaAs$[35,161,186], alkylsiloxanes on Si/SiO_2[35,160-166,168], and alkylsiloxanes on $Ti/TiO_2/GaAs$[166,187].

Figure 5(a) (from Craighead, Allara and workers[186]) shows an AFM image of a pattern generated by e-beam writing on a SAM of octadecyltrichlorosilane (ODT) on silicon oxide. The pattern was a 50 nm period grating with 25 nm wide features. Figure 5(b) (also from Craighead, Allara and workers[161]) shows a grating of 15–20 nm wide lines. This pattern was fabricated by STM writing using a SAM of octadecanethiol ($CH_3(CH_2)_{17}SH, ODT$) as a resist. At present, these features are the smallest fabricated in a patterned SAM by any method.

The detailed mechanism of pattern formation by electron beam or STM writing is not known, although damage to the SAM by secondary electrons (due to the

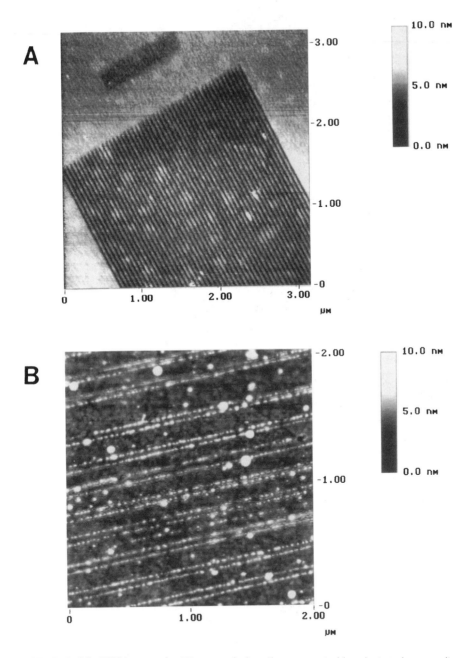

FIGURE 5. (a) AFM image of a 50 nm period grating generated by electron beam writing on a SAM of octadecyltrichlorosilane on SiO_2. (b) AFM image of a 15–20 nm period grating fabricated by writing with an STM on a SAM of octadecanethiol on Au. These images were provided by Craighead, Allara and co-workers. See text for citations.

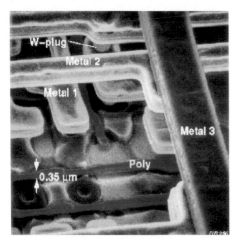

Plate 1 (Figure 2.7)

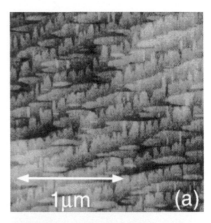

(a)

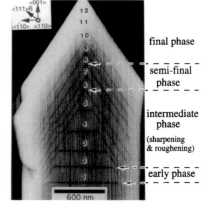

Plate 2 (Figure 2.13)

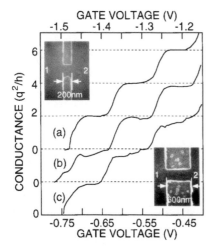

Plate 3 (Figure 2.19)

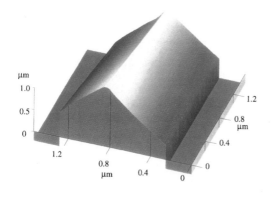

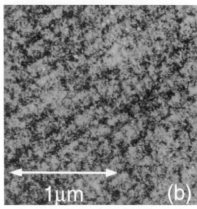

Plate 4 (Figure 5.20)

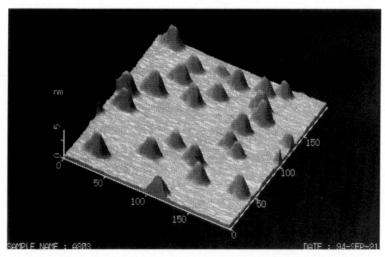

Plate 5 (Figure 5.22)

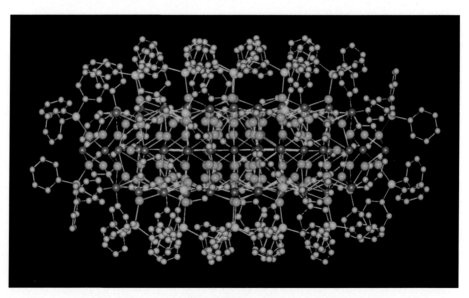

Plate 6 (Figure 6.1)

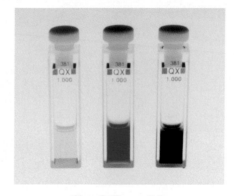

Plate 7 (Figure 6.4)

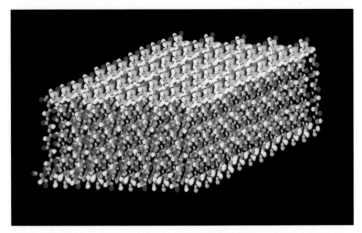

Plate 8 (Figure 9.4)

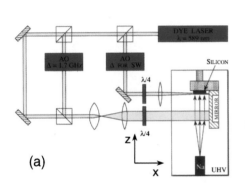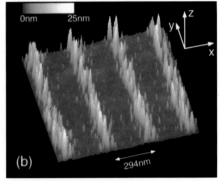

Plate 9 (Figure 10.3)

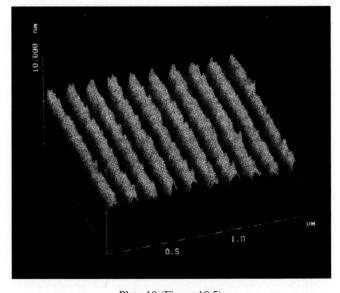

Plate 10 (Figure 10.5)

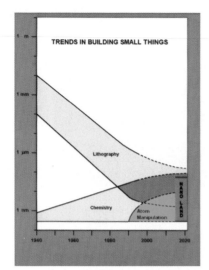

Plate 11 (Figure 11.1)

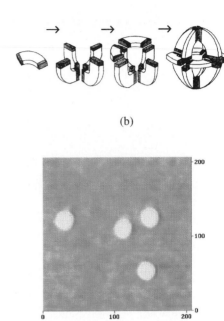

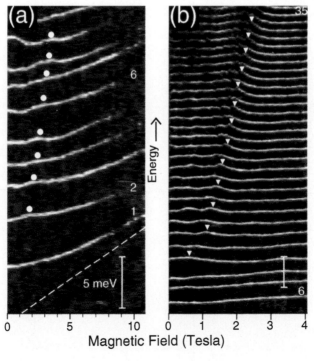

Plate 12 (Figure 12.7)

Plate 13 (Figure 13.19)

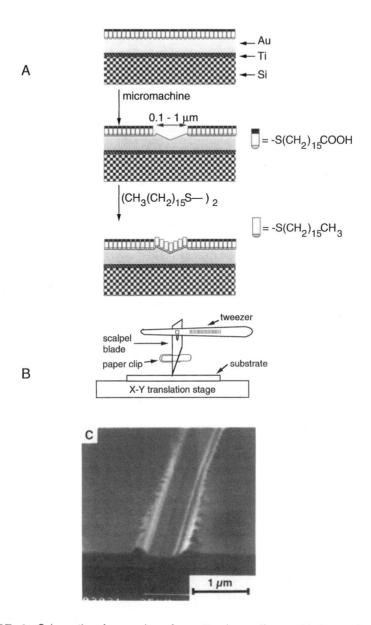

FIGURE 6. Schematic of procedure for patterning self-assembled monolayers by micromachining. (a) Micromachining removed a hydrophilic SAM (formed from $HOOC(CH_2)_{15}SH$) and exposed a region of bare gold. Exposing the substrate to $(CH_3(CH_2)_{15}S)_2$ formed hydrophobic SAMs on this bare gold. (b) Schematic of apparatus for micromachining. (c) Image of a line formed by micromachining.

large cross section for inelastic scattering of the low energy electrons in the monolayer[103,188]) is certainly involved. XPS measurements indicated that damage to the monolayer was incomplete: for monolayers of alkylsiloxanes on $Ti/TiO_2/GaAs$, the amount of carbon decreased by a maximum of 25% and the amount of Si (from the siloxane monolayer) remained constant, even for doses that were more than 10-fold the critical dose for damage[166,187]. The exact nature of the residual carbon was not determined unambiguously, although changes in the wettability by water (contact angle ~115° for unexposed regions, ~60° for exposed regions) suggested that it may be similar to graphite (contact angle ~50°)[166,187]. Although damage was not complete, it was sufficient to allow penetration of wet-chemical etchants in regions where the SAM was damaged (see below).

C. Other Techniques for Patterning SAMs

SAMs can be patterned by micromachining [i.e. scratching, see Figure 6] [93,150,151]. In micromachining, a scratch is formed on the SAM-coated surface of a substrate with a sharp object (a surgical scalpel or carbon fiber, for example): this process removes the SAM and exposes the bare substrate for further processing. The exposed surface can be rederivatized with a different SAM by washing with a second thiol or disulfide. Micromachining has produced patterned SAMs with features (the width of a line, for example) as small as 100 nm; smaller features could certainly be generated with better micromachining techniques.

Photolithographic techniques have also been used to pattern SAMs. Procedures have used conventional photolithography to make patterned alkyl siloxanes[152], UV photo-oxidation of alkanethiolate SAMs through a photolithographic mask[155-158,189] and photochemistry of alkanethiolate SAMs incorporating a photolabile terminal group[153,154].

It is also possible to "write" a patterned SAM of alkanethiolates on gold using a pen or capillary filled with alkanethiol; this technique can produce patterned SAMs with features > 1 μm[172,173].

V. APPLICATIONS OF SAMS

SAMs have been used in a number of applications. Here we focus on applications that are relevant to microfabrication. What advantages do SAMs offer (relative to conventional materials) for microfabrication? SAMs are: i) robust and simple to prepare; ii) generated with molecular-level control over their size and composition; iii) ultra-thin (~1-2 nm) and therefore plausible resists for nanometer-scale, high-resolution lithography; iv) compatible with a wide-variety of functional groups which allows control of surface properties such as wettability and surface free energy; and v) can be patterned by several techniques.

A. Passivation of Surfaces: Protection from Corrosion

Self-assembled monolayers can provide excellent protection of an underlying metal against corrosion. A study of SAMs of alkanethiols on copper[58] concluded that: i) SAMs protected the metal from oxidation; ii) increasing the length of the alkyl chain in the monolayer decreased the rate of oxidation (increased protection), and iii) the kinetics of oxidation was consistent with a model in which the SAM protected the copper by inhibiting transport of O_2 to the copper surface. SAMs of alkanethiolates also provide protection for *GaAs* against oxidation[35,190]. SAMs of octadecanethiol retarded the growth of oxide on *GaAs* significantly, although the oxide did slowly develop over a period of hours to days due to instability of the $As - S$ bond. Protection provided by other types of thin films against corrosion for *Fe* and *Ni*[191] suggests that SAMs will probably be effective in providing corrosion resistance to many other materials.

B. Fabrication Using Patterned SAMs as Resists

SAMs of alkanethiolates protect gold from corrosion for a variety of wet-chemical etchants, including basic solutions of CN^-[192] and ferricyanide[193]. Alkanethiolate SAMs terminated by CH_3 provide the best protection: SAMs with terminal groups such as *COOH, OH* or *CN* are less effective. The extent of protection depends on the length of the alkyl chain. For a C_{16} SAM of *HDT* on gold, the protection is sufficient for fabrication of gold structures using patterned SAMs as resists: in the time required to remove a 100 nm thick film of bare gold fully with an etchant solution, gold protected by SAM formed from *HDT* on *Au* showed only pinhole defects. Thinner SAMs are much less effective. SAMs of hexanethiol (C_6) on gold, for example, offer only marginal protection from the same etchant relative to bare gold.

Figure 7 shows gold and nickel microstructures fabricated by microcontact printing with *HDT* on gold to form patterned SAMs, and subsequently etching with an aqueous solution of ferricyanide[193] or a basic solution of CN^- to remove the gold film in regions not protected by these SAMs. Similar structures can be produced by μCP with alkanethiols on copper and silver and etching[185]. Although functional electronic circuits have not yet been fabricated by μCP and etching, fabrication of electrical components, such as micro-wires[146,148,150,192,194,195] and micro-electrodes, has been demonstrated[150,196].

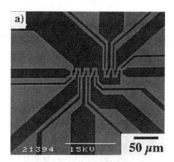

FIGURE 7. Gold and nickel structures formed by μCP and etching with a solution of ferricyanide (a) or a basic, oxygenated solution of cyanide (b–d). (a–b) SEM images of gold patterns typical of those used in microelectronics fabrication. (c) Fracture profile of gold lines. (d) The smallest features fabricated to date by standard μCP: 200 nm wide gold lines separated by 200 nm wide spaces of Si/SiO_2 (where the gold was removed by etching).

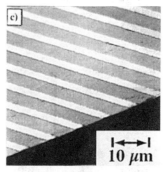

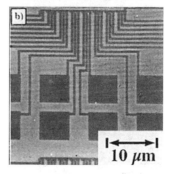

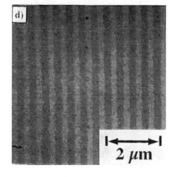

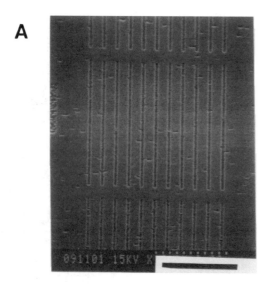

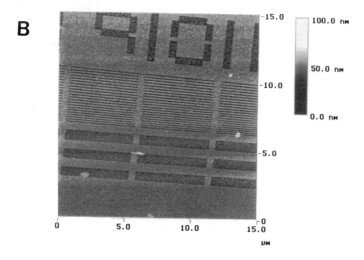

FIGURE 8. (a) A 400 nm period grating fabricated by electron beam writing on a SAM of octadecanethiol ($CH_3(CH_2)_{17}SH, ODT$) on $GaAs$ and etching. The scale bar (lower right) represents 2 μm. (b) Test pattern created by e-beam lithography using a SAM resist formed from the adsorption of octadecyltrichlorosilane ($CH_3(CH_2)_{17}SiCl_3, OTS$) on SiO_2/Si, and etching of SiO_2. These images were provided by Craighead, Allara and co-workers. See text for citations.

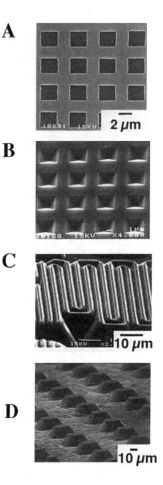

FIGURE 9. Structures of silicon formed by anisotropic etching using gold lines (formed by μCP and etching) as resists. (a) Grid of gold fabricated by μCP and etching of gold. (b) Etched *Si* using grid in (a) as mask. (c–d) etched silicon structures.

The smallest features fabricated to date using patterned SAMs as resists against wet-chemical etches have been produced by electron beam[35,160-168] and STM[161,165,169-171] writing. Figure 8 (from Craighead, Allara and co-workers[35,185,186]) shows nanostructures fabricated from *GaAs* and *SiO*$_2$. The 400 nm-period grating (feature size < 100 nm) in Figure 8(a) was fabricated by e-beam writing on a SAM of octadecanethiol ($CH_3(CH_2)_{17}SH,ODT$) on *GaAs* and etching for 30 min. an aqueous solution of ammonium hydroxide (NH_4OH)[35]. The test pattern in Figure 8(b) was generated by e-beam lithography using a SAM resist formed from the adsorption of octadecyltrichlorosilane ($CH_3(CH_2)_{17}SiCl_3,OTS$) on *SiO*$_2$/*Si*[166]. The pattern was developed by etching *SiO*$_2$ and *Si* with NH_4:HF:H_2O (6:1:4) and *KOH* (0.1 M) respectively. The depth of the etched pattern was 28 nm.

Patterned SAMs do not have the durability to serve as etch masks for conventional reactive ion etching (RIE). They can, however, be used to direct the deposition of a second thin film (usually an oxide or metal layer[155,165,168,187]). Features as small as 20–50 nm have been made using this method. A second procedure[187] uses a SAM damaged by electron beam writing as a *negative* resist, that is, regions of the monolayer damaged by the electron beam protect the underlying substrate from RIE etching while surrounding, undamaged regions do not. The details of this process are not well-understood, although a cross-lined carbonaceous material (either from the SAM, or more probably, from a contaminant in the vacuum chamber) is involved.

Patterned SAMs prepared by μCP with alkyltrichlorosilanes on SiO_2/Si are not sufficiently stable to be an effective resist in etching SiO_2 with aqueous *HF*. However, using a technique developed in other systems– assembling a second layer of liquid resist (polymethylmethacrylate or liquid hexadecane) on hydrophobic regions of the patterned SAMs[197-199]– they can make effective resists for etching of SiO_2[184].

Metal structures prepared by μCP and etching of the metal film can be used as masks for further processing[146,149,175,198,200]. Figure 9 shows SEM images of a patterned silicon surface fabricated as follows: i) a 50 nm thick gold film (with 25 nm of *Ti* used as an adhesion promoter) was deposited on a (100) silicon wafer; ii) microcontact printing formed a patterned SAM; iii) exposure to a CN^- solution created gold microstructures corresponding to the pattern of the SAM formed by μCP; iv) the *Ti* layer was removed by washing with dilute *HF*; v) anisotropic etching of the silicon wafer, in areas exposed by the removal of the gold and titanium layers, which was accomplished by submerging the substrate in a stirred solution of *KOH* (4 M) in isopropanol (15% by volume) at 60C for approximately 30 minutes, and vi) dissolving the remaining gold layer using aqua regia (1:1 HNO_3/HCl) which had previously protected the underlying silicon (in the patterned regions) from the chemical etch [Figure 9(b)]. Small variations in etching times resulted in slightly different features. Complex patterns can be fabricated using this procedure [Figure 9(c–d)] with features as small as 100 nm[175].

C. Fabrication on curved surfaces

Microcontact printing[146,148,149] can also be used to fabricate micron-scale features on curved surfaces[201]. Microcontact printing on curved substrates uses stamps fabricated from PDMS. Because they are elastomeric, these stamps can conform to a non-planar substrate without distorting the pattern on their surface. Figure 10 outlines a procedure, and Figure 11 shows examples of structures formed. At present, there is no comparable technique for microfabrication on curved surfaces. This work opens the door immediately to new optical and

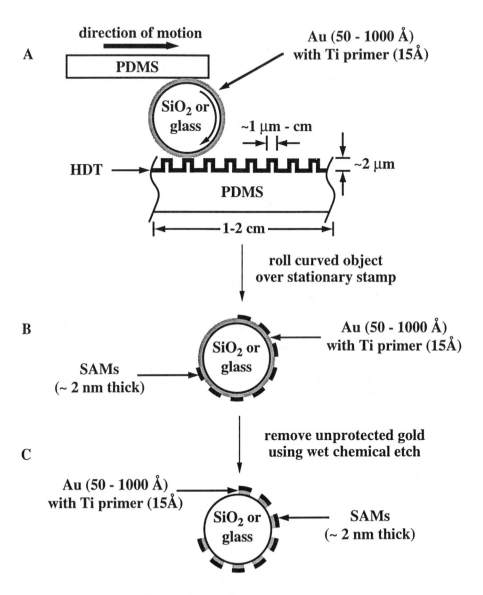

FIGURE 10. Scheme for μCP on curved surfaces.

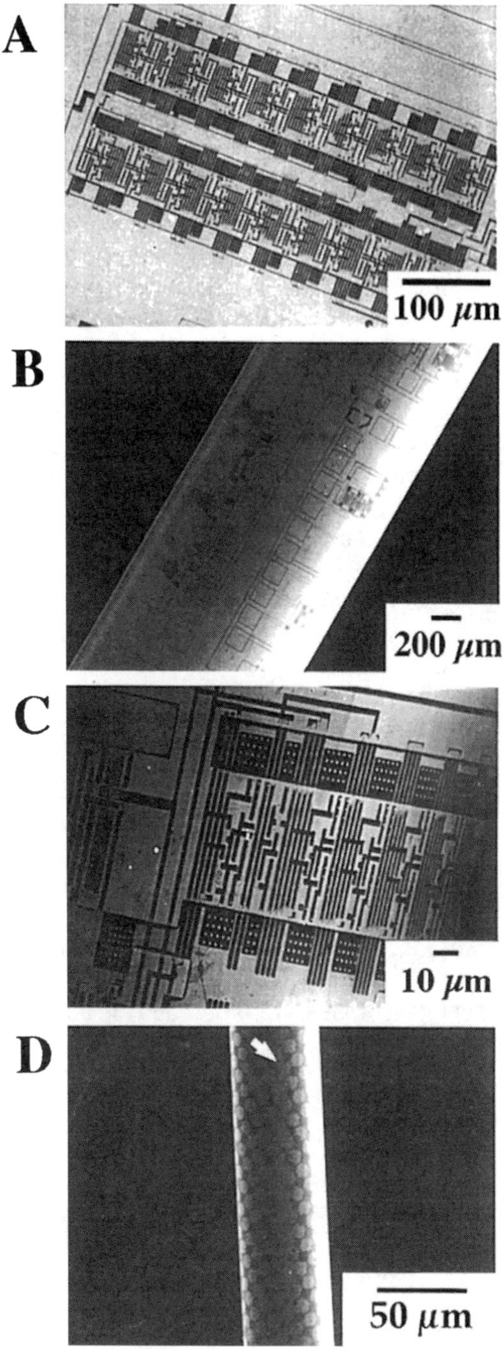

FIGURE 11. Images of gold microstructures fabricated by µCP with hexadecanethiol and etching on gold-coated curved substrates and etching of gold with a solution of ferricyanide. (a) a lens with r = 5 cm. (b–c) a glass capillary with r = 500 µm and (d) an optical fiber with r = 50 µm.

Fig.: 8-11

perhaps optoelectronic structures (see section V.E. below.)

D. Directed Assembly of Materials on the Surface of a Patterned SAM

Patterned SAMs can be used to extend the principle of molecular self-assembly fabrication in the plane *perpendicular* to the plane of the monolayer (Table 4) by using a SAM in which different regions of the surface are patterned in different organic functional groups. These functional groups make different regions of the surface have different values of surface free energy and different wettabilities. These differences in surface free energy and wettability can then be used to direct the assembly of certain material to specific regions of the surface. In one example, when patterned SAMs are exposed to water vapor (in air) at low temperatures (or high humidity), droplets of water condense on the surface selectively on hydrophilic regions [Figure 12(a)]. These so-called "condensation figures" have been used to image patterned SAMs, as optical diffraction gratings, and as sensors for humidity and temperature monitoring[146,173,194]. Passing a SAM with a pattern in wettability through an interface between water and an immiscible hydrocarbon liquid forms droplets of hydrocarbon selectively on hydrophobic regions of the surface[197,199]. Drops of hexadecane [Figure 12(b)] or UV-curable polymers [Figure 12(c)] formed micron-scale arrays of microlenses. Droplets of liquids containing dissolved organic salts can also be organized selectively on hydrophilic regions of patterned SAMs: evaporation of these droplets formed a periodic array of microcrystals [Figure 12(d)][146,202].

Patterned SAMs can also direct the assembly of material on a surface by inhibiting a deposition process. Patterned SAMs of alkanethiolates on gold have been used to direct the electrochemical deposition of metals[166,167] and polymers [Figure 12(e)][195], CVD deposition of metals[171,203-205], and electroless

TABLE 4. Directed assembly of materials on the surface of patterned SAMs.

MATERIAL ASSEMBLED ON PATTERNED SAMs	REFERENCE
water (condensation figures)	146,173,194
organic liquids	197,199
polymers	195,197,198
inorganic salts	146,163
metals (chemical vapor deposition, CVD)	171,203 – 205
metals (electrochemical deposition)	166,167

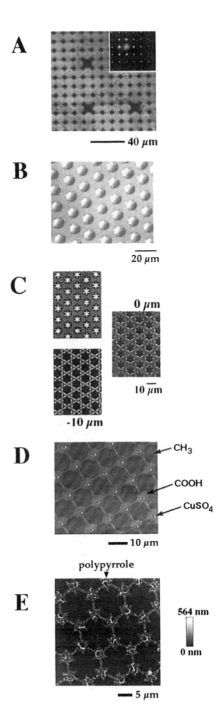

FIGURE 12. Images of materials patterned on the micron-scale by directed assembly using the surface of a patterned SAM as a template. (a) condensation figures (dark) of water on a patterned SAM. These condensation figures acted as optical diffraction gratings (inset). (b) Droplets of hexadecane assembled on a patterned SAM. (c) a microlens array formed by directed assembly of a UV-curable polymer on a patterned SAM. (d) AFM image of copper sulfate deposited in a micron-scale grid using a patterned SAM. (e) polypyrrole film deposited electrochemically using a patterned SAM as a template.

deposition of metals[146] in selected areas.

E. Optical Systems

Microcontact printing is well suited to the fabrication of optical elements. Metal diffraction gratings have been fabricated by μCP [146,148]. Optical waveguides have been prepared by directed assembly of polymethylmethacrylate (PMMA) and polyurethane on SAMs patterned by μCP [Figure 13][206]. Structures ranging in width from 3 μm to 100 μm, with lengths of 1–1.5 cm are easily accessible. Multi-mode waveguiding was demonstrated in structures with widths > 10 μm.

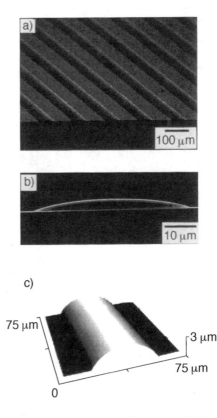

FIGURE 13. Scanning electron micrographs (a–b) and an AFM image (c) of polymeric waveguides prepared by directed assembly of polymethylmethacrylate on a patterned SAM.

F. Fabrication of Colloids[207]

Colloids[208] are used in catalysis, micro- and optoelectronics, seeding of film growth, and spectroscopy[209]. The electrical, chemical and thermodynamic properties of colloids depend sensitively on their size and shape[209]. Recently, methods have been developed to conveniently control the size of colloids. In these methods, the surfaces of colloidal clusters are covered with inorganic or organic functional groups during synthesis[12,98,210-215]. Typically, the alkanethiol is added to the solution from which the colloids are prepared; by controlling its concentration, the size of the colloids can be controlled. Alkanethiols are particularly effective for use with gold colloids since the resulting alkanethiolate SAM is very stable[98,99]. SAMs provide a useful technique for modifying the properties of dispersions of colloids[98,216,217].

G. Other applications

SAMs (both patterned and unpatterned) have found use in several areas not covered in this chapter. These applications are summarized in Table 5.

TABLE 5. Other applications of SAMs.

APPLICATION	REFERENCES
fundamental studies of wetting	29,30,38,39,41,59,69,97,100–104, 106,197,218–220
fundamental studies of electrochemistry	79,102,107,111,124–126,170,221,222
adhesion	223,224
tribology	174
biotechnology/biosurfaces	225–227
bioadhesion	228,229
attached cell culture	152,226,230,231

VI. CURRENT ISSUES

A number of issues will need to be addressed as self-assembly is developed for applications in the fabrication of micro- and nanosystems. These issues are:

- **Developing new systems of SAMs and optimizing existing systems.** SAMs of alkanethiolates on gold are presently the best developed system. Gold, however, is not a good substrate for many applications, particularly in silicon microelectronics. SAMs of alkanethiolates on gold are also too unstable thermally for many processing applications. Other systems of SAMs exist (see Table 1), but these systems need further development to be used generally for fabrication. Iron, nickel, chromium, titanium, aluminum, and stainless steel are possible targets for new substrates.

- **Decreasing the density of defects in the monolayer.** At present, the density of defects in metal structures fabricated by etching using patterned SAMs as resists is too high for applications in high-density microelectronics. Improved methods for evaluating and reducing the density of defects are needed.

- **Using SAMs to augment current technologies for microelectronics fabrication.** The most plausible application is surface passivation. Initial results that use SAMs to direct CVD processes are promising[171,203-205].

- **Developing SAMs as components of optical systems.** Self-assembly at the micron-scale seems particularly well-suited for use in optical systems. Initial work with waveguides[206] and microlens arrays[197] suggests wide application, but both types of systems are in very early stages of development.

- **Techniques for registration of multiple patterning steps.** Most complex systems involve multilayer processing and structures. For techniques that rely on elastomeric stamps (such as microcontact printing), developing methods for registration will be crucial.

- **Exploring techniques for self-assembly in three dimensions.** Most of the work to date using SAMs for fabrication has generated patterns in the plane of the monolayer. Directed assembly of materials[166,167,173,195,197,199] demonstrates the practicality of the fabrication of mesoscopic, 3-dimensional structures by self-assembly.

- **Extending current techniques to structures with nanometer-scale lateral resolution.** At present, only electron beam and STM writing have been used to fabricate patterned SAMs with features smaller than 100 nm. Many important phenomena involving electrons appear only in structures with lateral dimensions less than 50 nm[232]. To make an impact in nano-electronics, techniques that use self-assembly must provide reliable and convenient access to this regime of size.

- **Exploring non-organic self-assembled structures.** There is no intrinsic reason that self-assembly should be restricted to organic structures. Self-assembled inorganic multilayers[81,82,88,90] have been reported. Extension of this work to other materials such as ceramics is an important goal.

- **Developing SAMs showing useful electrical function.** Most SAMs are based on organic molecules, which are electrical insulators. Developing methods of building organic conductivity into organic self-assembled systems is an important issue. Several attempts have been made to enhance the

electrical properties of SAMs, including the incorporation of ferro-
cenes[124,126,127], the synthesis of thiol-terminated polythiophene and
polyphenylene chains[233,234], and the synthesis of polypyrrole-terminated
alkanethiolate SAMs[222,235].

VII. SUMMARY/CONCLUSIONS

Self-assembly is a new paradigm for micro- and nanofabrication. This chapter
has focused on self-assembled monolayers (SAMs) as model systems for self-
assembled materials. The successful development of SAMs from an intellectual
curiosity to a material used in several applications illustrates the promise of self-
assembly in fabrication.

Is self-assembly a concept for fabrication that warrants the substantial
development effort that will be required to convert laboratory processes into
manufacturing practices? New long-range strategies for microfabrication are certainly needed. The methods for microfabrication currently used may be reaching
their limits in terms of scale; other considerations (capital and processing costs,
waste management/environmental concerns, degree of perfection) are also becoming increasingly important. Self-assembly is one option that addresses some of
these problems. It offers an efficient route to complex structures ranging in size
from a few nanometeres to hundreds of microns, with a relatively low level of
defects and with molecular control over structure and composition; it requires
only simple equipment and facilities, has low capital costs and, in some cases,
generates only small quantities of wastes.

Several applications for self-assembly in the short term are possible. The use
of SAMs for passivation of surfaces (protection from corrosion or contamination)
is a plausible point of entry into several manufacturing processes; the use of patterned SAM as resists against wet and dry chemical etches or as materials to control the surface free energy of materials may also find applications. At present,
patterned SAMs are exceptionally convenient for fabrication at the micron-scale
and larger– a scale that is well-suited to applications in optics and biotechnology.
Patterned SAMs with dimensions as small as 200 nm have been prepared but are
not yet routine or reliable; techniques aiming for still smaller scales are being
developd.

Whether self-assembly will be effective in high-end microelectronics fabrication remains to be determined. At this early stage of development, patterned
SAMs may compete effectively as low-cost alternative techniques for micron-
scale printing such as silk-screening. At present, however, no method of self-
assembly can compete with photolithography for smaller (sub-micrometer) scale
microelectronics fabrication.

ACKNOWLEDGMENTS

We thank Dr. Gregory Poirier and Dr. Michael Tarlov for providing the images shown in Figure 2. We also thank Professor Harold Craighead and Michael Lercel for providing the images used in Figures 5 and 8. JLW gratefully acknowledges a postdoctoral fellowship from the NIH. The research in the GMW group was supported by the Office of Naval Research, ARPA and NSF (PHY 9312572).

REFERENCES

[1]   Whitesides, G. M., Mathias, J. P., Seto, C. T., *Science* **254**, 1312–19 (1991).

[2]   Lindsey, J.S., *New J. Chem.* **15**, 153–180 (1991).

[3]   Varner, J.E., *The 46th Symposium of the Society for Developmental Biology: Self-Assembling Architecture*, New York: Alan R. Liss, Inc., 1988, p. 276.

[4]   Kossovsky, N., *et al.*, *Bio/Technology,* **11**, 1534–1536 (1993).

[5]   McGrath, K.P., Kaplan, D.L., "Biological self-assembly: a paradigm for materials science," in *Mater. Res. Soc. Symp. Proc., (Biomaterials by Design)* **330**, 1994, pp. 61–68.

[6]   Creighton, T.E., *Proteins: Structure and Molecular Properties,* Freeman, New York: 1983.

[7]   Sanger, W., *Principles of Nucleic Acid Structure,* New York: Springer-Verlag, 1986.

[8]   Kim, E., Whitesides, G.M., *J. Am. Chem. Soc.,* (submitted).

[9]   For a review of hydrogen bonded aggregates see: Simanek, E.E., *et al.*, *Acc. Chem. Res.* **28**, 37–44 (1995).

[10] Archibald, D.D., Mann,S., *Nature* **364**, 430–433 (1993).

[11] Heywood, B., Mann, S., *Chem. Mat.* **6**, 311–318 (1994).

[12] Leff, D.V., *et al.*, *J. Phys. Chem.* **99**, 7036–7041 (1995).

[13] Menger, F.M., *Angew. Chem. Int. Ed. Engl.* **30**, 1086–1099 (1991).

[14] Dawson, K.A., *Nato Asi Ser.,* Ser. C (1992).

[15] Gompper, G., Schick, M., *Phase Transitions Crit. Phenom* (1994).

[16] de Gennes, P.-G. *The Physics of Liquid Crystals, New York:* 2nd ed. Oxford University Press, New York, 1993, Vol. **83**.

[17] De Santis, P., Morosetti, S., Rizzo, R., *Macromolecules* **7**, 52–58 (1974).

[18] Tomasic, L., Lorenzi, G.P., *Helv. Chim. Acta* **70**, 1012–1016 (1987).

[19] Ghadiri, M.R., *et al.*, *Nature* **366**, 324–7 (1993).

[20] Ghadiri, M.R., Granja, J.R., Buehler, L.K., *Nature* **369**, 301–4 (1994).

[21] Borra, E.F., *et al.*, *Astrophysical Journal* **393**, 829–847 (1992).

[22] Hubin, N., Noethe, L., *Science* **262**, 1390–1394 (1993).

[23] Borra, E.F., *Scientific American* 76–81 (February 1994).

[24] Popovic, Z. D., Sprague, R.A., Neville-Connell, G.A., *Appl. Opt.* **27**, 1281–1284 (1988).

[25] Georges, J. M., *et al.*, *Nature* **320**, 342–344 (1986).

[26] Noshay, A., McGrath, J.E., *Block Copolymers: Overview and Critical Survey,* New York: Academic Press, 1977.

[27] For reviews see: Bain, C.D., Whitesides, G.M., *Angew. Chem. Int. Ed. Engl.* **28**, 506–512 (1989); Whitesides, G.M., Laibinis, P.E., *Langmuir* **6**, 87–96 (1990); Ulman, A., *J. Mater. Educ* **11**, 205–80 (1989); Ulman, A., *An Introduction to Ultrathin Organic Films,* San Diego: Academic Press, 1991; Dubois, L.H., Nuzzo, R.G., *Ann. Rev. Phys. Chem.* **43**, 437–463 (1992). Bard, A.J., *et al.*, *J. Phys. Chem.* **97**, 7147–7173 (1993). Whitesides, G.M., Gorman, C.B., in *Handbook of Surface Imaging and Visualization,* Hubbard, A.T., ed., Boca Raton: CRC Press, (in press).

[28] For a review of LB films see: Ulman, A., An Introduction to Ultrathin Organic Films, San Diego, CA.: Academic Press, 1991; Ulman, A., *J. Mater. Educ* **11**, 205–280 (1989), and references therein.

[29] Bain, C.D., Evall, J., Whitesides, G.M., *J. Am. Chem. Soc.* **111**, 7155–64 (1989).

[30] Bain, C.D., Whitesides, G.M., *J. Am. Chem. Soc.* **111**, 7164–75 (1989).

[31] Wasserman, S.R., Tao, Y.T., Whitesides, G.M., *Langmuir* **5**, 1074–87 (1989).

[32] Wasserman, S.R., *et al.*, *J. Am. Chem. Soc.* **111**, 5852–61 (1989).

[33] Nakagawa, O.S., *et al.*, *Jpn. J. Appl. Phys., Part* **30**, 3759–62 (1991).

[34] Bain, C.D., *Adv. Mat.* **4**, 591–4 (1992).

[35] Tiberio, R.C., *et al.*, *Appl. Phys. Lett.* **62**, 476–8 (1993).

[36] Li, T.T.T., Waever, M.J., *J. Am. Chem. Soc.* **106**, 6107–6108 (1984).

[37] Porter, M.D., *et al.*, *J Am. Chem. Soc.* **109**, 3559–3568 (1987).

[38] Bain, C.D., Whitesides, G.M., *Science* **240**, 62–63 (1988).

[39] Bain, C.D., Whitesides, G.M., *J. Am. Chem. Soc.* **110**, 3665–3666 (1988).

[40] Bain, C.D., *et al.*, *J. Am. Chem. Soc.* **111**, 321–335 (1989).

[41] Bain, C.D., Biebuyck, H.A., Whitesides, G.M., *Langmuir* **5**, 723–7 (1989).

[42] Bain, C.D., Whitesides, G.M., *Angew. Chem. Int. Ed. Engl.* **28**, 506–512 (1989).

[43] Nuzzo, R.G., Allara, D.L., *J. Am. Chem. Soc.* **105**, 4481–4483 (1983).

[44] Troughton, E.B., *et al.*, *Langmuir* **4**, 365–385 (1988).

[45] Sagiv, J., *J. Am. Chem. Soc.* **102**, 92–98 (1980).

[46] Netzer, L., Sagiv, J., *J. Am. Chem. Soc.* **105**, 674–676 (1983).

[47] Maoz, R., Sagiv, J., *J. Colloid Interface Sci.* **100**, 465–496 (1984).

[48] Hoffmann, H., Mayer, U., Krischanitz, A., *Langmuir* **11**, 1304–1312 (1995).

[49] Linford, M.R., Chidsey, C.E.D., *J. Am. Chem. Soc.* **115**, 12631–12632 (1993).

[50] Linford, M.R., *et al.*, *J. Am. Chem. Soc.* **117**, 3145–3155 (1995).

[51] Fenter, P., *et al.*, *Langmuir* **7**, 2013–16 (1991).

[52] Walczak, M.M., *et al.*, *J. Am. Chem. Soc.* **113**, 2370–2378 (1991).

[53] Laibinis, P.E., *et al.*, *J. Am. Chem. Soc* **113**, 7152–67 (1991).

[54] Laibinis, P.E., *et al.*, *Langmuir* **7**, 3167–73 (1991).

[55] Laibinis, P.E., Lewis, N.S., *Chemtracts: Inorg. Chem* **4**, 49–51 (1992).

[56] Chang, S.C., Chao, I., Tao, Y.T., *J. Am. Chem. Soc.* **116**, 6792–805 (1994).

[57] Li, W., Virtanen, J.A., Penner, R.M., *J. Phys. Chem* **98**, 11751–5 (1994).

[58] Laibinis, P.E., Whitesides, G.M., *J. Am. Chem. Soc.* **114**, 9022–8 (1992).

[59] Laibinis, P.E., Whitesides, G.M., *J. Am. Chem. Soc.* **114**, 1990–5 (1992).

[60] Smith, E.L., *Report* **18**, (1992).

[61] Bigelow, W.C., Pickett, D.L., Zisman, W.A., *J. Colloid Interface Sci.,* **1**, 513–538 (1946).

[62] Timmons, C.O., Zisman, W.A., *J. Phys. Chem.* **69**, 984–990 (1965).

[63] Golden, W.G., Snyder, C.D., Smith, B., *J. Phys. Chem.* **86**, 4675–4678 (1982).

[64] Allara, D.L., Nuzzo, R.G., *Langmuir* **1**, 45–52 (1985).

[65] Schlotter, N.E., *et al.*, *Chem. Phys. Lett.* **132**, 93–98 (1986).

[66] Laibinis, P. E., *et al.*, *Science* **245**, 845–7 (1989).

[67] Chen, S.H., Frank, C.W., *Langmuir* **5**, 978–987 (1989).

[68] Chau, L.-K., Porter, M.D., *Chem. Phys. Lett.* **167**, 198–204 (1990).

[69] Allara, D.L., *et al.*, *J. Am. Chem. Soc* **113**, 1852–4 (1991).

[70] Tao, Y.T., Lee, M.T., Chang, S.C., *J. Am. Chem. Soc.* **115**, 9547–55 (1993).

[71] Samant, M.G., Brown, C.A., Gordon, J.G.I., *Langmuir* **9**, 1082–5 (1993).

[72] Smith, E., Porter, M.D., *J. Phys. Chem.* **97**, 8032–8038 (1993).

[73] Ahn, S.J., MIrzakhojaev, D.A., Son, D.H., Kim, K., *Bull. Korean Chem. Soc.* **15**, 369–74 (1994).

[74] On acidic or neutral metal oxides, (e.g. TiO_2), the major species bound to the surface is the hydroxamic acid; on basic metal oxides (e.g. Cu(II) oxide) the ligand binds as a hydroxamate. See next reference.

[75] Folkers, J.P., *et al.*, *Langmuir* **11**, 813–824 (1995).

[76] Lee, T.R., *Pure Appl. Chem.* **63**, 821–8 (1991).

[77] Pemberton, J.E., Bryant, M.A., Joa, S.L., Garvey, S.D., *Proc. SPIE Int. Soc. Opt. Eng.* (1992).

[78] Black, A.J., *et al.*, *J. Am. Chem. Soc.* **115**, 7924–5 (1993).

[79] Shimazu, K., *et al.*, *Bull. Chem. Soc. Jpn* **67**, 863–5 (1994).

[80] Hines, M.A., Todd, J.A., Guyot, S.P., *Langmuir* **11**, 493–7 (1995).

[81] Lee, H., Kepley, L.J., Hong, H.-G., Mallouk, T.E., *J. Am. Chem. Soc.* **110**, 618 (1988).

[82] Putvinski, T.M., *et al.*, *Langmuir* **6**, 1567–1571 (1990).

[83] Katz, H.E., *et al.*, *Science* **254**, 1485–1487 (1991).

[84] Schilling, M.L., *et al.*, *Langmuir* **9**, 2156–2160 (1993).

[85] Byrd, H., Pike, J.K., Talham, D.R., *Chem. Mater.* **5**, 709–15 (1993).

[86] Frey, B.L., Hanken, D.G., Corn, R.M., *Langmuir* **9**, 1815–1820 (1993).

[87] Byrd, H., *et al.*, *J. Am. Chem. Soc.* **116**, 295–301 (1994).

[88] Kleinfeld, E.R., Ferguson, G.S., *Science* **265**, 370–3 (1994).

[89] Feng, S., Bein, T., *Nature* **368**, 834–836 (1994).

[90] Keller, S.W., Kim, H.-N., Mallouk, T.E., *J. Am. Chem. Soc.* **116**, 8817–8818 (1994).

[91]   Maoz, R., Sagiv, J., *Thin Solid Films* 1985.

[92]   Tillman, N., Ulman, A., Penner, T.L., *Langmuir* **5**, 101–111 (1989).

[93]   Abbott, N.L., Folkers, J.P., Whitesides, G.M., *Science* **257**, 1380–2 (1992).

[94]   Krysinski, P., Chamberlin, R.V., II, Majda, M., *Langmuir* **10**, 4286–4294 (1994).

[95]   For reviews that consider the detailed structure of SAMs see: Dubois, L.H., Nuzzo, R.G., *Annu. Rev. Phys. Chem.* **43**, 437–463 (1992); Ulman, A., *An Introduction to Ultrathin Organic Films,* San Diego, CA.: Academic Press, 1991.

[96]   Fenter, P., Eberhardt, A., Eisenberger, P., *Science* **266**, 1216–18 (1994).

[97]   DiMilla, P.A., *et al.*, *J. Am. Chem. Soc.* **116**, 2225–6 (1994).

[98]   Brust, M., *et al.*, *J. Chem. Soc., Chem. Comm.* 801–802 (1994).

[99]   Sondag-Huethorst, J.A.M., Schonenberger, C., Fokkink, L.G.J., *J. Phys. Chem.* **98**, 6826–6834 (1994).

[100]   For a review see: Whitesides, G.M., Laibinis, P.E., *Langmuir* **6**, 87–96 (1990).

[101]   Laibinis, P.E., *et al.*, *J. Am. Chem. Soc.* **112**, 570–9 (1990).

[102]   Chidsey, C.E.D., Loiacono, D.N., *Langmuir* **1990**, 682–691 (1990).

[103]   Tidswell, I.M., *Phys. Rev. B* **44**, 10869–79 (1991).

[104]   Hautman, J., Klein, M.L., *Mater. Res. Soc. Symp. Proc.* (1992).

[105]   Nuzzo, R.G., Dubois, L.H., Allara, D.L., *J. Am. Chem. Soc.* **112**, 558–569 (1990).

[106]   Folkers, J.P., Laibinis, P.E., Whitesides, G.M., *Langmuir* **8**, 1330–41 (1992).

[107]   Kim, J.H., Cotton, T.M., Uphaus, R.A., *J. Phys. Chem.* **92**, 5575–8 (1988).

[108]   Bryant, M.A., Pemberton, J.E., *J. Am. Chem. Soc.* **113**, 8284–93 (1991).

[109]   Bryant, M.A., Pemberton, J.E., *J. Am. Chem. Soc.* **113**, 3629–37 (1991).

[110]   Bryant, M.A., Pemberton, J.E., *J. Am. Chem. Soc.* **113**, 8284–93 (1991).

[111]   Matsuda, N., *et al.*, *Chem. Lett.* (1992).

[112]   Evans, S.D., *et al.*, *J. Am. Chem. Soc.* **113**, 4121–31 (1991).

[113]   Thompson, W.R., Pemberton, J.E., *Chem. Mater* **5**, 241–4 (1993).

[114]   Caldwell, W.B., *et al.*, *Langmuir* **10**, 4109–15 (1994).

[115]   Tang, X., Schneider, T., Buttry, D.A., *Langmuir* **10**, 2235–40 (1994).

[116]   Strong, L., Whitesides, G.M., *Langmuir* **4**, 546–58 (1988).

[117]   Chidsey, C.E.D., *et al.* *Langmuir* **6**, 1804–6 (1990).

[118]   Camillone, N.I., *et al.* *J. Chem. Phys.* **98**, 3503–11 (1993).

[119]   Camillone, N., III, *et al.*, *J. Chem. Phys.* **101**, 11031–11036 (1994).

[120]   Fenter, P., *et al.*, *Mater. Res. Soc. Symp. Proc.* 1992.

[121]   Arndt, T., Schupp, H., Schrepp, W., *Thin Solid Films* (1989).

[122]   Sun, L., Kepley, L.J., Crooks, R.M., *Langmuir* **8**, 2101–3 (1992).

[123]   Chidsey, C.E.D., Loiacono, D.N., *Langmuir* **6**, 709–12 (1990).

[124]   Uosaki, K., Sato, Y., Kita, H., *Langmuir* **7**, 1510–14 (1991).

[125]   De, L.H.C., Donohue, J.J., Buttry, D.A., *Langmuir* **7**, 2196–202 (1991).

[126]   Hickman, J.J., *et al.*, *J. Am. Chem. Soc.* **113**, 1128–32 (1991).

[127]   Chidsey, C.E.D., *et al.*, *Chemtracts: Inorg. Chem.* **3**, 27–30 (1991).

[128]   Sabatani, E., *et al.*, *Langmuir* **9**, 2974–81 (1993).

[129]   Sun, L., Crooks, R.M., *Langmuir* **9**, 1951–4 (1993).

[130]   Creager, S.E., Hockett, L.A., Rowe, G.K., *Langmuir* **8**, 854–61 (1992).

[131]   Kim, Y.T., Bard, A.J., *Langmuir* **8**, 1096–102 (1992).

[132]   Gregory, B.W., Dluhy, R.A., Bottomley, L.A., *Proc. Spie Int. Soc. Opt. Eng.* 1993.

[133]   Delamarche, E., *et al.* *Langmuir* **10**, 2869–71 (1994).

[134]   Schoenenberger, C., *et al.*, *Langmuir* **10**, 611–14 (1994).

[135]   Stranick, S.J., *et al.* *J. Vac. Sci. Technol., B* **12**, 20004–7 (1994).

[136]   Stranick, S.J., *et al.*, *J. Phys. Chem.* **98**, 7636–46 (1994).

[137]   Delamarche, E., *et al.*, *Langmuir* **10**, 4103–8 (1994).

[138]   Bucher, J.P., Santesson, L., Kern, K., *Langmuir* **10**, 979–83 (1994).

[139]   Gregory, B.W., Dluhy, R.A., Bottomley, L.A., *J. Phys. Chem.* **98**, 1010–21 (1994).

[140]   Sondag, H.J.A.M., Schonenberger, C., Fokkink, L.G.J., *J. Phys. Chem.* **98**, 6826–34 (1994).

[141]   Wolf, H., *et al.*, *J. Phys. Chem.* **99**, 7102–7107 (1995).

[142]   (a) Poirier, G.E., Tarlov, M.J., *Langmuir* **10**, 2853–6 (1994). (b) Poirier, G.E., Tarlov, M.J., *J. Phys. Chem.* **99**, 10966–10970 (1995).

[143]   McCarley, R.L., Dunaway, D.J., Willicut, R.J., *Langmuir* **9**, 2775–7 (1993).

[144]   Poirier, G.E., Tarlov, M.J., Rushmeier, H.E., *Langmuir* **10**, 3383–6 (1994).

[145]   Dubois, L.H., Nuzzo, R.G., *Annu. Rev. Phys. Chem.* **43**, 437–463 (1992).

[146]   Kumar, A., Biebuyck, H.A., Whitesides, G.M., *Langmuir* **10**, 1498–1511 (1994).

[147]   Zhao, M., Wilbur, J.L., and Whitesides, G.M., (unpublished results).

[148]   Kumar, A., Whitesides, G.M., *Appl. Phys. Lett.* **63**, 2002–4 (1993).

[149]   Wilbur, J.L., *et al.*, *Adv. Mat.* **7–8**, 600–604 (1994).

[150]   Abbott, N.L., Rolison, D.R., Whitesides, G.M., *Langmuir* **10**, 2672–82 (1994).

[151]   Lopez, G.P., *et al.*, *J. Am. Chem. Soc.* **115**, 10774–81 (1993).

[152]   Kleinfeld, D., Kahler, K.H., Hockberger, P.E., *J. Neurosci.* **8**, 4098–4120 (1988).

[153]   Rozsnyai, L.F., Wrighton, M.S., *J. Am. Chem. Soc.* **116**, 5993–4 (1994).

[154]   Wollman, E.W., *et al.*, *J. Am. Chem. Soc.* **116**, 4395–404 (1994).

[155]   Calvert, J.M., *et al.*, *Thin Solid Films* **211**, 359–63 (1992).

[156]   Tarlov, M.J., Burgess, D.R.F.J., Gillen, G., *J. Am. Chem. Soc.* **115**, 5305–6 (1993).

[157]   Dressick, W.J., Calvert, J.M., *Jpn. J. Appl. Phys., Part 1,* 5829–39 (1993).

[158]   Huang, J., Dahlgren, D.A., Hemminger, J.C., *Langmuir* **10**, 626–8 (1994).

[159]   Gillen, G., *et al.*, *J. Appl. Phys. Lett.* **65**, 534–6 (1994).

[160]   Sondag-Huethorst, J.A.M., van Helleputte, H.R.J., Fokkink, L.G., *J. Appl. Phys. Lett.* **64**, 285–7 (1994).

[161]   Lercel, M.J., *et al.*, *Appl. Phys. Lett.* **65**, 974–6 (1994).

[162]   Marrian, C.R.K., *et al.*, *Appl. Phys. Lett.* **64**, 390–2 (1994).

[163]   Rieke, P.C., *et al.* *Langmuir* **10**, 619–22 (1994).

[164]   Mino, N., *et al.*, *Thin Solid Films* **243**, 374–7 (1994).

[165]   Perkins, F.K., *et al.*, *J. Vac. Sci. Technol., B* **12**, 3725–30 (1994).

[166]   Lercel, M.J., *et al.*, *J. Vac. Sci. Technol. B* **12**, 3663–7 (1994).

[167]   Sondag-Huethorst, J.A.M., van Helleputte, H.R.J., Fokkink, L.G., *J. Appl. Phys. Lett.* **64**, 285–287 (1994).

[168]   Lercel, M.J., *et al.*, *Microelec. Eng.* **27**, 43 (1995).

[169]   Kim, Y.T., Bard, A.J., *Langmuir* **8**, 1096–102 (1992).

[170]   Ross, C.B., Sun, L.i., Crooks, R.M., *Langmuir* **9**, 632–6 (1993).

[171] Schoer, J.K., *et al.*, *Langmuir* **10**, 615–18 (1994).

[172] López, G.P., Biebuyck, A., Whitesides, G.M., *Langmuir* **9**, 1513–1516 (1993).

[173] López, G.P., *et al.*, *Science* **260**, 647–649 (1993).

[174] Wilbur, J.L., *Langmuir* **11**, 825–831 (1995).

[175] Wilbur, J.L., *et al.*, *Adv. Mat.*, (in press).

[176] Chaudhury, M.K., Whitesides, G.M., *Langmuir* **7**, 1013–25 (1991).

[177] Bain, C.D., Whitesides, G.M., *Langmuir* **5**, 1370–8 (1989).

[178] Hare, E.F., Zisman, W.A., *J. Phys. Chem.* **59**, 335–340 (1995).

[179] de Gennes, P.-G., *Rev. Mod. Phys.* **57**, 827–863 (1985).

[180] Holmes-Farley, S.R., *et al.*, *Langmuir* **1**, 725–740 (1995).

[181] Biebuyck, H.A., Whitesides, G.M., *Langmuir* **10**, 4581–4587 (1994).

[182] Xia, Y., Whitesides, G.M., *Adv. Mater.* **7**, 471–473 (1995).

[183] Xia, Y., Whitesides, G.M., *J. Am. Chem. Soc.* **117**, 3274–3275 (1995).

[184] Xia, Y., *et al.*, *Langmuir*, (in press).

[185] Xia, Y., *et al.*, (unpublished results).

[186] Lercel, M.J., *et al.*, *J. Vac. Sci. Technol. B* **11**, 2823–2828 (1993).

[187] Lercel, M.J., *et al.*, *J. Vac. Sci. Technol. B*, (in press).

[188] Laibinis, P.E., *et al.*, *Science* **254**, 981–3 (1991).

[189] Huang, J., Hemminger, J.C., *J. Am. Chem. Soc.* **115**, 3342–3 (1993).

[190] Sheen, C.W., *et al.*, *J. Am. Chem. Soc.* **114**, 1514–15 (1992).

[191] Stratmann, M., *et al.*, *Bull. Electrochem.* **8**, 52–6 (1992).

[192] Kumar, A., *et al.*, *J. Am. Chem. Soc.* **114**, 9188–9189 (1992).

[193] We used an aqueous ferricyanide etch $(0.001M\ K_4Fe(CN)_6, 0.01M\ K_3 Fe(CN)_6,\ 0.1M\ K_2S_2O_3,\ 1M\ KOH)$. Xia, Y., Zhao, M., and Whitesides, G.M. .

[194] Kumar, A., Whitesides, G.M., *Science* **263**, 60–62 (1994).

[195] Gorman, C.B., Biebuyck, H.A., Whitesides, G.M., *Chem. Mat.* **7**, 526–9 (1995).

[196] Abbott, N.L., *et al.*, *NATO ASI Ser.*, Ser. E (1993).

[197] Biebuyck, H. A., Whitesides, G.M., *Langmuir* **10**, 2790–2793 (1994).

[198]   Kim, E., Kumar, A., Whitesides, G.M., *J. Electrochem. Soc.* **142**, 628–633 (1995).

[199]   Gorman, C.B., Biebuyck, H.A., Whitesides, G.M., *Chem. Mat.* **7**, 252–4 (1995).

[200]   Whidden, T.K., *et al. Nanotechnology,* (in press).

[201]   Jackman, R.J., Wilbur, J.L., Whitesides, G.M., *Science,* (in press).

[202]   Wilbur, J.L., *et al.*, *Nanotechnology,* (in press).

[203]   Potochnik, S.J., *et al.*, "Advanced Metallization for Devices and Circuits: Science, Technology and Manufacturability," in *Mater. Res. Soc. Symp. Proc.:* **337**, 1994, pp.429–34.

[204]   Potochnik, S.J., *et al. Langmuir* **11**, 1841–1845 (1995).

[205]   Jeon, N.L., *et al.*, *Langmuir,* (in press).

[206]   Kim, E., *et al.*, *Adv. Mat.,* (in press).

[207]   The rich field of colloids and nanocrystals is discussed in more detail in Chapter 6.

[208]   Reiss, H.J., *Chem. Phys.* **19**, 482–487 (1951).

[209]   Schmid, G., *Chem. Rev.* **92**, 1709–1727 (1992).

[210]   Steigerwald, M.L., *et al.*, *J. Am. Chem. Soc.* **110**, 3046–3050 (1988).

[211]   Wang, Y., Herron, N., *J. Phys. Chem.* **95**, 525–532 (1991).

[212]   Fenske, D., Krautscheid, H., *Angew. Chem. Int. Ed. Engl.* **29**, 1452–1454 (1990).

[213]   Krautscheid, H., *et al.*, *Angew. Chem. Int. Ed. Engl.* **32**, 1303–1305 (1993).

[214]   Herron, N., *et al.*, *Science* **259**, 1426–1428 (1993).

[215]   Murray, C.B., Norris, D.J., Bawendi, M.G., *J. Am. Chem. Soc.* **115**, 8706–8715 (1993).

[216]   Colvin, V.L., Goldstein, A.N., Alivisatos, A.P., *J. Am. Chem. Soc.* **114**, 5221–30 (1992).

[217]   Maeda, Y., Yamamoto, H., Kitano, H., *J. Phys. Chem.,* **99**, 4837–4841 (1995).

[218]   Whitesides, G.M., Laibinis, P.E., *Langmuir* **6**, 87–96 (1990).

[219]   Lee, T.R., *et al.*, *Langmuir* **10**, 741–749 (1994).

[220]   Abbott, N.L., Gorman, C.B., Whitesides, G.M., *Langmuir* **11**, 16–18 (1995).

[221]   Taniguchi, I., *et al.*, *Microchem. J.* **49**, 340–54 (1994).

[222]   Willicut, R.J., McCarley, R.L., *Langmuir* **11**, 296–301 (1995).

[223]   Ferguson, G.S., *et al.*, *Science* **253**, 776–8 (1991).

[224] Chaudhury, M.K., Whitesides, G.M., *Science* **255**, 1230–2 (1992).

[225] Prime, K.L., Whitesides, G.M., *Science* **252**, 1164–7 (1991).

[226] Mrksich, M., Whitesides, G.M., *Trends in Biotechnology,* (in press).

[227] Pritchard, D.J., Morgan, H., Cooper, J.M., *Angew. Chem. Int. Ed. Engl.* **34**, 91–93 (1995).

[228] López, G.L.A., *et al., J. Am. Chem. Soc.* **115**, 5877–5878 (1993).

[229] Amador, S.M., *et al., Langmuir* **9**, 812–17 (1993).

[230] Singhv, R., *et al., Science* **264**, 696–698 (1994).

[231] Sukenik, C.N., *et al., Biomed. Mater. Res.* **24**, 1307–23 (1990).

[232] Kirk, W.P., Reed, M.A., *Nanostructures and Mesoscopic Systems,* Academic Press, 1992.

[233] Tour, J.M., *Adv. Mat.* **6**, 190–8 (1994).

[234] Pearson, D.L., *et al., Polym. Prepr.* **35**, 202–3 (1994).

[235] Sayre, C.N., Collard, D.M., *Langmuir* **11**, 302–6 (1995).

Chapter 9

Biocatalytic Synthesis of Polymers of Precisely Defined Structures

Timothy J. Deming

Department of Materials
University of California at Santa Barbara
Santa Barbara, CA 93106

Vincent P. Conticello

Department of Chemistry
Emory University
Atlanta, GA 30322

David A. Tirrell

Department of Polymer Science and Engineering
University of Massachusetts at Amherst
Amherst, MA 01003

I. INTRODUCTION

The fabrication of functional nanoscale devices requires the construction of complex architectures at length scales characteristic of atoms and molecules. Currently microlithography and micro-machining of macroscopic objects are the preferred methods for construction of small devices, but these methods are limited to the micron scale. An intriguing approach to nanoscale fabrication involves the association of individual molecular components into the desired architectures by supramolecular assembly. This process requires the precise specification of intermolecular interactions, which in turn requires precise control of molecular structure.

Organic polymers offer several advantages as materials for the construction of small-scale devices, including ease of synthesis and fabrication, well-delineated structure-property correlations, and thermal and mechanical stability. However the ability to precisely define polymer architecture is severely restricted for all but the most simple sequences. Macromolecular structure is defined in terms of four architectural variables: molecular size, composition, sequence, and stereochemistry. The degree to which these variables can be controlled depends on the method of polymer synthesis; conventional methods of polymer synthesis afford only statistical control of each. Consequently most synthetic polymers are not pure substances but instead comprise heterogeneous populations of molecular species, and significant advances in the solution and solid state properties of polymeric materials have been associated with increased degree of control over one or more of the architectural variables. For example, the stereoregular polymerization of α-olefins by the Ziegler-Natta procedure revolutionized polymer materials science by providing high melting, crystalline materials from simple, inexpensive building blocks[1]. In general, the synthesis of complex architectures is limited by the few available methods for the synthesis of precisely defined macromolecular architectures.

Polymers of uniform structure are synthesized by either of two techniques: iterative coupling of selectively activated monomers[2], or template-directed polymerization[3]. The former process consists of stepwise assembly of the desired material via a repetitive sequence of intermolecular coupling and activation steps (Scheme I). Two reactive end-groups are coupled intermolecularly, and one of the remaining end-groups is selectively deprotected. The process is repeated with addition of new reactants and the desired molecule is assembled sequentially. The serial nature of this process ensures that the microstructure of the polymer is determined by the identity of the individual reactants at each step. This process has found application in the assembly of complex structures both in solution and on solid supports. The former has been used, for example, in the synthesis of dendritic macromolecules[4], and the latter procedure forms the basis for the Merrifield synthesis of polypeptides[5]. However, owing to the linear nature of the process, the degree of synthetic difficulty increases geometrically with the length (and thereby the complexity) of the sequence. This feature, coupled with the lack of an intrinsic proof-reading mechanism, limits this process to the synthesis of moderate length polymers, e.g., sixty residues for a polypeptide sequence.

In contrast to the iterative coupling procedure, template-directed synthesis can provide an intrinsic proof-reading capacity, and therefore a self-correction mechanism. In this procedure, a master template is used to specify the exact sequence of the target polymer, which is then assembled in a parallel process from the component monomers (Scheme II). The template can direct the synthesis of many copies of the target polymer by dissociation of the complementary polymer and repetition of the process. This procedure is currently limited in scope by the availability of suitably designed templates, which ideally must be uniform polymers themselves, and by formulation of appropriate mechanisms for the transfer of sequence information from the template to the reactive monomers. The best realized example of this process is protein biosynthesis, in which a DNA sequence serves as a template for the synthesis of the polypeptide chain through the intermediacy of a complementary mRNA sequence [Figure 1].

Scheme I. Iterative Step Growth

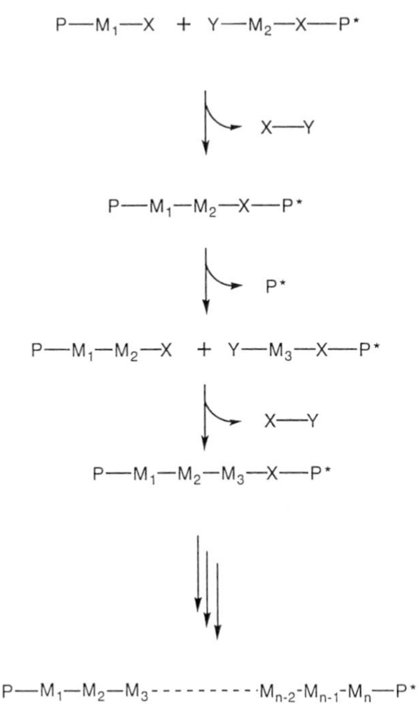

P—M1—X + Y—M2—X—P*
X—Y
P—M1—M2—X—P*
P*
P—M1—M2—X + Y—M3—X—P*
X—Y
P—M1—M2—M3—X—P*
P—M1—M2—M3---------Mn-2-Mn-1-Mn—P*

Scheme II. Template Polymerization

A—B—C-------------X—Y—Z
M1
M2
M3,......,Mn-2,Mn-1,Mn
M1—M2—M3------------Mn-2—Mn-1—Mn

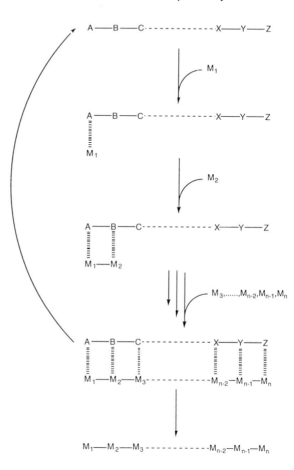

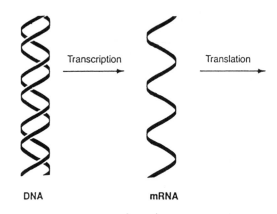

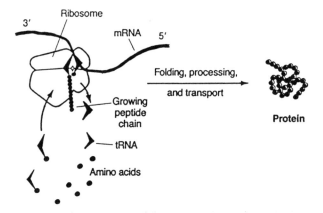

FIGURE 1. Schematic representation of protein biosynthesis from a DNA template. Reproduced with permission from King, J., *Chem. Eng. News* **32**, April 10, 1989.

The sequence information specifying any given protein is encoded in a particular DNA sequence in the host organism. Under appropriate conditions, a copy of the coding sequence is transcribed as a messenger RNA (mRNA). The triplet codons of the mRNA, each specifying a particular amino acid residue, are then sequentially decoded on the ribosome by specific amino-acylated transfer RNA molecules. This procedure serves as the mechanism for coupling the DNA sequence information to that of the polypeptide. Near-absolute specificity of all of the aforementioned structural parameters, e.g., size, composition, sequence, and stereochemistry, is ensured for the nascent polypeptide not only by the template polymerization mechanism but also by simultaneous proof-reading steps that occur during the transcription/translation process. Thus polypeptides derived from genetic templates may be considered as model uniform polymers.

Polypeptides also differ from most synthetic polymers in that they adopt ordered three-dimensional structures in solution and in the solid state. Most naturally occurring proteins adopt globular structures consistent with their roles in substrate recognition and transformation, for example, as enzymes. However a number of naturally occurring proteins have fibrous structures and serve as structural components *in vivo*. These proteins are essentially polymeric in the classical sense, consisting of tandem repeats of oligomeric peptide sequences. Yet these proteins are synthesized by the same mechanism and exhibit the same uniformity of structure as their more complex globular counterparts. In addition, these polypeptides also adopt regular and persistent secondary structures in both the solution and solid states, consistent with the geometrical requirements of their oligopeptide repeats.

By utilizing the principles of protein structure and the concepts of material science as guides, unnatural protein-based materials can be designed that are capable of self-assembly into unique two- or three-dimensional shapes on the basis of their primary structures. The sequence of a polypeptide chain contains sufficient information to direct its specific assembly into a particular native conformation. The driving force for this process is the non-covalent, primarily hydrophobic interaction occurring between segments of the polypeptide chain. These interactions act cooperatively to direct the formation and association of secondary structure elements, e.g., α-helix, β-sheet, and reverse turn, into the native structure. The known preferences for the formation of specific secondary structures by particular amino acid sequences provide a pattern for the *de novo* design of protein materials. Novel combinations of secondary structure can be programmed into the polypeptide sequence to generate unique geometries in the folded protein. These dimensionally-defined protein materials may serve as subunits for the hierarchical construction of nanoscale objects and devices. In addition, a diverse repertoire of functional groups may be incorporated into these materials, comprising the side chain functionalities of both natural and unnatural amino acid residues, providing geometrically-defined sites for further interaction between individual protein subunits.

Artificial protein sequences can be synthesized with absolute uniformity of structure by utilizing the techniques of recombinant DNA technology and bacterial protein expression [Figure 2][6,7]. A fundamental oligopeptide repeat, designed for its ability to adopt a particular fold in the final material, is encoded into a double-stranded DNA sequence. The oligonucleotide repeat is generated by solid-state synthesis and then sequenced to verify its integrity. Ligation of the ends of the oligonucleotide in a specific head-to-tail fashion affords a population of DNA multimers that contain tandem repeats of the coding sequence. Fractionation generates the target length multimer (the artificial gene), which is inserted into a plasmid, an autonomously replicating, extrachromosomal circular genetic element. The plasmid also contains a gene that confers resistance to a particular antibiotic and thus provides a method for selection of transformed cells. The recombinant plasmid is then introduced into a bacterial strain (most typically *Escherichia coli*) that is capable of expression of the target protein. When cultured in the presence of an appropriate antibiotic, only bacteria that produce the products of the recombinant plasmid survive and grow. The transformed bacteria are grown to a particular cell density and target protein production is induced.

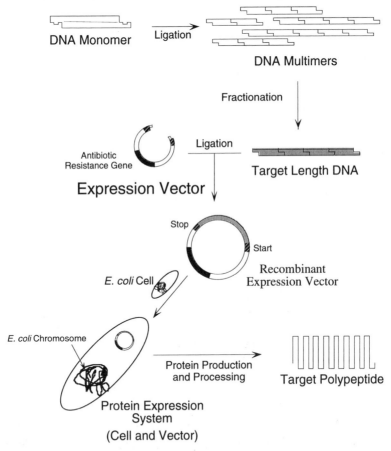

FIGURE 2. Synthesis, cloning, and expression of artificial genes encoding rRepetitive polypeptides.

The cells are harvested and lysed, and the target protein is then purified from the host proteins. Yields of up to grams of protein per liter of culture have been obtained in this manner.

Methods of this kind have now been used to prepare natural structural proteins as well as artificial proteins that have no direct parallel in nature. The remaining sections of this chapter summarize the state of the art of bacterial expression of protein-like polymers of potential utility in nanoscale fabrication of structures and devices.

II. EXPRESSION OF STRUCTURAL PROTEINS

Natural structural proteins, e.g. silk, collagen, elastin, and keratin, are ubiquitous and essential to the proper function of all organisms. Natural evolutionary processes have yielded proteins that surpass the performance of man-made materials, e.g. mammalian elastin in the cardiovascular system that lasts half a century without loss of function, and spider webs composed of silk threads that are tougher than any synthetic fiber[6]. There has been much recent interest in understanding and producing these natural polymers to make products for commercial applications. Purification of the biological materials from natural sources can be complicated or impossible in some cases. For example, adhesive proteins used by marine organisms to attach themselves to surfaces are irreversibly cross-linked upon secretion and cannot then be processed or reused[8]. Isolation of structural proteins from biological sources (e.g. collagen) is practiced on a large scale commercially, yet the physical properties of those extracted materials are often less than optimal. Furthermore, problems in processing these materials combined with inconsistent quality and limited sources act to restrict commercial use of natural structural proteins to a few applications (e.g. photographic gelatin and cosmetic fillers).

As an alternative method to achieve production of large quantities of well-defined structural proteins, many researchers have turned to biocatalytic polymer synthesis using genetic engineering[6,7]. Two approaches to the synthesis of natural polymers through this technique are (i) to clone the complementary DNA which encodes the protein of interest from the source organism and then express the resulting gene and (ii) to utilize or create a consensus repeat from highly repetitive proteins to construct a corresponding DNA sequence ("monomer") which is described in the previous section. This second procedure has been widely employed since it avoids the need to obtain cDNA clones and since it permits selection of codons to avoid unstable repetitive sequences and adaptation of codon usage to the requirements of the host organism. Many natural structural proteins or artificial proteins designed to mimic natural materials have been successfully expressed in microorganisms and one such product has been commercialized[6].

A. Silks

The silk fibroin of the common silk caterpillar *Bombyx mori* consists of extensive β-strands which hydrogen bond to form sheets[9]. As such, the protein chains are very densely packed and are also highly oriented in silk fibers resulting in high tensile strength. *B. mori* silk fibroin primarily consists of tandem repeats of the sequence (*GlyAlaGlyAlaGlySer-*) and this has been the repeat sequence targeted for production of materials having the properties of silk fibers. Joseph Cappello and coworkers at Protein Polymer Technologies, Inc. (PPTI) have reported the expression in *E. coli* of high molecular weight (40–100 kDa) proteins that incorporate blocks of the *B. mori* silk hexapeptide repeat[6]. These recombinant proteins were shown to be crystalline with structures similar to the β-sheet structures found in natural silk. Fibers prepared from these samples were found to contain β-sheet structures but no molecular orientation or alignment was seen for any of the samples. This is a limitation which must be overcome to attain and assess the ultimate physical properties of these silk-like polymers.

Sequences derived from elastin (*ValProGlyValGly-*) or from the cell-binding portion of fibronectin (*-ArgGlyAspSer-*) have also incorporated into silk-like polymers[6]. The result is that the elastin blocks lower the crystallinity of the polymers and enhance their solubility and processability. The incorporation of the fibronectin sequence into silk-like polymers creates a useful substrate material for cell culture. This polymer can be coated on plastic or glass surfaces and subjected to sterilization at high temperature and pressure without loss of activity. This material has been prepared in quantities greater than 500g and has been commercialized by PPTI as ProNectin F®[6]. Progress has also been made in the cloning and expression of fragments of the silk proteins from midges (*Chironomus tentans*) and spiders. Midge larvae spin insoluble fibers consisting in part of a family of silk proteins designated spI, which contain *ca.* 150 copies of an 82 residue core repeat. The spI are very large proteins, *ca.* 1MDa. Steven Case and his colleagues at the University of Mississippi have prepared synthetic oligonucleotides which encode the 82 residue core repeat, and have reported successful expression of these oligonucleotides in *E. coli* under control of a bacteriophage T7 promoter[10]. The protein has been purified to homogeneity and is being used to study the mechanisms of fiber assembly. Spider dragline silk from the golden orb weaver (*Nepila clavipes*) is produced as fibers which are stronger than those of high-strength synthetic aromatic polyamides such as Kevlar®[11]. Randolph Lewis and coworkers at the University of Wyoming have cloned genes encoding various spider silk proteins and attempts to express these proteins in *E. coli* are in progress[12].

B. Collagen

Collagen makes up as much as one fourth of all the proteins in mammals. It is the major constituent of connective tissues and thus acts as the main load bearing, soft tissue component. Collagen fibers consist of bundles of triple helical proteins held together through extensive hydrogen bonding. The individual strands of the triple helices are constructed from many repeats of the tripeptide sequence (-*GlyProXxx*-) where *Xxx* can be any amino acid, but is usually proline or trans-4-hydroxyproline. The proline residues favor extended conformations of the polypeptide chains which are further stabilized by the intertwining of two additional protein strands. The glycine residues allow close packing of the three strands favoring formation of intrastrand hydrogen bonds and close packing of triple helical bundles.

Ina Goldberg and coworkers at Allied Signal, Inc. have prepared artificial genes which encode repetitive proteins (-*GlyProPro*-)$_n$ analogous to natural collagen sequences[13]. The synthetic genes have been cloned in *E. coli* using the thermally inducible promoter, λpL. A 22 kDa polypeptide was successfully synthesized, but unless the heat-shock response of the *E. coli* host is altered, the protein is proteolytically degraded. The protein can be stably produced in mutants that have impaired protein degradation systems. Cappello and coworkers at PPTI have also created artificial genes based on human collagen for preparation of collagen-like polymers[6]. In *E. coli*, they have successfully expressed these genes to prepare polymers ranging in size from 40 to 70 kDa. The properties of these materials have not been extensively explored to date.

C. Elastin

The 72 kDa protein tropoelastin is a precursor of the protein elastin, which provides the elastic properties of mammalian arteries, lungs, skin, and other tissues[14]. The tropoelastin sequence consists of tandem repetitions of a number of oligopeptide blocks which are four to nine amino acids in length. A synthetic analog of one of these blocks has been elaborated into a polymeric material which exhibits extraordinary elasticity. This material contains the block (-*ValPro-GlyValGly*-) which forms a type II β-reverse turn around the proline-glycine dipeptide[15]. The amino acids, not participating in the turn, span the distance between turns with glycine acting as a spacer of high flexibility. When the turns are linked together they wind into a spiral that functions as a molecular spring, resulting in a material that can be extended to over 300% of its resting length with no deformation[15].

A tropoelastin cDNA has been cloned and expressed in *E. coli* by Joel Rosenbloom and his colleagues at the University of Pennsylvania[16]. Since the free polypeptide is rapidly degraded in *E. coli*, it is necessary to express a fusion protein of tropoelastin linked at its N-terminus to an 81-amino acid fragment of influenza virus NS 1 protein. CNBr digestion can then be used to isolate the tropoelastin portion if a methionine codon is included at the junction of tropoelastin with the virus protein. (Tropoelastin contains no internal methionine residues.) The purified recombinant tropoelastin cross reacts with antibodies to elastin-like peptides and exhibits chemotactic activity toward fetal calf ligament fibroblasts[16]. Recently McPherson, *et al.*[17] have described the biosynthesis of an analogue of the core repeat of mammalian elastin, [*Gly*-(*Val*-*Pro*-*Gly* -*Val*-*Gly*)$_{19}$-*Val*-*Gly*-*Pro*-*Gly*], in *E. coli*[18]. The target protein was isolated in high purity and exhibited spectroscopic properties identical to those of chemically synthesized samples.

D. Adhesive Proteins

Many marine organisms (e.g. barnacles, mussels, and tube worms) secrete adhesive proteins for surface attachment or for physical protection[8]. These adhesives are remarkable in that they can be spread on a surface immersed in salt water and form a permanent bond within seconds. The bond will also withstand a wide range in temperature and large fluctuations in tidal currents. Since many man-made adhesives are ineffective in wet environments, the development of commercial adhesives based on these proteins is an attractive option. Although marine adhesive proteins from different organisms have different amino acid sequences, they all have a high percentage of DOPA (3, 4–dihydroxyphenylalanine) residues which are present in repetitive sequences of 7 to 10 amino acids in length[8]. The DOPA residues are created by post-translational hydroxylation of tyrosine in precursor proteins and are thought to be primarily responsible for the adhesive properties of the polymers. *In situ* enzymatic oxidation of the DOPA residues to quinones is thought to result in cross-linking of the materials and their adhesion to surfaces[18].

Robert Strausberg and coworkers at Genex Corporation have isolated a cDNA clone from the marine mussel *Mytilus edulis* that consists of 20 repeats of (-*AlaLysProSerTyrProProThrTyrLys*-), the putative adhesive sequence. They have expressed this protein and several of its multimers, e.g., dimers, trimers and tetramers, in *Saccharomyces cerevisiae* by using an expression vector containing a promoter comprised of portions of the *S. cerevisiae* GAL1 and MF-α1 promoters[19]. This expression system is efficient enough to produce the recombinant proteins in amounts corresponding to about 3-5% of the total cell protein. These proteins are all insoluble; the largest having a molecular weight of *ca.* 96 kDa. The purified precursor proteins do not show adhesive properties until they are treated with a bacterial tyrosinase, which converts tyrosine residues to DOPA.

A synthetic gene, encoding an analog of the mussel adhesive protein, has been constructed by Anthony Salerno and Ina Goldberg of Allied Signal. A DNA sequence encoding 20 copies of the decapeptide adhesive sequence was constructed by polymerization of a 30-base pair DNA monomer; the codons were chosen to correspond to *E. coli* use patterns and expression was controlled by the bacteriophage T7 promoter[20]. The resulting recombinant protein accumulated in intracellular inclusion bodies and could be isolated in amounts approaching 60% of total cellular protein. When this material is mixed with mushroom tyrosinase, it forms an adhesive which can be spread on surfaces and used for cell attachment and growth. The adhesive has also been shown to be effective in forming a gas permeable seal when used as a cement for wet ocular tissue samples[21].

E. Viral Proteins

Human adenoviruses are believed to attach to host cells via fibrillar protein spikes located on the surface of the viral capsid[22]. The 200 Å long spikes consist of a repetitive 62 kDa protein organized into a trimeric array which is about 30 Å in diameter. The secondary structure of the protein has not yet been determined although triple helical and cross-β models have been proposed. The potential for producing polymers which could be processed into high performance fibers through self-assembly makes the synthesis of analogues of these viral spike proteins an attractive prospect.

The 35 kDa fibrous protein from adenovirus serotype 3 has been expressed in *E. coli* by Corinne Albiges-Rizo and Jadwiga Chroboczek of the European Molecular Biology Laboratory. The product is a fusion protein containing at its N-terminus a short leader sequence derived from the cloning vector based on bacteriophage T7[23]. The recombinant protein was insoluble and appeared as a trimer when subjected to gel electrophoresis under denaturing conditions. Since the naturally occurring viral protein and the recombinant protein exhibit similar gel filtration behavior, Albiges-Rizo and Chroboczek have suggested that the native viral fiber also consists of a trimeric protein assembly[23].

John O'Brien and coworkers at E. I. du Pont de Nemours and Co. have expressed synthetic analogs of viral spike proteins in bacterial systems[24]. Oligonucleotide sequences encoding three synthetic analogs consisting of polypeptide repeats of 15 residues were prepared, multimerized and cloned in *E. coli*. The polypeptide repeat sequences were designed to contain reverse-β turns of either 4 or 5 residues and two short (three residue) β-strands. The resulting expressed proteins formed inclusion bodies and ranged in molecular weight from 20 to 100 kDa. All of the proteins were soluble in hexafluoroisopropanol (HFIP). Most interesting is the observation that HFIP solutions of several of these polymers were birefringent, suggesting that liquid crystalline phases were formed.

Fibers were drawn from the anisotropic solutions and were shown to have mechanical properties similar to those of textile fibers in commercial use.

F. Coiled-coil Proteins

A family of polymers analogous to the protein transcription factors GCN4, Fos and Jun has been expressed in bacterial systems by Kevin McGrath and David Kaplan of the U. S. Army Natick Research, Development and Engineering Center[25]. The transcription factors form coiled-coil structures consisting of two intertwined helices with a superhelical pitch of *ca.* 140 Å[26]. The amphiphilic helices assemble with the hydrophilic residues facing the solvent and helping to stabilize the assemblies by forming interchain electrostatic interactions; the hydrophobic residues are located on the inside of the coil. Heterodimerization of the protein chains is encouraged by the specific design of the artificial polypeptide sequences. Although these studies are at an early stage, the project has the potential for creating materials capable of self-assembly into complex structures and intermolecular recognition.

III. *DE NOVO* DESIGN AND SYNTHESIS OF WELL-DEFINED POLYPEPTIDES

A. Design Methodology

Previous research in this area initially had two major goals. The first was to evaluate the feasibility of using genetic engineering to create proteins with materials potential. Natural proteins known to have excellent materials properties were purposely set aside in favor of testing the potential of creating new proteins *ab initio*. These proteins would be designed from first principles utilizing structural elements found in natural proteins and their simpler analogs, namely β-strands, reverse turns and α-helices. The second key objective was to assess the extent to which protein chain folding and supramolecular organization could, in fact, be controlled at the molecular level. Judicious utilization of the compositional and sequential control available through biocatalytic synthesis should yield materials with precisely-defined structures.

Two classes of proteins have been featured in this work (Table 1). One consists of periodic chain-folding proteins designed to form crystalline lamellar solids. Nearly all flexible, stereoregular polymers crystallize in the form of lamellar aggregates in which the chain is oriented normal (or nearly normal) to the lamellar plane and folds regularly at the lamellar surface[27]. The thickness of such crystals is determined primarily by the kinetics of crystallization. Through proper sequence design, it has become possible to construct polymer chains from

TABLE 1. Repeating Units of Periodic Polypeptides Currently Under Study

REPEATING UNIT	INTEREST	STATUS[a]
(GlyAla)$_x$ GlyProGlu	Sequence-dependent chain folding	A (x = 3, 4, 5, 6,∞)
(GlyAla)$_3$ XY	Turn requirement for folded-chain lamellae	A (XY = GlyLeu. GlySer, GlyVal, GlyMet, SerGly, GlyAsp,GlyTyr, GlyAsn, GlyPhe, GlyThr)
		B (XY = 10 additional pairs)
(GlyAla)$_x$ GlyGlu	Correlation of chemical and spatial periodicity	A (x = 3-6)
(GlyAla)$_3$ GlyZ[b] (GlyAla)$_x$ GlyZGlu[c]	Incorporation of non-natural amino acids	A
(GlyAla)$_3$ GlyGlu(AlaGly)$_3$ GlyGlu	Effect of stem sequence polarity on crystal structure and morphology	A
(GlyAla)$_3$ GlyGlu(GlyAla)$_3$ GlyVal (GlyAla)$_3$ GlyGlu(GlyAla)$_3$ GlyMet	Differentiation of lamellar surface functionality	B B
Glu$_{17}$Asp	Monodisperse poly(glutamic acid) and mesogenic poly(γ-benzyl-L-glutamate)	A

[a] A: Protein expressed and isolated;
B: Coding sequence prepared and cloned.
[b] Z = L-selenomethionine, p-fluorophenylalanine, 5,5,5-trifluoroleucine, 3-thienylalanine.
[c] Z = azetidine-2-carboxylic acid, thiazolidine-4-carboxylic acid, 3,4-dehydroproline.

elements that are known to form antiparallel β-sheets and that allow control of the thickness and surface chemistry of lamellar polymer crystals. The second class of materials under study consists of proteins predicted to form α-helices. Poly(α,L-glutamic acid) (PLGA) and related polymers have played a central role in investigations of the physical chemistry and materials science of chain molecules[28]. Studies of this rod-like molecule are complicated, however, by the polydispersity of chain lengths obtained when this polymer is prepared synthetically. Our research efforts have concentrated on the production of monodisperse variants of PLGA and poly(α, L-aspartic acid) to overcome these synthetic limitations.

More recent research in this field has branched into an additional area which serves to extend the original goals. This new focus, and the subject of the next section, concerns the use of intact cellular protein synthesis machinery to incorporate *unnatural* amino acids into some of the artificial proteins described above. Unnatural amino acids (or amino acid analogs) allow the incorporation of unusual atoms and functional groups (e.g. fluorine and electroactive groups) into proteins, and extend the range of properties obtainable for protein materials.

B. Folding β-Sheet Proteins

The formation of folded-chain lamellar crystals is common among polymers of regular chemical structure. For most such materials, however, the folded-chain structure is determined by the kinetics, rather than the thermodynamics, of the crystallization process, and the observed lamellar organization is metastable. The opportunity to design sequences of controlled periodicity offers the prospect of thermodynamic control of regular chain folding. Our approach has been to attempt to exploit known sequence-dependent secondary structures (e.g., β-strands and reverse turns) to design folded-chain lamellar crystals of controlled thickness and surface functionality. As shown schematically in Figure 3, the thickness of the crystal is controlled by the length of the extended element (the β-strand), while the surface functionality is determined by the composition of the turn sequences.

Glycine-alanine (*Gly-Ala*) dyads are featured in the stem segments of our periodic proteins as this motif is known to form β-strands. *GlyAla* repeats are responsible for the β-sheets characteristic of the silk-II crystal morphology as well as the β-strands in type-I poly-*AlaGly*[29]. One of the design variables under study is the effect of the stem length on lamellar dimensions (see below). A variety of turn structures is also under study.

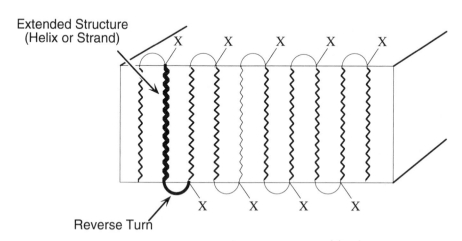

FIGURE 3. Schematic representation of a folded chain lamellar crystal of a periodic polypeptide built up from alternating extended and folded secondary structural elements.

The first proteins to be constructed contained proline and glutamic acid in the turn sequences since proline is a well known turn-former and the bulky, polar glutamic acid residue should be excluded from the crystalline interior by both steric interactions and solvent-binding effects[30]. Crystalline solids have been obtained for several sequences of this type (Table 1), but the formation of regular lamellae is dependent on the size of the basic repeat unit (see below). A library of genes encoding several different two-amino acid turn sequences has been constructed, based on sequences reported for the type-I' and type-II' reverse turns found in natural proteins. The anticipated solid-state organization of these polymers would thus be characterized by regular folding of the chain at the turn sequences, with the glycylalanine β-sheets stacked in the lamellar interior and the turn sequences exposed at the surface. The first repetitive copolypeptides that we designed were multimers of either the nonapeptide [(*GlyAla*)$_3$ *GlyProGlu*][7] (**1**) or the undecapeptide [(*GlyAla*)$_4$ *GlyProGlu*][31], (**2**). Chain-length variants containing 10–54 repeats of the above sequences have been efficiently produced in *E. coli* using a bacteriophage T7-based expression system. Although several lines of evidence indicate that the expressed proteins have the correct sequence, such materials do not readily adopt regularly folded chain arrangements, but instead form amorphous glasses at room temperature.

In a second attempt at sequence-controlled crystallization, a gene encoding thirty-six repeats of the octapeptide [(*GlyAla*)$_3$ *GlyGlu*] (**3**) was expressed in *E. coli* as a fusion protein containing short flanking sequences derived from the cloning and expression vectors[32]. The target protein was confined to the soluble fraction of the whole cell lysate, and a simple procedure involving sequential pH adjustments with glacial acetic acid afforded a substantial enrichment in the supernatant. The product was isolated by ethanol precipitation at -10 C, and the periodic portion of the target protein was liberated from the flanking sequences by cleavage with cyanogen bromide. The predominance of the antiparallel β-sheet structure in the crystalline polypeptide is revealed in the key features of the infrared (IR), Raman, and cross polarization magic angle spinning nuclear magnetic resonance (CP/MAS NMR) spectra of samples crystallized from 70% formic acid[32]. The IR spectrum of (**3**) exhibits amide I, II, and III vibrational modes at 1623, 1521, and 1229 cm^{-1}, respectively, characteristic of the β-sheet conformation, and the weak amide I component at 1698 cm^{-1} indicates the regular alternation of chain direction that defines the antiparallel β-sheet architecture. The Raman spectrum of (**3**) exhibits the amide I band at 1664 cm^{-1} and the splitting of the amide III band into two components at 1260 and 1228 cm^{-1}, characteristic of the antiparallel β-sheet conformation. Likewise, the CP/MAS ^{13}C NMR spectrum of (**3**) showed chemical shifts analogous to those obtained for poly(*L*-*AlaGly*) in the β-form[32].

Wide-angle x-ray diffraction patterns of crystal mats of (**3**) were obtained with the x-ray beam parallel to the plane of the mat. The diffraction patterns exhibit discrete Bragg reflections consistent with a crystalline polymer and the reflections index on an orthorhombic unit cell with dimensions $a = 0.948$ nm, $b = 1.060$ nm, and c (chain axis) $= 0.695$ nm. These unit cell parameters are commensurate with previously published x-ray diffraction results from various silk fibroins and synthetic polypeptides that exhibit similar crystalline structures[33]. In addition, well-defined meridional reflections were observed, which correspond to a long period spacing of 3.6 nm. This result appears to be completely consistent with the expected lamellar thickness, since molecular modeling predicts a spacing of *ca.* 3.3 nm between the termini of glutamic acid side chains arrayed on opposite faces of crystals built up from stacked, folded β-sheets of (3) [Figure 4]. A long-period spacing of 3.6 nm would thus correspond to an interlamellar separation of ca. 0.3 nm, which appears plausible. Electron microscopy of (3) reveals individual lamellae *ca.* 1 μm in length, generally similar in appearance to crystals grown from conventional nylons[32]. All of the experimental evidence collected on the solid-state structure of (3) [and other stem length variants of (3)] strongly supports a crystalline antiparallel β-sheet architecture involving a chain-folded lamellar structure as the basic crystalline unit.

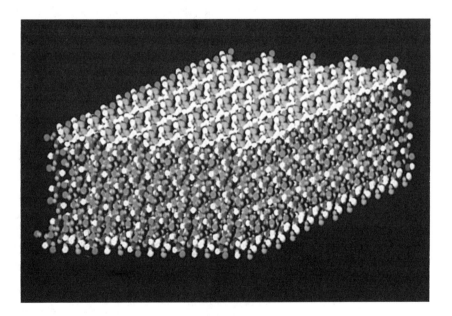

FIGURE 4. Molecular graphics representation of a folded-chain lamella crystal formed from a repeating polymer of sequence (3). The blue and red "core" of the aggregate comprises the repeating alanylglycine β-sheet portion of the chain; the green surfaces present the reactive side chains of the periodic glutamic acid residues. The essential elements of the structure are supported by the results of spectroscopic and scattering studies. Reprinted with permission from reference[32]. (See color plate.)

Other stem-length and turn sequence variants of **3** are also under investigation. One variant in particular is the sequence $[(AlaGly)_3 GluGly(GlyAla)_3 GluGly]$, (**4**), which was synthesized to probe the chain folding pattern in the polymeric crystalline domains[34]. None of the diffraction analyses mentioned above directly addresses the nature of chain folding, yet successful assembly of the predicted lamellar structures depends on the ability of the chains to fold and crystallize in an adjacent reentry fashion. Polymers comprising 10 repeats of sequence (**4**) were expressed in *E. coli* under conditions where ^{13}C-enriched glycine and alanine could be incorporated into the chains. Infrared spectra of blends of natural-abundance and ^{13}C-enriched polymers of sequence (**4**) showed two bands in the polypeptide amide I region (1629 and 1596 cm^{-1}) assignable to hydrogen-bonded β-sheets of the isotopically distinct samples. It has been suggested that the frequency shift of the unperturbed amide I vibration is due to transition dipole coupling between adjacent carbonyls within the β-sheets. Hence, the presence of separate ^{13}C and ^{12}C derived amide I bands (instead of a composite band) led us to propose that a majority of the chains are folding in an adjacent reentry fashion as predicted[34].

C. Monodisperse Alpha Helical Proteins

Polymers that adopt a persistent rodlike structure in solution and in the solid state have been a topic of much current research in materials science[35]. These materials should form ordered solutions due to the shape anisotropy of the individual rods, thereby facilitating their fabrication into oriented solid materials. Polypeptides that adopt a persistent α-helical secondary structure should display this behavior for structures that have a critical aspect ratio of rod length to diameter. Indeed, studies of poly (α,L-glutamic acid) (PLGA) and its ester derivatives have played a central role in elucidating the physical chemistry and materials science of rodlike polymers[28]. PLGA itself has been used in fundamental studies of the helix-coil transition of polyelectrolytes[36], and ester derivatives of PLGA most notably poly(γ-benzyl-α,L-glutamate) (PBLG) form organic liquid crystalline solutions[37] and oriented monolayer films[38]. In addition, α-helical rods should have a large dipole moment associated with the peptide bond that is essentially oriented along the helical axis[39]. The dipoles of individual PBLG helices can be oriented in solution under the influence of an electric or a magnetic field affording ferroelectrically ordered mesophases and films[40]. These materials display properties associated with oriented acentric materials such as piezoelectric behavior and second order non-linear optical properties.

The traditional synthetic route to PLGA and its esters involves the ring-opening polymerization of *N*-carboxy-α-amino acid anhydrides. This technique affords a heterogeneous population of chains that have relatively broad distributions of molecular sizes. This heterogeneity of molecular lengths complicates the assessment of the critical structural criteria for the observation of liquid crystallinity in rodlike polymers, which remains an open question. The utility of these materials in the design and construction of complex macromolecular architectures also is limited by this structural heterogeneity. The incorporation of more complex sequences of residues in a rational manner is also clearly precluded via this synthetic approach due to the inability to accurately specify the positions of individual residues.

Uniform length α-helical rods of a PLGA analogue $[GluAsp(Glu_{17}Asp)_4 GluGlu]$, (5), have been prepared via a genetically directed synthesis[41]. The periodic aspartic acid residues provide recognition and cleavage sites for the restriction enzyme *BbsI*, which is used to liberate the coding sequence for head-to-tail oligomerization of the oligonucleotides. Aspartic acid was chosen as the second residue because of its similarity to glutamic acid, which is expected to reduce to a minimum any perturbation of the chemical and physical properties of the chain. The gene for (5) was expressed as a fusion protein to the affinity label glutathione-S-transferase, which enabled a convenient separation from the bacterial host proteins by chromatography. The target polypeptide was liberated from the fusion protein by either enzymatic proteolysis or cyanogen bromide cleavage. The PLGA analogue (5) displays chemical and spectroscopic properties consistent with its expected structure, in particular, a pH dependence of its circular dichroism spectrum indicative of a helix to coil transition upon deprotonation at alkaline pH. In addition, the electrophoretic behavior of (5) compares favorably with commercial samples of narrow polydispersity (polydispersity index ≈ 1.2), chemically synthesized PLGA. The latter samples appear as broad smears while the former appears as a single, tightly focussed band, thus illustrating the power of the biosynthetic strategy for the control of macromolecular architecture.

Monodisperse (5) has been cleanly converted to its benzyl ester derivative (6) by alkylation of the carboxylate groups with phenyldiazomethane. GPC analysis of the product indicates that the uniformity of (5) is retained during the benzylation. Spectroscopic analysis of (6) is consistent with chemically synthesized samples of PBLG. The uniform polyglutamates should generate unique liquid crystalline phases because the polymer's uniform chain length should allow it to stack into a phase with both orientational and positional order. In contrast, the polydisperse polymer lacks positional order and forms a more limited range of liquid crystalline structures. Solutions of (6) in benzyl alcohol form optically birefringent phases characteristic of lyotropic liquid crystalline solutions of rodlike polymers. The effect of the uniformity of the polypeptide on the degree of order displayed by the liquid crystalline phases is currently under investigation.

IV. INCORPORATION OF ARTIFICIAL AMINO ACIDS INTO PROTEINS

A. Methodology

While biological synthesis offers superior control over polymer chain architecture, including purity of sequence, size and stereochemistry, its versatility is normally limited to the twenty amino acids that appear in natural proteins. Thus, an important aim for the future will be to expand the number of amino acids that can be utilized in protein biosynthesis. First efforts in this direction would logically start with amino acid analogs known to be incorporated during translation[42]. The unnatural amino acids (or amino acid analogs) employed are those which are in many ways similar to a natural amino acid, yet differ slightly in structure (thienylalanine vs. phenylalanine), functionality (hydroxyproline vs. proline), or constituent atoms (selenomethionine vs. methionine) (Tables 2 and 3).

The biosynthesis of proteins containing unnatural amino acids has been approached in three different ways. One approach involves attaching a desired substituent to a suppressor tRNA that decodes a nonsense codon in an appropriate location in the gene of interest[43]. This approach allows the usual restrictions on enzymatic charging of tRNA to be by-passed; it also permits the unnatural monomer to be introduced into a specific site. However, only sub-milligram quantities of protein are likely to be obtained, since chemical acylation is experimentally difficult and the tRNA is not recycled. Furthermore, since the efficiency of suppression generally is 50% or less, multi-site replacement (for example, in a repetitive protein) is impractical. In a second method, the amino acid analogue is added to a cell-free protein synthesis system[44]. Only substrates that can be enzymatically activated and charged can be used, but the method avoids the difficulties of low suppression efficiency and chemical synthesis of the aminoacyl tRNA. The third approach, which provides the potential for synthesizing proteins containing many modified sites, is *in vivo* incorporation of unnatural amino acids.

It has been known for many years that certain analogs of the natural amino acids can be incorporated efficiently *in vivo* into bacterial proteins[43]. Success of this process is entirely dependent on the ability of the protein synthesis machinery of the cell to accept the analog in place of the natural amino acid. The artificial amino acid must first enter the cell; however, because the relevant transport processes are usually not very specific, transport is not normally limiting to analog incorporation[45]. Much more important are the aminoacyl-tRNA synthetases, which are responsible for recognition of the 20 natural amino acids and attachment ("charging") to their respective tRNAs[46]. The unnatural analogs must be similar enough to their natural counterparts to be recognized and activated by these enzymes. In addition, a certain lack of specificity is desirable in the aminoacyl-tRNA synthetase enzyme for the amino acid of interest: the less

TABLE 2. Unnatural Amino Acid Analogs Incorporated into [(GlyAla)$_3$-XxxGly].

Code	Structure	Name	Previous *E. coli* Incorporation
SeMet		selenomethionine	yes
p-FP		*p*-fluorophenylalanine	yes
TFL		5,5,5-trifluroleucine	yes
3-TA		3-thienylalanine	no

TABLE 3. Analogs of Proline Under Study.

Code	Structure	Name	Previous *E. coli* Incorporation
Azc		azetidine-2-carboxylic acid	yes
Pip		piperidine-2-carboxylic acid	no
Dhp		3,4-dehydroproline	yes
Thp		thiazolidine-4-carboxylic acid	yes
c-Hyp		*cis*-4-hydroxyproline	yes
t-Hyp		*trans*-4-hydroxyproline	no
Sar		sarcosine	no
Nip		piperidine-3-carboxylic acid	no
Pyr		pyrrolidone-5-carboxylic acid	no

Pro

specific the enzyme, the greater the range of analogs which can be successfully incorporated.

Since the synthetase enzymes are nearly always much more effective in activating the natural amino acid than the analog, it is often crucial to decrease the amount of natural amino acid available to the cells. This requirement is met through the use of host strains which are amino acid auxotrophs, i.e., cells which require a certain amino acid (the one of interest for analog incorporation) in order to grow. Thus, our methodology for incorporating unnatural amino acids into artificial proteins begins with the preparation of a suitable *E. coli* auxotroph[47] followed by transformation of this strain with an expression vector containing the gene encoding the target protein of interest with the natural amino acid encoded in the positions where the analogs are to be inserted. If the amino acid analog supports cell growth, the cells can then be grown in medium lacking the natural amino acid, but supplemented with the analog of interest. Expression of the target protein is then induced and the protein is isolated by standard methods. If the analog is toxic to the cells or does not support growth, an alternate procedure is employed; the cells are grown in medium containing the natural amino acid, they are then pelleted, washed, and resuspended in medium containing the analog and lacking the natural amino acid, and protein expression is induced. This procedure is necessary to grow large numbers of cells for production of sufficient quantities of target proteins.

B. Incorporation of Selenomethionine

In a first attempt to generate artificial proteins containing unnatural amino acids, we produced a β-sheet forming protein containing selenomethionine[48]. The protein was expressed in *E. coli* from plasmid DNA that encoded a lamellar protein of sequence $[(GlyAla)_3 GlyMet]_{11}$ (**7**). Selenomethionine was selected for two reasons: first, it is known to support growth of *E. coli* methionine auxotrophs in the absence of externally supplied methionine[39]; second, the presence of selenomethionine endows product proteins with useful sites for chemical modification and a ^{77}Se NMR probe useful for mapping local folding properties. In addition, the selenium containing side chains can be utilized selectively for addition of structurally or functionally important new substituents or for site-specific linkage to other materials.

The target protein used in our study contains methionines in each putative β-turn segment, offering the potential of selective placement of selenomethionines on the surfaces of a resulting lamellar material. This prediction is supported (but certainly not conclusively demonstrated) by x-ray scattering data obtained with the methionine containing product, showing that this material assumes a crystalline morphology[48]. An *E. coli* expression strain was first converted to a strict methionine auxotroph and shown to grow in medium in which methionine was replaced with selenomethionine; although the growth rate was slower with the analog, similar saturating cell densities were observed[48]. Electrophoretic analysis showed that the target protein was produced in cells grown on selenomethionine and that no inhibitory effects on protein production were observed. The level of selenomethionine substitution was estimated indirectly using a competition assay where the loss of radiolabeled methionine in product protein was correlated with the ratio of methionine and selenomethionine in the growth medium. The results were consistent with perfect or near-perfect replacement of methionine with selenomethionine. Plans for future work include characterization of the crystalline morphology of the analog containing material and chemical modification studies.

C. Fluorine Containing Polypeptides

Another area of interest is the incorporation of fluorinated amino acid analogs. Natural polymers lack halogens, yet many synthetic polymers which incorporate halogens have proven to be extremely useful. In particular, fluorine containing polymers display the useful characteristics of excellent solvent resistance, good hydrolytic stability, low surface energy, and low coefficient of friction[50]. We reasoned that it would be advantageous to transfer some of these properties to polypeptides to create unique materials which might prove useful for the fabrication of novel membranes, low-friction surface treatments, and biomedical materials. Furthermore, since fluorine is only slightly larger than hydrogen it is not surprising that many fluorinated amino acids are recognized and utilized by the bacterial protein synthesis machinery[42].

The analogs chosen for the *in vivo* synthesis of fluorinated polypeptides in our laboratory have included *para*-fluorophenylalanine (*p*fF) and 5,5,5-trifluoroleucine (TFL), since both had been shown previously to be incorporated into *E. coli* proteins[42]. The strategy was similar to that used for selenomethionine, with the exception that the host auxotroph strains were first grown in the presence of the natural amino acids and then protein synthesis was induced in medium containing the analog. This step was necessary because these analogs did not support growth of the *E. coli* strains used for protein expression. The artificial proteins into which the analogs were inserted were, as with selenomethionine, lamellar proteins of the type $[(GlyAla)_3 GlyXxx]_n$ (**8**), where *Xxx* was encoded as Phe for *p*fF (n = 13) and Leu for TFL (n = 12).

Proteins containing *p*fF[51] or TFL[52] were expressed and purified through standard techniques. The extent of analog incorporation was found to be very high for both materials (>95% for *p*fF and 88% for TFL) as determined by ^{1}H NMR spectra and amino acid analyses. Both materials were crystallizable and were found to adopt antiparallel β-sheet structures similar to those obtained for other proteins of this type (see below). Thus, it is expected that the fluorinated residues, since they occupy turn positions, will be exposed at the surfaces of the polymer crystals and should affect the bulk material properties of the samples. Advancing contact angle measurements with hexadecane for the TFL polypeptides gave increasing contact angles with increasing analog content relative to the natural residues. The presence of fluorine in the samples was found to decrease the surface energy as determined by decreased wettability toward hydrocarbon liquids[52].

D. Electroactive Substituents

The scope of functional groups that can be incorporated into the protein backbone by genetic engineering has recently been extended to residues with electroactive substituents, such as 3-thienylalanine (3-TA)[53]. 3-Alkylthiophenes, which are structurally similar to 3-TA, can be oxidatively polymerized through the 2- and 5-positions of the ring to produce polymers with extended p conjugation. Poly(3-alkylthiophene)s can be electrochemically doped to form conducting polymers with conductivities (2000 S cm^{-1}) that are among the highest recorded for organic materials[54]. Copolymerization of the 3-TA residues with 3-alkylthiophenes can provide a method for the crosslinking (or grafting) of polypeptides onto electrodes via a layer of conducting polymer that may be useful for the fabrication of enzyme-based sensory elements.

Prior to this work, 3-TA had not been shown to be utilized by the protein synthesis machinery of *E. coli*, although the isomeric 2-TA had been incorporated into native bacterial proteins in place of phenylalanine. Therefore, the target protein selected for the incorporation assay was the sequence, [(*GlyAla*)$_3$*GlyPhe*]$_{13}$ (**9**) which contains a unique *Phe* residue in each repeat, which can serve as an accepting site for 3-TA. The chain-folded structure of the polypeptide could potentially control the spatial orientation of the electroactive substituents and facilitate conversion of these groups to an oriented conducting polymer. *In vivo* radiolabelling experiments demonstrated that the analogue-containing target protein was expressed at levels similar to the natural variant (**9**) in a *Phe* auxotroph. The extent of substitution, quantitated by UV and ^{1}H NMR spectroscopy, and amino acid analysis, is about 80% suggesting that 3-TA is efficiently utilized by the *E. coli* protein synthesis machinery under these conditions. Preliminary electrochemical investigations indicate that the 3-TA-containing polypeptide can be oxidatively copolymerized with 3-methylthiophene yielding conductive films.

E. Structural Modifications of Polypeptides

It was mentioned earlier that a key to successful *in vivo* analog incorporation is lack of specificity in the aminoacyl tRNA synthetases. The prolyl-tRNA synthetase has been reported to be the least specific of all[55]. Not only does this mean that a wide variety of proline analogs should be incorporated into proteins, but proline, the only imino acid, is also a key residue for structural modification studies. The lack of a hydrogen atom on nitrogen to serve as a hydrogen bond donor, coupled with the steric bulk of the prolyl ring, in almost all cases excludes proline from involvement in α-helical and β-sheet structures[56]. Proline therefore usually occupies the key positions of helix starter or helix terminator, and plays an active role in protein folding. It has also been reported that proline can act to bend the axis of a helix by *ca.* 30°. Finally, although it is a hydrophobic residue, proline is usually exposed at the hydrophilic surfaces of proteins by virtue of its localization in turns. This is an ideal location if reactive analogs are to be accessible for post-translational modification.

It was thought that analogs of proline might cause significant perturbations in the solid-state structure of polymers of the sequence $[(GlyAla)_3 GlyProGlu]$ (**1**) (see below). Although these polypeptides crystallize with difficulty, it was proposed that substitution of an analog for proline might facilitate crystallization of the *AlaGly* repeats into a lamellar β-sheet architecture. Of key interest was incorporation of homologs of proline, where a simple change in the size of the ring structure would cause a significant change in the conformation of the residue and thus allow chain folding. As precedent for this, it has been reported that incorporation of L-azetidine-2-carboxylic acid in place of proline at levels as low as 4% is sufficient to prevent triple helix formation in collagen[57].

Many analogs of proline have been reported and some have been demonstrated to be incorporated *in vivo* into cellular proteins[58]. These results are briefly summarized in Table 3. The bulk of studies on analogs of proline have focused on collagen synthesis[59], although a few reports exist on the behavior of these analogs in bacteria[60]. The proline analogs used in our studies began with some of those which had been previously incorporated into proteins, namely: L-azetidine-2-carboxylic acid (*Azc*)[57], L-γ-thiaproline (*Thp*)[61], and 3,4–dehydroproline (*Dhp*)[62]. As can be seen in Table 3, *Azc*, *Thp*, and *Dhp* have all been incorporated into proteins in *E. coli* cells *in vivo*.

A proline auxotroph of *E. coli* containing a gene encoding sixteen repeats of (**1**) was used for the expression of proteins containing the analogs[63]. Proteins containing varying degrees of *Azc*, *Thp*, and *Dhp* were expressed and purified. *Dhp* appeared to be utilized most efficiently with incorporation levels approaching 100%. *Azc* was incorporated to a lesser extent (*ca.* 40%) and *Thp* was incorporated to a level of around 10%. The physical properties of the *Dhp* variant were found to be very similar to those of the proline form: both formed amorphous solids and were very soluble in water. While the material containing *Dhp* is amorphous, it is an intriguing material which contains olefinic functionality which can serve as a site of specific reactivity not present in natural proteins.

Both the *Thp* and *Azc* variants were found to be water insoluble, and the *Azc* material fractionated differently from the parent material during purification. The decreased solubility of these polypeptides is an indication that they may be crystalline; a hypothesis supported for the *Azc* material by preliminary structural studies. These proteins illustrate the use of small specific variations in monomer structure to control and vary the overall conformation of the chain.

V. HYBRID ARTIFICIAL PROTEINS

A. Design Criteria

The coupling of natural and artificial domains to produce hybrid artificial proteins offers the prospect of new classes of functionalized nanoscale architectures. In preliminary experiments directed toward these objectives, enzymes have been linked at the genetic level to crystalline artificial proteins to promote the assembly of catalytic arrays and sensors [Figure 5]. In designing materials of this type many issues relating to the composition of the natural and artificial domains must be addressed. Choice of the artificial domain in our initial experiments was simple: we sought a polypeptide which would present functional groups useful in surface assembly and attachment. Such functional groups might include the ammonium groups of lysine (which can attach to anionic surfaces), the sulfhydryl groups of cysteine (which can bind to gold surfaces), and the carboxylate

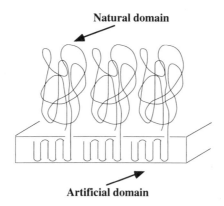

Natural domain

Artificial domain

FIGURE 5. Schematic representation of a new class of reactive polymers, hybrid artificial proteins, which combine the catalytic or recognition properties of natural proteins with the materials properties of artificial proteins. In the authors' laboratories, hybrid artificial proteins are being evaluated as elements of enzyme-based sensors and reactors.

groups of glutamate and aspartate (which can be attached to cationic surfaces). A polypeptide which fulfilled all of these requirements was the repeating octapeptide (**3**) which is expressed in high yield in *E. coli.*

The enzyme comprising the natural domain of the hybrid protein must satisfy a more complex set of requirements. The enzyme should function as a single-subunit, so that post-translational association of enzyme subunits, common in many enzymes, is not essential. It must express well in appropriate cellular hosts, and it must tolerate modification (i.e., fusion to the artificial domain) without loss of function. Our first candidate was a single chain phosphotriesterase from *Pseudomonas diminuta* which hydrolyzes many phosphorus based pesticides and nerve agents[64]. This enzyme should be useful in detection and decontamination of these toxins, it has been cloned and expressed in *E. coli*, and it was available in plasmid-encoded form at the outset of our experiments[65].

B. Results

A gene encoding both the 27 kDa artificial protein of sequence (**3**) and the 39 kDa phosphotriesterase domain was constructed and expressed in *E. coli* using a bacteriophage T7-based expression system[66]. The enzyme itself was also expressed in *E. coli* and was found to form inclusion bodies which segregate into the insoluble portion of the cell lysate. The fusion protein was found to segregate into the soluble portion of the cell lysate which should help to reduce problems involved in renaturing the enzymatic portion of the protein. Fractionation experiments on an ion-exchange column showed that the fusion protein could be effectively separated from the free enzyme based on the excess acidic sites present in the artificial domain. Furthermore, these sites were sufficient to bind the fusion protein to the column under conditions where the free enzyme is not bound, and preliminary results suggest that enzymatic activity is retained. Current work is aimed at immobilizing the fusion protein onto optical glass fibers to create biosensor devices.

REFERENCES

[1]   Pino, P., Moretti, G., *Polymer* **28**, 683 (1987).

[2]   Tour, J.M., *Trends in Polymer Science* **2**, 332 (1994).

[3]   Tan, Y.Y., *Prog. Polym. Sci.* **19**, 561 (1994).

[4]   Fréchet, J.M.J., *Science* **263**, 1710 (1994).

[5]   Merrifield, R.B., *Science* **232**, 241 (1986).

[6]   Cappello, J., Ferrari, F., *Plastics from Microbes: Microbial Synthesis of Polymers and Polymer Precursors,* Mobley, D.P., ed., Munich: Hanser/Gardner Publications, 1994, pp. 35–92.

[7]   McGrath, K.P., Fournier, M.J., Mason, T.L., Tirrell, D.A., *J. Am. Chem. Soc.* **114**, 727 (1992).

[8]   Waite, J.H., *Biol. Rev.* **58**, 209 (1983).

[9]   Tsujimoto, Y., Suzuki, Y., *Cell* **18**, 591 (1979).

[10] Case, S.T., Powers, J., Hamilton, R., Burton, M.J., "Silk Polymers" in *ACS Symposium Series* **544**, American Chemical Society, Kaplan, D., Adams, W.W., Farmer, B., Viney, C., eds., Washington, DC, 1994, pp. 81–90.

[11] Gosline, J.M., DeMont, M.E., Denny, M.W., *Endeavor* **10**, 37 (1986).

[12] (a) Hinman, M.B., Lewis, R.V., *J.Biol. Chem.* **10**, 1 (1992). (b) Xu, M., Lewis, R.V., *Proc. Natl. Acad. Sci. USA,* **87**, 7120 (1990).

[13] Goldberg, I., Salerno, A.J., "Materials Synthesis Utilizing Biological Processes," in *Mater. Res. Soc. Proc.* **174**, Materials Research Society, Reike, P.C., Calvert, P.D., Alper, M., eds., Pittsburg, PA, 1990, pp 229–236.

[14] Sandberg, L.B., Soskel, N.T., Leslie, J.B., *N. Engl. J. Med.* **304**, 566 (1981).

[15] Urry, D.W.J., *Prot. Chem.* **7**, 1 (1988).

[16] Indik, Z., Abrams, W.R., Kucich, U., Gibson, C.W., Mecham, R.P., Rosenbloom, J., *Arch. Biochem. Biophys.* **280**, 80 (1990).

[17] McPherson, D.T., Morrow, C., Minehan, D.S., Wu, J., Hunter, E., Urry, D.W., *Biotechnol. Prog.* **8**, 347 (1992).

[18] (a) Waite, J.H., Tanzer, M.L., *Science* **212**, 1038 (1981). (b) Waite, J.H., Housely, T.J., Tanzer, M.L., *Biochemistry* **24**, 5010 (1985). (c) Jensen, R.A., Morse, D.E., *J. Comp. Physiol. B* **158**, 317 (1988).

[19] Filipula, D.R., Lee, S.-M., Link, R.P., Strausberg, S.L., Strausberg, R.L., *Biotechnol. Prog.* **6**, 171 (1990).

[20] Salerno, A.J., Goldberg, I., *Appl. Microbiol. Biotechnol.* **58**, 209 (1993).

[21] Benedict, C.V., Picciano, P.T., "Adhesives From Renewable Resources," in *ACS Symposium Series* **385**, American Chemical Society, Hemingway, R.W., Conners, A.H., Branham, S.J., eds., Washington, DC, 1989, pp. 465–483.

[22] Valentine, R.C., Pereira, H.G., *J. Mol. Biol.* **13**, 13 (1965).

[23] Albiges-rizo, L., Chroboczek, J., *J. Mol. Biol.* **212**, 247 (1990).

[24] O'Brien, J.P., *et al.* "Silk Polymers" in *ACS Symposium Series* **544**, American Chemical Society, Washington, DC, Kaplan, D., Adams, W.W., Farmer, B., Viney, C., eds., 1994, pp. 104–117.

[25] McGrath, K.P., Kaplan. D.L., "Biomolecular Materials" in *Mater. Res. Soc. Proc.* **292**, Materials Research Society, Viney, C., Case, S.T., Waite, J.H., eds., Pittsburg, PA., 1993, pp 83–92.

[26] O'Shea, E.K., Kim, P.S., *Science* **254**, 539 (1991).

[27] Bassett, D.C., *Principles of Polymer Morphology,* Cambridge: Cambridge University Press, 1981.

[28] Block, H., *Poly(g-benzyl-L-glutamate) and Other Glutamic Acid Containing Polymers*, New York: Gordon and Breach, 1983.

[29] Fraser, R.D.B., McRae, T.P., *Conformations of Fibrous Proteins,* New York: Academic, 1973.

[30] Chou, P.Y., Fasman, G.D., *Biochemistry* **13**, 211 (1974).

[31] Creel, H.S., Fournier, M.J., Mason, T.L., Tirrell, D.A., *Macromolecules* **24**, 1213 (1991).

[32] Krejchi, M.T., Atkins, E.D.T., Waddon, A.J., Fournier, M.J., Mason, T.L., Tirrell, D.A., *Science* **265**, 1427 (1994).

[33] Warwicker, J.O., *J. Mol. Biol.* **2**, 350 (1960).

[34] Parkhe, A.J., Fournier, M.J., Mason, T.L., Tirrell, D.A., *Macromolecules* **26**, 6691 (1993).

[35] Finkelman, H., *Angew. Chem. Int. ed. Engl.* **26**, 816 (1987).

[36] Doty, P., *J. Polymer Sci.* **23**, 851 (1957).

[37] Horton, J.C., Donald, A.M., Hill, A., *Nature* **346**, 44 (1990).

[38] McMaster, T.C., Carr, H.J., Miles, M.J., Cairns, P., Morris, V.J., *Macromolecules* **24**, 1428 (1991).

[39] Hol, W.G.J., *Nature* **273**, 443 (1978).

[40] (a) Go, Y., Ejiri, S., Fukuda, E., *Biochim. Biophys. Acta* **175**, 454 (1969). (b) Iizuka, E., *Biochim. Biophys. Acta* **175**, 456 (1969).

[41] Zhang, G., Fournier, M.J., Mason, T.L., Tirrell, D.A., *Macromolecules* **25**, 3601 (1992).

[42] (a) Wilson, M.J., Hatfield, D.L., *Biochim. Biophys. Acta* **781**, 205 (1984). (b) Richmond, M.H., *Bacteriol. Rev.* **26**, 398 (1962). (c) Hortin, G., Boime, I., *Methods in Enzymology,* Fleischer, S., Fleischer, B., eds., New York: Academic, 1983, Vol. **96**,

pp 777–784.

[43]   (a) Noren, C.J., Anthony-Cahill, S.J., Griffith, M.C., Schultz, P.G., *Science* **244**, 182 (1989). (b) Bain, J.D., Glabe, C.G., Dix, T.A., Chamberlin, A.R., *J. Am. Chem. Soc.* **111**, 8013 (1989). (c) Mendel, D., Ellman, J.A., Schultz, P.G., *J. Am. Chem. Soc.* **113**, 2758 (1991). (d) Mendel, D., Ellman, J.A., Chang, Z., Veenstra, D.L., Kollman, P.A., Schultz, P.G., *Science* **256**, 1798 (1992).

[44]   Zubay, G., *Ann. Rev. Gen.* **7**, 267 (1973).

[45]   Saks, M.E., Sampson, J.R., Abelson, J.N., *Science* **263**, 191 (1994).

[46]   Schimmel, P., *Ann. Rev. Biochem.* **56**, 125 (1987).

[47]   Miller, J.H., *Experiments in Molecular Genetics;* Cold Spring Harbor, New York, 1972.

[48]   Dougherty, M.J., Kothakota, S., Fournier, M.J., Mason, T.L., Tirrell, D.A., *Macromolecules* **26**, 1779 (1993).

[49]   Tuve, T., Williams, H., *J. Am. Chem. Soc.* **79**, 5830 (1957).

[50]   Gangal, S.V., in *Encyclopedia of Polymer Science and Engineering,* 2nd ed., New York: Wiley-Interscience, 1990, Vol. **16**, pp. 577–648.

[51]   Yoshikawa, E., Fournier, M.J., Mason, T.L., Tirrell, D.A., *Macromolecules* **27**, 5471 (1994).

[52]   Kothakota, S., Fournier, M.J., Mason, T.L., Tirrell, D.A., Manuscript in preparation.

[53]   Kothakota, S., Fournier, M.J., Mason, T.L., Tirrell, D.A., *J. Am. Chem. Soc.* **117**, 536 (1995).

[54]   Roncali, J., Yassar, A., Garnier, F., *J. Chem. Soc. Chem. Commun.* **581** (1988).

[55]   Papas, T.S., Mehler, A.H., *J. Biol. Chem.* **245**, 1588 (1970).

[56]   Fasman, G.D., *Prediction of Protein Structure and the Principles of Protein Conformation,* New York: Plenum, 1989, pp. 48–53.

[57]   (a) Grant, M.M., Brown, A.S., Corwin, L.M., Troxler, R.F., Franzblau, C., *Biochim. Biophys. Acta* **404**, 180 (1975). (b) Fowden, L., Richmond, M.H., *Biochim. Biophys. Acta* **71**, 459 (1963). (c) Takeuchi, T., Prockop, D.J., *Biochim. Biophys. Acta* **175**, 142 (1969).

[58]   Mauger, A.B., Witkop, B., *Chem. Rev.* **66**, 47 (1966).

[59]   (a) Rosenbloom, J., Prockop, D.J., *J. Biol. Chem.* **245**, 3361 (1970). (b) Uitto, V.-J., Uitto, J., Kao, W.W.-Y., Prockop, D.J., *Arch. Biochim. Biophys.* **185**, 214 (1977). (c) Jimenez, S., Rosenbloom, J., *Arch. Biochim. Biophys.* **163**, 459 (1974). (d) Uitto, J., Prockop, D.J., *Biochim. Biophys. Acta* **336**, 234 (1974). (e) Gottlieb, A.A., Fujita, Y., Udenfriend, S., Witkop, B., *Biochemistry* **4**, 2507 (1965). (f) Cleland, R.,

Olson, A.C., *Biochemistry* **7**, 1745 (1968). (g) Rosenbloom, J., Prockop, D.J., *J. Biol. Chem.* **246**, 1549 (1971).

[60]   (a) Busiello, V., Di Girolamo, M., Cini, C., De Marco, C.,   *Biochim. Biophys. Acta* **564**, 311 (1979). (b) De Marco, C., Busiello, V., Di Girolamo, M., Cavallini, D., *Biochim. Biophys. Acta* **478**, 156 (1977).

[61]   Busiello, V., Di Girolamo, M., Cini, C., De Marco, C., *Biochim. Biophys. Acta* **606**, 347 (1980).

[62]   (a) Fowden, L., Neale, S., Tristam, H., *Nature* **199**, 35 (1963). (b) Smith, L.C., Ravel, J.M., Skinner, C.G., Shive, W., *Arch. Biochim. Biophys.* **99**, 60 (1962).

[63]   Deming, T.J., Fournier, M.J., Mason, T.L., Tirrell, D.A., *ACS Polym. Mat. Sci. Engr.* **71**, 673 (1994).

[64]   Dumas, D.P., Rauschel, F.M., *J. Biol. Chem.* **265**, 21498 (1990).

[65]   McDaniel, C.S., Harper, L.L., Wild, J.R., *J. Bacteriol.* **170**, 2306 (1988).

[66]   Dong, W., Fournier, M.J., Mason, T.L., Tirrell, D.A., *Polym. Prepr.* **35(2)**, 419 (1994).

Chapter 10

Atom Optics:
Using Light to Position Atoms

Jabez J. McClelland

Electron Physics Group
National Institute of Standards and Technology
Gaithersburg, Maryland 20899-0001

Mara Prentiss

Physics Department
Harvard University
Cambridge, MA 02138

I. INTRODUCTION

In most conventional lithography techniques, a light-sensitive resist is used to transfer a pattern from a mask to a substrate. The process is massively parallel because one mask can be used to expose millions of features at the same time. The minimum size of the features that can be created with conventional lithography is of the order of half the wavelength of the light being used. Extensions of optical lithographic techniques, involving deep ultraviolet light, are predicted to reach a limit of approximately $0.10 \, \mu m$[1] by the year 2001. Other exposure technologies such as electron beam and x-ray, that have been proposed for producing smaller features than ultraviolet light, are problematic. For example, in conventional electron beam systems the writing process is serial, so large complicated patterns take a very long time to write. Moreover, in both electron beam and x-ray lithography, the masks and the substrates can be damaged by the high energy beams. Thus, there is great interest in developing parallel techniques for creating nanometer-scale features without damage and without masks. The new field of atom optics offers various massively parallel fabrication techniques which hold the promise for creating nanometer-scale features with good contrast and high resolution using low energy atomic beams. These advantages are augmented by the "direct-write" nature of atom optical fabrication, that is, structures can be deposited directly on a substrate without the intermediate process steps involved with using a resist.

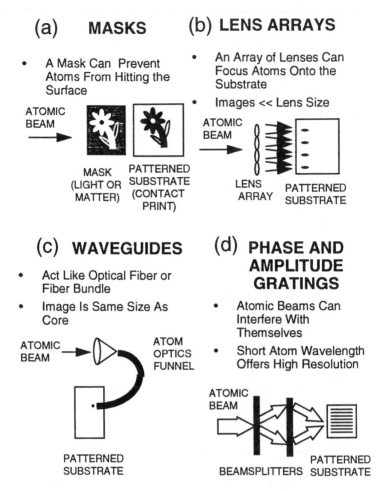

FIGURE 1. Atom optics as a tool for nano-fabrication.

Figure 1 shows a number of conceptual ways in which atom optics can be used to fabricate nanostructures. Generally, a beam of atoms propagates toward a substrate, and a pattern of light beams is used to manipulate the atoms, controlling their deposition. The light can deflect the atoms from parts of the substrate, a process which uses the light as a mask or a stencil for the atoms as shown in Figure 1(a). The light can also be used as a lens to focus atoms onto particular areas of the substrate, producing high flux of atoms at the focal points of the lens, and diminished or zero flux over the remaining areas, as shown in Figure 1(b). In addition, light can serve as a waveguide which confines atoms in narrow spatial regions, so that the atoms are delivered to the substrate in only narrow selected regions. Each waveguide is analogous to an optical fiber, which contains light within the core and can be used to deliver the light to a particular chosen area of a substrate. The complete atom optics process is then equivalent to a bundle of optical fibers which can be used to simultaneously deliver small points of atoms to many narrowly confined regions of a substrate, as shown in Figure 1(c). Finally, amplitude or phase holograms can be used to create interference patterns between atomic wave functions, so that atoms are deposited on the substrate where the interference pattern is constructive, and no atoms are deposited at the surface where the interference pattern is destructive, as shown in Figure 1(d).

In addition to the processes shown in Figure 1, it is also possible to simply take advantage of the direct-write feature of atom optics, using a material mask to shadow regions of the substrate during deposition. As a final note, we mention that with direct-writing, the eventual shape of the permanent structure that remains on the surface could in some cases depend strongly on interactions which occur after the atoms are deposited on the substrate. For example, depending on the atomic species and surface material, degree of surface contamination, and ambient temperature, atoms can sometimes migrate across the surface after deposition, resulting in surface distributions which can be radically different from those originally induced by the light.

In the next section we begin by discussing some general principles in order to explain how light can manipulate matter, and then we discuss some of the history of how these principles have been used so far. In section III we discuss some of the current research as it relates directly to fabrication, and some future prospects.

II. ATOMIC MANIPULATION

A. Fundamentals of the Light Force

We have mentioned how light can control the deposition of atoms onto a substrate by acting as a lens, a mask, a waveguide, or a hologram for the atoms. In order to understand how light can act in these ways, we will discuss how the interaction between an atom and light results in a force on the atom.

Traditionally, the force exerted by light on atoms is divided into two categories: (1) stimulated or coherent processes in which the energy of the atoms is conserved during the interaction; and (2) spontaneous or incoherent processes where the energy of the atom is not conserved[2,3]. Most of the coherent processes lead to motion of the atoms that can be easily modeled using analogies with conventional optics. The incoherent processes, however, find no analogy in conventional optics. These processes can cool atoms, compressing their phase space distribution, something that cannot be accomplished by the conservative

forces of conventional optics. Thus atomic beams can be simultaneously collimated and brightened. This is an important difference between atom optics and
conventional optics.

Both coherent and incoherent forces can be explained by considering the
interaction between the electric field associated with the light field, and the dipole
moment induced in the atom by pppthis electric field. Consider a simple model
for the atom which consists of a massive positive charge at the origin surrounded
by a circular cloud of very light negative charges. In the presence of an external
electric field **E**, the positive and negative charges will be displaced from each
other, resulting in a net dipole moment **p** for the atom, where **p** is proportional to
E and can be written $\mathbf{p} = \alpha \, \mathbf{E}$. The energy of the dipole in the field is then $-\mathbf{p}\cdot\mathbf{E}$. If
the field is oscillating at a frequency ω, then the dipole moment will also oscillate
at a frequency ω, though there will generally be some phase shift between the two
oscillations. If, as is the case for light, the oscillations are very fast in comparison
with the time scale for external motion, the relevant energy is not the instantaneous energy $-\mathbf{p}\cdot\mathbf{E}$, but the average of that energy over an optical cycle, i.e.,
$-|\mathbf{p}| \, |\mathbf{E}| \cos\phi = -|\alpha| \, |\mathbf{E}|^2 \cos\phi \, \alpha -|\alpha| I \cos\phi$, where I is the laser intensity
and ϕ is the phase difference between the dipole and the field.

The stimulated or coherent force, which is sometimes also referred to as the
dipole force, can be associated with the change in this average energy as a function of position, and hence arises from a gradient in the intensity. The sign of the
force depends on the phase difference ϕ. If the field is detuned below resonance,
($\Delta < 0$, where Δ is the detuning of the laser from the transition), the phase difference is between $0°$ and $90°$ and $\cos\phi > 0$; therefore, if $\Delta < 0$ atoms are attracted to
the intensity maxima. Conversely, for fields detuned above resonance, ϕ is
between $90°$ and $180°$, so $\cos\phi < 0$ and atoms are attracted to the intensity
minima.

While the above derivation of the behavior of the stimulated force is essentially correct, it must be noted that there are a number of quantum mechanical
subtleties that must be taken into account in order to properly model all
phenomena. Discussion of these effects can be found, for example, in reference
[4]. For the present we simply note that the stimulated force can also be
described as the result of stimulated emission and absorption of photons, which
produces a coherent redistribution of photons among modes of the laser field. The
force is considered coherent because the coherence of the atomic wavepacket is
preserved during the interaction. In the limit of large detuning and/or low intensity, it can be shown that the potential associated with this force is given by

$$U = \frac{\hbar\gamma^2}{8\Delta} \frac{I}{I_s}, \tag{1}$$

where γ is the natural line width of the atomic transition (in rad/s), Δ is the detuning of the laser frequency from the atomic resonance (also in rad/s), I is the laser intensity, and I_s is the saturation intensity associated with the atomic transition. We note that U is sometimes expressed in terms of the Rabi frequency $\Omega = \gamma\sqrt{I/2I_s}$ as $U = \hbar\Omega^2/4\Delta$. For smaller detunings, and higher intensities, the potential can be written as

$$U = \frac{\hbar\Delta}{2} \ln(1+p), \tag{2}$$

where

$$p = \frac{I}{I_s} \frac{\gamma^2}{\gamma^2 + 4\Delta^2}. \tag{3}$$

This expression is valid provided the atom moves slowly enough in a spatially varying field to maintain equilibrium between its internal and external degrees of freedom. If this is not the case, velocity-dependent forces arise and the motion of the atoms can no longer be derived from a conservative potential.

The spontaneous or incoherent force cannot easily be associated with a change in energy as a function of position, although it also results from the induced atomic dipole. The incoherent force results from the induced dipole moment which is in quadrature with the electric field, and results in a force proportional to the gradient of the *phase* of the field, rather than the gradient of the intensity. This results in a net push in the direction of propagation of the field. The spontaneous force is non-conservative and can therefore result in compression in phase space. For example, a geometry where two weak counter-propagating traveling wave fields are detuned below resonance results in very strong cooling, and is referred to as optical molasses. From a quantum mechanical point of view, this force can be viewed as the result of a stimulated absorption followed by a spontaneous emission. The force is considered incoherent because the coherence of the atomic wavepacket is not preserved during the interaction. In addition, the spontaneous emission occurs in a random direction, so there is a heating process associated with this force, even when the net average force results in cooling.

B. First Approaches to Atom Focusing with Lasers

The first atom optics element to be demonstrated was the single thick lens. In this experiment, Bjorkholm et al.[5] focused an atomic beam using the stimulated force generated by a strong traveling wave field with a Gaussian intensity distribution $I \propto \exp(-2r^2/\sigma^2)$, where σ is the $1/e^2$ radius of the laser beam, which was about 100 μm. The laser field was detuned below resonance, so the potential minimum was at the intensity maximum. The researchers allowed an atomic beam of atoms to co-propagate with the light, and showed that the atomic beam was focused by the light to a spot approximately 200 μm in diameter. The lens is considered thick because the atomic position changed significantly during the time that the atoms interacted with the lens.

Another type of lens, making use of the spontaneous force, was demonstrated by Balykin et al.[6]. In this configuration, four diverging laser beams were incident transversely upon an atomic beam. The laser beams were tuned exactly on resonance with the atoms, so a spontaneous force was exerted in the direction of the light propagation. Since the beams were diverging, the light intensity, and hence the force, increased as a function of distance away from the beam axis, resulting in a lensing effect. The concentration of an atom beam into a point and the imaging of a two-slit atomic beam source were both observed with this lens.

In 1992, focusing of an atomic beam by a large-period standing wave was observed by Sleator et al.[7]. This experiment consisted of passing a beam of metastable helium atoms through a laser beam ($\lambda = 1.083\,\mu m$) that was reflected at grazing incidence from a glass surface. The optical standing wave that resulted from this grazing reflection had a period of about 45 μm, and the atoms were apertured to 25 μm, so all the atoms passed through a single period of the light field. The intensity variation within this single period resulted in a dipole force potential that formed a lens for the atoms. Imaging at a magnification of unity of both a single 2 μm slit and a 8 μm-period grating were observed with this lens.

The first suggestion that lasers could be used to focus atoms into the nanometer regime was made by Balykin and Letokhov[8], who proposed that a focused TEM^*_{01} ("doughnut"-mode) laser beam with atoms propagating axially would be a particularly good arrangement for this. Further analysis by Gallatin and Gould[9] and McClelland and Scheinfein[10] showed that indeed, even considering all the possible aberrations, spot sizes of a few nanometers could be expected.

C. Lens Arrays, Channeling and Optical Lattices

Another important development in atom optics has been the lens array[11]. In the experiments discussed above, the light produced a single positive or negative lens for atoms, which focused the atoms into a single spot. It is also possible to use light to produce an array of lenses. These lenses can simultaneously focus atoms into arrays of lines or dots. This allows parallel lithography over large areas, which is a key factor in making atom optics a desirable approach for fabrication.

A simple example of a lens array is the array of parallel cylindrical lenses which can be generated using a standing-wave field. Consider an optical standing wave formed by two counter-propagating traveling plane waves with wave vectors $\pm k$ along $\hat{x}$. The x-dependence of the intensity is then given by $I = I_o \cos^2(kx)$, which results in a potential proportional to $\cos^2(kx)$ for positive detuning. Let this standing wave interact with an atomic beam propagating in the $\hat{z}$-direction. The force associated with the potential variation in x will tend to focus the atoms into a series of lines along potential minima, which are at the nodes of the standing wave, found at $kx = (2n+1)\pi/2$, where n is an integer.

Thus, the plane standing wave will act like an array of cylindrical lenses focusing a plane wave of atoms into a series of parallel lines. It is important to note that each of the lenses in the array is separated from each other lens by a distance which is an exact half integral multiple of the atomic wavelength, which means that the resulting atomic distributions have excellent registration.

The discussion above describes the case where the plane wave forms a thin lens array, so that the atoms do not move significantly during the interaction. Most of the early experiments were done in the opposite limit where the atoms interacted with the lenses for a long time and were "channeled," that is, periodically focused and defocused as they passed through the lens array. The conventional optical analogy would be an array of graded index lenses, which could also be considered a waveguide.

The wave-guiding or channeling associated with atomic motion in thick cylindrical lens arrays was observed using spectroscopic techniques: Prentiss and Ezekiel[12] observed the distortion in the fluorescence due to atomic motion, and Salomon et al.[13] measured the distortion in absorption from a weak probe. The channeling was also demonstrated more directly by Balykin et al.[14] who showed that the atoms could be gently guided around curves.

The atomic waveguiding discussed above is analogous to the guiding of light by a multi-mode fiber or waveguide, where the initial atomic distribution includes many eigenmodes of the waveguide. It is also possible to do experiments in a regime where only one or two of the lowest waveguide modes are populated. An array of these waveguides are used to form an optical lattice. This is potentially good for lithography because these low order modes can be very well separated spatially allowing for deposition of well separated features with little or no background, whereas the multimode waveguides discussed above can have a significant background connecting the features. In the following we will discuss how these lowest order modes can be very efficiently populated.

As mentioned in section II A above, one of the important differences between conventional optics and atom optics is that in atom optics it is possible to make use of the spontaneous force to cool atoms and compress them in phase space. In the channeling experiments described above, a planar optical standing wave acted as an array of thick cylindrical graded index lens which periodically focused and defocused an atomic beam. Atoms are originally distributed uniformly in the potential and always have the opportunity to return to their initial value in the potential. Adding cooling to the potential allows the motion of the atoms in the wells to be damped so that they are eventually confined to a small region near the bottom of the potential wells, even if they started at a potential energy much higher than the potential minimum. The potential wells are periodically spaced and separated by half an optical wavelength, with spacings as small as a quarter of a wavelength possible in some geometries. Experiments which observe simultaneous cooling and channeling have been done in 1, 2, and 3 dimensions[15,16,17,18,19,20,21]. Indirect spectroscopic measurements indicate that the atoms have been confined to a region approximately $\lambda/20$ in diameter. The remaining regions of space are not populated, and the resulting distributions are often described as "optical crystals".

D. Atom Mirrors as Masks

In the sections above, we discussed early experiments which showed that light can form lens arrays or waveguides that can control atomic deposition. In this section, we will discuss early experiments that showed that light can be used as a mask to selectively prevent deposition. Basically, a light field detuned above resonance can form a potential barrier that will reflect atoms back in their initial direction of propagation. If such a light field is present above a substrate, there will be no deposition onto the substrate in the regions where the light is present. Thus, a patterned light field detuned above resonance will act as a mask that prevents deposition in regions where the light is present, and allows deposition in regions without light.

The early experiments discuss atom mirrors or ''trampolines'' rather than atom masks because they were more interested in the atoms that bounced off the light than in those that hit the surface. Consider a light field with an intensity given by $I = I_o \exp(-\alpha z)$, for $z > 0$. An atom of mass m will bounce off the light if it has a velocity, v_z, such that $\frac{1}{2}mv_z^2$ is less than the potential barrier due to the light. This atom will then freely propagate with velocity $-v_z$. This was first demonstrated by Balykin[22], who created the appropriate intensity dependence by using internal reflection from a prism. Similar experiments have since been done by other groups [23,24], who used curved prism surfaces to make a focusing mirror for the atoms, and have enhanced the reflection by coating the prism with a dielectric cavity that enhances the intensity of the light at the surface to almost 1000 times the initial intensity.

E. Diffraction of Atoms From a Laser Beam
or a Microfabricated Grating

In the sections above, we discussed early experiments which showed that light can act as a lens array or a mask which controls atomic deposition. These experiments can be described by analogy with classic geometric optics, or ray tracing, where the atom can be modeled as a billiard ball. This is not unreasonable since the typical De Broglie wavelength of the atoms is 0.01 nm. In the following section, we will consider experiments where the wave-like nature of the atoms is important, so the billiard ball model will not give correct results. In particular, we will discuss experiments which demonstrated that the atomic wave function can be modified by phase and amplitude gratings, creating atomic interference patterns that can also be used to control deposition onto a substrate.

A series of pioneering experiments have been done which used the diffraction from phase and amplitude holograms to produce atomic interference patterns. In the first experiment Gould *et al.*[25] used the potential energy associated with an optical standing wave to produce periodic variation on the phase of an atomic wavepacket. Thus, the light acted as a phase grating for the atoms. The grating produced an atomic distribution with a 300 nm period. The period of the pattern was independent of the intensity of the standing wave, but the intensity did control the depth of the phase modulation and thus the amplitude of various orders of the diffraction pattern. Later, Ekstrom *et al.*[26] from the same group did an experiment with an amplitude hologram composed of a matter grating with a 200 nm period. This grating produced an interference pattern with a 100 nm period. Any of these diffraction patterns could have been deposited onto a substrate, creating periodic arrays of lines.

Atomic diffraction effects can also be used to focus atomic beams. Carnal *et al.*[27] demonstrated this by focusing a beam of metastable helium atoms with a microfabricated Fresnel zone plate with diameter 210 μm and innermost zone diameter 18.76 μm. They showed that a 10 μm source of atoms could be imaged to a spot with a diameter of 18 μm, and also that the image of a double slit of width 22 μm and spaced 49 μm could be clearly resolved.

III. CURRENT RESEARCH IN NANOFABRICATION WITH ATOM OPTICS

While a fairly broad range of atom manipulation techniques has been evolving over the past decade or so, and possibilities for fabrication have been suggested, actual demonstration of the fabrication of nanostructures has only recently begun to appear. To date, efforts have concentrated mainly on the standing-wave lens array, in which a standing-wave laser field is used to focus atoms into a series of nanometer-scale lines as they deposit onto a surface [see Figure 2]. The standing wave is directed across the surface of the substrate, usually as close as possible, and atoms are focused by a stimulated, or dipole force that concentrates them into the nodes. This configuration has a number of important advantages for surface fabrication of structures. A laser standing wave, by its nature, consists of very small features, of the order of the wavelength of light, which repeat with extreme regularity. Thus small scale as well as massive parallelism are built into the lens from the beginning. Furthermore, the geometry allows for easy arrangement of the laser and atomic beams; many other configurations, such as focusing in a "doughnut"-mode laser beam[8,9,10], require making the atoms and the laser coaxial, which could pose practical challenges.

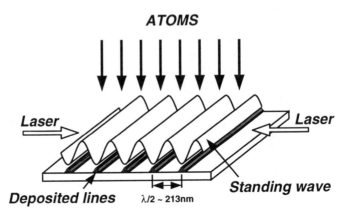

FIGURE 2. Focusing of atoms in a standing wave laser field.

A. Sodium

In a pioneering experiment using the standing-wave configuration, Timp
et al.[11] were able to deposit an array of lines of sodium atoms, forming a grating on a silicon substrate. The grating covered an area of about 0.2 cm^2, and had
a pitch of 294.3 ± 0.3 nm.

The manipulation and focusing of the sodium atoms proceeded in two stages
in this experiment [see Figure 3(a)]. First, an atom beam emerging from an
effusive oven was optically collimated using optical "molasses" [30] to cool the
beam transversely from two sides. This was achieved by illuminating the beam
transversely with circularly-polarized light from a CW single-frequency, stabilized ring-dye laser tuned slightly below the sodium $(3s - {}^2S_{1/2}$ F=2 $) \rightarrow$
$(3p - {}^2P_{3/2}$ F=3 $)$ transition at 589.0 nm. (The $\lambda/4$ plate shown in the schematic
was used to ensure that the light beam entering the deposition chamber was circularly polarized.) An additional beam, shifted in frequency by 1.7 GHz, was combined with this beam to repump those atoms which had fallen into the other
ground-state hyperfine level $(F = 1)$. The collimation of the atomic beam with
optical molasses was necessary to reduce the transverse kinetic energy of the
atoms to the level at which their trajectories could be significantly modified by the
standing wave potential, which had a depth of about 1 μeV.

After collimation, the atom beam passed through a standing wave positioned
very close to the substrate. This laser beam was also circularly polarized, and
also had a small 1.7 GHz-shifted repumping component. Its main frequency was
shifted by 60MHz-10GHz relative to the atomic line so that there would be little

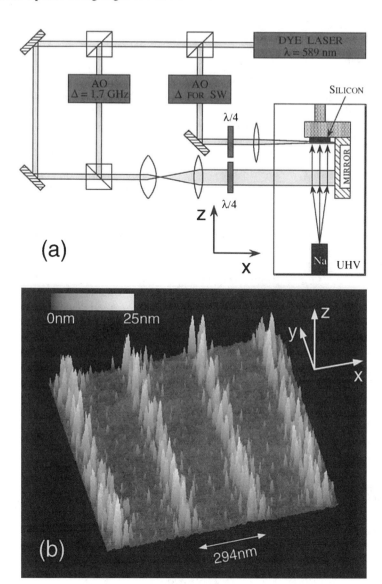

FIGURE 3. (a) Experimental apparatus for focusing sodium atoms with a standing wave into a grating with 294 nm pitch and (b) a scanning tunneling micrograph of a typical grating (ref. [11]). In the schematic the optical molasses is used to cool the transverse kinetic energy in the atomic beam, and subsequently, the optical standing wave is used to focus the cool atoms. The resulting lines shown in the micrograph are comprised of grains of sodium about 10-20nm in diameter. The average linewidth is estimated to be about 45nm. (See color plate.)

excited-state population, and hence little possibility for spontaneous emission and the associated velocity-dependent forces. Both positive and negative detunings were tested, and both gave comparable results, though in one case the atoms were attracted to the nodes (positive detuning) and in the other they were attracted to the antinodes (negative detuning).

An important component of this experiment was ensuring that the standing-wave nodes did not move relative to the substrate on which the atoms were being deposited, as this would result in a smearing of the pattern. Since the node positions are determined solely by the position of the surface of the standing-wave mirror, it was necessary to stabilize this mirror relative to the substrate. This was accomplished using two different techniques. In early experiments the spatial stabilization was done by forming an optical cavity between a small mirror on the sample mount and a standing wave mirror external to the deposition chamber. Using standard optical cavity-locking techniques, it was possible to monitor the transmission of this cavity and servo the position of the standing-wave mirror using a piezoelectric transducer, keeping the mirror in registry with the substrate within $\lambda/45$, or 13 nm. Not only does this configuration reject mechanical vibrations between the substrate and the surface mirror, but also potentially provides a mechanism for doing lithography since the nodes of the standing wave can be moved relative to the substrate during deposition by changing the frequency of light in the cavity. In subsequent experiments, the mirror was mounted in direct contact with the substrate, as it is represented in the schematic of Fig. (3a), to provide even better registration, but sacrificing control of the relative motion between the substrate and the standing wave.

The sodium gratings were observed while under vacuum by illuminating them with a laser beam with a wavelength between 560 and 580 nm, and by using an ultra-high vacuum scanning tunneling microscope.[11] Diffraction was observed, and the periodicity of the grating was inferred from the diffraction angle. A number of detunings and laser powers were used, and each resulted in a grating with the expected pitch of $\lambda/2 = 294.5$ nm within an uncertainty of 0.4 nm. Figure 3(b) shows a typical scanning tunneling micrograph of a $1\,\mu m \times 1\,\mu m$ area of the grating. The lines are comprised of grains of sodium less than 10nm in diameter. The average linewidth is less than 20nm.

B. Chromium

The experiments on sodium were closely followed by work reported by McClelland *et al.*[28], in which chromium atoms were focused into an array of lines on a substrate using a laser standing wave. With chromium substituted for sodium, two important advances were possible. The deposited structures could be removed from vacuum for direct examination with a variety of microscopy techniques, and for the first time applications involving a hard, ultra-high vacuum

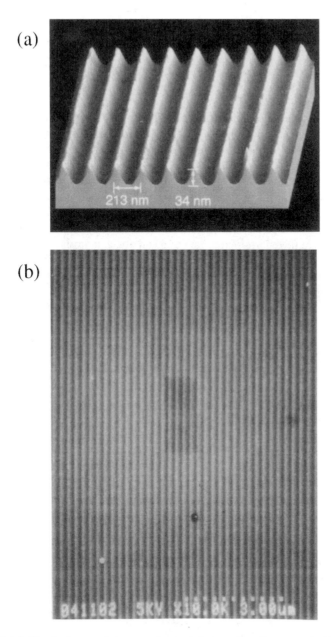

FIGURE 4 (a) Atomic force microscope image of chromium lines created by focusing chromium atoms in a standing wave (ref. [28]). (b) Scanning electron microscope image of chromium lines created by focusing chromium atoms in a standing wave (ref. [29]).

compatible material with many useful properties could be envisioned.

As with sodium, the chromium experiments consisted of an effusive atomic beam that was first collimated in a region of optical molasses and then focused into a series of lines by a standing wave grazing along the surface of a silicon wafer. The laser wavelength used in this case was 425.43 nm, which corresponds to the $(4s - {}^7S_3) \rightarrow (4p - {}^7P_4^o)$ transition in the chromium atom, and no additional laser frequencies were required because the predominant isotope of chromium (^{52}Cr at 84%) has no hyperfine structure. The optical molasses for the chromium atoms was also set up in a polarization-gradient configuration[30], in order to achieve a sub-Doppler transverse kinetic energy spread. And finally, the silicon wafer was held in direct contact with the standing-wave mirror which was mounted in the vacuum. This provided registry between the standing wave and the substrate to within a few nanometers without the need for active stabilization.

Because chromium forms a very thin (~1 nm) layer of very tough, passive oxide when exposed to air, the chromium lines formed by laser-focused deposition could be removed from the vacuum and imaged in air with atomic force microscopy (AFM), or examined by scanning electron microscopy (SEM). Figure 4(a) shows an AFM image of a 2 μm by 2 μm section of the lines, which cover a 0.4 mm by 1 mm region of the substrate. Figure 4(b) shows an SEM image of the lines with a larger field of view, illustrating their uniformity[29].

The lines on the substrate whose image is shown in Figure 4 had an average line width of 65±6 nm (full-width at half-maximum), a pitch of 212.78 nm, and an average height of approximately 20 nm. The width of the lines depends on details of the deposition, such as the degree of collimation and velocity spread of the atomic beam, and the laser beam power, detuning and waist size. The pitch of the lines was not measured directly, but rather inferred from the known wavelength of the laser light. Because of the geometry of the experiment, the pitch of the lines must be equal to the periodicity of the standing wave, i.e., half the laser wavelength, with very small corrections (of order parts per million). The height of the lines was of course dependent on the length of the deposition, which was 20 minutes in the case of Figure 4.

C. Aluminum

While chromium has a number of potentially useful applications, such as nanowires, etch resists, and structured magnetic materials, other materials are also worth pursuing for atom-optical applications. Very recently McGowan and Lee[31] have reported one-dimensional laser cooling and focusing of an aluminum atomic beam. The advantages of laser focusing aluminum are twofold: aluminum is currently the material of choice for interconnects in microcircuits, so it is a desirable atom to manipulate, and the wavelength of the laser used is shorter than in the case of sodium or chromium, so the resolution is potentially

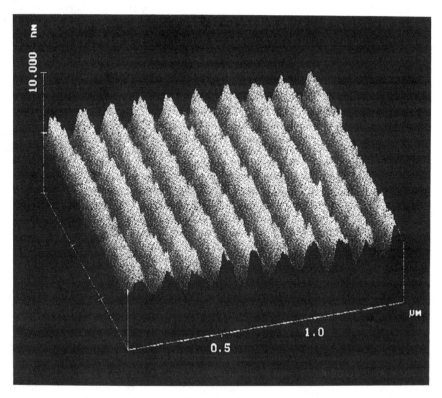

FIGURE 5. Atomic force microscope image of aluminum lines created by focusing aluminum atoms in a standing wave (ref. [31]). (See color plate.)

higher. For this experiment the laser was tuned near the $(3p - {}^2P^o_{3/2}) \rightarrow$ $(3d - {}^2D_{5/2})$ transition in atomic aluminum, which has a wavelength of 309.3 nm. The UV laser light was produced by frequency doubling of a CW dye laser in an external build-up cavity. While the $(3p - {}^2P^o_{3/2})$ state is not the ground state of aluminum, it is sufficiently low-lying to be thermally populated by roughly 20% of the atoms emerging from the atomic oven. Significant Doppler cooling and focusing of the atomic beam was observed as shown in Figure 5. A feature size of about 180nm was reported.

D. Modeling of the Process

Along with the experiments described above, there has also been work done on understanding the behavior of laser focusing in a standing wave from the point of view of being able to predict the line shape of deposited structures and hence the ultimate resolution of the process. Berggren et al.[32] and McClelland[33] have taken a semiclassical approach, solving the equation of motion of the atoms as they pass through a single node of the standing wave assuming a conservative potential given by the light shift induced by the laser. Any effects due to quantization of the atomic motion are ignored in this approach, mainly because these effects are small for a wide range of parameters. The advantage of this approach is that rays can be traced rather simply and the influence of experimental parameters can be explored.

One outcome of this work has been the realization that the standing wave node can act as a true lens, and focal lengths and aberrations can be discussed in complete analogy to optics, in particular particle optics. For example, McClelland[33] has shown that the essential behavior of the lens is governed by a single parameter a, given by:

$$a = \frac{\hbar \Delta}{2E_0} \frac{I}{I_0} \frac{\gamma^2}{\gamma^2 + 4\Delta^2} k^2 \sigma_z^2, \qquad (4)$$

where Δ is the laser detuning from the atomic resonance, E_0 is the kinetic energy of the atoms, $k = 2\pi/\lambda$ (λ being the atomic wavelength), σ_z is the $1/e^2$ radius of the laser beam, I is the laser intensity, I_0 is the saturation intensity of the atomic transition, and γ is the atomic line width. In the paraxial, thin-lens limit of a weak lens, the focal length is simply given by:

$$f = \sqrt{2/\pi} \, \sigma_z / a. \qquad (5)$$

When the lens is thick, such as when the substrate plane is within the laser beam, the focal length and principal plane location are still entirely determined by a, though a numerical calculation must be carried out to obtain the dependence. These parameters scale with σ_z, so once the numerical calculation is done, the results can be used to predict the behavior of any lens.

A central result of the semiclassical calculations is that line widths of the order of ten nanometers are possible with a thermal atomic beam having sub-Doppler-cooled transverse velocity spread. This is generally borne out in the experiments so far; however sufficient discrepancies still exist, suggesting that further theoretical work is needed. Along these lines, Marksteiner et al.[34] have performed general quantum calculations of the localization (or quantum "squeezing") of atoms in a standing wave potential. While no numerical comparisons can be made at present, this is clearly an area where fruitful comparisons can be made to enhance our understanding of the role played by quantum and non-adiabatic effects in the focusing of atoms in a standing wave.

IV. FUTURE PROSPECTS

In the research that has been performed so far, it has been clearly demonstrated that nanometer-scale patterns can be made on a surface by utilizing the techniques of atom optics. The new approaches introduced by these techniques, with the associated improvement in resolution and parallelism, could have a very significant impact on the field of nanostructure fabrication. Atom optics, however, is in its infancy, and the extent of its impact on nanostructure fabrication depends on progress in two broad areas: (a) the techniques should be extended to allow working with as many different materials as possible, especially those of interest to specific applications, and (b) methods should be devised to allow the possibility of fabricating more general patterns. Already, a number of possible approaches are being discussed to overcome these hurdles, and it is expected that many new developments will be forthcoming in the near future.

A. Other Materials

One way that atom optics can be extended to other materials is simply to follow the approach taken already to some extent, i.e., look for appropriate atomic transitions and the associated lasers in the species of interest. The requirements can be somewhat stringent in this case, but nevertheless it has led to the introduction of chromium, and possibly aluminum as candidates. For an atomic species to be considered for atom optics, it must first of all have a transition that is accessible with a CW, single-frequency laser with enough power to saturate the transition. This transition must originate in the ground state of the atom, or in a state that has enough population such that most or all of the atoms are affected. In this regard the atomic beam must not have too high a concentration of other isotopes, dimers, ions or excited species. Second, if an optical molasses stage is necessary for collimation, then it is necessary for the atomic transition to be a cycling one; that is, atoms excited by the laser must return to the same state so that they can repeatedly absorb photons. Sometimes the cycling requirement is difficult to meet because atoms have hyperfine structure or other complicated level structure. This difficulty can often be overcome, however, by the introduction of other laser frequencies, as in the case of sodium. While all these requirements seem rather restrictive, it is still quite possible that atom optics techniques will be applied to new species in the future. Laser technology is constantly improving, and as new CW tunable sources emerge, additional atomic species will probably become accessible to atom optics.

Another perhaps more fruitful approach to generalizing the applicability of atom optics is to make use of some sort of pattern-transfer mechanism. In this approach, a pattern is generated with an atomic species that is compatible with atom optical techniques, and then the pattern is transferred to the desired material by some sort of etching process. At present, two possible routes exist toward this; in the future, other possibilities may evolve.

The first technique involves making use of the etch resist properties of chromium. For example, if a pattern in gold is desired, an atom-optically generated pattern of chromium can be deposited on the surface, and then the sample is etched in a solution of KCN. The chromium prevents the gold from being etched and the pattern is transferred to the gold. If desired, the chromium can then be removed with HCl or commercial chrome etch. Dry, or reactive-ion etching can also be used in a similar way to transfer the pattern to *GaAs* or other materials, because chromium provides excellent resistance to most of these etch processes[35].

The second technique uses a resist layer to transfer the pattern. The resist might be one of the standard high resolution electron-beam resists used for conventional lithography, or it could consist of a thiol or siloxane-based self-assembled monolayer (SAM). SAM-based resists offer some unique advantages, promising very high sensitivity and resolution. The desired material is coated with the resist, and then the resist is exposed by an atom-optically focused pattern of atoms. Subsequent etching transfers the pattern to the substrate.

Exposure of the resist can occur either through a chemical or an energetic mechanism. In the first case, a reactive atom such as one of the alkalis (all of which can be used for atom optics) strikes the surface and reacts with the resist, making it susceptible to removal by a solvent or etch. In the second, metastable rare-gas atoms (again, all accessible to atom optics) strike the surface and liberate their internal energy in a shower of secondary electrons that can go on to damage the resist and render it soluble or no longer resistant to an etch. Preliminary investigations on both of these processes have indicated that they are real possibilities and should be pursued further.

B. More General Patterns

While atom optics has been shown to have the potential for very high resolution and massive parallelism, a pattern more complicated than a grating has yet to be demonstrated. Before discussing approaches to the general pattern problem, however, it should be noted that simple one-dimensional atom-optically generated artifacts have already found a unique, critical application. As discussed above, the pitch of the gratings fabricated in a standing wave can be directly derived from the wavelength of the laser light. Since the laser light is locked to a well-known atomic transition, the pitch is traceable through the speed of light in vacuum to a well-characterized frequency. Thus the grating can be used as a length standard on the nanometer scale, a regime where accurate calibration artifacts are sorely needed.

But in order to explore the full range of possibilities for atom-optical nanofabrication, it is useful to contemplate the extension of the techniques to more general patterns. A number of concepts are being discussed, and these range from simple extensions of current techniques to whole new concepts.

Perhaps the simplest generalization of a one-dimensional deposition experiment involves adding another standing wave perpendicular to the first one, so that a two-dimensional pattern is deposited. In this case a pattern of dots with spacing $\lambda/2$ can be produced. These dots could already be useful for the fabrication of a quantum dot array, or a patterned magnetic medium. However, if the dots can be made with sufficiently small size, i.e. 10–20 nm, then the possibility exists for scanning the substrate within the unit cell of the array as they deposit, thereby "painting" a pattern of any desired shape. Because of the natural periodicity of the standing wave, this pattern would be reproduced precisely in each well of the standing wave across the entire substrate. The result would be the massively parallel fabrication of an arbitrary structure, with obvious implications for device applications.

Another approach to generalizing the pattern concerns cases when the periodicity of the standing wave is too high, i.e., when the dots or lines are too close together. In this situation, one can imagine arranging a mask in registry with the standing wave that allows atoms to deposit only in certain areas. The mask needs only be fabricated with a resolution corresponding to $\lambda/2$, or a few hundred nanometers, that is, just enough to select the desired periods of the standing wave. Once the atoms enter the standing wave, they are focused to the limits of atom optics, i.e., of order 10 nm.

Even more general patterns can in principle be created by recognizing that a standing wave is only a very simple example of what is essentially a broad continuum of possible intensity patterns that can be generated on a surface. The challenge of this approach is to decide on the desired pattern, and then design an optical field that will concentrate atoms into this pattern using our knowledge of atom optics. In order to take advantage of the small sizes (or high spatial frequencies) attainable with optical fields, it is probably most fruitful to consider patterns that can be made as a result of some form of interference. The standing wave is the simplest example of such a pattern, but far more complicated ones are possible when more beams from other directions with controllable phase are introduced. While there are some limitations on what optical fields can be generated because of the laws of diffraction and the need to use only a single wavelength, a wide range of patterns are still possible.

At present, it is difficult to predict in what exact form atom optics could prove the most useful for nanolithography, although it is clear that a large variety of possibilities exist for the future. Eventually, any or all of the possible processes outlined in Figure 1, from masks to lenses to waveguides to atom holograms, may break some fundamental barriers in the ongoing search for new ways to fabricate with increasingly smaller dimensions at ever higher efficiency.

REFERENCES

[1]   *Working Group Reports, Semiconductor Technology Workshop,* Semiconductor Industry Association, San Jose, CA (1993).

[2]   Gordon, J.P., and Ashkin, A., *Phys. Rev. A* **21**, 1606 (1980).

[3]   Cook, R.J., *Phys. Rev. A* **20**, 224 (1979).

[4]   Cohen-Tannoudji, C., Dupont-Roc, J., Grynberg, G., *Atom-Photon Interactions: Basic Processes and Applications,* New York: John Wiley and Sons, 1992.

[5]   Bjorkholm, J.E., Freeman, R.R., Ashkin, A., and Pearson, D.B., *Phys. Rev. Lett.* **41**, 1361 (1978); Bjorkholm, J.E., Freeman, R.R., Ashkin, A., and Pearson, D.B., *Opt. Lett.* **5**, 111 (1980).

[6]   Balykin, V.I., *et al., J. Mod. Opt.* **35**, 17 (1988).

[7]   Sleator, T., Pfau, T., Balykin, V., and Mlynek, J., *Appl. Phys. B* **54**, 375 (1992).

[8]   Balykin, V.I., and Letokhov, V.S., *Opt. Commun.* **64**, 151 (1987).

[9]   Gallatin, G.M., and Gould, P.L., *J. Opt. Soc. Am. B* **8**, 502 (1991).

[10] McClelland, J.J., and Scheinfein, M.R., *J. Opt. Soc. Am. B* **8**, 1974 (1991).

[11] Timp, G.L., Behringer, R.E., Tennant, D.M., Cunningham, J.E., Prentiss, M., and Berggren, K.K., *Phys. Rev. Lett.* **69**, 1636 (1992); and Natarajan, Vasant, Behringer, R.E. and Timp, G., *J.Vac. Sci. Tech. B* to be published Nov./Dec. (1995).

[12] Prentiss, M.G., and Ezekiel, S., *Phys. Rev. Lett.* **56**, 46 (1986).

[13] Salomon, C., *et al., Phys. Rev. Lett.* **59**, 1659 (1987).

[14] Balykin, V.I., *et al., Opt. Lett.* **13**, 958 (1988).

[15] Jessen, P.S., Gerz, C., Lett, P.D., Phillips, W.D., Rolston, S.L., Spreeuw, R.J.C., and Westbrook, C.I., *Phys. Rev. Lett.* **69**, 49 (1992).

[16] Verkerk, P., *et al., Phys. Rev. Lett.* **68**, 3861 (1992).

[17] Hemmerich, A., and Hänsch, T., *Phys. Rev. Lett.* **70**, 1410 (1993).

[18] Marte, P., Dum, R., Taïeb, R., Lett, P.D., and Zoller, P., *Phys. Rev. Lett.* **71**, 1335 (1993).

[19] Taïeb, R., Marte, P., Dum, R., and Zoller, P., *Phys. Rev. A* **47**, 4986 (1993).

[20] Grynberg, G., *et al., Phys. Rev. Lett.* **70**, 2249 (1993).

[21] Verkerk, P., Meacher, D., Coates, A., Courtois, J.-Y., Guibal, S., Lounis, B., Saloman, C., and Grynberg, G., *Europhys. Lett.* **26**, 171 (1994).

[22]   Balykin, V.I., Letokhov, V.S., Ovchinnikov, Yu. B., and Sidorov, A.I.,   *Phys. Rev. Lett.* **60**, 2137 (1988).

[23]   Kasevitch, M.A., Weiss, D.S., and Chu, S.,   *Opt. Lett.* **15**, 607 (1990).

[24]   Kaiser, R., *et al.*,   *Opt. Commun.* **104**, 234 (1994).

[25]   Gould, P.L., Ruff, G.A., and Pritchard, D.E.,   *Phys. Rev. Lett.* **56**, 827 (1986).

[26]   Ekstrom, C.R., Keith, D.W., and Pritchard, D.E.,   *Appl. Phys. B* **54**, 369 (1992).

[27]   Carnal, O., Sigel, M., Sleator, T., Takuma, H., and Mlynek, J.,   *Phys. Rev. Lett.* **67**, 3231 (1991).

[28]   McClelland, J.J., Scholten, R.E., Palm, E.C., and Celotta, R.J.,   *Science* **262**, 877 (1993).

[29]   Scholten, R.E., McClelland, J.J., Palm, E.C., Gavrin, A., and Celotta, R.J.,   *J. Vac. Sci. Technol.* **12**, 1847 (1994).

[30]   Cohen-Tannoudji, C., and Phillips, W.D.,   *Physics Today* **43**, 33 (October, 1990).

[31]   McGowan, R.W., and Lee, S.A., in *Abstracts of Contributed Papers,* ICAP XIV, Boulder, CO, July 31–August 5, (1994), p. 2H-7; and McGowan, R.W. *et al.*, to be published in *Optics Letters.*

[32]   Berggren, K.K., Prentiss, M., Timp, G.L., and Behringer, R.E.,   *J. Opt. Soc. Am. B* **11**, 1166 (1994).

[33]   McClelland, J.J., (submitted to *J. Opt. Soc. Am. B*).

[34]   Marksteiner, S., Walser, R., Marte, P., and Zoller, P.,   *Appl. Phys. B* **60**, 145 (1995).

[35]   Maluf, N.I., Chou, S.Y., McVittie, J.P., Kuan, S.W.J., Allee, D.R., and Pease, R.F.W.,   *J. Vac. Sci. Technol. B* **7**, 1497 (1989).

Chapter 11

From The Bottom Up: Building Things With Atoms

Don Eigler[1]

IBM Research Division, Almaden Research Center
San Jose, CA 95120-6099, USA

Throughout much if not all of his existence, man has been motivated to build things. The first objects built by man: weapons, shelters and tools, were certainly motivated by the need to survive. Man is not alone in this endeavor. Birds build nests. Beavers build dams. Chimpanzees build and use tools. The advance of man's ability to build objects of increasing sophistication has enabled him to satisfy motivations beyond survival, motivations such as bettering the quality of life and expansion of knowledge. What I want to talk about today is the achievement of a milestone in man's ability to build things. That milestone is the ability to build things using individual atoms as the building blocks; the ability to build things *from the bottom up*, by placing the atoms where we want them[2].

In the second half of the twentieth century we have witnessed technology revolutions in the electronics industry and the chemical and drug industries. These revolutions have been brought about by an ability to build two kinds of small things: molecules and electronic circuits. The challenge faced by the electronics industry is opposite to that faced by the chemical and drug industries. While the electronics industry strives to build ever smaller circuits, the chemical and drug industries strive to build ever larger and more complex molecules. In addition, the approach that each industry takes is opposite to that of the other. The electronics industry makes its circuits in a top-down approach using lithographic techniques to whittle circuits out of a block of silicon, while the chemical and drug industries use the bottom-up approach of swirling together chemical subunits in such a way that huge numbers of the desired molecule or product are ultimately created (thanks to the miracles of thermodynamics). In the last decade of this century we are witnessing a sort of collision of these two manufacturing approaches. What we see happening is that the size of the objects that we can construct in our laboratories using a top-down approach is approaching the size of the largest molecules that chemists construct. As shown in Figure 1, that size is about five nanometers.

The construction of things one-atom-at-a-time is a blending of concepts from both electronic and chemical construction techniques. Similar to the way electronic circuits are built, atomic scale construction embodies the idea that the structure is exact in the sense that (within manufacturing tolerances) we build just

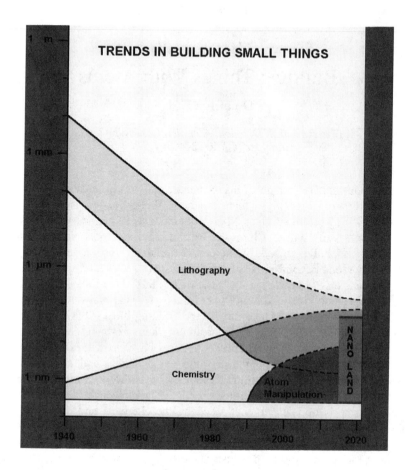

FIGURE 1. The scale of the smallest things built using the top-down approach of the electronics industry is approaching the scale of the largest molecules built by the bottom-up approach of chemistry. (See color plate.)

what we want and nothing else. However, unlike the electronics manufacturing processes, atomic scale construction techniques more closely resemble those of chemistry, in which the desired product is created from assembly of smaller units - the bottom-up approach.

The technological driving force to build things using atomic construction techniques comes from the anticipated economic and technological benefits that it might offer. It must be emphasized that achievement of an *economically viable* atomic construction technology is, at present, no more than a dream. An equally important and more achievable motivation to build things with atoms is that it allows us to do science in an arena we have never before entered – and that, at least for the scientist, is reason enough. The ability to fabricate on the atomic scale allows us to conduct experiments in a way which here-to-fore was impossible. This ability can best be viewed as a tool which the scientist can use to explore intellectual *terra incognita*. It is the excitement that comes from this exploration that ultimately is the real driving force for building things with atoms today.

Atom manipulation was made possible by the invention in the early 1980's of a most remarkable and versatile instrument: the scanning tunneling microscope, or STM for short[3]. For this invention, Gerd Binnig and Heinrich Rohrer of IBM's Research Division were awarded the 1986 Nobel Prize in physics. The STM is an instrument capable of creating atomic resolution images of the surface of electrically conducting materials such as metals and semiconductors. However, unlike optical microscopes and their close cousins the electron microscopes, the STM does not rely upon wave optics to form an image. Incongruously, it forms an image in a way which is similar to the way a blind person can form a mental image of an object by feeling the object. The STM achieves this feat by scanning a metal needle, called the "tip", across a surface while maintaining the tip within a few atomic diameters of the surface. This is done by making an electric current flow between the tip and the surface to be imaged. The magnitude of this electric current is very sensitive to the separation between the tip and the surface and thus can be used as a signal for stabilizing the height of the tip as the tip is moved laterally across the surface. By scanning the tip in a raster pattern over the surface and recording the height of the tip, an image of the topography, or shape, of the surface can be recorded.

Just as the blind person can sense not only the shape of the object, but also its texture, compressibility and thermal conductivity, the STM can tell us much more about a surface than just its shape. We can use it to study how the electrons in the near-surface layer have arranged themselves not only spatially, but energetically. Even more remarkable, just as the blind person can push, pull, pick up and put down objects on a table top, so too can we push, pull, pick up and put down surface atoms using the tip of the microscope. Here is how it is done.

In Figure 2(A) we see a schematic illustration of an atom which lies on top of a metal surface. The forces which hold the atom to the surface are due to the chemical binding of the atom to the nearby atoms of the surface. These forces are schematically represented by springs shown in Figure 2. To image this atom the tip is scanned over the surface, following a trajectory shown by the dashed line in Figure 2(A). Under these conditions the separation between the tip and the atom on top of the surface (called an "adatom") are great enough so that any forces between the tip and the adatom are negligible, the result being that the adatom will hold still to have its picture taken.

A: IMAGING MODE   B: MANIPULATION MODE

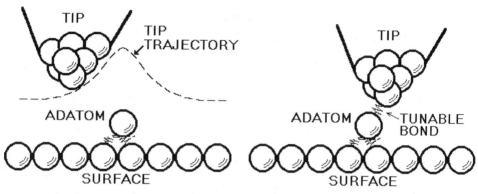

FIGURE 2. (A) In the imaging mode the tip is stabilized far enough above the surface so that the interaction between the tip and the adatom is negligible. (B) In the manipulation mode the tip is brought close enough to the adatom to drag the adatom along the surface. The force exerted on the adatom by the tip is due to the partially formed chemical bond between the tip and the adatom.

Just as there are chemical binding forces between the adatom and the nearby atoms of the surface, there will be a chemical binding force between the adatom and the outermost atom(s) of the tip. In Figure 2(B) we have schematically represented the chemical bonding interaction between the tip and the adatom as a spring. I like to call this the "tunable bond" because we can tune both the direction and the magnitude of the force we exert on the adatom simply by adjusting the position of the tip. Now it turns out that for a wide range of adatoms, and even for groups of adatoms or molecules, it is possible to adjust the height of the tip so that the in-plane force exerted by the tip on the adatom is great enough to overcome the in-plane forces between the adatom and the underlying surface, yet at the same time the out-of-plane force exerted on the adatom by the tip is not so great as to overcome the out-of-plane forces between the surface and the adatom. When these conditions are achieved, it is possible to move the tip sideways and drag the adatom along the surface. This process, called the sliding process, allows one to position adatoms with atomic-scale precision. The trick is to be able to successfully switch between operating the microscope in an imaging mode with the tip at a height where its' interaction with the adatom is negligible, to operating the microscope in the manipulation mode with the force between the tip and the adatom sufficient to drag the adatom along the surface under the tip. The sliding process[4] is indicated in Figure 3.

In order for the sliding process to work correctly, it is best to use combinations of adatoms and surfaces where the lateral, or in-plane, interaction between the adatom and the surface is not too great. Fortunately, this can be achieved for a wide range of adatoms on top of of metal surfaces, but at a price. The week in-plane interaction between adatom and surface means that very little thermal shaking of the adatom would be required in order for the adatom to spontaneously hop from site to site across the surface, that is, to undergo thermal diffusion. Thermal diffusion is generally undesirable because it causes the randomization of the locations of adatoms, thus destroying the work you do in placing the adatoms at particular locations. The solution to this problem – the price you have to pay – is to conduct the experiment at low enough temperature that the rate of thermal diffusion is very slow compared to the duration of the experiment. This can be achieved by cooling the sample and the tip of the microscope to the temperature of liquid helium, some 269 degrees centigrade below the freezing point of water. Cooling to liquid helium temperature introduces some complications in the design of the apparatus, but it brings with it a variety of advantages related to the overall performance of the STM. All of the experiments I will discuss were conducted with an STM I built at IBM's Almaden Research Center. That STM is contained in a vacuum vessel which is designed to provide a degree of vacuum sufficient to maintain the surface of a sample clean for weeks at a time (a metal surface exposed to atmospheric pressure would be contaminated with a monolayer of gas

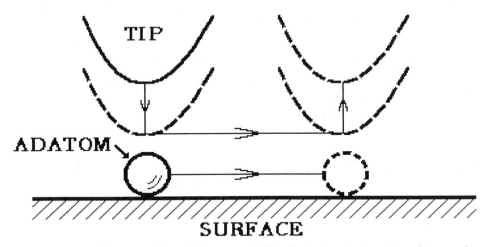

FIGURE 3. Schematic of the sliding process. The tip is placed above the adatom and then lowered to an empirically determined height at which the attractive interaction between the tip and the adatom is sufficient to pull the adatom along the surface. Once the adatom is moved to its final location, the tip is raised back to the height used for imaging, effectively terminating the tip-adatom interaction.

FIGURE 4. Left: Image of the first atom to be moved under control by the STM. This xenon atom is bound to a platinum surface. The streak located near the lower left corner of the atom is a partial image of the atom during the first line-scan of the STM. The forces on the xenon atom due to the tip caused the atom to slightly change its position during imaging. Streaks such as these made us suspect that the STM could be used to move atoms under control. Right: Image of the same xenon atom after it had been purposefully moved to the defect site which is apparent in the left hand image. The in-plane interactions of the xenon atom with the bare platinum surface are so weak that the xenon atom will always move under the influence of the tip, even when we operate the microscope with the greatest possible tip-sample separation. When xenon is bound to a defect site on the platinum surface the in-plane interactions are greater and it is possible to image the xenon atom without causing it to move. For this reason, the xenon atom had to be moved from defect site to defect site on the platinum surface to obtain these images. On other metal surfaces, or different crystalline orientations of the platinum surface, this is not a problem.

molecules in about one microsecond, thus the need for so called "Ultra-High-Vacuum"). In addition, the chamber which houses the microscope, and everything within that chamber, is cooled to liquid helium temperature for the above mentioned reasons.

Atom manipulation came about almost by accident. Paul Weiss and I had spent some time studying the adsorption and ordering of xenon on a platinum surface. During these studies we had noted that when the tip was brought close enough to an island of xenon atoms, the island would change its shape. In September of 1989, Erhard Schweizer and I were continuing these studies when we noted some unusual streaks across our STM images which were clearly due to tip induced motion of the xenon adatoms. Rather than ignore these streaks, we investigated the experimental conditions that caused them. Doing so, we learned that the presence of the streaks depended upon how we operated the microscope. If we operated the microscope with the tip close enough to the sample then we saw the streaks, otherwise we did not. This immediately suggested that we could use the tip to control the position of the xenon. Trying out our ideas required modifications to the software we used to operate the microscope. Within a day the necessary modifications were made. These modifications allowed us to switch from an imaging mode where the tip executed a raster scan of the surface, to a mode in which we could move the tip of the microscope along any desired path across the surface, and with a tunnel current different from that used for imaging. With these modifications in place, I began by imaging an isolated xenon atom which was bound to a defect site on the platinum surface as shown in Figure 4-Left. I then stopped imaging, moved the tip directly over the xenon atom, increased the magnitude of the tunneling current in order to bring the tip a little closer to the xenon atom, and then I had the computer move the tip from the location where the xenon atom originally was to a new location not too far away.

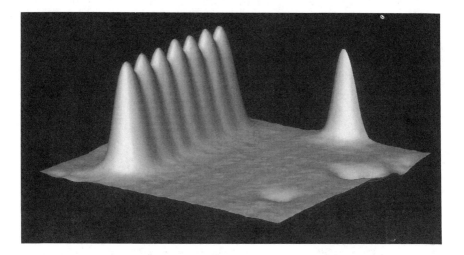

FIGURE 5. A row of seven xenon atoms constructed with the STM. The xenon atoms are spaced apart every other atom of the underlying nickel surface. The xenon atom cannot be packed together any tighter and remain in a single row. From building structures like this we learn about the strength of the xenon-xenon interaction relative to the strength of the in-plane interaction between the xenon atoms and the underlying nickel atoms.

Once the tip reached the new location, I reduced the magnitude of the tunnel current in order to increase the separation between the tip and the xenon atom and thus return to the imaging mode. Next, I re-imaged the surface to find [Figure 4-Right] that the xenon atom had been successfully moved to the location of my choice. I then repeated the same experiment four times, and it worked each time. With this xenon atom, the milestone was achieved.

The utility of being able to move around atoms on a surface is that this freedom allows us to learn new things. One of the first things we did once we learned to move xenon atoms was to construct the row of seven xenon atoms on top of the nickel surface shown in Figure 5[5]. By doing this we learn about the stability of the structure. In addition, we learn about the strength of forces between xenon atoms coadsorbed on a surface in relation to the lateral forces exerted on the xenon atoms by the atoms of the metal surface. The real importance of this experiment was not the knowledge we gained about adsorbed xenon atoms. It was the demonstration that atom manipulation was a useful tool for gaining knowledge.

We have come a long way in the sophistication of the structures that we can build. Figure 6 shows a structure we call a "Quantum Corral" made from 96 iron atoms carefully positioned on a copper surface[6]. The reason for building this ring shape structure was to trap some electrons on the inside. The circular wave structure exhibited on the inside is the density distribution due to three of the quantum states of the corral. The wave pattern, which is closely related to the wave pattern on the head of a drum, provides us a very vivid demonstration of quantum mechanics.

The xenon atoms in Figure 5 and the iron atoms in Figure 6 are both assemblies of atoms which are bound to the surface, but not directly to each other. Another way to say this is that the separation between nearby xenon or iron atoms is still so great that the xenon-xenon or iron-iron chemical bonds have not been formed. Ultimately, we want to build structures where the atoms are bonded to one another just as in any lump of stuff. Ultimately, we want to build molecules and three dimensional solids. Recently, we[7] have taken a big step in this direction by demonstrating the ability to assemble molecules from their constituent atoms. Figure 7 shows a molecule formed from eight cesium and eight iodine atoms. The individual iodine atoms are not visible in this image because their apparent height is so little compared to that of the cesium atoms. This molecule is held together so well by the forces between the various cesium and iodine atoms that we can slide and rotate the molecule over the copper surface on which it rests without breaking the molecule apart. The immediate utility of assembling molecules with the STM is that it enables us to learn more about how chemical reactions occur on surfaces.

Finally, it is worthwhile to consider where atom manipulation with the STM is likely to lead us. The application of atom manipulation for technological purposes is, at the moment, far beyond sensible consideration. The speed and

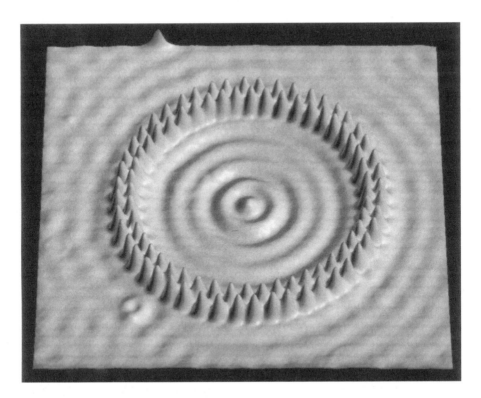

FIGURE 6. A "Quantum Corral" made from 96 iron atoms on a copper surface. The copper surface supports surface states which form a two dimensional gas of electrons. These surface state electrons are scattered by the iron atoms and as a result some are trapped in the corral's interior. The wave structure seen inside the corral is the density distribution of three quantum states of the corral. The wave structure outside the corral is the standing wave pattern due to the reflection of surface state electrons on the exterior of the corral. Corrals let us study the wave properties of electrons in very small confining structures (something that the electronic device designers of the future will need to know).

reliability that we achieve make any idea of mass manufacturing, now, or in the *foreseeable* future, completely ridiculous. (I will of course be delighted to find myself wrong!) What we cannot predict is *when*, by *whom*, and even *if* the key discovery(ies) will be made that would enable a bottom-up construction *technology* which uses atoms as the building blocks. On the other hand, it is likely that in the years ahead atom manipulation will be heavily exploited as a scientific tool. One of the severe limitations that we face is that it is particularly difficult, at least so far, to build three dimensional structures from arbitrary atoms. Our level

FIGURE 7. This molecule, made of eight cesium and eight iodine atoms, was constructed one atom at a time with an STM. The electrons of the molecule do not protrude
nearly as far into the vacuum at the site of the iodine atoms as they do at the cesium
atoms. As a result, the iodine atoms cannot be individually discerned in this image.
Assembling molecules with the STM lets us learn about chemical reactions on a metal
surface.

of control and knowledge is not yet that advanced. Even if we restrict our considerations to two dimensional structures we find that our atom manipulation abilities are not universal: it only works for certain combinations of surfaces and adatoms. This-not-withstanding, the realm of experiments opened up to us by being
able to move around a few atoms is vast, largely untapped and ripe for exploration.

REFERENCES

[1]   The work described in this paper was carried out in collaboration with C.P. Lutz, and
      a string of postdocs: P.W. Weiss, E.K. Schweizer, P. Zeppenfeld, M.F. Crommie
      and A. Hopkinson. Each of these individuals has played a crucial role in the
      achievement of the work presented here. I am deeply indebted to them all.

[2]   The possibility of building things "from the bottom up" using atoms as the building
      blocks was discussed by Feynman in what has now become a famous article:
      "There's Plenty of Room at the Bottom," Engineering and Science, pg. 22, February
      1 960. Feynman considered whether there were physical limits that would prevent
      the construction of objects by "putting the atoms where you want." He concluded

that there was no fundamental reason why this could not be done. From this consideration, he predicted that atomic scale construction was inevitable.

[3]   Binnig, G. and Rhorer, H., *Sci. Am.* **253**, 50 (1985). Hansma, Paul K., *J. Appl. Phys.* **61 (2)**, R1 (1987).

[4]   The sliding process is just one of several processes that may be used to manipulate atoms on surfaces with the STM. See Stroscio, J.A., and Eigler, D.M., *Science* **254**, 1319 (1991).

[5]   Eigler, D.M., and Schweizer, E.K., *Nature* **344**, 524 (1990).

[6]   Crommie, M.F., Lutz, C.P., and Eigler, D.M., *Science* **262**, 218 (1993); Heller, E.J., Crommie, M.F., Lutz, C.P., and Eigler, D.M., *Nature* **369**, 464 (1994); Crommie, M.F., Lutz, C.P., Eigler, D.M., and Heller, E.J., *Physica D* **83**, 98 (1995); Crommie, M.F., Lutz, C.P., Eigler, D.M., and Heller, E.J., *Surf. Rev. & Lett.* **2**, 127 (1995).

[7]   Hopkinson, A., Lutz, C.P., and Eigler, D.M., submitted for publication.

Chapter 12

Physical Properties of Nanometer-Scale Magnets

David D. Awschalom

Department of Physics
University of California
Santa Barbara, CA 93106

Stephan von Molnár

Department of Physics
Materials Research and Technology Center
Florida State University
Tallahassee, FL 32308

I. MESOSCOPIC MAGNETISM

A. Introduction

Contemporary needs for increasing information storage and high density magnetic recording have stimulated new research in mesoscopic magnetism. In analogy to semiconductor physics, the process of miniaturizing magnetic materials has revealed a variety of new and unexpected classical and quantum mechanical phenomena. On a small enough length scale the interactions between individual atomic spins cause their magnetic moments to be aligned in the ordered pattern of a single domain, without the complication of domain walls separating regions of varying orientation. For particle sizes at or below that of a single domain, standard theoretical models of dynamical behavior predict simple stable magnets with controllable properties. However, recent experimental studies of micron-scale ferromagnetic particles show them to be far less stable and to switch more easily than expected from traditional classical theories of thermal activation. In even smaller systems, new theoretical predictions and experimental observations suggest that at low enough temperature, quantum mechanical tunneling may ultimately control the particle's magnetization.

It is the purpose of this chapter to sketch the salient features of these theories, review various methods for fabricating small magnetic structures, and describe several different experimental measurements aimed at exploring these phenomena[1]. A section is devoted to novel time-resolved instrumentation for micromagnetism, and the chapter will conclude with some comments on future directions of research in nanometer-scale structures and possible implications for technology. Finally, we note that a discussion of the research activity in giant magnetoresistance (GMR) is not included. Although this area had its renaissance in the study of thin film magnetic structures, it deserves a full account in its own right.

B. Thermally Activated Dynamics: Shape Engineering

Understanding the behavior of individual magnetic domains has become an important issue in magnetics technology. Lorentz electron microscopy images of magnetic recording "bits" often show ripples, vortices, and feather-like patterns in the magnetic structure. At this detailed level the structure deviates significantly from a simple recording pattern: strong interactions between the domains cause the magnetization vector to splay away from the desired direction inside each bit, and also make the transition from one bit to another somewhat unpredictable. These effects contribute to "media noise" which is one of the major current problems in magnetic storage. Such issues are important in present recording technology of ~100 Mbits/in^2 , and they will be critical to the future development of high density storage now expected to be ~10 Gbits/in^2. At this density a single bit is expected to be 1 μm wide and only 70 nm long, with a film thickness approaching ~30 nm. For conventional magnetic materials, this length approaches the regime of single domain magnetism[2]. The standard picture of magnetic dynamics of small magnetic particles is the Stoner-Wohlfarth (SW) model[3] and is an ideal starting point for a discussion of this problem. This model is based on the following assumptions:

1) The particle size is smaller than the width of a domain wall, and the orientation of the spins is strictly the same throughout the particle. The magnitude of the magnetization remains constant for all applied magnetic fields, and the only change is the direction of the magnetization with respect to the particle shape.

2) The magnetic energy of the particle is a function only of the collective orientation of the spins, and is determined by a combination of the external magnetic field, and the demagnetizing energy (i.e., the magnetic-dipole energy of the magnetic fields created by the spins) or the magneto-crystalline anisotropy (i.e., the energy arising from the spin-orbit interaction between the spins and the orbital motion of the electrons in the crystal lattice). For example, if the particle is ellipsoidal, then the magnetic energy as a function of the magnetic orientation θ is [2]:

$$E = \frac{1}{4}(D_a + D_b)M^2 V - \frac{1}{4}(D_b - D_a)M^2 V\cos 2\theta - V\vec{H}\cdot\vec{M}, \qquad (1)$$

where D_a and D_b are demagnetizing factors along the major and minor axes of the ellipsoid, M is the magnetization, V is the volume, and H is the external

magnetic field. Alternatively, the SW model may be applied to a spherical particle with crystalline anisotropy in which case, for uniaxial anisotropy, the first two
terms in energy would be replaced by $KVsin^2\theta$, where K is the anisotropy constant.

3) The particle orientation will always be such that this energy, E, is at a
stable or metastable minimum. If a configuration becomes unstable as the applied
field is varied, then the particle orientation switches at the coercive field H_c,
immediately seeking the nearest available energy minimum.

A modification of this premise by Néel[4] and Brown[5] is that the escape
from a metastable state at finite temperature is governed by an Arrhenius thermal
activation rate exp(-E/k$_B$T), where E is the energy barrier and T is the temperature. The field-dependent hysteretic properties of a particle at any given temperature depend on Γt where $1/\Gamma$ is the relaxation time for the system to reach thermodynamic equilibrium from the saturated state and t is the time after the removal of the external field. The rate for this process is given by:

$$\Gamma = \Gamma_o e^{(-E/k_B T)},\qquad(2)$$

where Γ_o is the microscopic attempt frequency $\sim 10^{10} sec^{-1}$. For $\Gamma t \gg 1$, the particles behave superparamagnetically and the magnetization is described by
independent spins (i.e. there is no hysteresis). If one assumes Γ^{-1} equal to an
experimental time of typically 100 seconds, and the anisotropy is magnetocrystalline, then Eq. (2) yields the condition that the upper particle volume limit, V_p, for
superparamagnetic behavior is approximately $25 k_B T/K$[6,7]. The dependence of
the coercive field (that field for which the magnetization is zero) for constant K is
directly related to the volume of the particle, increasing monotonically from
$H_c = 0$ at V_p with increasing volume[7]. It is also possible to predict that the
coercive field increases monotonically with uniaxial anisotropy, i.e.
$H_c \sim 2K/M$[2]. Similarly, if one chooses to neglect magnetocrystalline anisotro-
pies and only confines the discussion to the demagnetizing effects, then
$H_c \sim (D_b - D_a)M$[2]. This expression predicts a monotonic increase in H_c with
increased aspect ratio of the prolate ellipsoidal particle.

An earlier study[7] of H_c for spherical iron magnets found that the predictions
for a monotonic increase of H_c with increasing volume are not correct even for
dimensions small enough for the particles to be considered single domain. Moreover, H_c as a function of increasing particle volume was seen to achieve a maximum value and subsequently decrease - a result that could not be explained using
the SW and Néel-Brown arguments. One of the motivations for producing the
well-characterized magnets of varying aspect ratio described below was to search
for a similar deviation from classical predictions.

C. Quantum Tunneling of Magnetization

The classical picture of an ideal single domain magnet implies that the magnetic orientation of a small particle will remain stable indefinitely below a given temperature. However, identical energy states in two potential wells separated by a small barrier can be coupled by quantum mechanical tunneling, leading to strikingly different dynamics normally forbidden by classical physics. In the case of a microscopic magnet (~ 10 nm), the number of atoms remains very large ($\sim 10^4$), and the macroscopic degree of freedom can be a magnetization direction. This phenomenon, macroscopic quantum tunneling (MQT), is the transition between two magnetization states, each consisting of ordered lattices of many spins[8]. While large systems are generally unsuitable for the study of quantum phenomena, mainly because they have many interacting modes of excitation, small magnetic structures are believed to couple only weakly to the environment. Early suggestions of such quantum effects were observed through relaxation of the magnetization in particulate media[9]. Although the reversal of the magnetization in such systems is complicated because of the range of particle sizes and the range of interaction strengths between the particles, the decay of magnetization upon reversal of the field could be modeled by an Arrhenius activation over a range of energy barriers. However, at low temperature it is found that the magnetic relaxation does not freeze out, but exhibits behavior independent of temperature. In these cases, it is believed that the rotation of the magnetic domains is accomplished by a quantum tunneling process in an asymmetrical potential.

One-dimensional barrier penetration is a misleading depiction of quantum tunneling of the magnetization; there is no Schrodinger equation which describes this process, since it is not an elementary particle which is tunneling, but a collective coordinate, the magnetization direction of a collection of spins. However, the classical micromagnetic theory of magnetic dynamics can be quantized to produce WKB-type equations for the expected zero-temperature tunneling rate for ferromagnetic and simple antiferromagnetic particles[8]:

$$\Gamma_F = \Gamma_o e^{-N(K_{\parallel}/K_{\perp})^{1/2}}, \tag{3a}$$

$$\Gamma_{AF} = \Gamma_o e^{-N(K_{\parallel}/J)^{1/2}}, \tag{3b}$$

Here N is the number of spins in the particle, $K_{\perp}$ and $K_{\parallel}$ are anisotropy constants for the easy and hard axes respectively, and J is the exchange interaction energy. Note that the number of spins appearing in the exponent means that, for both the ferromagnetic and antiferromagnetic cases, this process becomes unobservably small when the magnetic particles contain a large number of spins. The difference in the nature of these two exponents arises from the different spin wave spectra associated with ferromagnetic and antiferromagnetic excitations. By analogy to the demagnetizing factors in Eq. (1), the anisotropy constants determine the energy landscape for the magnetization. In cases 3(a) and 3(b), the tunneling rate is predicted to be exponentially suppressed as a function of size of the particle; however, the antiferromagnetic rate is typically much higher than the ferromagnetic rate because the ratio $(K_\parallel /J)$ can easily be $\sim 10^{-5}$ in the magnetic materials which have been studied, while $K_\parallel /K_\perp$ is typically of order unity.

There is great interest in the production of small magnetic structures and in understanding and controlling many of their dynamical properties. Although one normally thinks of technology as being driven by basic science, the reverse applies as well: researchers have realized and explored the theoretical concepts of classical micromagnetics and MQT in real systems by exploiting the most recent advances in the technology of low temperature instrumentation, materials preparation, and magnetic structures.

II. MICROFABRICATED MAGNETS BY SCANNING TUNNELING MICROSCOPY

A. Fabrication and Structural Characterization

An important goal in the study of nanometer-scale magnets is to understand the magnetic behavior of individual particles, the interactions between particles in various array configurations, and possible interactions between magnetic structures and their supporting substrate. By controlling the size, separation and proximity of the particles to the supporting substrate, one expects to enter regimes where characteristic dimensions such as the surface-to-volume ratio or the aspect ratio play an important role in the dynamical properties of the magnets. Furthermore, it is known both from experiment and theory that the characteristic length scales for single domain magnetic structures are nominally ~ 50 nm[2,7]. These requirements suggest a scheme which is complementary to optical or even electron beam lithographic techniques.

Scanning tunneling microscopy (STM) offers atomic scale spatial resolution that, in combination with metallo-organic chemical vapor deposition (MOCVD) techniques, is a promising technology for the direct deposition of magnetic nanostructures. The process requires that a vapor containing the species of interest be injected into an otherwise clean space and locally decomposed upon a surface. This method has been used to produce microscopic iron-containing deposits having diameters as small as 10 nm and aspect ratios as large as 100:1 from an iron pentacarbonyl precursor onto *Si*, *GaAs* or *Au* substrates as well as onto the tunneling tip itself[10]. Based on earlier work[11], arrays of iron particles and freestanding iron fibers have been fabricated and are shown in Figure 1. The fibers are particularly important for use as TEM samples to determine the material microstructure[10].

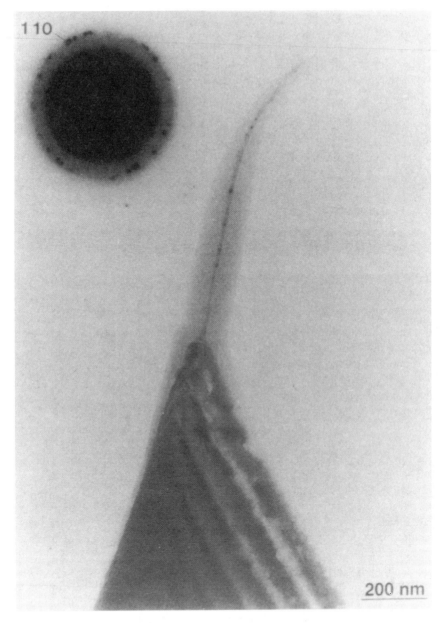

FIGURE 1. Transmission electron micrograph of a high aspect ratio filament on the apex of a silicon tip. The inner filament shows diffraction contrast, has a diameter of 9 nm, and is 880 nm high. Inset shows a ring of diffraction spots indexed to (110) planes of bcc iron. (From ref. 10).

The growth procedures require a STM head which is compatible with an ultra-high vacuum environment capable of operation at pressures $\sim 10^{-10}$ Torr, having a wide range of x-y positioning, and chemically passive in the presence of reactive gases. The STM tips are tungsten or silicon. The size, morphology, and cleanliness of the deposits depends strongly on the value and sign of the applied potential, which varies in value from approximately 5-30 V and produces currents up to approximately 50 pA. Growth rates and the concomitant variation in debris, due to diffusion of material during the growth of the iron particles along the substrate, depend critically on the ambient gas pressure during growth. We speculate that there are two modes of dissociation of the carbonyl, one due to bombardment by energetic electrons, the other by physically splitting apart the molecule due to the large electric field acting on the molecular dipoles. The former process, electron bombardment, produced the structure shown in Figure 1. This fiber is composed of an amorphous outer shell surrounding a granular core which consists of serial microcrystallites. As many as seven rings are observable in the original diffraction data, and by indexing the pattern [inset, Figure 1], it is possible to determine that the crystallites are body-centered cubic iron. This result is important as it also demonstrates that the deposit is very pure, containing less than 0.4% carbon; larger amounts would yield the equilibrium phase of $Fe - C$ which is face-centered cubic. Thus, the presence of the bcc phase is evidence for relatively pure iron and nearly complete decomposition of the precursor[10].

In addition to TEM, various other analytical tools have been applied to characterize the magnetic nanostructures including the STM itself. While the STM is useful for investigating some of the smallest structures (~ 10 nm), the most powerful tool is the scanning electron microscope (SEM). It is capable of examining a relatively large field of view, thus giving detailed information on the structures and their uniformity, the particle aspect ratio, as well as the regularity of patterned arrays. Although the internal structure of the dots and fibers as seen through TEM contains both an outer shell and a magnetic core which is not visible by SEM, the SEM is often the only tenable electron microscopy for arrays on flat surfaces. Finally, Auger electron spectroscopy (AES) is used to determine the chemical composition of the deposits. Despite the fact that the resolution of this technique, both because of electron beam focusing and scattering within the material, is approximately 50 nm and much larger than a single dot, it has been possible to determine the average composition of the deposits. In the case of growth through energetic electron bombardment, the carbon content is large and the deposited material is spread over large areas, since the reaction products are mobile on the surface. The experiments thus far have shown that the diffusion rate of reaction products is limited by the adsorption of the precursors on the surface, and, hence, on the pressure. The higher the pressure, the faster the growth, the smaller in diameter are the dots thus produced. (The speculation is that the mobility will decrease for lower substrate temperatures.) In the case of field-induced decomposition, the morphology is quite different. The iron dots are confined to areas of maximum field and thus to sharp boundaries either on tip or substrate. However, the resulting structures are amorphous, carbonatious filaments with only a small percentage of iron[10].

B. Magnetic Measurements

1. Miniature Hall Detectors

Measurements of small elongated and spherical single domain particles are not consistent with the Néel-Brown theory of thermally assisted magnetization reversal over a simple potential barrier[7,12]. Consequently, these studies have motivated a variety of new theoretical investigations on the stability of magnetic nanostructures[13]. Of fundamental interest is the reversal mechanism in real particles and whether it is due to classically coherent (Néel-Brown) or incoherent rotations, or heterogeneous nucleation and growth. Using STM nanolithography, small particle systems of various aspect ratios have been produced and their magnetic properties investigated using a novel Hall effect magnetometer[14]. Deposits are fabricated directly upon the active area of a Hall bar gradiometer. With this device, hysteresis loops of dilute particle arrays comprising 100–600 particles, each having magnetic signals of less than 10^{-14} emu, have been measured at low temperatures. The measurements are made using a novel high sensitivity design based on the Hall response within a semiconductor heterostructure. In contrast to integrated superconducting quantum interference device (SQUID) micro-susceptometers, this device allows for a systematic investigation over a wide range of applied fields and temperatures. As described later, however, dc SQUIDs are presently more sensitive by several orders of magnitude.

A high mobility $GaAs/Ga_{0.7}Al_{0.3}As$ two-dimensional hole system (carrier concentration $n_{2D} = 3 \times 10^{11} \, cm^{-2}$, carrier mobility $\mu(5K) = 10^5 \, cm^2/Vs$) is chemically etched into the form of a Hall gradiometer as shown in Figure 2(b). The device is composed of two Hall crosses, in which the ac currents are opposite in direction and the sum of the induced Hall voltages are measured in the presence of a perpendicularly applied magnetic field. In principle, there should not be a Hall signal in the absence of a sample as long as the material is uniform. However, if one of the crosses is penetrated by a magnetic field different from the other, there will be a net signal sensitive to the differential magnetic field. Figure 2(c) shows a SEM micrograph of the STM-fabricated array placed in a 10μm x 2.5 μm active area of one of a pair of Hall crosses in the Hall magnetometer. It is interesting to note that a comparable $2\mu m^2$ device has been constructed and seen to have a spin sensitivity of approximately 10^{-14} emu. Figure 3(a) shows the results of measurements on two different samples. These data have demonstrated an ability to detect a field change of approximately 1 Gauss, which translates into a total moment of 2×10^{-13} emu. Magnetizations of this magnitude, approximately

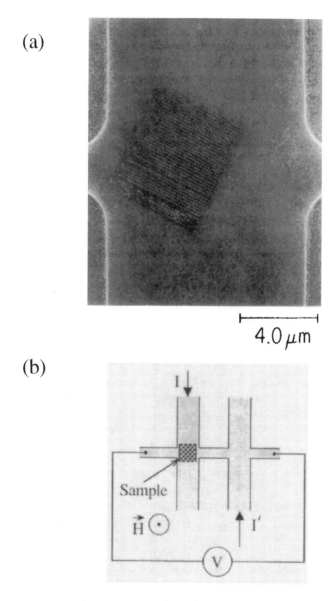

FIGURE 2. (a) A SEM image of a magnetic array composed of 30 nm diameter by 66 nm diameter high structures spaced 100 nm apart, (b) Schematic of the Hall magnetometer showing the device layout and bridge measurement circuit. The sample is deposited into one Hall cross while the other serves as a reference. I and I' are independent current sources that float with respect to one another.

(c)

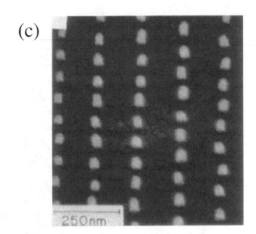

FIGURE 2. (c) Scanning electron microscope micrograph of a STM-fabricated array of iron particles (aspect ratio 2.2+/-0.3) placed in the 10 μm x 2.5 μm active area of a 2D *GaAs/AlGaAs* semiconductor heterostructures Hall bar magnetometer. (From ref. [14]).

10^7 spins, would be difficult to detect using conventional means. Finally, it should be noted that despite an increase in aspect ratio, the coercive field determined from a series of measurements similar to those shown in Figure 3(a) initially increases but then appears to show a decrease. This effect is not predicted by SW theory, and more detailed experiments appear to support the initial observation that there is a maximum in H_c for increasing values of the aspect ratio[14]. The temperature-dependence of the coercive fields can be clearly seen in Figure 3(b), systematically decreasing with increasing temperature.

2. Integrated DC SQUID Microsusceptometry

The experiments necessary to observe the magnetic dynamics in nanometer-scale particles are demanding, especially in view of the ultimate quest to observe a single microscopic structure. State-of-the-art superconducting quantum interference devices (SQUIDs) are uniquely suited for making these observations. Recent advances in superconducting VLSI (very large scale integration) technology have led to the development of miniature SQUID gradiometers and susceptometers configured for the detection of small magnetic fields with nearly quantum limited sensitivity[15,16]. These devices have proven to be valuable tools in the study of both local and global magnetic behavior of low dimensional structures[16]. As shown below, they have been used to measure the frequency-dependent magnetic susceptibility and magnetic noise of nanometer-scale magnets using a fully

(a)

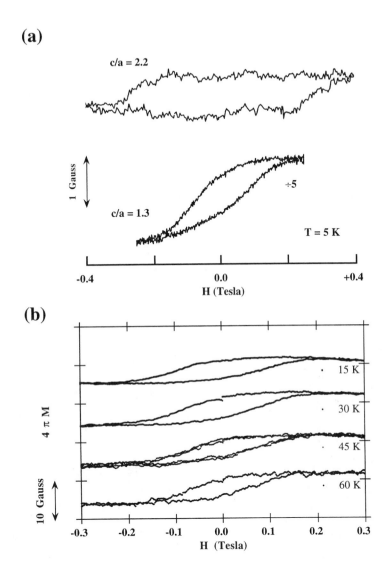

FIGURE 3. (a) Hysteresis loops of STM-deposited Fe particles as a function of particle dimension. Measurements on sample A (aspect ratio 1.3+/-0.3) at T = 5K, 250 G/min, and sample B (aspect ratio 2.2+/-0.3) at T = 15K, 1kG/min. These variations in measurement time and temperature do not affect the curves on the scale presented (From reference [14]). (b) Temperature-dependence of the hysteresis behavior in sample A.

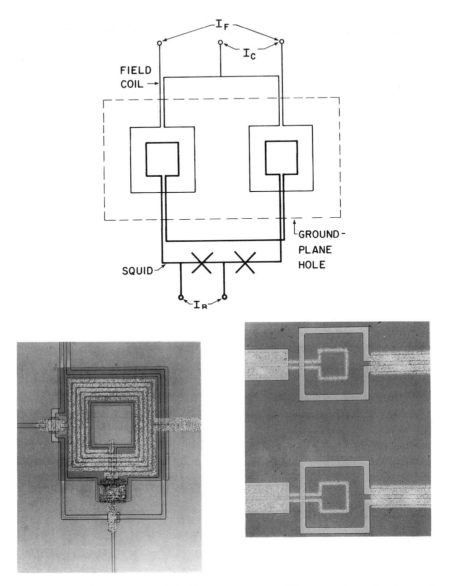

FIGURE 4. An integrated dc SQUID susceptometer. The top schematic shows the design configuration of the device built using niobium and lead lithographic processes; I_F, current in the field coil; I_C, current in the center-tapped coil; I_B, SQUID bias current. The bottom two photographs display the dc SQUID with planar input transformer, and the square pickup coil structure with coplanar field coils, respectively. The linewidths are 2.0 μm.

integrated thin film dc SQUID susceptometer. This miniature planar superconducting circuit is shown in Figure 4.

In order to reduce the effects of stray magnetic fields, two remotely located
series of Josephson junctions are connected in a low inductance fashion to a
microfabricated gradiometer detector. This planar loop structure consists of two
square counterwound Nb pickup loops 15.0 μm on a side over a superconducting
ground plane, one of which contains the sample. In addition, a center-tapped field
coil, which takes a single square turn around each pickup loop, is used to apply an
ac magnetic field (10^{-5} G) for susceptibility measurements as well as to apply
static dc fields. The single chip experiment and associated electronics are
attached to the mixing chamber of a liquid helium dilution refrigerator and cooled
to temperatures as low as 20mK. In addition, the susceptometer assembly is elec-
tronically and magnetically shielded within a radio-frequency tight superconducting chamber and cooled within mu-metal cylinders to achieve a magnetic flux
noise $< 10^{-7} \phi_0/\sqrt{Hz}$, where ϕ_0 is the Josephson flux quantum.

C. Stability of Mesoscopic Magnetic Structures

Changes in the coercive field with dimensionality, notably the decrease in
high aspect ratios and large diameter particles, are not consistent with conventional coherent or incoherent modes of spin reversal. The standard theory of magnetization reversal is the mean field treatment where, in order to avoid an energy
barrier due to exchange interactions between atomic moments with unlike orientations, Néel-Brown theory assumes a uniform rotation of all atomic moments in
the system[4,5]. For uniform coherent reversals at zero field, the barrier is set by
the anisotropy volume product KV, and is expected to increase with volume.
Other modes of reversal such as buckling, fanning, and curling also predict lifetimes which increase with some power of the dimension[17]. Thus, none of these
models can explain the observed maximum, including recent theories which
attempt to explain a reduction in the magnitude of the coercivity at finite temperatures in elongated particles due to non-uniform magnetization fluctuations. However, for highly anisotropic materials, there exists another mode of relaxation
which is expected to lead to shorter lifetimes[18]. This relaxation process is based
on droplet theory, in which small regions of the magnetization along the applied
magnetic field are continuously created and destroyed in the background of opposite magnetization. Monte Carlo simulations have succeeded in obtaining the
qualitatively observed experimental features for the coercivity as a function of
particle volume (and particle shape) and also give expressions for the expected
relaxation times in various regions of parameter space defined by the particle
volume and the applied magnetic field. These predictions will be a powerful
motivation for reexamination and continued study of the switching behavior of
independent small particles. Presently available dynamical results on

lithographically prepared magnets will be reviewed in the next section.

III. LITHOGRAPHICALLY PREPARED MAGNETS

A. Magnetic Recording Technology

There exist continuing efforts to produce magnetic recording media with ever higher bit density in order to store greater amounts of information in even smaller spaces. One approach has been to push the limits of optical and electron beam lithography and produce structures on the 10-100 nm scale. Recent efforts have demonstrated the ability to fabricate arrays of nickel pillars on silicon with uniform diameters of 35 nm and heights of 120 nm[19]. The density of such arrays are at least two orders of magnitude greater than any other state-of-the-art magnetic storage devices. However, on these scales, the surface to volume ratio becomes ever more important, as does the challenge to try to measure arrays or even single particles. Below, we discuss some of the problems inherent in structures at these small scales and will review some of the critical experiments which have been performed over the last few years, often using newly developed or refined techniques.

B. Effects of Surfaces

Small particles can have properties quite different from bulk materials due to the large surface/volume ratio. For example, it has been found that $\gamma-Fe_2O_3$ particles approximately 6 nm in dimension do not have a collinear magnetization[20]. The arrangements of the particle spin structure can be modified by binding those spins at the surface to other molecules. Thus, the atomic scale surface environment which surrounds the particle is critical to its intrinsic physical properties. Similar particles having acicular shapes, used for magnetic recording and having an aspect ratio of 6:1 and typical lengths of 600 nm, can have their coercive fields changed by as much as 50% simply by a surface treatment[21]. The physical arguments put forward suggest that a sodium phosphate treatment binds preferentially to the ferrous ions, thereby increasing the surface anisotropy and, consequently, H_c.

Theoretical arguments suggest that strong modifications of surface spins can occur depending on the strength of their binding to the external molecular environment, and there are predictions of anisotropy changes by a factor of 100[22]. Thus, depending on the surface to volume ratio, the bulk will rotate completely only if the surface anisotropy can be overcome, which means that reversals no longer depend on a simple anisotropy energy density and the volume of the particle. Clearly, the energy barrier as a function of magnetic field will be far more complex, most likely leading to magnetic relaxation times which are not simple exponentials with a single constant.

C. Switching

How does one precisely study the effects of shape and surface on the static and dynamic magnetic properties of nanoparticles? Typically this is done by collecting data from a large ensemble of structures using standard techniques and then attempting to mathematically derive the average properties of isolated particles from the results. This assumes that the particles are relatively uniform: as this is rarely the case, the subsequent analysis is often difficult and occasionally speculative. It is, therefore, the challenge of the experimentalists to try and measure ever smaller collections of magnetic particles, with the intent of ultimately addressing a lonely single particle. Several special techniques have been devised for this purpose. The Hall gradiometer magnetometer and integrated DC SQUID devices have already been discussed above. Here we mention a select set of innovative experiments on lithographically prepared magnets and commercial recording media. While by no means a complete or exclusive list, these studies have led to a more detailed understanding of the magnetic properties of nanostructures.

Over the last few years, the field of magnetic force microscopy (MFM) has seen substantial progress. Whereas the signal derived from MFM is proportional to the force gradient, and thus the second spatial derivative with respect to the magnetic field produced by the magnet under investigation, it has been possible to relate the signals to the measured moment[23,24]. Recent measurements on magnetotactic bacteria demonstrated that moments as small as 10^{-13} emu can be measured using mechanical cantilever techniques[25]. Furthermore, as seen in Figure 5, magnetic structures of bits 1 μm in diameter have been examined with lateral resolution finer than 10 nm[26]. All of these measurements show that magnetic force microscopy is rapidly evolving into a powerful experimental tool for micromagnetic studies.

It has also been possible to perform dynamical studies and observe the switching of the magnetic state of one or a small group of particles as a function of time and magnetic field using MFM. Systems studied include permalloy, with which MFM was compared to other independent dynamical measurements including an alternating gradient magnetometer[23], as well as single Fe_2O_3 particles[27]. Very recent switching studies using isolated nanoscale nickel bars of varying width and separation confirmed the earlier work that the coercive field does not increase monotonically with aspect ratio[28]. This latter work also showed the unexpected result that all larger width bars were multi-domain with the exception of a 15 nm wide bar. All of these observations confirm that the statistics of reversal are not simply describable as an activated process over a single valued barrier, and that reversal, even in single domain particles, does not occur via coherent

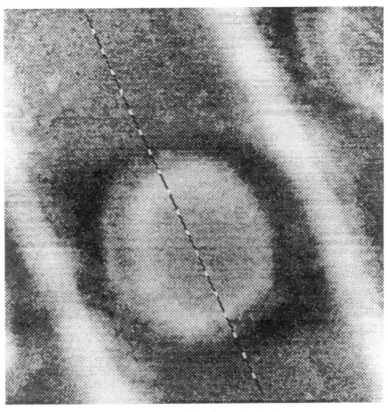

2 μm

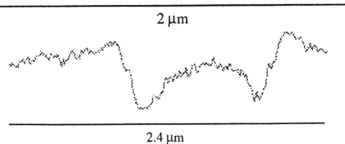

2.4 μm

FIGURE 5. Magnetic image of one bit from a commercial magneto-optical disk. The image was measured approximately 25 nm above the surface of the sample. The reflector layer was removed so that only a 10 nm dielectric layer was on top of the MO layer. Below the image, the profile of the MFM response along the diameter of the bit is plotted. The dotted line indicates the direction along which the profile below was taken. The image, measured 35 nm above the magnetically-active layer, resolves the fine structure of the bit. (From ref. [26]).

rotation. For high aspect ratio samples, in particular, the process is more complex, including nucleation of magnetic configurations, possibly of the "droplet" type mentioned earlier. Observation of switching, using Lorentz microscopy, in which the magnetic state of individual particles can be detected via the (magnetic) deflection of the electron beam, also support the idea that coherent rotation does not describe the entirety of the reversal process[29].

It is worthwhile mentioning two other unusual experimental techniques in this context although they may not have quite the spatial resolution of MFM. One of these is the use of the anomalous Hall coefficient. In this case, the signal is proportional to the magnetization of the material being measured and was used to investigate switching properties[30] in cobalt-chromium films, a popular medium for vertical recording. Hall crosses of linear dimension 0.7 μm achieved sensitivities of 4 x 10^{-14} and comparable to approximately 4 x $10^{6} \mu_B$. This technique makes it possible to look at the switching of single particles by observing the Barkhausen noise in magnetic hysteresis and magnetic relaxation. The decrease in magnetization occurs in minute discrete jumps characteristic of particles, not domains, as does the time dependence of the decay. Detailed observations of the distribution of jump sizes for varying field conditions led to the conclusion that individual switching events do not obey the Néel model.

The alternating gradient magnetometer is another technique which makes use of the resonance frequency of a lever arm to sample an alternating gradient in an external field and measures the force on a magnetic structure attached to the lever. This method was used to study the particle size and aspect ratio dependence of the hysteresis for a series of particles with total saturation magnetization of approximately $7 x 10^{-6}$ emu[31]. The coercivity of lithographically patterned permalloy particles, having thicknesses of 0.05 μm and lengths which varied from 0.5 μm to 9 μm was studied in detail and compared to theory. As can be seen from Figure 6[31] the agreement between experiment and theory is qualitative, although a peak in coercivity is apparent in the experimental data which is not predicted theoretically. Additional theoretical simulations[32] show the type of magnetization distribution resulting from micromagnetic computation. These pictures suggest that coherent rotation does not give a complete description of the reversal and more complex modes are favored.

A variation of the cantilever technique described earlier has been employed to investigate the behavior of small magnetic particles over an extremely wide range of magnetic fields. Single crystal *Si* cantilevers have been used as magnetometer devices sensitive to both torque and force in order to directly measure the temperature-dependent magnetization of the nanostructures[33]. This method relies on the capacitive detection of the motion of a micromachined semiconductor platform, upon which microscopic magnets are deposited. While traditional torque meters require mechanical contact between the sample and a measurement detection apparatus at room temperature, this scheme replaces the mechanical coupling with electrical sensing. In particular, the measurement of torque has

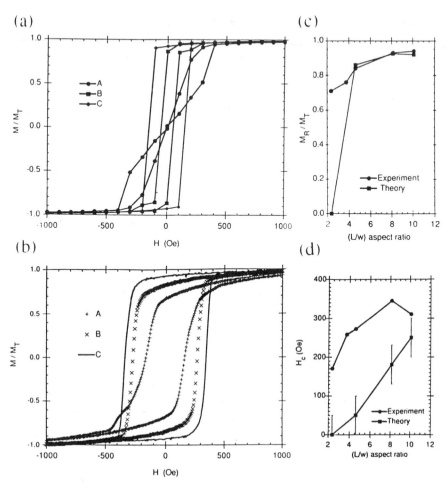

FIGURE 6. Hysteresis of permalloy particle with various aspect ratios (L/w). (a) Experimental loops for (A) L=0.41 μm, w=0.175 μm, (B) L=0.83 μm, w=0.18 μm, and (C) L=1.47 μm, w=0.18 μm. (b) Loops from the numerical calculation for the same size particles as in (a). Comparison between experiment and theory for (c) "squareness" (MR/MT) and (d) coercivity versus particle aspect ratio. (From ref. [34]).

proven to be an extremely useful technique in the study of microscopic single crystals and thin film superconductors[34]. Further mechanical enhancements and electronic improvements have already led to considerable advances in small signal detection using cantilevers for NMR measurements[35].

(a)

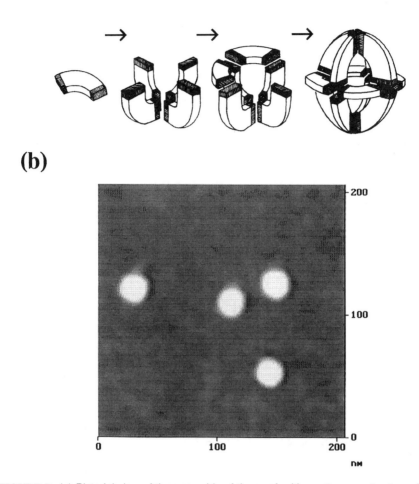

(b)

FIGURE 7. (a) Pictorial view of the assembly of the apoferritin quaternary structures into the complete protein shell which contains the magnetic material. (From reference [36].) (b) An atomic force microscope image of the antiferromagnetic protein. (See color plate.)

It is clear from the foregoing brief discussion of some of the salient results that single particle dynamics are not satisfactorily described by simple Arrhenius relaxation, and that the rotation occurs by means other than the (assumed) coherent mode. These discoveries will have a profound effect on any device design, should particle sizes in such devices approach the dimensions we are discussing here. On a more basic level, these discoveries also mean that the MQT predictions, which have assumed simple potential barriers, may not describe fully the tunneling modes. It will, therefore, be necessary to include these new findings in future studies of MQT and thermal switching dynamics.

IV. BIOLOGICAL MAGNETS

A. Natural and Artificially Engineered Proteins

While current work on creating lower dimensional structures is largely focused on manipulation of condensed matter systems through a variety of growth and processing techniques, small length scales already exist in many biological systems. One example of this novel "in situ" fabrication is the protein complex ferritin, a natural constituent of the metabolic system in most animals; it provides one of the variety of ways in which the organism stores Fe^{3+} for physiological needs[36]. This protein is of interest to researchers in a range of fields: in the biology of mammals and bacteria, ferritin serves as the main form of iron storage; in the chemical synthesis of organic-inorganic nanostructures, the protein shell (apoferritin) can serve as a host for inorganic materials; and in the physics of magnetism, ferritin is one of the smallest realizable magnets and displays a variety of classical and quantum spin phenomena. As seen in Figure 7, it consists of a segmented protein shell which has the shape of a hollow sphere, with 12.5 nm OD and 7.5 nm ID. *In vivo* the inner space is normally filled with a crystal of iron oxide similar to ferrihydrite ($5Fe_2O_3 - 9H_2O$) that is antiferromagnetic below T = 240K. Measurements of the lattice magnetization using a cantilever magnetometer in fields up to 27 T[37] yield additional indications of antiferromagnetic interactions in natural magnetic proteins. The observed size of the signal suggests that only a small fraction of the spins are involved. With a large surface to volume ratio, the properties must be strongly dependent on the size of the proteins, but the size of available samples is largely determined by natural processes.

Recent synthetic techniques in biochemistry have offered a means of extensively manipulating the size and magnetic state (antiferromagnetic or ferrimagnetic) of the mineral core[37]. The iron oxide can be synthesized *in vitro* within empty protein shells, and the reaction controlled to give particles with desired iron loading from a few to a few thousand ferric ions per protein (the maximum is nearly 4500). This makes possible a more systematic study of the magnetic properties of these nanometer-scale proteins. It is worth noting that apoferritin can also serve as a vessel for the synthesis of other materials such as magnetite (Fe_3O_4), maghemite ($\gamma-Fe_2O_3$), $MnOOH$, and UO_3. Magnetite and maghemite are ferrimagnetic. Thus, we have an extremely small system in which we can vary both the size and the nature of the magnetic interactions.

B. Classical Properties and Interactions

A series of artificial ferritin samples is synthesized beginning with the empty apoferritin shell. In the "magnetoferritin" sample, a *ferrimagnetic* core is synthesized, and in the other samples the same antiferromagnetic material which constitutes the natural ferritin core is synthesized with loading from 100 to 4000 iron ions[1,37]. The dramatic range in the magnetic properties of artificial ferritin can be seen in Figure 8(a), which shows the magnetization of the ferrimagnetic ferritin as compared to natural antiferromagnetic horse spleen ferritin. The magnetic moment of artificially-engineered ferrimagnetic particles is found to be nearly ten times larger than the comparable antiferromagnetic ones. In the latter case, the net moment of the antiferromagnetic protein arises only from uncoupled surface spins. Moreover, the stability of these ferrimagnetic particles is also more sensitive to externally applied fields than their antiferromagnetic counterparts. Along with the absence of hysteresis, the sharp reversal in magnetization indicates that the anisotropy of the magnetoferritin must be weak. In contrast, the natural ferritin displays a slight hysteresis and remanence which disappear when the temperature is raised to 30K. Thus, above a certain temperature, the anisotropy barrier is overcome by thermal fluctuations leading to classical superparamagnetic behavior.

These synthetic magnets therefore have potentially more valuable applications, first in providing a system in which to study the volatility of memory as the size of the particles is reduced. Due to the large surface to volume ratio in these nanometer-scale structures, the surface spins are also likely to play a significant role and may be used to microscopically tune their magnetic properties. Magnetic dynamics experiments used to characterize the behavior of these artificial nanomagnets have determined the blocking temperature T_B as a function of their iron content in one series of artificially-engineered ferritin [Figure 8(b)]. In the Néel-Brown picture described earlier[4,5], T_B is the temperature below which thermally-assisted hopping between different magnetic orientations becomes frozen out:

$$\Gamma = \Gamma_0 e^{\left(-\dfrac{KV}{k_B T_B} \right)} \tag{4}$$

Here K is the anisotropy and V is the magnetic volume. The particle is effectively blocked if a measurement is performed on a time scale short compared to Γ^{-1}. This expression would predict that T_B is linearly proportional to the iron loading. The data in Figure 8(b) shows that this agrees with the overall trend, but not the extrapolation to zero. Again, some non-ideal classical behavior, perhaps akin to the surface pinning effects, must be invoked. We will see that ferritin deviates from SW behavior in a much more exotic way, which arises from quantum mechanical effects.

(a)

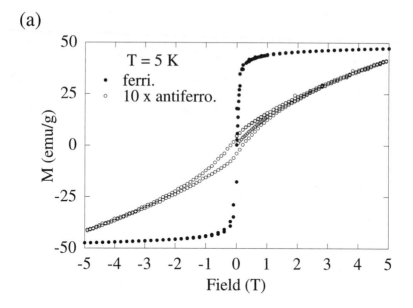

(b)

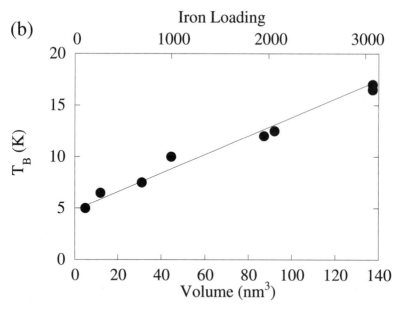

FIGURE 8. (a) The hysteresis loops from two artificially-engineered magnetic proteins at T=4.5K showing the behavior of antiferromagnetic and ferrimagnetic samples. (b) The measured blocking temperature T_B for artificial antiferromagnetic proteins of various sizes (Γ_{LAB} = 1/8 sec). (From ref. [37]).

C. Quantum Magnetic Phenomena

At temperatures far below the blocking temperature, the net magnetization is classically forbidden from fluctuating to the opposite direction because there is not enough thermal energy to overcome the anisotropy barrier. Due to its antiferromagnetic structure and small size, natural ferritin is a good candidate for the quantum tunneling predicted in Eq. 3(b)[8]. In zero applied magnetic field, the symmetric double well potential between "spin up" and "spin down" states leads to the quantum mechanical prediction that the magnetization should tunnel coherently back and forth between the two wells. While the particles are antiferromagnetic, they have a relatively high number of uncompensated surfaces spins owing to their large surface to volume ratio. This permits a monitoring of their dynamics using the coupling of this excess moment to an external detection apparatus. These excess spins serve as tracers for the antiferromagnetic particle dynamics. In contrast to magnetic relaxation experiments, this leads to a prediction for the appearance of a resonance line in the magnetic susceptibility and noise spectra at the frequency given by Eq. 3(b).

Figure 9 shows an experimental observation of this resonance and its dependence on protein volume. This resonance is interpreted as the tunnel splitting between two macroscopic states of the ferritin particles, namely, the net sublattice magnetization pointing up and pointing down. The sharpness of the line indicates that coupling to the environment is weak, one of the important requirements for MQT. In addition, the frequency of the line, which appears at temperatures below 200mK, corresponds reasonably well to the prediction of Eq. 3(b), particularly the exponential dependence of the frequency on the number of spins, N, in the particle. It was also confirmed experimentally[38] that the resonance disappears rapidly as the temperature is increased, or as a magnetic field is applied and the symmetry of the double-well potential is broken, consistent with theoretical expectations for the occurence of MQT in antiferromagnets. These results have stimulated a number of theoretical investigations on the effects of dissipation, including the role of nuclear spins on magnetic quantum tunneling, as well as on the feasibility of seeing quantum effects in larger magnetic structures.

V. SPATIOTEMPORAL STUDIES OF MICROMAGNETISM

A. Time-Resolved and Spatially-Resolved Information

The dynamics of small structures tend to be very fast, generally requiring time-resolved optical techniques for direct observation. However, many of the structures presently of interest are too small to be resolved by diffraction-limited optical microscopy. Similarly, scanning probe microscopes, while offering tremendous spatial resolution, generally have been of very limited temporal

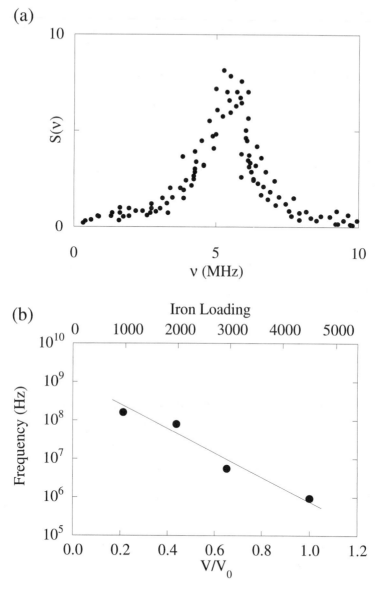

FIGURE 9. (a) A typical magnetic noise resonance curve observed for an artificially-engineered magnetic protein containing 3000 Fe spins at T = 24.3mK. (b) The dependence of this resonance frequency on the particle size at this temperature is shown using a series of manufactured magnetic proteins. Note the logarithmic scale in frequency in accord with quantum mechanical predictions. (From ref. [37]).

frequency bandwidth. This situation has generated substantial recent interest in combined scanning probe/ultrafast optical methods for the investigation of small-scale dynamics, using tunneling and other local interactions[39,40,41]. In addition to exploring the intrinsic behavior of nanometer-scale magnets, there is a great deal of interest in understanding the interaction of such structures with electronic charge carriers. The underlying physics controlling spin-dependent phenomena in magnetic, metallic, superconducting, and granular layers remains a lively issue in condensed matter and materials science. In many of these media, the microscopic interaction between the electronic and magnetic spins is believed to occur at the interface between the two subsystems. Moreover, recent developments (theoretical and experimental) in new magnetic and superconductor/semiconductor nanostructures require a detailed understanding of these dynamics at the local level. To this end, it is important to develop and apply spectroscopic experimental techniques which have a high degree of spatial resolution.

Micromagnetism is, at present, a particularly fruitful arena for the development and application of such methods. The resolution requirements in this case, while beyond the limits of conventional microscopy, are orders of magnitude more relaxed than those necessary to study true atomic dynamics. In fact, a combined resolution on the order of 50 ps / 50 nm is sufficient to uncover many of the remaining details of the dynamics (in the classical regime, at least) of magnetic switching in small structures, either by rotation or nonlinear wall propagation. This resolution is now accessible through near-field scanning optical microscopy (NSOM)[42], a technique that is rapidly emerging as a powerful tool for studying nanometer-scale phenomena. With applications ranging from physics and biology to molecular chemistry, NSOM combines the power of optical techniques with the potential high resolution of other near atomic-scale probes such as atomic force microscopy and scanning tunneling microscopy.

B. Ultrafast Probes of Dynamical Magnetism

1. Picosecond Pulsed Magnetic Fields

Of course, any scanned image of a dynamical process is necessarily a stroboscopic one, requiring many repetitions of the process to complete the picture. It is therefore necessary, in addition, to have complete control of the timing of the onset of the dynamics with respect to the narrow interval in which an instantaneous image is captured. For many magnetic systems, optically triggered (e.g., via a photoconductive switch) picosecond pulsed magnetic fields are a convenient means of initiating dynamics with a fast perturbation from equilibrium[43]. Very fast transition times are achieved with lithographic field coil and transmission line structures to apply the transient field to the sample. This is also an instance of an approach that works better for smaller structures. Higher peak fields at the same rise-time are achieved as the dimensions of coil structure are reduced. For example, 1 T transient fields switching in 1 ps are within reach at the micrometer length scale.

As an example, this approach has been applied to the investigation of the ferromagnetic resonance instability in lithographic structures fabricated from bismuth-doped yttrium-iron-garnet films. In this case, snapshots of the magnetization are recorded by the Faraday rotation of picosecond laser pulses transmitted through a near-field optical probe. A threshold in the pulsed-field amplitude is found above which nonequilibrium (dynamic) domain structure is observed [see Figure 10]. Such measurements tend to be "photon-starved" by the low optical throughput (on the order of 10^{-6}) of the near-field probes, and a more efficient approach is still highly desirable.

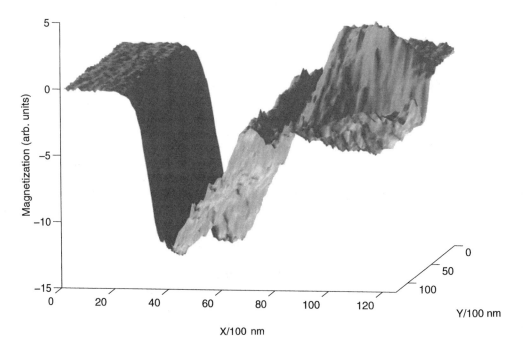

FIGURE 10. Dynamic domains in a bismuth-doped yttrium-iron-garnet film induced by a 500 Oe, 1 ns duration spin tipping pulse, which drives the system beyond the ferromagnetic resonance instability. This image is recorded 2 ns after the onset of the tipping pulse. The film is masked at the left where no signal is seen; the magnetization to the right is uniform below the instability.

2. *Femtosecond Near-field Scanning Optical Microscopy*

A pervasive question which arises while exploring magneto-electronic interactions in confined and interfacial geometries is precisely where electronic spin-scattering from magnetic moments takes place. By introducing magnetic ions into semiconductor quantum structures, one can obtain a magnetically-tunable system in which to study spin interactions using optical spectroscopy techniques[44]. The high degree of electronic confinement in quantum wells provides a strong overlap between photoexcited charge carriers and the neighboring magnetic ions, leading to large Zeeman energy splittings in relatively modest magnetic fields (~ 25 meV/T). Following optical excitation, radiative recombination of the spin-up and spin-down charge carriers in the semiconductor produces left- and right-circularly polarized luminescence. Polarization analysis of the emitted light yields the carrier spin orientation which, in turn, carries an imprint of the local magnetic state. Thus, this class of structures serves as a template for investigating the role of magnetic interactions on electronic spin behavior.

Conventional optical techniques have spatial resolutions which are diffraction-limited to the wavelength of light, λ. However, this limit may be surpassed by performing measurements in the near-field, at distances small compared to λ. In practice, scanning a small collecting aperture very close (~10 nm) to the surface of a sample allows an optical image to be obtained with spatial resolution as high as λ /40[42]. This technique has recently been used to study the role of patterned defects on spin-scattering in magnetic semiconductor quantum structures, where equally spaced planes of paramagnetic *MnSe* are introduced within 12 nm wide *ZnSe/ZnCdSe* quantum wells during growth by molecular beam epitaxy[45]. The lattice is then locally damaged using a focused beam of 140 keV Ga^+ ions to implant a series of horizontal and vertical lines spaced 2.0 micrometeres apart. Upon excitation with a Helium-Cadmium laser (λ=442 nm), spatially-resolved luminescence is acquired at λ=474 nm in a magnetic field of 0.2T normal to the patterned sample surface at T=5K. An example of an image taken from a sample containing 12 magnetic planes is shown in Figure 11, revealing modulated optical structure. The detected luminescence intensity drops smoothly from maxima in the unimplanted areas to minima in the implanted regions [Figure 11(a)], and is attributed to an increase of nonradiative carrier recombination channels due to lattice damage from the Ga^+ ions. In marked contrast to the intensity, the spatially resolved luminescence polarization remains constant in the unimplanted areas [Figure 11(b)], and drops sharply at the regions of implantation. The data reveal that the spin-scattering of charge carriers does not simply scale with their optical recombination. As seen in Figure 11(c, d), time-resolved information from the same structure directly indicates that the electronic *lifetime* drops quickly within the implanted regions, and is largely responsible for the image seen in Figure 11(a). Taken together with the polarization picture of Figure 11(b), these data suggest that carrier relocation does not

involve spin-scattering from the magnetic ions. These results provide a demonstration of the unique capability of near-field techniques for locating physical
interactions within composite magnetic structures which would otherwise be
difficult to ascertain by spatially-averaged methods.

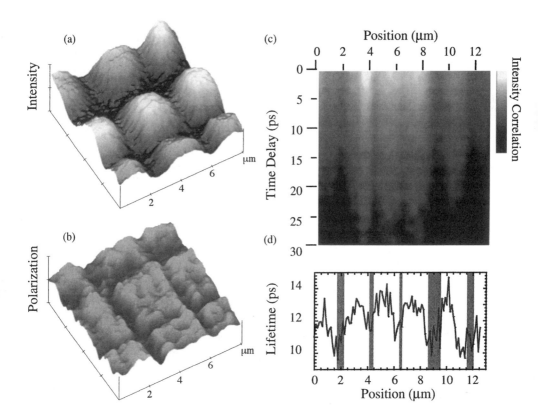

FIGURE 11. (a) Spatially-resolved luminescence from a patterned magnetically-doped
semiconductor quantum well at B = 0.2 T and T = 5.0K, obtained using near-field scanning optical microscopy. Modulation of the strong luminescence intensity arises from an
increase in nonradiative relaxation channels due to ion implantation. (b) Spatially-
resolved polarization image reveals uniform spin-scattering in the implanted regions.
(c) Time- and spatially-resolved luminescence taken across a series of vertically
implanted regions in a similar sample. Near t=0, the intensity is modulated by the localized lattice damage. The luminescence decays exponentially in time, yielding a spatially-
dependent lifetime shown in (d).

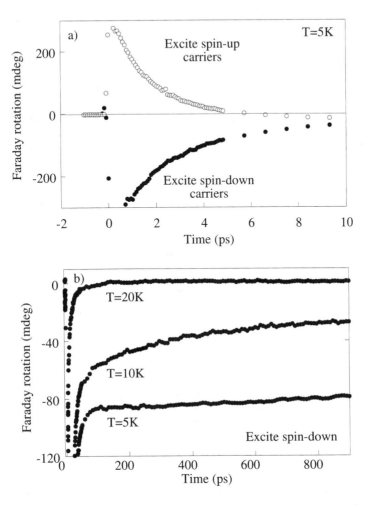

FIGURE 12. Time-resolved Faraday rotation (TRFR) at T = 5K of a transmitted probe beam through 24 isolated submonolayer magnetic *MnSe* planes embedded within a 12 nm *ZnSe*/*ZnCdSe* semiconductor quantum well. Magnetization is induced in the sample from the scattering of spin-polarized carriers created by an exciting laser pulse. Data is shown over two different time scales. (a) Direct observation of electronic carrier spin-flip scattering from the magnetic planes and recombination, (b)Temperature-dependent TRFR in B=2T long after the carriers have recombined. This optically-induced magnetization is generated by 120 femtosecond optical pulses yet persists for several microseconds. (From reference [45]).

3. *Femtosecond Faraday Spectroscopy*

In contrast to detecting the polarization of emitted light from magnetic sem-iconductors, it is possible to optically induce a magnetization by illuminating the lattice with circularly polarized light. The magnetic d-like orbitals of the man-ganese ions are strongly coupled to the s- and p- like conduction (electron) and valence (hole) bands, and give rise to a large Zeeman spin-splitting of the bands in the presence of an applied magnetic field. Circularly polarized light of the appropriate sense couples selectively to either the spin up or spin down carrier eigenstates, allowing for the net transfer of spin angular momentum, and thus magnetization, from the optically excited carriers to the paramagnetic manganese sublattice.

With ultrafast lasers in a pump-probe configuration it is, thus, possible to monitor, with femtosecond resolution, the dynamical effects on the magnetic moments due to the injection of spin polarized carriers into the lattice. The tech-nique of time-resolved Farady rotation reveals the onset, transfer, and eventual decay of magnetization optically induced by a 100fs pump pulse of circularly polarized light[46]. The Faraday rotation experienced by a time delayed linearly polarized probe pulse is a measure of the magnetization of the sample. This enables studies of the field- and temperature-dependence of magnetic relaxation in monolayer nanostructures, as seen in Figure 12. On short timescales the meas-ured change in magnetization is sensitive to the presence of the injected spin-polarized carriers. Spin-up (spin-down) carriers generate an initial increase (decrease) in the measured magnetization, which subsequently decays as the car-riers spin-scatter and recombine. Long timescale measurements (hundreds of picoseconds) reveal the magnetization induced to the paramagnetic *Mn* sublattice which persists long after the electronic carrier have recombined. The magnitude of the induced magnetic signal is highly temperature dependent, and decays on microsecond timescales.

VI. CONCLUSIONS AND FUTURE DIRECTIONS

It might appear that all of these studies point to limitations and restrictions of the use of small magnetic structures for the storage of information. The switching of magnetic domains depends on a myriad of detailed features of the particles, and quantum effects ultimately limit the length of time that a magnetic bit can remain stable. Nevertheless, it seems equally possible that these investigations will provide fundamentally new ways of using magnetic structures in technology. For example, theoretical investigations of magnetic quantum tunneling led to the surprising discovery that a selection rule quite generally forbids quantum tunnel-ing for particles with an odd number of electrons[47]. Thus, limitations imposed by quantum effects may be overcome. Another area for new investigations concerns the use of magnetic systems, not for memory, but for logic. It is now known that some computational problems can be greatly accelerated in a "quantum computer" in which bits and gates are implemented at the level of individual spins, allowing for calculations which exploit quantum interference effects within the computer itself[48]. It seems fair to say that "the book is not closed" on the discovery of fundamental directions for magnetic phenomena at the mesoscopic or microscopic level, nor will it be for some time to come.

ACKNOWLEDGMENTS

We would like to thank Savas Gider, Jeremy Levy, Scott Crooker, Mark Freeman, and Andrew Kent for their help with this chapter. This work is supported in part by grant #F49620-93-1-0117 from the Air Force Office of Scientific Research, and the UCSB MRL Central Facilities which are supported by the NSF under award #DMR-9123048.

REFERENCES

[1] This article is an expansion of the summary description by D.D. Awschalom and D. DiVincenzo, *Physics Today* **48**, 43 (April 1995).

[2] Morrish, A.H., *The Physical Principles of Magnetism,* New York: John Wiley & Sons, 1965.

[3] Stoner, E.C., and Wohlfarth, E.P., *Phil. Trans. Roy. Soc. (London), A* **240**, 599 (1948).

[4] Néel, L., *Compt. Rend.* **228**, 664 (1949); *Ann. Geophys.* **5**, 99 (1949).

[5] Brown, W.F., *J. Appl. Phys.* **30**, 1305 (1959).

[6] Bean, C.P., and Livingston, J.D., *J. Appl. Phys.* **30**, 1205 (1959).

[7] Kneller, E.F., and Luborsky, F.E., *J. Appl. Phys.* **34**, 656 (1963).

[8] Stamp, P.C.E., Chudnovsky, E.M., Barbara, B., *Int. J. Mod. Phys.B* **6**, 1355 (1992).

[9] Barbara, B., *et al.*, *J. Appl. Phys.* **73**, 6703 (1993).

[10] Kent, A.D., Shaw, T.M., von Molnár, S., and Awschalom, D.D., *Science* **262**, 1249 (1993).

[11] Silver, R.M., Ehrichs, E.E., de Lozanne, A.L., *Appl. Phys. Lett.* **51** , 247 (1987); Ehrichs, E.E., Smith, W.F., de Lozanne, A.L., *Ultramicroscopy* **42–44B**, 1438

(1991); McCord, M.A., Kern, D.P., Chang, T.H.P., *J. Vac. Sci. Technol. B* **6**, 1877 (1988); McCord, M.A., and Awschalom, D.D., *Appl. Phys. Lett.* **57**, 2153 (1990).

[12]   Lederman, M., Schultz, S., and Ozaki, M., *Phys. Rev. Lett.* **73**, 1986 (1994).

[13]   Braun, H.B., *Phys. Rev. Lett.* **71**, 3557 (1993); H.B. Braun and H. Neal Bertram, to be published.

[14]   Kent, A.D., von Molnár, S., Gider, S., and Awschalom, D.D., *J. Appl. Phys.* **76**, 6656 (1994).

[15]   Clarke, J., *Sci. Am.* **271**, 46 (1994).

[16]   Ketchen, M.B., Awschalom, D.D., Gallagher, W.J., Kleinsasser, A.W., Sandstrom, R.L., Rozen, J.R., and Bumble, B., *IEEE Trans. Mag.* **25**, 1212 (1989); Awschalom, D.D., and Warnock, J., *IEEE Trans. Mag.* **25**, 1186 (1989).

[17]   Köstner, E., and Arnoldussen, T.C., in *Magnetic Recording*, Mee, C.D., and Daniel, E.D., eds., New York: McGraw-Hill, 1987, Vol. 1, p. 98; Kneller, E., in *"Magnetism and Metallurgy"*, Berkowitz, A.E., and Kneller, E., eds., New York: Academic, 1969, Vol. 1.

[18]   Richards,   H.L.,   Sides,   S.W.,   Novotny,   M.A.,   and   Rikvold,   P.A., *J. Mag. Magn. Mat.,* submitted for publication.

[19]   Chou, S.Y., Wei, M.S., Krauss, P.R., and Fischer, P.B., *J. Appl. Phys.* **76**, 6673 (1994).

[20]   Coey, J.M.D., *Phys. Rev. Lett.* **27**, 1140 (1971).

[21]   Spada, F.E., Berkowitz, A.E., and Prokey, N.T., *J. Appl. Phys.* **69**, 4475 (1991).

[22]   Slonczewski, J.C., *J. Mag. Magn. Mat.* **117**, 368 (1992).

[23]   Gibson, G.A., Schultz, S., Smyth, J.F., and Kern, D.P., *IEEE Trans. Magn.* **27**, 5000 (1991).

[24]   Grütter, P., Rugar, D., and Manin, H. J., *Ultramicroscopy* **47**, 393 (1992).

[25]   Dahlberg, E. Dan, Proksch, R.B., Moskowitz, B.M., Bazylinski, D.A., and Frankel, R.B., *Proc. of ICM'94,* to be published.

[26]   Foss, S., Proksch, R., and Dahlberg, E. Dan, *Proc. of MORIS'94,* to be published.

[27]   Proksch, R.B., Foss, S., and Dahlberg, E. Dan, *IEEE Trans. Mag.* **30** , 4467 (1994).

[28]   Wei, M.S., and Chou, S.Y., *J. Appl. Phys.* **76**, 6679 (1994).

[29]   Salling, C., O'Barr, R., Schultz, S., McFadyeu, I., and Ozaki, M., *J. Appl. Phys.* **75**, 7989 (1994).

[30] Webb, B.C., and Schultz, S., *Solid State Commun.* **68**, 437 (1988).

[31] Smyth, J.F., Schultz, S., Fredkin, D.R., Kern, D.P., Rishton, S.A., Schmid, H., Cali, M., and Koehler, T.R., *J. Appl. Phys.* **69**, 5692 (1991).

[32] Fredkin, D.R., Koehler, P.R., Smyth, J.F., and Schultz, S., *J. Appl. Phys.* **69**, 5276 (1991).

[33] Chaparala, M., Chung, O.H., Naughton, M.J., in *AIP Conference Proceedings No. 273: Superconductivity and Its Applications* (Buffalo, NY, 1992) Kwok, H.S., Shaw, D.T., Naughton, M.J., eds. (American Institute of Physics, 1992), p. 407.

[34] Farrell, D.E., Rice, J.P., and Ginsberg, D.M., *Phys. Rev. Lett.* **67** , 1165 (1991).

[35] Rugar, D., Zuger, O., Hoen, S., and Yannoni, C.S., *Science* **264**, 1560 (1994).

[36] Ford, G.C., *et al.*, *Phil. Trans. R. Soc. Lond. B* **304**, 551 (1984); Gerl, M., and Jaenicke, R., *Biochem.* **27**, 4089 (1988).

[37] Gider, S., Awschalom, D.D., Douglas, T., and Mann, S., *Science* (in press, 1995).

[38] Awschalom, D.D., DiVincenzo, D.P., and Smyth, J.F., *Science* **258**, 414 (1992).

[39] Nunes, Jr., G., and Freeman, M.R., *Science* **262**, 1029 (1993).

[40] Weiss, S., Ogletree, D.F., Botkin, D., Salmeron, M., and Chemla, D.S., *Appl. Phys. Lett.* **63**, 2567 (1993).

[41] Hamers, R.J., and Cahill, D.G., *Appl. Phys. Lett.* **57**, 2031 (1990).

[42] Betzig, E., and Trautman, J.K., *Science* **257**, 189 (1992).

[43] Freeman, M.R., *J. Appl. Phys.* **75**, 6194 (1994).

[44] Awschalom, D.D., and Samarth, N., in *Optics of Semiconductor Nanostructures,* Hennenberger, F., and Schmitt-Rink, S., eds., (Akademie Verlag, Berlin, 1993), Chapter II, pp. 291–312.

[45] Crooker, S.A., *et al*, "Enhanced Spin Interactions in Digital Magnetic Heterostructures", submitted for publication.

[46] Baumberg, .J.J., Crooker, S.A., Awschalom, D.D., Samarth, N., Luo, H., and Furdyna, J.K., *Phys. Rev. B* **50**, 7689 (1994).

[47] Loss, D., DiVincenzo, D.P., and Grinstein, G., *Phys. Rev. Lett.***69** , 3232 (1992); von Delft, J., and Henley, C.L., *Phys. Rev.Lett.* **69**, 3236 (1992).

[48] Shor, P., *Proceedings of the 35th Annual Symposium on the Foundations of Computer Science,* (IEEE Comput. Soc. Press, Los Alamitos, CA, 1994), p. 124.

Chapter 13

Single Electron Transport
Through a Quantum Dot

Leo P. Kouwenhoven

Department of Applied Physics
Delft University of Technology
P.O. Box 5046, 2600 GA Delft, The Netherlands

Paul L. McEuen

Department of Physics
University of California and Materials Science Division
Lawrence Berkeley Laboratory
Berkeley, CA 94720

I. INTRODUCTION

The push toward miniaturization has been a dominant technological force for the past 30 years. In addition to revolutionizing the computer world, this trend toward ever-smaller systems has profoundly influenced solid state physics. Using the fabrication technologies of the semiconductor industry, scientists can now routinely design electronic structures on a size scale where quantum effects are important. Their properties are regulated by quantization phenomena previously observable only in naturally occurring small systems. For example, current flow in these devices can be regulated by the fact that charge is quantized in units of e, just as quanta of charge regulated the falling of Millikan's oil drops in 1909. Furthermore, electrons in these structures behave like the quantum-mechanical waves that Schrödinger, Bohr, and others used to describe the properties of atoms in the 1920s. The rapidly advancing fabrication techniques of the semiconductor industry have thus allowed physicists and engineers to make structures that probe the quantum properties of single electrons, but in novel and controlled geometries unavailable to previous generations of scientists.

But insights into the esoteric and puzzling underpinnings of our physical world is not all that this field promises. There are hopes for practical applications of this work. The quantum, or single-electron, transistor is an example of where abstract and applied science can productively intersect. This transistor consists of a small region, or "dot", through which electrons must tunnel to get from the source to drain. This device, in its incarnation as a semiconductor quantum dot[1], is the subject of this chapter.

The name "dot" suggests an infinitely small region. Solid state quantum dots, however, have dimensions much larger than the atomic scale of an Å. Although a semiconductor quantum dot is made out of roughly a million atoms, virtually all electrons are solidly bound to the nuclei of the material. The number of unbound, free electrons in the dot can be very small; between one and a few hundred. The deBroglie wavelength of these electrons is comparable to the size of the dot, so they can express themselves as waves. The electrons occupy discrete quantum levels (akin to atomic orbitals in atoms), and have a discrete excitation spectrum. A quantum dot also has an energy scale analogous to the ionization energy of an atom. This is the energy required to add or remove an electron from the dot. It is typically called the charging energy because it is dominated by the Coulomb interaction required to add/remove a charge e from the dot.

Because of the analogies to real atoms, quantum dots are sometimes referred to as artificial atoms[2]. The atomic-like physics of dots is studied not via their interaction with light, however, but instead by measuring their current–voltage characteristics. Quantum dots are therefore artificial atoms with the intriguing possibility of attaching current and voltage leads to probe their atomic states.

A. Outline

This chapter reviews many of the main experimental and theoretical results reported to date on electron transport through semiconductor quantum dots. We note that other reviews also exist[3]. For theoretical reviews we refer to Averin and Likharev[4] for detailed transport theory; Ingold and Nazarov[5] for the theory of metallic and superconducting systems; and Beenakker[6] and van Houten, Beenakker and Staring[7] for the single electron theory of quantum dots. Collections of single electron papers can be found in references [8] and [9]. For reviews in popular science magazines see references [1,2,10-12].

The outline of this chapter is as follows. In the remainder of this section we review the conditions under which charge and energy quantization effects are observable in quantum dots. In Section II we review the history of quantum dots and describe fabrication and measurement methods. A simple theory of electron transport through dots is outlined in Section III. Section IV presents basic single electron experiments which will be compared with the theory of Section III. Section V addresses the behavior of dots in high magnetic fields, and Section VI describes ac transport through dots. Finally, applications and future directions are summarized in Section VII.

B. Quantized Charge Tunneling

In this section we examine the circumstances under which Coulomb charging effects are important. In other words, we answer the question, "How small and how cold should a piece of solid be so that adding or subtracting a single electron has a measurable effect?" To answer this question, let's consider the electronic

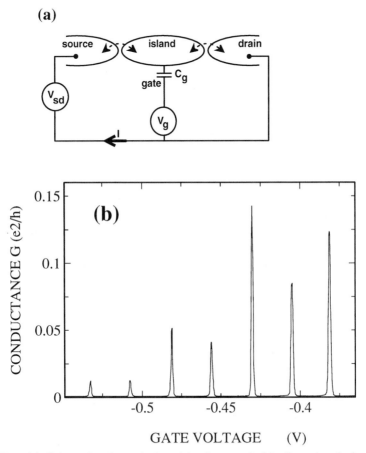

(a)

(b)

FIGURE 1. (a) Schematic of an electron island connected to three terminals. (b) An example of a measurement of Coulomb oscillations to illustrate the effect of single electron charges on the macroscopic conductance The conductance is the ratio I/V_{sd} and the period in gate voltage V_g is about e/C_g. (from Nagamune et al. [13]).

properties of the small conductor depicted in Figure 1(a), which is coupled to three terminals. Particle exchange can occur with only two of the terminals, as indicated by the arrows. These source and drain terminals connect the small conductor to macroscopic current and voltage meters. The third terminal provides an electrostatic or capacitive coupling and can be used as a gate electrode. If we first assume that there is no coupling to the source and drain contacts, then our small conductor acts as an island for electrons. The number of electrons on this island is an integer N and the charge on the island is quantized and equal to Ne. If we now allow tunneling to the source and drain electrodes, then the number of electrons N adjusts itself until the energy of the total system is minimized.

When tunneling occurs, the charge on the island suddenly changes by the quantized amount e. The associated change in the Coulomb energy is conveniently expressed in terms of the capacitance C of the island. An extra charge e changes the electrostatic potential by the charging energy $E_C = e^2/C$. This charging energy becomes important when it exceeds the thermal energy $k_B T$. A second requirement is that the barriers are sufficiently opaque such that the electrons are located either in the source, in the drain, or on the island. This means that quantum fluctuations in the number N due to tunneling through the barriers is much less than one. This requirement translates to a lower bound for the tunnel resistances R_t of the barriers. To see this, consider the typical time to charge or discharge the island $\Delta t = R_t C$. The Heisenberg uncertainty relation: $\Delta E \Delta t = (e^2/C) R_t C > h$ implies that R_t should be much larger than the resistance quantum $h/e^2 = 25.813 k\Omega$ in order for the energy uncertainty to be much smaller than the charging energy. This summarizes the two main conditions for observing effects due to the discrete nature of charge[3,4]:

$$R_t \gg h/e^2 \qquad (1)$$

$$e^2/C \gg k_B T \qquad (2)$$

The first criterion can be met by weakly coupling the object to the source and drain leads. The second criterion can be met by making the dot small. Recall that the capacitance of an object scales with its radius R. For a sphere, $C = 4\pi\varepsilon_r\varepsilon_o R$, while for a flat disc, $C = 8\varepsilon_r\varepsilon_o R$, where ε_r is the dielectric constant of the material surrounding the object. For a 0.5 μm radius sphere in free space, for example, the charging energy is ~3 meV, which is easily resolvable at cryogenic temperatures.

While tunneling changes the electrostatic energy of the island by a discrete value, a voltage V_g applied to the gate capacitor C_g can change the island's electrostatic energy in a continuous manner. In terms of charge, tunneling changes the island's charge by an integer while the gate voltage induces an effective continuous charge $q = C_g V_g$ that represents, in some sense, the charge that the dot would like to have. This charge is continuous even on the scale of the elementary charge e. If we sweep V_g the build up of the induced charge will be compensated in periodic intervals by tunneling of discrete charges onto the dot. This competition between continuously induced charge and discrete compensation leads to so-called *Coulomb oscillations* in a measurement of the current as a function of gate voltage at a fixed source-drain voltage. An example of a measurement[13] is shown in Figure 1(b). In the valley of the oscillations, the number of electrons on the dot is fixed and necessarily equal to an integer N. In the next valley to the right the number of electrons is increased to $N+1$. At the crossover between the two stable configurations N and $N+1$ a "charge degeneracy"[14] exists where the number can alternate between N and $N+1$. This allowed fluctuation in the number (i.e. according to the sequence $N => N+1 => N => \cdots$) allows current to flow and results in the observed peaks.

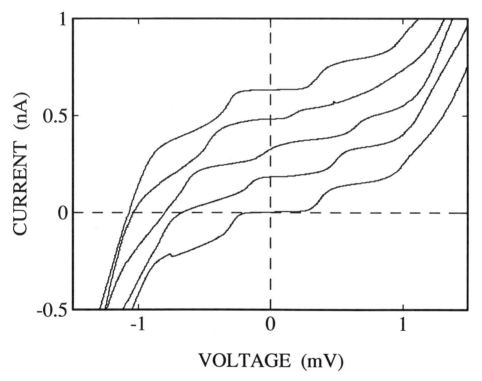

VOLTAGE (mV)

FIGURE 2. An example of a measurement of the Coulomb staircase in current-voltage characteristics. The different curves have an offset for clarity ($I=0$ occurs at $V_{sd}=0$) and are taken for five different gate voltages to illustrate periodicity in accordance with the oscillations shown in Figure 1(b) (from Kouwenhoven *et al.*[15]).

An alternative measurement is performed by fixing the gate voltage, but varying the source-drain voltage. The observation as shown in Figure 2[15] is a nonlinear current-voltage characteristic exhibiting a *Coulomb staircase*. A new current step occurs at a threshold voltage $(\sim e^2/C)$ at which an extra electron is energetically allowed to enter the island. It is seen in Figure 2 that the threshold voltage is periodic in gate voltage, in accordance with the Coulomb oscillations of Figure 1(b).

C. Energy Level Quantization

Electrons residing on the dot occupy quantized energy levels. To be able to resolve these levels, the energy level spacing $\Delta E \gg k_B T$. The level spacing at the Fermi energy E_f for a box of size L is given by the inverse of the density of states, $\Delta E(E_f) = 1/(L^d g^d(E_f))$ where d is the dimensionality. Including spin degeneracy, we have:

$$\Delta E = (N/4)\,\hbar^2\pi^2/mL^2. \qquad\qquad 1D \qquad (3a)$$

$$= (1/\pi)\,\hbar^2\pi^2/mL^2. \qquad\qquad 2D \qquad (3b)$$

$$= (1/3\pi^2 N)^{1/3}\,\hbar^2\pi^2/mL^2. \qquad 3D \qquad (3c)$$

The characteristic energy scale is thus $\hbar^2\pi^2/mL^2$, which is (approximately) the level spacing for the lowest 2 levels in the box. For a 1D box, the level spacing grows for increasing N, in 2D it is constant, while in 3D it decreases as N increases.

Lithographic dots can be made as small as ~ 100 nm. For a 100 nm dot, this yields $\hbar^2\pi^2/mL^2 = 0.08$ meV, which corresponds to a temperature of about 1K. Metallic dots are 3-dimensional (the Fermi wavelength of electrons in metals is on the order of 0.1 nm, which is much less than the 100 nm dot size), so we must use Eq. (3c) to find the level spacing at E_f. A 100 nm metallic dot has approximately 10^7 electrons, so the level spacing is about 5 µV, which is too small to be observed at obtainable temperatures.

The level spacing of a 100 nm 2D dot, on the other hand, is ~0.03 mV, which is large enough to be observable at dilution refrigerator temperatures. Electrons confined at a semiconductor heterointerface are effectively 2-dimensional, as will be discussed in the next section. In addition, they have a small effective mass that further increases the level spacing. As a result, dots made in semiconductor heterostructures are true artificial atoms, with both observable quantized charge states and quantized energy levels.

The fact that the quantization of charge and energy can drastically influence transport through a quantum dot is demonstrated by the Coulomb oscillations in Figure 1(b) and the Coulomb staircase in Figure 2. Although we have not yet explained these observations in detail (see Section III), we note that one can obtain spectroscopic information about the charge state and energy levels of the dot by analyzing the precise shape of the Coulomb oscillations and the Coulomb staircase. In this way, single electron transport can be used as a spectroscopic tool. The bulk of this chapter is devoted to a discussion of such measurements.

II. HISTORY, FABRICATION, AND MEASUREMENT TECHNIQUES

In this section, we briefly review some of the history and fabrication aspects of quantum dots. We start in Section II.A by discussing the history of single electron effects in metallic systems. We then turn to semiconductor structures in Section II.B. A brief history of quantum effects in semiconductors is given, followed by a discussion of one of the most popular "artificial atoms", the *GaAs/AlGaAs* lateral quantum dot. This type of dot will be the focus of most of the experiments presented in Sections IV.–VI.

A. Single Electron Effects in Metals

Single electron effects are really nothing new. In his famous 1911 experiments, Millikan[16] observed the effects of single electrons on the falling rate of oil drops. Single electron tunneling was first studied in solids in 1951 by Gorter[17], and later by Giaever and Zeller in 1968[18], and Lambe and Jaklevic in 1969[19]. These pioneering experiments investigated transport through thin films consisting of small grains. A detailed transport theory was developed by Kulik and Shekhter in 1975[20]. Much of our present understanding of single electron charging effects was already developed in these early works. However, a drawback was the averaging effect over many grains and the limited control over device parameters. Rapid progress in device control was made in the mid 80's when several groups began to fabricate small systems using nanolithography and thin-film processing. The new technological control, together with new theoretical predictions by Likharev[21] and Mullen *et al.*[22], boosted the interest in single electronics and led to the discovery of many new transport phenomena. The first clear demonstration of controlled single electron tunneling was performed by Fulton and Dolan in 1987[23] in an aluminum structure similar to the one in Figure 1(a). They observed that the macroscopic current through the two junction system was extremely sensitive to the charge on the gate capacitor. These are the so-called Coulomb oscillations. This work also demonstrated the usefulness of such a device as a *single – electrometer*, i.e. an electrometer capable of measuring single charges.

Electrometer-type experiments in metals have been performed with both normal and superconducting devices[24,25]. In the latter case, under certain conditions, charge can only be added in units of the Cooper pair charge, $2e$[26-29]. Also, other multi-junction configurations, such as 1D and 2D arrays, have been investigated in the normal and superconducting state[3]. Somewhat surprisingly, the apparently simplest system, a single junction, has turned out to be the most complicated. In single junctions the coupling between the electron that tunnels and collective fluctuations in the environment is found to be crucial[30-36]. Although this is particularly interesting from a fundamental point of view, this chapter is restricted to systems with at least one island, i.e. *two* tunnel junctions.

The advent of the scanning tunneling microscope (STM)[37] has renewed interest in the Coulomb blockade in small grains. STMs can both image the topography of a surface and measure the current-voltage characteristic of a single grain. The charging energy of a grain of size ~10 nm can be as large as 100 meV, so that single electron phenomena can occur up to room temperature[38]. These charging energies are 10 to 100 times larger than those obtained in artificially fabricated Coulomb blockade devices. However, the absence of a third gate electrode in these naturally formed structures limits their usefulness.

B. Quantum Confinement in Semiconductors

As discussed in Section I.B, quantization of energy can be important in semiconductor dots. The effects of quantum confinement on the electronic properties of semiconductor heterostructures were well known prior to the study of quantum dots. The chapter by Sakaki[39] reviews growth techniques such as molecular beam epitaxy, which allows fabrication of quantum wells and heterojunctions with energy levels that are quantized along the growth (z) direction. For proper choice of growth parameters, the electrons are fully confined in the z-direction (i.e. only the lowest 2D eigenstate is occupied by electrons). The electron motion is free in the x-y plane. This forms a two dimensional electron gas (2DEG).

Quantum dots emerge when this growth technology is combined with electron-beam lithography to produce confinement in all three directions. Some of the earliest experiments were on *GaAs/AlGaAs* resonant tunneling structures etched to form sub-micron pillars. These pillars are called vertical quantum dots because the current flows along the z-direction. Reed *et al.*[40] found that the I-V characteristics reveal structure that they attributed to resonant tunneling through quantum states arising from the lateral confinement. Results on similar vertical tunneling structures have been reported more recently in references [41-46].

At the same time as the early studies on vertical structures, gated *GaAs/AlGaAs* devices were being developed in which the transport is entirely in the plane of the 2DEG. The starting point for these devices is a 2DEG at the interface of a *GaAs/AlGaAs* heterostructure (see Figure 3 and reference [39]). The only mobile electrons at low temperature are confined at the *GaAs/AlGaAs* interface, which is typically ~ 100 nm below the surface. Typical values of the 2D electron density are $n_s \sim (1-5) \cdot 10^{15} \, m^{-2}$. To define the small device, metallic gates are patterned on the surface of the wafer using electron beam lithography[47]. Gate features as small as 50 nm can be routinely written. Negative voltages applied to metallic surface gates define narrow wires or tunnel barriers in the 2DEG.

Such a system is very suitable for quantum transport studies for two reasons. First, the wave length of electrons at the Fermi energy is $\lambda_F = (2\pi/n_s)^{\frac{1}{2}}$ ~(80–30) nm. This is 100 times larger than in metals. Second, the mobility of the 2DEG can be as large as 100 $m^2 V^{-1} s^{-1}$, which corresponds to a transport elastic mean free path larger than 10 μm. This technology thus allows fabrication of devices which are much smaller than the mean free path; electron transport through the device is ballistic. In addition, the device dimensions can be comparable to the electron wavelength, so that quantum confinement is important. The observation of quantized conductance steps in short wires, or quantum point contacts, demonstrated quantum confinement in two directions[48,49]. Later work on different gate geometries led to the discovery of a wide variety of mesoscopic transport phenomena[50]. For instance, coherent resonant transmission was demonstrated through a quantum dot[51] and through an array of quantum dots[52]. These early dot experiments were performed with barrier conductances of order e^2/h or larger, so that the effects of charge quantization were relatively weak.

The effects of single-electron charging were first seen in semiconductors in experiments on narrow wires by Scott-Thomas *et al.*[53]. With an average conductance of the wire much smaller than e^2/h, their measurements revealed a periodically oscillating conductance as a function of a voltage applied to a nearby gate. It was pointed out by van Houten and Beenakker[54], along with Glazman and Shekhter[14], that these oscillations arise from single electron charging of a small segment of the wire, delineated by an impurity. This pioneering work on "accidental dots"[53,55-58] stimulated the study of more controlled systems.

The most widely studied type of device is a lateral quantum dot defined by metallic surface gates. Figure 3 shows a schematic and an SEM micrograph of a device used by Staring *et al.*[59]. The tunnel barriers between the dot and the source and drain 2DEG regions can be tuned using the left and right pair of gates. The dot can be squeezed to smaller size by applying a potential to the center pair of gates. Similar gated dots, with lithographic dimensions ranging from 1.5 μm down to ~0.3 μm, have been studied by a variety of groups. The size of the dot formed in the 2DEG is somewhat smaller than the lithographic size, since the

(a)

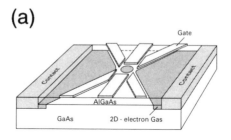

(b)

FIGURE 3. (a) Schematic of a typical $GaAs/AlGaAs$ heterostructure device and (b) a scanning electron microscope photo of the device (from Staring *et al.*[59]) A two dimensional electron gas (2DEG) is formed at the interface between the $AlGaAs$ and the $GaAs$, which is about 100 nm below the surface. Negative voltages applied to the fabricated metallic gates at the surface deplete the 2DEG underneath. The resulting structure is a dot containing a few electrons which are coupled via tunnel barriers to the large 2DEG regions to the left and right. The tunnel barriers and the size of the dot can be tuned individually with the voltages applied to the left/right pair of gates and to the center pair, respectively.

2DEG is typically depleted 100-200 nm away from the gate. This yields effective dot diameters from ~1.3 μm to ~0.1 μm for the devices studied.

We can estimate the charging energy e^2/C and the quantum level spacing ΔE from the dimensions of the dot. The total capacitance C – the capacitance between the dot and all other pieces of metal around it, plus contributions from the self-capacitance – should in principle be obtained from self-consistent calculations[60-62]. A quick estimate can be obtained from the formula given previously for an isolated 2D metallic disk given previously, yielding $e^2/C = e^2/(8\varepsilon_r\varepsilon_o R)$ where R the disk radius and $\varepsilon_r = 13$ in $GaAs$. For example, for a dot of radius 200 nm, this yields $e^2/C = 1$ meV. This is really an upper limit for the charging energy, since the presence of the metal gates and the adjacent 2DEG increases C.

An estimate for the single particle level spacing can be obtained from Eq. (3b), $\Delta E = 2/m^*R^2$, where $m^* = 0.067m_e$ is the effective mass in $GaAs$, yielding $\Delta E = 0.03$ meV. To observe the effects of these two energy scales on transport, the thermal energy k_BT must be well below the energy scales of the dot. This corresponds to temperatures of order 1K ($k_BT = 0.086$ meV at 1K). As a result, most of the transport experiments have been performed in dilution refrigerators with base temperatures in the $10 - 50$mK range. The measurement techniques are fairly standard, but care must be taken to avoid spurious heating of the electrons in the device. Since it is a small, high resistance object, very small noise levels can cause significant heating. With reasonable precautions (e.g. filtering at low temperature, screened rooms, etc.), effective electron temperatures in the 50–100mK range can be obtained.

It should be noted that other techniques like far-infrared spectroscopy on arrays of dots[64] and capacitance measurements on arrays of dots[65] and on single dots[66] have also been employed. Infrared spectroscopy probes the collective plasma modes of the system, yielding very different information than that obtained by transport. Capacitance spectroscopy, on the other hand, yields nearly identical information, since the change in the capacitance due to electron tunneling on and off a dot is measured. Results from this single-electron capacitance spectroscopy technique will be presented in Section V.

III. TRANSPORT THROUGH QUANTUM DOTS - THEORY

This section presents a theory of transport through quantum dots that incorporates both single electron charging and energy level quantization. We have chosen a rather simple description which still explains most experiments. We follow Korotkov et al.[67], Meir et al.[68], and Beenakker[6], who generalized the charging theory for metal systems to include 0D-states. This section is split up into parts that separately discuss the period of the Coulomb oscillations, III.A; the amplitude and lineshape of the Coulomb oscillations, III.B; the Coulomb staircase, III.C; and related theoretical work, III.D.

A. Period of Coulomb Oscillations

Figure 4(a) shows the potential landscape of a quantum dot along the transport direction. The states in the leads are filled up to the electrochemical potentials μ_{left} and μ_{right} which are connected via the externally applied source-drain

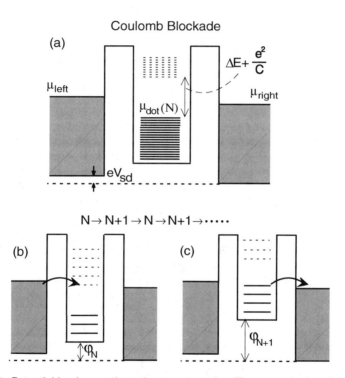

FIGURE 4. Potential landscape through a quantum dot. The states in the 2D reservoirs are filled up to the electrochemical potentials μ_{left} and μ_{right} which are related via the external voltage $V_{sd} = (\mu_{left} - \mu_{right})/e$. The discrete 0D-states in the dot are filled with N electrons up to $\mu_{dot}(N)$. The addition of one electron to the dot would raise $\mu_{dot}(N)$ (i.e. the highest solid line) to $\mu_{dot}(N+1)$ (i.e. the lowest dashed line). In (a) this addition is blocked at low temperature. In (b) and (c) the addition is allowed since here $\mu_{dot}(N+1)$ is aligned by means of the gate voltage with the reservoir potentials μ_{left}, μ_{right}. (b) and (c) show two parts of the sequential tunneling process at the same gate voltage. (b) shows the situation with N and (c) with $N+1$ electrons in the dot.

voltage $V_{sd} = (\mu_{left} - \mu_{right})/e$. At zero temperature (and neglecting co-tunneling[69]) transport occurs according to the following rule: current is (non) zero when the number of available states on the dot in the energy window between μ_{left} and μ_{right} is (non) zero. The number of available states follows from calculating the electrochemical potential $\mu_{dot}(N)$. This is, by definition, the minimum energy for adding the *Nth* electron to the dot: $\mu_{dot}(N) \equiv U(N) - U(N-1)$, where $U(N)$ is the total ground state energy for N electrons on the dot at zero temperature.

To calculate $U(N)$ from first principles is nearly impossible. To proceed, we make several assumptions. First, we assume that the quantum levels can be calculated independently of the number of electrons on the dot. Second, we parameterize the Coulomb interactions among the electrons in the dot and between electrons in the dot and those somewhere else in the environment (as in the metallic gates or in the 2DEG leads) by a capacitance C. We further assume that C is independent of the number of electrons on the dot. This is a reasonable assumption as long as the dot is much larger than the screening length (i.e. no electric fields exist in the interior of the dot). We can now think of the Coulomb

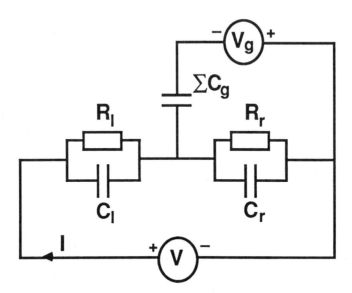

FIGURE 5. Circuit diagram in which the tunnel barriers are represented as a parallel capacitor and resistor. The different gates are represented by a single capacitor ΣC_g. The charging energy in this circuit is $e^2/(C_l + Cr + \Sigma C_g)$.

interactions in terms of the circuit diagram shown in Figure 5. Here, the total capacitance $C = C_l + C_r + C_g$ consists of capacitances across the barriers, C_l and C_r, and a capacitance between the dot and gate, C_g. This simple model leads in the linear response regime (i.e. $V_{sd} \ll \Delta E/e$, e/C) to an electrochemical potential $\mu_{dot}(N)$ for N electrons on the dot[15]:

$$\mu_{dot}(N) = E_N + \frac{(N - N_0 - 1/2)\, e^2}{C} - e\frac{C_g}{C} V_g. \qquad (4)$$

This is of the general form $\mu_{dot}(N) = \mu_{ch}(N) + e\phi_N$, i.e. the *electrochemical* potential is the sum of the *chemical* potential, $\mu_{ch}(N) = E_N$, and the *electrostatic* potential, $e\phi_N$. The single-particle state E_N for the Nth electron is measured from the bottom of the conduction band and depends on the characteristics of the confinement potential. The electrostatic potential ϕ_N contains a discrete and a continuous part. In our definition the integer N is the number of electrons at a gate voltage V_g and N_0 is the number at zero gate voltage. The continuous part in ϕ_N is proportional to the gate voltage and represents the induced polarization charge $q = C_g V_g$ discussed in Section I.A At fixed gate voltage, the number of electrons on the dot N is the largest integer for which $\mu_{dot}(N) < \mu_{left} \cong \mu_{right}$. When, at fixed gate voltage, the number of electrons is changed by one, the resulting change in electrochemical potential is:

$$\mu_{dot}(N+1) - \mu_{dot}(N) = \Delta E + \frac{e^2}{C}. \qquad (5)$$

The *addition* energy, $\mu_{dot}(N+1) - \mu_{dot}(N)$, is large for a small capacitance and/or a large energy splitting, $\Delta E = E_{N+1} - E_N$, between 0D-states. It is important to note that the many-body contribution e^2/C to the energy gap of Eq. (5) exists only at the Fermi energy. Below $\mu_{dot}(N)$, the energy states are only separated by the single particle energy differences ΔE [see Figure 4(a)]. These energy differences ΔE are the *excitation energies* of a dot with constant number N.

A non-zero addition energy can lead to a blockade for tunneling of electrons on and off the dot, as depicted in Figure 4(a), where N electrons are localized on the dot. The $(N+1)th$ electron cannot tunnel on the dot, because the resulting electrochemical potential $\mu_{dot}(N+1)$ is higher than the potentials of the reservoirs. So, for $\mu_{dot}(N) < \mu_{left}, \mu_{right} < \mu_{dot}(N+1)$ the electron transport is blocked, which is known as the *Coulomb blockade*.

The Coulomb blockade can be removed by changing the gate voltage, to align $\mu_{dot}(N+1)$ between μ_{left} and μ_{right}, as illustrated in Figure 4(b) and (c). Now an electron can tunnel from the left reservoir on the dot [since, $\mu_{left} > \mu_{dot}(N+1)$].

The electrostatic increase $e\phi_{N+1} - e\phi_N = e^2/C$ is depicted in Figure 4(b) and (c) as a change in the conduction band bottom. Since $\mu_{dot}(N+1) > \mu_{right}$, one electron can tunnel off the dot to the right reservoir, causing the electrochemical potential to drop back to $\mu_{dot}(N)$. A new electron can now tunnel on the dot and repeat the cycle $N \to N+1 \to N$. This process, whereby current is carried by successive discrete charging and discharging of the dot, is known as *single electron tunneling*, or SET.

On sweeping the gate voltage, the conductance oscillates between zero (Coulomb blockade) and non-zero (no Coulomb blockade), as illustrated in Figure 6. In the case of zero conductance, the number of electrons N on the dot is fixed. Figure 6 shows that upon going across a conductance maximum (a), N changes by one (b), the electrochemical potential μ_{dot} shifts by $\Delta E + e^2/C$ (c), and the electrostatic potential $e\phi$ shifts by e^2/C (d). From Eq. (4) and the condition $\mu_{dot}(N, V_g) = \mu_{dot}(N+1, V_g + \Delta V_g)$, we get for the change in gate voltage ΔV_g between oscillations[69]:

$$\Delta V_g = \frac{C}{eC_g}\left[\Delta E + \frac{e^2}{C}\right]; \tag{6}$$

and for the position of the Nth conductance peak:

$$V_g(N) = \frac{C}{eC_g}\left[E_N + (N - 1/2)\frac{e^2}{C}\right]. \tag{7}$$

For vanishing energy splitting $\Delta E \cong 0$, the classical capacitance-voltage relation for a single electron charge $\Delta V_g = e/C_g$ is obtained; the oscillations are periodic. Non-vanishing energy splitting results in quasi-periodic oscillations. For instance, in the case of spin-degenerate states two periods are, in principle, expected. One corresponds to electrons N and $N+1$ having opposite spin and being in the same spin-degenerate 0D-state, and the other to electrons $N+1$ and $N+2$ being in different 0D-states.

B. Amplitude and Lineshape of Coulomb Oscillations

We now wish to consider the detailed shape of the oscillations and, in particular, the dependence on temperature. We assume that the temperature is greater than the quantum mechanical broadening of the 0D energy levels $h\Gamma \ll k_B T$. We return to this assumption later. We distinguish three temperature regimes:

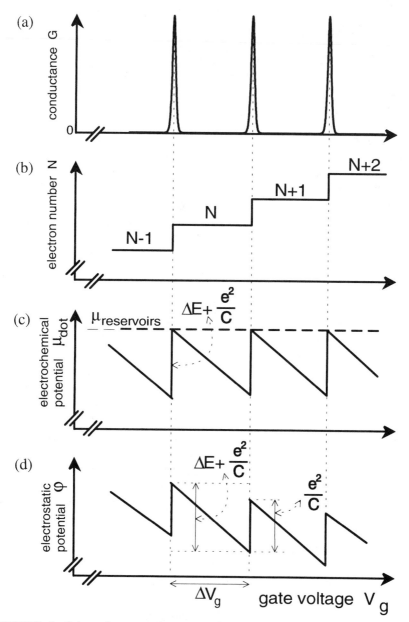

FIGURE 6. Schematic comparison, as a function of gate voltage, between (a) the Coulomb oscillations in the conductance G, (b) the number of electrons in the dot $(N+i)$, (c) the electrochemical potential in the dot, $\mu_{dot}(N+i)$, and (d) the electrostatic potential ϕ.

1) $e^2/C \ll k_B T$, where the discreteness of charge cannot be discerned.

2) $\Delta E \ll k_B T \ll e^2/C$, the *classical* or *metallic Coulomb blockade* regime, where many levels are excited by thermal fluctuations.

3) $k_B T \ll \Delta E \ll e^2/C$, the *quantum Coulomb blockade* regime, where only one or a few levels are relevant.

In the high temperature limit where $e^2/C \ll k_B T$, the conductance is independent of the electron number and is given by the Ohmic sum of the two barrier conductances $1/G = 1/G_\infty = 1/G_{left} + 1/G_{right}$. (Note that this requires equilibration in the dot, which may not occur in the quantum Hall regime[70].) This high temperature conductance G_∞ is independent of the size of the dot and is characterized completely by the two barriers.

The classical Coulomb blockade regime can be described by the so-called "orthodox" Coulomb blockade theory[4,5,20]. Figure 7(a) shows a calculated plot of Coulomb oscillations at different temperatures for energy-independent barrier conductances and an energy-independent density of states. The Coulomb oscillations are visible for temperatures $k_B T < 0.3 e^2/C$ (curve c). The lineshape of an individual conductance peak is given by[6,20]:

$$\frac{G}{G_\infty} = \frac{2\delta/k_B T}{\sinh(\delta/k_B T)} \approx 2\cosh^{-2}\left(\frac{\delta}{2.5 k_B T}\right)$$

$$\text{for } h\Gamma, \Delta E \ll k_B T \ll e^2/C. \tag{8}$$

The parameter δ measures the distance to the center of the peak: i.e. $\delta = e(C_g/C) \cdot |V_{g,res} - V_g|$, with $V_{g,res}$ the gate voltage at resonance. The width of the peaks are linear in temperature as long as $k_B T \ll e^2/C$. The peak maximum G_{max} is independent of temperature in this regime [curves a and b in Figure 7] and equal to half the high temperature value $G_{max} = G_\infty/2$. This conductance is half the Ohmic addition value because of the effect of correlations. Since an electron must first tunnel off before the next can tunnel on, the probability to tunnel through the dot decreases.

In the quantum Coulomb blockade regime, tunneling occurs through a single level. The temperature dependence calculated by Beenakker[6] is shown in Figure 7(b). The single peak conductance is given by:

$$\frac{G}{G_\infty} = \frac{\Delta E}{4 k_B T} \cosh^{-2}\left(\frac{\delta}{2 k_B T}\right) \quad \text{for } h\Gamma \ll k_B T \ll \Delta E, e^2/C, \tag{9}$$

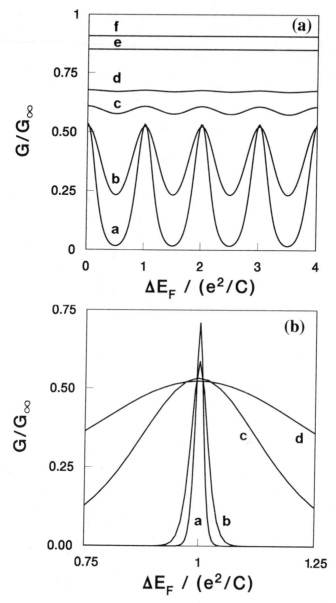

FIGURE 7. Calculated temperature dependence of the Coulomb oscillations as a function of Fermi energy in the classical regime (a) and in the quantum regime (b). In (a) the parameters are $\Delta = 0.01\, e^2/C$ and $k_B T/(e^2/C) = 0.075$ [a], 0.15 [b], 0.3 [c], 0.4 [d], 1 [e], and 2 [f]. In (b) the parameters are $\Delta = 0.01\, e^2/C$ and $k_B T/(\Delta) = 0.5$ [a], 1 [b], 7.5 [c], and 15 [d]. (From van Houten, Beenakker and Staring [7].)

with the assumption that ΔE is independent of E and N. The lineshape in the classical and quantum regimes are virtually the same, except for the different 'effective temperatures'. However, the peak maximum $G_{max} = G_\infty \cdot (\Delta E/4k_B T)$ decreases linearly with temperature in the quantum regime, while it is constant in the classical regime. This distinguishes a quantum peak from a classical peak.

The temperature dependence of the peak height is summarized in Figure 8. On decreasing the temperature, the peak maximum first decreases down to half the Ohmic value. On entering the quantum regime, the peak maximum increases and starts to exceed the Ohmic value. Thus, at intermediate temperatures, Coulomb correlations reduce the conductance maximum below the Ohmic value, while at low temperatures, quantum phase coherence results in a resonant conductance exceeding the Ohmic value.

Above we discussed the temperature dependence of an individual conductance peak and how it can be used to distinguish the classical from the quantum regimes. Comparing the heights of different peaks at a single temperature (i.e. in a single gate voltage trace) can also distinguish classical from quantum peaks. Classical peaks all have the same height $G_{max} = G_\infty/2$. (In semiconductor dots the peak heights slowly change since the barrier conductances change with gate voltage.) On the other hand, in the quantum regime, the peak height depends sensitively on the coupling between the levels in the dot and in the leads. This

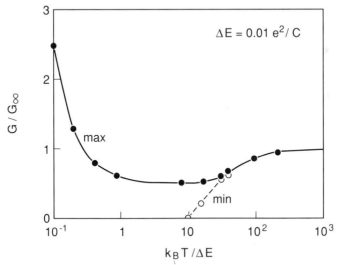

FIGURE 8. Calculated temperature dependence of the maxima and minima of the Coulomb oscillations for $h\Gamma \ll k_B T$ and $\Delta = 0.01 e^2/C$. (From van Houten, Beenakker and Staring[7].)

coupling can vary strongly from level to level. Also, as can be seen from $G_{max} = G_\infty \cdot (\Delta E / 4 k_B T)$, the Nth peak probes the specific excitation spectrum around $\mu_{dot}(N)$ when the temperatures $k_B T \sim \Delta E$[60]. The quantum regime therefore usually shows randomly varying peak heights. An example of such behavior was already seen in Figure 1(b).

An important assumption for the above description of tunneling in both the quantum and classical Coulomb blockade regimes is that the barrier conductances are small: $G_{left,right} \ll (e^2/h)$. This assumption implies that the broadening $h\Gamma$ of the energy levels in the dot due to the coupling to the leads is much smaller than $k_B T$, even at low temperatures. The charge is well defined in this regime and quantum fluctuations in the charge can be neglected (i.e. the quantum proba-bility to find an electron in the dot is either zero or one). For tunneling it means that only first order tunneling has to be taken into account and higher order tun-neling via virtual intermediate states can be neglected[69]. A treatment of the regime $k_B T \sim h\Gamma$ involves the inclusion of higher order tunneling processes. Such complicated calculations have recently been performed[72-74]. For simplicity, we discuss this regime by considering non-interacting electrons and equal bar-riers. Then the zero temperature conductance is given by the well-known Breit-Wigner formula[75]:

$$G_{BW} = \frac{2e^2}{h} \frac{(h\Gamma)^2}{(h\Gamma)^2 + \delta^2} \quad \text{for } T = 0, \; e^2/C \ll h\Gamma, \; \Delta E. \quad (10)$$

The on-resonance peak height (i.e. for $\delta=0$) is equal to the conductance quantum $2e^2/h$; the factor 2 results from spin-degeneracy. The finite temperature conduc-tance follows from $G = \int dE \cdot G_{BW} \cdot (-\partial f/\partial E)$. Although the electron-electron interactions are ignored, it will be shown in the experimental section that Coulomb peaks in the regime $k_B T \sim h\Gamma$ have the Lorentzian lineshape of Eq. (10). However, the peak maximum of $2e^2/h$ in the non-interacting case of Eq. (10) is reduced by Coulomb interactions to e^2/h; i.e. despite spin-degenerate lev-els at zero magnetic field, electrons can only tunnel one-by-one through the dot.

C. Non-linear Transport

In addition to the linear-response Coulomb oscillations, one can obtain infor-mation about the relevant energy scales of the dot by measuring the non-linear dependence of the current on the source-drain voltage V_{sd}. Following the rule that the current depends on the number of available states in the window $eV_{sd} = \mu_{left} - \mu_{right}$, one can monitor changes in the number of available states when increasing V_{sd}. To discuss non-linear transport it is again helpful to distin-guish the classical and the quantum Coulomb blockade regime.

Coulomb Staircase

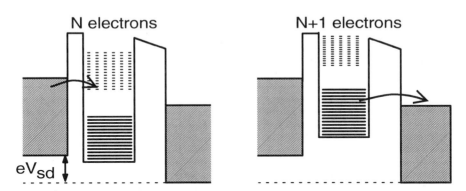

FIGURE 9. Energy diagram to indicate that for larger source-drain voltage V_{sd} the empty states above the Coulomb gap can be occupied. This can result in a Coulomb staircase in current–voltage characteristic [see Figure 2].

In the classical regime, the current is zero as long as the interval between μ_{left} and μ_{right} does not contain a charge state [i.e. when $\mu_{dot}(N) < \mu_{right}$, $\mu_{left} < \mu_{dot}(N+1)$ as in Figure 4(a)]. On increasing V_{sd}, current starts to flow when either $\mu_{left} > \mu_{dot}(N+1)$ or $\mu_{dot}(N) > \mu_{right}$, depending on how the voltage drops across the two barriers. One can think of this as opening a charge channel, corresponding to either the $N \rightarrow (N+1)$ [this example is shown in Figure 9] or the $(N-1) \rightarrow N$ transition. On further increasing V_{sd}, a second channel will open up when *two* charge states are contained between μ_{left} and μ_{right}. The current then experiences a second rise.

For a highly asymmetric quantum dot, where the barriers are unequal, the voltage will mainly drop across one of the barriers. This keeps the electrochemical potential of one of the reservoirs fixed relative to the charge states in the dot, while the electrochemical potential of the other reservoir moves in accordance with V_{sd}. In this asymmetric case, the current changes are expected to appear in the $I - V_{sd}$ characteristics as pronounced steps, the so-called *Coulomb staircase*[3,4,20]. The current steps ΔI occur at voltage intervals $\Delta V \approx e/C$. In a region of constant current, the topmost charge state is nearly always full or nearly always empty depending on whether the reservoir with the higher electrochemical potential is coupled to the dot via the small barrier or via the large barrier, respectively. In the example depicted in Figure 9 the $(N+1)$ charge state is nearly always occupied. Reversing the sign of V_{sd} would leave the $(N+1)$ charge state nearly always empty.

In the quantum regime a finite source-drain voltage can be used to perform spectroscopy on the discrete energy levels[67]. On increasing V_{sd} we can get two types of current changes. One corresponds to a change in the number of charge states in the source-drain window, as discussed above. The other corresponds to changes in the number of energy levels which electrons can choose for tunneling on or off the dot. The voltage difference between current changes of the first type measures the addition energy while the voltage differences between current changes of the second type measures the excitation energies. More of this spectroscopy method will be discussed in the experimental Section IV.B.

D. Overview of other Theoretical Work

The theory outlined above greatly simplifies the way that the electrons in the dot interact with each other and with the reservoirs. We have made two simplifications. First, we have assumed that the coupling to the leads does not perturb the levels in the dot. Second, we have represented the electron-electron interactions by a constant capacitance parameter. This paragraph briefly comments on these crude simplifications and discusses more advanced theories.

A non-zero coupling between dot and reservoirs is included by assuming an intrinsic width $h\Gamma$ of the energy levels. A proper calculation of $h\Gamma$ should not only include direct elastic tunnel events but also tunneling via intermediate states at other energies. Such higher order tunneling processes are referred to as co-tunneling events[69]. They become particularly important when the barrier conductances are not much smaller than e^2/h. Experimental results on co-tunneling have been reported by Geerligs et al.[76] and Eiles et al.[77] for metallic structures and by Pasquier et al.[78] for semiconductor quantum dots. In addition to higher order tunneling mediated by the Coulomb interaction, the effects of spin interaction between the confined electrons and the reservoir electrons have been studied theoretically[79-84]. A quantum dot coupled to reservoirs with a net spin —for instance, when the dot has an odd number of electrons— resembles a magnetic impurity coupled to the conduction electrons in a metal. "Screening" of the localized magnetic moment by the conduction electrons leads to the well-known *Kondo effect*[79-84]. This is particularly interesting since parameters like the exchange coupling and the Kondo temperature should be tunable with a gate voltage. However, given the size of present day quantum dots, the Kondo temperature is hard to reach, and no experimental results have been reported to date. Experimental progress has been made recently in somewhat different systems[85,86].

The second simplification is that we have modeled the Coulomb interactions with a constant capacitance parameter, and we have treated the single particle states as independent of these interactions. More advanced descriptions calculate the energy spectrum in a self-consistent way. In particular, for small electron number $(N < 10)$ the capacitance is found to depend on N and on the particular confinement potential[60-62,87]. In this regime, screening within the dot is poor and the capacitance is no longer a geometric property. It is shown in Section V that the constant capacitance model also fails dramatically when a high magnetic field is applied. Calculations beyond the self-consistent Hartree approximation have also been performed. Several authors have followed Hartree-Fock[88-90] and exact[91,92] schemes in order to include spin and exchange effects in few-electron dots[93]. One prediction is the occurrence of spin singlet-triplet oscillations by Wagner *et al.*[94] of which evidence has been given recently[66,95]. This will be discussed further in Section V.

There are other simplifications as well. Real quantum dot devices do not have perfect parabolic or hard wall potentials. They usually contain many potential fluctuations due to impurities in the substrate away from the 2DEG. Their 'thickness' in the z-direction is not zero but typically 10 nm. And as a function of V_g the potential bottom not only rises, but also the shape of the potential landscape changes as well. Theories virtually always assume effective mass approximation, zero thickness of the 2D gas, and no coupling of spin to the lattice nuclei. In discussions of delicate effects, these assumptions may be too crude for a fair comparison with real devices. In spite of these problems, however, we point out that the constant capacitance model and the more advanced theories yield the same, important, qualitative picture of having an excitation and an addition energy. The experiments in the next section will clearly confirm this common aspect of the different theories.

IV. Transport through Quantum Dots - Experiment

This section presents experiments which can be understood with the theory of the previous section. Section IV.A covers linear response measurements while Section IV.B addresses experiments with a finite source-drain voltage.

A. Linear Response Coulomb Oscillations

Figure 10 shows a measurement of the conductance through a quantum dot of the type shown in Figure 3 as a function of a voltage applied to the center gates[59,96]. As in a normal field-effect transistor, the conductance decreases when the gate voltage reduces the electron density. However, superimposed on this decreasing conductance are periodic oscillations. As discussed in Section III, the oscillations arise because, for a weakly coupled quantum dot, the number of electrons can only change by an integer. Each period seen in Figure 10 corresponds to changing the number of electrons in the dot by one. The period of the oscillations is independent of magnetic field. The peak height is close to

e^2/h. Note that the peak heights at $B=0$ show a gradual dependence on gate voltage. This indicates that the peaks at $B=0$ are classical (i.e. the single electron current flows through many 0D-levels). The slow height modulation is simply due to the gradual dependence of the barrier conductances on gate voltage. A close look at the trace at $B=3.75\,T$ reveals a quasiperiodic modulation of the peak amplitudes. This results from the formation of Landau levels within the dot and will be discussed in detail in Section V. It does not necessarily mean that tunneling occurs through a single quantum level.

The effect of increasing barrier conductances from increases in the gate voltage can be utilized to study the effect of an increased coupling between dot and macroscopic leads. This increased coupling to the reservoirs [i.e. from left to right in Figure 10] results in broadened, overlapping peaks with minima which do not go to zero. Note that this occurs despite the constant temperature during the measurement. In Figure 11 the coupling is studied in more detail in a different, smaller dot where tunneling is through individual quantum levels[97,98]. The peaks in the left part are so weakly coupled to the reservoirs that the intrinsic width is negligible: $h\Gamma \ll k_B T$. The expanded peak in (b) confirms that the

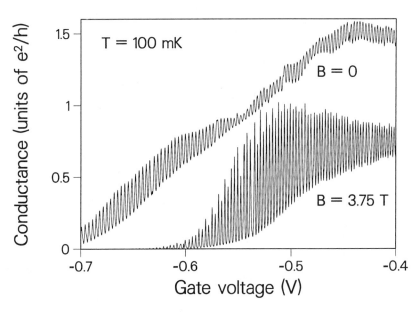

FIGURE 10. Coulomb oscillations in the conductance as a function of center gate voltage measured in the device of Figure 3 at zero magnetic field and in the quantum Hall regime. (From Williamson *et al.*[96]).

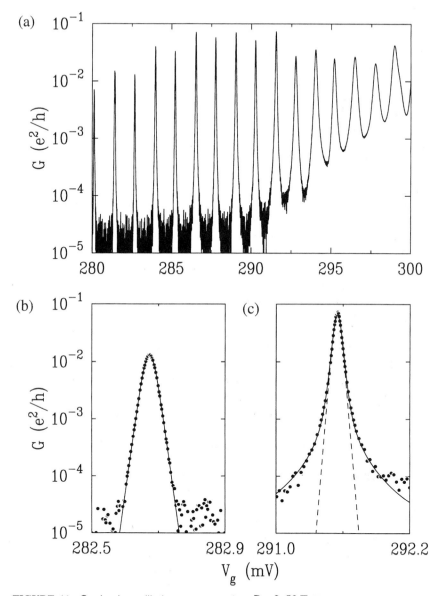

FIGURE 11. Coulomb oscillations measured at $B = 2.53\,T$. The conductance is plotted on a logarithmic scale. The peaks in the left region of (a) have a thermally broadened lineshape as shown by the expansion in (b). The peaks in the right region of (a) have a Lorentzian lineshape as shown by the expansion in (c). (From Foxman *et al.*[97].)

lineshape in this region is determined by the Fermi distribution of the electrons in the reservoirs. On a logarithmic scale, the finite temperature Fermi distribution leads to linearly decaying tails for the peaks. The solid line in (b) is a fit to Eq. (9) with a temperature of 65mK and fit parameter $e^2/C = 0.61$ meV. The peaks in the right part of (a) are so broad that the tails of adjacent peaks overlap. The peak in (c) is expanded from this strong coupling region and clearly shows that the tails have a slower decay than expected for a thermally broadened peak. In fact, a good fit is obtained with the Lorentian lineshape of Eq. (10) with the inclusion of a temperature of 65mK (so, $k_B T = 5.6 \mu eV$). In this case the fit parameters give $h\Gamma = 5 \mu eV$ and $e^2/C = 0.35$ meV. The Lorentzian tails are still clearly visible despite the fact that $k_B T \approx h\Gamma$. The fits also reveal that the charging energy decreases significantly on increasing the coupling to the reservoirs. An important conclusion from this experiment is that for strong Coulomb interaction the lineshape for tunneling through a discrete level is approximately Lorentzian, similar to the non-interacting Breit-Wigner formula (3.6).

Figure 12(a) shows the temperature dependence of a set of selected conductance peaks[68]. These peaks are measured for barrier conductances much smaller than e^2/h where $h\Gamma \ll k_B T$. At approximately 1K, the peak heights increase monotonically, but at low temperature, a striking randomness in peak heights is observed. Moreover, some peaks decrease and others increase on increasing temperature. Random peak behavior is not seen in metallic Coulomb islands. It is due to the discrete density of states in quantum dots. The randomness from peak to peak is usually ascribed to the variations in the nature of the energy levels in the dot. The observed behavior is reproduced in the calculations of Figure 12(b), where variations are included in the form of a random coupling of the quantum states to the leads[68]. This randomness can arise from disorder, or, in clean systems, from chaotic character of electron trajectories inside of dots[99,100].

The temperature dependence of a single quantum peak at $B = 0$ is shown in Figure 13[98]. The upper part shows that the peak height decreases as inverse temperature up to about 0.4K. Beyond 0.4K the peak height is independent of temperature up to about 1K. We can compare the temperature behavior with the theoretical temperature dependence of classical peaks in Figure 7(a) and the quantum peaks in Figure 7(b). The height of a quantum peak first decreases until $k_B T$ exceeds the level spacing ΔE, where around 0.4K it crosses over to the classical Coulomb blockade regime. This transition is also visible in the width of the peak. The lower part of Figure 13 shows the full-width-at-half-maximum (FWHM) which is a measure of the lineshape. The two solid lines differ in slope by a factor 1.25 which corresponds to the difference in "effective temperature" between the classical lineshape of Eq. (8) and quantum lineshape of Eq. (9). At about 0.4K a transition is seen from a quantum to classical temperature dependence, which is in good agreement with the theory of Section III.B.

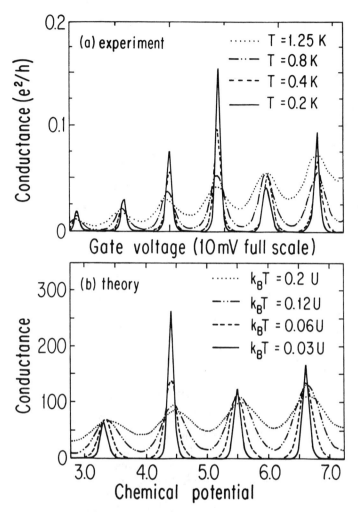

FIGURE 12. Comparison between (a) measured and (b) calculated Coulomb oscillations in the quantum regime for different temperatures at $B = 0$. For the calculation, the level spacing was taken to be uniform: $\Delta = 0.1 U = 0.1 e^2/C$, but the coupling of successive energy levels was varied to simulate both an overall gradual increase and random variations from level to level (From Meir, Wingreen and Lee[68].)

The randomness in peak heights at low magnetic fields reflects that each quantum level has a specific overlap with the levels in the reservoirs. As we discuss in the next section, a high magnetic field suppresses this randomness. In particular,

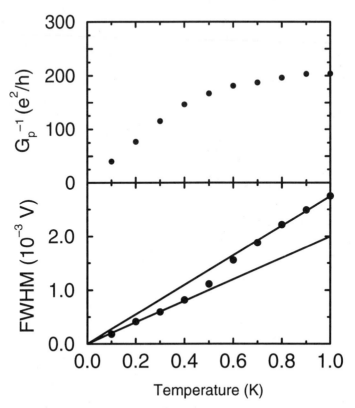

FIGURE 13. Inverse peak height and full-width-half-maximum (FWHM) versus temperature at $B = 0$. The inverse peak height shows a transition from linear T dependence to a
constant value. The FWHM has a linear temperature dependence and shows a transition to a steeper slope. These measured transitions agree well with the calculated transition from the classical to the quantum regime. (From Foxman *et al.*[98].)

when only one Landau level is occupied there is only one type of quantum level at
the Fermi energy (i.e. confined edge states). For such regular quantum levels we
can study the dependence of the conductance on the barrier transmission probabil-
ities. In particular, we can study *coherent resonant tunneling*, which occurs
when a quantum particle of appropriate energy propagates without loss of phase
memory across two barriers. In analogy to an optical Fabry-Perot cavity, multiply
reflected partial waves in an electron cavity can constructively interfere. The total
transmission probability can approach 1, even if each barrier alone is highly
reflecting. At zero temperature, in one dimension, and in the absence of charging
effects, the (spin-resolved) conductance of this interferometer is given by[51]:

$$G_{1D} = \frac{e^2}{h} \frac{t_1 t_2}{1+(1-t_1)(1-t_2)-2[(1-t_1)(1-t_2)]^{1/2}\cos\phi}. \quad (11)$$

Eq. (11) explicitly depends on the barrier transmissions, t_1 and t_2, or via the Landauer relations, $G_1 = t_1 e^2/h$ and $G_2 = t_2 e^2/h$, on the two barrier conductances[101]. The phase, ϕ, is acquired by an electron wave during one round trip between the barriers. For the case $t_1 = t_2 \ll 1$, Eq. (11) reduces to the Lorentzian lineshape of the Breit-Wigner Eq. (10). Finite temperature can be included as described below Eq. (10).

Coherent resonant tunneling is studied[101] in a gate geometry similar to Figure 3 where the barrier transmissions can be varied independently by separate gate voltages. Figure 14(a) shows the measured conductance of the first barrier as a function of gate voltage V_1 with the second barrier fully conducting (i.e. $V_2 = 0$) The measurements were taken at $B = 7T$ such that only one Landau level is occupied. The irregular structure in $G(V_1)$ is probably due to potential fluctuations in or near the barrier. When the second barrier is set in the tunnel regime ($t_2 \approx 0.02$) a dot is formed. Now sweeping V_1 produces the Coulomb oscillations of Figure 14(b). The peak height of the oscillations shows a striking modulation that is correlated with the transmission, t_1, of Figure 14(a), albeit, in a non-classical manner. For example, near $V_1 = -770mV$ and $-850mV$, the peak conductance is strongly suppressed, even though t_1 is at a maximum of 0.6. The classical, sequential tunneling prediction for the conductance maxima is given by[101]: $G_{cl} = (e^2/h)t_1 t_2/(t_1 + t_2 - t_1 t_2)$ and shown by the dashed line in Figure 14(b). The fit to the data is poor, with the measured conductance exceeding the classical prediction by as much as a factor of 15.

In contrast, the peak conductance ($\cos(\phi) = 1$) from the quantum transmission formula of Eq. (11) agrees well with the data, when thermal averaging of 40mK is taken into account [Figure 14(b), solid line]. In a coherent quantum system, the barrier transmissions must match in order to have total transmission well above the sequential value. Since $t_2 \approx 0.02$ in the experiment of Figure 14(b), increasing the transmission of the first barrier above this value actually decreases the total transmission predicted by Eq. (11). This is precisely the effect seen in the data. This experiment demonstrates that despite the strong Coulomb interactions, on-resonance transport is well described by a quantum mechanical wave formula. This is because transport of the Nth electron is an elastic process, so the electron's phase memory is maintained, even though the other $(N-1)$ electrons undergo a large Coulomb energy change. (Note that Meir and Wingreen[102] have theoretically shown that on-resonance transport through an interacting system can be described by a non-interacting Landauer-type formula.) Off-resonance, in the valleys of the oscillations, the fluctuations in the number of electrons in the dot is suppressed by the Coulomb blockade. Here, the electron number N characterizes classical, charged particles. So the oscillations in Figure 14(b) can be interpreted

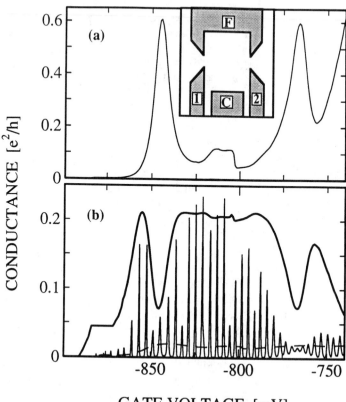

FIGURE 14. (a) Conductance versus V_1, the gate voltage associated with the first barrier, the second barrier fully conducting (i.e. $t_2 = 1$). (b) Coulomb oscillations versus gate voltage V_1 with changes in $t_1 \cdot t_2$ fixed at about 0.02. The maximum peak height calculated with Eq. (11) and an effective temperature of 40mK is shown with a solid line. The dashed line is the classical prediction (see text). The magnetic field is 7 T. (From Johnson *et al.*[101].)

as particle-wave oscillations or, following a recent experiment in superconductors[103], as demonstrating the Heisenberg uncertainty principle $\Delta N \Delta \phi > 1$, meaning that either the electron phase or the electron number is not well-defined.

The conclusions of this paragraph are that quantum tunneling through discrete levels leads to conductance peaks with the following properties: The maxima increase with decreasing temperature and can reach e^2/h in height; the lineshapes gradually change from thermally broadened for weak coupling to Lorentzian for strong coupling to the reservoirs; the peak amplitudes show random variations at low magnetic fields and regular behavior at high magnetic fields.

B. Non-Linear Transport Regime

In the non-linear transport regime one measures the current I (or differential conductance dI/dV_{sd}) while varying the source-drain voltages V_{sd} or the gate voltages V_g. The rule that the current depends on the number of states in the energy window $eV_{sd} = (\mu_{left} - \mu_{right})$ suggests that one can probe the energy level distribution by measuring the dependence of the current on the size of this window.

The example shown in Figure 15 demonstrates the presence of two energies: the *addition energy* and the *excitation energy*[101]. The lowest trace is measured for small $V_{sd}(V_{sd} << \Delta E)$. The two oscillations are regular Coulomb oscillations where the change in gate voltage ΔV_g corresponds to the energy necessary for adding one electron to the dot. In the constant capacitance model this addition energy is expressed in terms of energy in Eq. (5) and in terms of gate voltage in Eqs. (6) and (7). The excitation energy is discerned when the curves are measured with a larger V_{sd}. A larger source-drain voltage leads not only to broadened

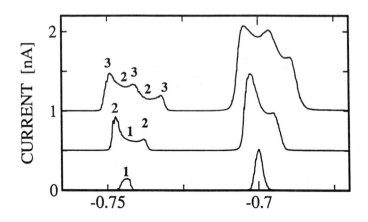

CENTER GATE VOLTAGE [V]

FIGURE 15. Coulomb peaks at $B = 4T$ measured for different source-drain voltage V_{sd}. From the bottom curve up $V_{sd} = 0.1$, 0.4, and 0.7 meV. The peak-to-peak distance corresponds to the addition energy $e^2/C + \Delta E$. The distance between shoulders within a peak corresponds to the excitation energy ΔE for constant number of electrons on the dot. (From Johnson *et al.*[101].)

oscillations, but also to additional structure. Single peaks clearly develop into double peaks, and then triple peaks. Below we explain in detail that a single peak corresponds to tunneling through a single level (when $eV_{sd} < \Delta E$), a double peak involves tunneling via two levels (when $\Delta E < eV_{sd} < 2\Delta E$), and a triple peak involves three levels (when $2\Delta E < eV_{sd} < 3\Delta E$). The spacing between mini-peaks is therefore a measure of the excitation energy with a constant number of electrons in the dot. These measurements yield to an energy splitting $\Delta E \approx 0.3$ meV (i.e. the excitation energy) which is about ten times smaller than the Coulomb charging energy in this structure.

To explain these results in more detail, we use the energy diagrams [104] in Figure 16 for five different gate voltages. The thick vertical lines represent the tunnel barriers. The source-drain voltage V_{sd} is somewhat larger than the energy separation $\Delta E = E_{N+1} - E_N$. In (a) the number of electrons in the dot is N and transport is blocked. When the potential of the dot is increased via the gate

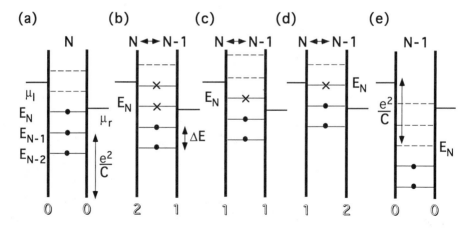

FIGURE 16. Energy diagrams for five increasingly negative gate voltages. The thick lines represent the tunnel barriers. Below the barriers the number of available states for transport are given. Horizontal lines with a bold dot denote occupied 0D-states, E_N. Dashed horizontal lines denote empty states. Horizontal lines with crosses may be occupied. In (a) transport is blocked with N electrons in the dot. In (b) there are 2 states available for tunneling on the dot and 1 state for tunneling off. The crosses in level E_N and E_{N+1} denote that only one of these states can get occupied. In (c) there is just one state that can contribute to transport. In (d) there is one state to tunnel to and there are 2 states for tunneling off the dot. In (e) transport is again blocked, but now with $N-1$ electrons in the dot. (From van der Vaart et al.[104].)

voltage, the 0D-states move up with respect to the reservoirs (see b). The *Nth* electron can now tunnel to the right reservoir. The non-zero source-drain voltage gives two possibilities for tunneling from the left reservoir into the dot. A new electron, which brings the number from $N-1$ back to N, can tunnel to the ground state E_N or to the first excited state E_{N+1}. This is denoted by the crosses in the levels E_N and E_{N+1} which are drawn for having N electrons in the dot. So there are two available states for tunneling onto the dot of which only one will get occupied.

If the potential of the dot is increased further, the electrons from the left reservoir can only tunnel to level E_N as shown in (c). Now, only the ground state level can contribute to the current. Continuing to move up the potential of the dot results in the situation depicted in (d). It is still possible to tunnel to level E_N, but in this case the electrons in both levels E_N and E_{N-1} can tunnel off the dot. Increasing the potential further brings the dot back to the Coulomb blockade (e).

Going through the cycle from (a) to (e) we see that the number of available states for tunneling changes like 0-2-1-2-0. This is also observed in the experiment of Figure 15. The second curve shows two maxima which correspond to cases (b) and (d). The minimum corresponds to case (c). For a source-drain voltage which is somewhat larger than $2\Delta E$ the sequence of contributing states is 0-3-2-3-2-3-0 as the center gate voltage is varied. Here the structured Coulomb oscillation will show three maxima and two local minima. This is observed in the third curve in Figure 15.

Figure 17 shows data related to that in Figure 15, but now in the form of dI/dV_{sd} for different values of V_g[97]. The quantized excitation spectrum of a quantum dot leads to a set of discrete peaks in the differential conductance– a peak in dI/dV_{sd} occurs every time a level in the dot aligns with the electrochemical potential of one of the reservoirs. (Similar observations have been reported on vertical quantum dot structures[42].) Many such traces like Figure 17(a), for different values of V_g, have been collected in (b). Each dot represents the position of a peak from (a) in $V_g - V_{sd}$ space. The vertical axis is multiplied by a parameter that converts V_{sd} into energy[97]. Figure 17(b) clearly demonstrates the Coulomb gap at the Fermi energy (i.e. around $V_{sd} = 0$) and the discrete excitation spectrum of the single particle levels. The level separation $\Delta E \approx 0.1$ meV in this particular device is much smaller than the charging energy $e^2/C \approx 0.5$ meV.

The transport measurements described in this section clearly reveal the addition and excitation energies of a quantum dot. These two distinct energies arise from both quantum confinement and from Coulomb interactions. Overall, agreement between the simple theoretical model of Section III and the data is quite satisfactory. There are other issues that can regulate tunneling through these dots that we have not discussed, however. For example, it was recently argued[105] that the electron spin can result in a spin-blockade. The observation of a negative-differential conductance by Johnson *et al.*[101] and by Weis *et al.*[106] provides evidence for this spin-blockade theory[105]. Other effects will be

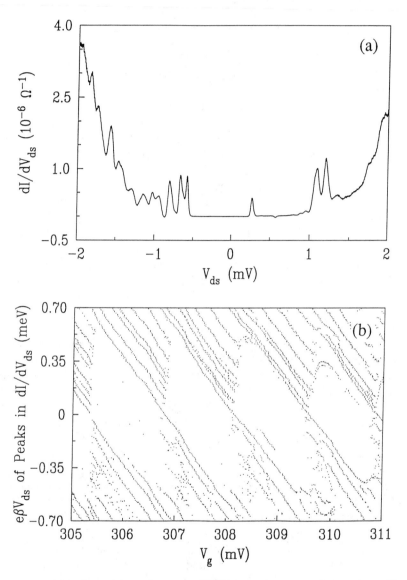

FIGURE 17. (a) Differential conductance dI/dV_{sd} versus source-drain voltage V_{sd} at $B = 3.35T$. The positions in V_{sd} of peaks from traces as in (a) taken at many different gate voltages are plotted in (b) as a function of gate voltage. A factor $e\beta$ has been used to convert peak positions to electron energies. The oscillating Coulomb gap and the discrete excitation spectrum are clearly seen. (From Foxman *et al.*[97]).

discussed at the end of Section V.

V. Quantum Dots in High Magnetic Fields

In this section, we will examine the addition spectra of quantum dots when a magnetic field is applied and compare the experimental results to theoretical predictions. In Section V.A, we address few electron dots, where exact calculations can be performed[91,92]. Interesting predictions, such as singlet-triplet oscillations in the spin state of the 2-electron dot[94], are compared with experiment. In section V.B., we discuss many-electron dots, where spectra are more difficult to calculate exactly. At low magnetic fields, the B-dependence of the energy spectrum is quite complex, due to both the Coulomb interactions and the complicated nature of the electron trajectories inside the dot. This regime is discussed in detail in the chapters on chaos in nanostructures[107]. In Section V.B, we concentrate instead on high magnetic fields, where the quantization of the electron orbits into Landau levels is important. Finally, Section V.C discusses open questions.

A. Few-electron Dots at High Magnetic Fields

The simplest model of a quantum dot consists of noninteracting electrons residing in a parabolic confining potential. The classical motion is then a periodic oscillation with a characteristic frequency ω_o. The addition of a magnetic field alters the motion, leading to orbits of the type shown in Figure 18. An electron at the center of the dot rotates in a cyclotron-like orbit, which becomes the cyclotron frequency $\omega_c = eB/m^*$ at high magnetic fields, where m^* is the effective mass in *GaAs*. Electrons away from the center slowly precess around the dot as they perform their cyclotron motion. This is due to the drift velocity $\mathbf{v_D} = \mathbf{ExB}$ of the cyclotron orbit in the electric field of the confinement potential. Quantum mechanically, this model can be easily solved[108,109]. At high magnetic fields ($\omega_c \gg \omega_o$) the expression simplifies to:

$$E(n,m,S_z) = (n + 1/2)\hbar\omega_c + (m + 1/2)\hbar\omega_o^2/4\omega_c + g\mu_B BS_z. \quad (12)$$

where $n = 0,1,2,...$ is the radial or Landau level (*LL*) index, $m = 0,1,2,...$ labels the angular momentum of the drifting cyclotron orbit, and $S_z = \pm 1/2$ is the spin index. Roughly speaking, the *LL* index n labels the number of magnetic flux quanta h/e enclosed by the electron orbit during its cyclotron motion, while m labels the number of flux quanta enclosed by the drifting orbit. Since each successive m-state encloses one more flux quantum, each (spin-resolved) *LL* within the dot can be occupied by one electron per flux quantum penetrating the area of

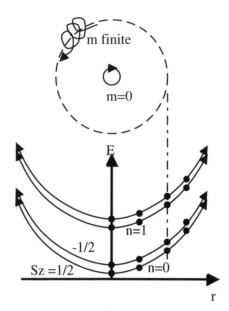

FIGURE 18. Top: Classical electron orbits inside a parabolically confined quantum dot. The orbit in the center exhibits cyclotron motion, while the orbit away from the center also drifts in the electric field of the confining potential. Bottom: Schematic energy level diagram of a quantum dot in a high magnetic field. The $n=0$ and $n=1$ orbital *LLs* are shown, each of which is spin-split. The dots represent quantized states within a *LL* that encircle m flux quanta in their drifting cyclotron motion.

the dot. Increasing B causes both types of orbits to shrink in order to encircle the same number of magnetic flux quanta, making more states fit in the same area and increasing the *LL* degeneracy.

Equation (12) ignores electron-electron interactions. Nevertheless, it should be valid for the first electron occupying a dot, since there are no other electrons with which to interact. The solution from Eq. (12) with $n=m=0$ should thus describe the ground-state addition energy of the first electron. At $B=0$ this is the zero-point energy of the harmonic oscillator, $\hbar\omega_o/2$. At high B it is the energy of the lowest *LL* $\hbar\omega_c/2$, which grows linearly with B. An electric to magnetic field crossover occurs when $\omega_c \approx \omega_o$.

Measuring a one-electron dot in the lateral gated geometry has proven to be difficult. Vertical dots with as few as one electron have been studied, however, by both nonlinear I-V measurements[42,95] and by capacitance spectroscopy[66]. Results from the latter technique are shown in Figure 19 taken from Ashoori, *et al.*[66]. The change in the capacitance due to a single electron tunneling on

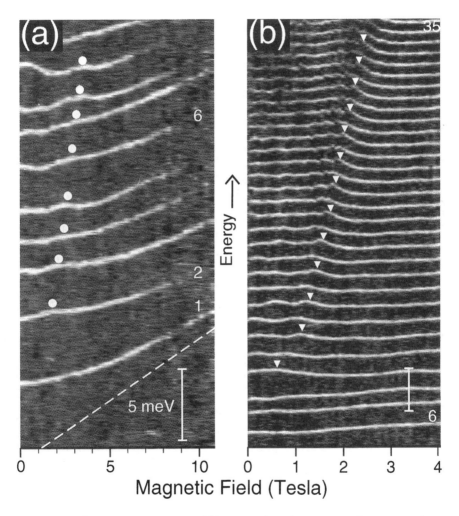

FIGURE 19. Grey scale plot of the addition energies of a quantum dot measured as a function of magnetic field. Each successive light colored line corresponds to the energy for adding an additional electron to the dot. (a) Addition spectrum for the first few electrons. The dot on the curve for the second added electron marks the singlet-triplet transition discussed in the text. (b) Addition spectrum for 6 through 35 electrons. The triangles mark the filling factor $v = 2$, (From Ashoori *et al.*[66]). (See color plate.)

and off a dot is plotted in grey scale as a function of gate voltage along the y-axis and magnetic field along the x-axis. The first line at the bottom of Figure 19 represents the addition energy for the first electron as a function of B. The addition energy is constant for low B and grows linearly for high B. Fitting to Eq. (12) allows the determination of the bare harmonic oscillator frequency: $\hbar\omega_0 = 5.4$ meV.

The situation gets more interesting for more than one electron on the dot. To describe the addition energy for larger numbers of electrons, the simplest approach is to use the non-interacting electron spectrum, Eq. (12), combined with the Coulomb-blockade model for the interactions. This model is discussed in Sections III and IV. In this approximation, the second electron would also go into the $n = m = 0$ state, but with the opposite spin, creating a spin singlet state. This spin singlet state remains the ground-state configuration until the Zeeman energy is large enough to make it favorable for the second electron to flip its spin and occupy the $n = 0, m = 1$ state. From Eq. (12), this occurs when $\hbar\omega_0^2/4\omega_c = g\mu_B B$. The 2 electron ground state is then an $S_z = 1$ spin-triplet state. For *GaAs* the spin splitting is quite small ($g = -0.4$), and the Zeeman-driven singlet-triplet transition would occur at a very large B of around $25T$ for the dot in Figure 19. The data, however, shows something quite different. The addition energy for the second electron has a feature at a much lower field (marked by a dot) that has been attributed to the singlet-triplet transition[66].

A more realistic model of the Coulomb interactions can explain this discrepancy[94]. The size of the lowest spatial state (i.e. $n = 0, m = 0$) shrinks in size with increasing B. As a result, the Coulomb interaction between the two spin-degenerate electrons grows. At some point, it becomes favorable for the second electron to occupy the $m = 1$ single-particle state, avoiding the first electron and reducing the Coulomb interaction energy. The dot thus switches to a triplet state. This transition is driven mostly by Coulomb interactions, since the spin splitting is still quite small.

Many other features are also observed in the addition energies of the first few electrons as a function of B, as seen in Figure 19(a). These features can also be interpreted by comparison with microscopic calculations[110]. The agreement between experiment and theory is far from perfect, however, which indicates the need for further study.

B. Many-electron Dots in the Quantum Hall Regime

At larger number of electrons on the dot ($N > 20$), the capacitance spectroscopy measurements begin to show very organized behavior. This large N regime has been extensively explored by transport spectroscopy in lateral structures[111,112]. An example is shown in Figure 20, where the addition energy for the *Nth* electron ($N \sim 50$) is measured as a function of B[112]. This plot is made

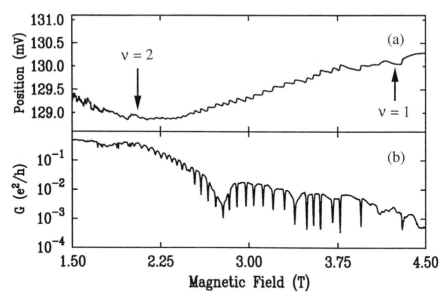

FIGURE 20. (a) Position in gate voltage and (b) peak height of a conductance peak measured as a function of magnetic field. The filling factors, v, in the dot are as marked. The quasi-periodic structure reflects single-electron charge rearrangements between the two lowest *LLs*. (Adapted from McEuen, *et al.*[112].)

by measuring a Coulomb oscillation and plotting the position in gate voltage [Figure 20(a)] and height [Figure 20(b)] of the peak as a function of B. The behavior is very regular in the regime between 2 T and 4 T. The peak positions drop slowly, and then rise quickly, with a spacing between rises of approximately 60 mT. At the same time that the peak position is rising, the peak amplitude drops suddenly. Regularities can also be seen in the peak amplitudes measured at a fixed B, but with changing V_g, i.e. for adding successive electrons. For example, the data presented in Figure 11 is a plot of a series of peaks in the ordered regime above 2 T[112]. A close examination reveals that the peak heights show a definite modulation with a period of every-other peak.

To understand these results, a theoretical model of the many-electron dot is needed. Unfortunately, for dots containing more than ~10 electrons, exact calculations cannot easily be performed and approximation schemes must be used. Again, the simplest approach is to assume the electrons fill up the noninteracting electron states, given by Eq. (12), and to use the Coulomb blockade model to describe the Coulomb interactions[7,111]. This model was used to interpret early experiments[111], but later work showed it to be seriously inadequate[112], for essentially the same reasons that we discussed above for the two-electron dot. In a high magnetic field, Coulomb interactions cause rearrangements among the states that cannot be understood from the behavior of non-interacting levels.

An improved description of the addition spectrum treats the Coulomb interactions in a self-consistent manner[112-115]. This proto-Hartree approach is essentially the Thomas-Fermi model, but with the *LL* energy spectrum replacing the continuous density of states that is present at $B = 0$. In this model, one views the quantum dot as a small electron gas with a nonuniform electron density. Classically, this density profile would be determined by the competition between the Coulomb interactions and the confinement potential. For example, for a parabolic confinement potential, the result is an electron density that is maximal at the center and decreases continuously on moving away from the center, as shown in Figure 21(a).

We now include the effects of Landau level quantization in this picture. In a first approximation, the electrons fill up the requisite number of Landau levels to yield the classical electrostatic distribution. For simplicity, we concentrate exclusively on the case where only two *LL*s are occupied ($n = 0$; $S_z = \pm 1/2$). This is shown in Figure 21(a). Note, however, that the states in the second (upper) *LL* have a higher spin energy than those in the first (lower) *LL*. As a result, some of these electrons will move to the lower *LL*. This continues until the excess electrostatic energy associated with this charge re-distribution cancels the gain from lowering the *LL* energy. The resulting (self consistently determined) charge distribution for the island is shown in Figure 21(b), and the electrochemical potentials for electrons added to the 2 *LL*s are shown in Figure 21(c). Note that partial occupation of a *LL* implies that there are states at the Fermi energy available to screen the bare potential. If we assume perfect screening, then the resulting self-consistent potential is flat. This is analogous to the fact that in the interior of a metal no electric fields are present. For example, in the center of the island, where the second *LL* is partially occupied the self-consistent electrostatic potential is flat. Similarly, near the edge, where the first *LL* is partially occupied, the potential is also flat. In between, there is an insulating region where exactly one *LL* is occupied.

The result is that we have two metallic regions, one for each *LL*, separated by an insulating strip. Electrons added to the dot will be added to one of these two metallic regions. If the insulating strip is wide enough, tunneling between the two metallic regions is minimal; they will effectively act as two independent electron gases. The charge is separately quantized on each *LL*. Not only is the total number N of electrons in the dot an integer, but also the numbers of electrons N_1 in LL_1 and N_2 in LL_2 are integers. In effect, we have a two-dot, or "dot-in-dot" model of the system.

This schematic picture of a quantum dot in high magnetic fields is supported by a number of simulations[112-116]. Figure 22 shows a contour map of the electrostatic potential for a quantum dot with two occupied *LL*s, as calculated by

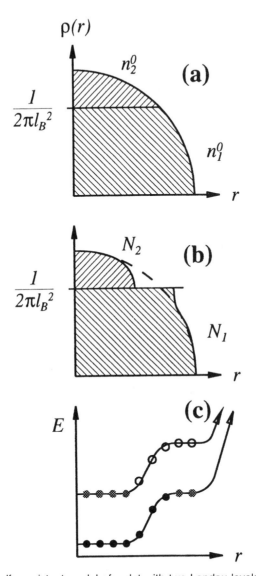

FIGURE 21. A self-consistent model of a dot with two Landau levels occupied. (a) Filling of the *LLs* that would yield the classical electrostatic charge distribution. (b) Electrons redistribute from the higher to the lower *LL* to minimize their *LL* energy. (c) Resulting self-consistent level diagram for the dot. Solid circles: fully occupied *LL* – an "insulating" region. Shaded circles: partially occupied *LL* – a "metallic" region. (Adapted from McEuen, *et al.*[112].)

M. Stopa[116]. In the center of the dot (region #2) where the second *LL* is partially occupied, the potential is flat. Similarly, the first *LL* creates a ring of constant potential where it is partially occupied (region #1). Electrons tunneling onto
the dot will go to either one of these metallic regions.

We now discuss the implications of this model for transport measurements.
First, as additional electrons are added to the dot, they try to avoid each other. As
a result, successively added electrons tend to alternate between the two metallic
regions. Note, however, that electrons will most likely tunnel into the outer *LL*
ring, as it couples most effectively to the leads. Peaks corresponding to adding

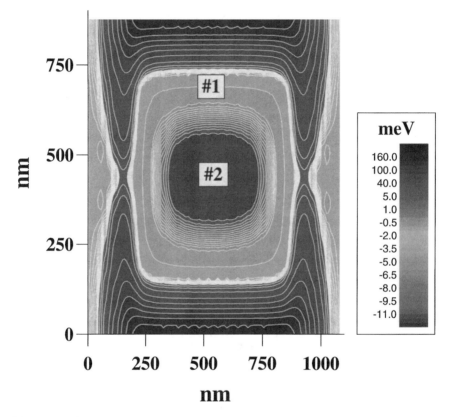

FIGURE 22. Contour plot of the self-consistent electrostatic potential for a quantum dot
in a high magnetic field. In the regions labeled #1 and #2, the first and second *LLs* are
partially occupied. The electrons can thus rearrange themselves to screen the external
potential, and the resulting self-consistent potential is constant. In between, where one
LL is fully occupied and no screening occurs, the potential rises sharply. (From M.
Stopa[116].)

an electron to the inner *LL* should thus be smaller. If electrons are alternately added to the inner and outer *LLs* with increasing gate voltage, the peaks should thus alternate in height. The measurements of Figure 11 show this behavior. Measurements[117] for higher numbers of *LLs* occupied give similar results (i.e. a periodic modulation of the peak amplitudes), with a repeat length determined (approximately) by the number of *LLs* occupied[104,118].

To understand the peak-position structure in Figure 20(a), we again note that, as B increases, the electrons orbit in tighter circles to enclose the same magnetic flux. In the absence of electron redistribution among the *LLs*, the charge density therefore rises in the center of the dot and decreases at the edges. This bunching causes the electrostatic potential of the second *LL* to rise and that of the first *LL* to drop. Therefore, the energy for adding an electron to the first *LL*, $\mu_1(N_1, N_2)$, and hence the peak position, decreases with increasing B. This is illustrated schematically in Figure 23. This continues until it becomes energetically favorable for an electron to move from the second to the first *LL*. This electron redistribution —which we call the internal Coulomb charging— causes the electrostatic

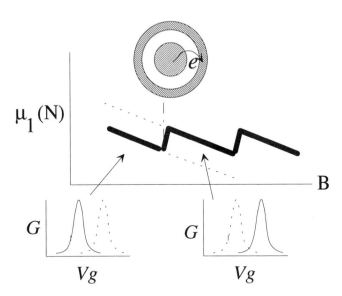

FIGURE 23. Schematic illustration of charge redistribution within a dot with increasing magnetic field. When a single electron moves from the 2nd to the 1st *LL*, the electrochemical potential for adding an additional electron to the 1st *LL* increases. As a result, the peak position shifts.

potential of the first *LL* to jump from $\mu_1(N_1, N_2)$ to $\mu_1(N_1+1, N_2-1)$ with $N = N_1 + N_2$. These jumps are clearly observable in the data of Figure 20, occurring every 60 *mT*. Note that these electron redistributions are a many-electron version of the two-electron singlet-triplet transition. In both cases, Coulomb interactions push electrons into states at larger radii with increasing *B*.

The peak height data shown in Figure 20(b) can be similarly explained. The peak amplitude for adding the *Nth* electron is strongly suppressed at *B* fields where it is energetically favorable to add the electron to the inner *LL*. This corresponds to the magnetic field where the peak position is rising. A dip in the peak amplitude thus occurs at every peak position where an electron is transferred from the second to the first *LL*. The period of the oscillation, 60 mT, roughly corresponds to the addition of one flux quantum to the area of the dot. This period implies an area of $(0.26 \ \mu m)^2$, a size which is consistent with the dimensions of the dot.

Care must be taken in interpreting the peak height, however. Other experiments show[111,117,119] that the heights of the smaller peaks do not directly reflect the tunneling rate into the inner *LL*. The tunneling rate into the inner *LL* is typically too small to produce a significant current. The observed peak is actually due to thermally-activated transport through the outer (first) *LL*. Since all of the observed current corresponds to tunneling through the first *LL*, the position of a peak is proportional to the electrochemical potential $\mu_1(N_1, N_2)$ for adding an electron to the first *LL*. This potential is a function of both N_1 and N_2, the number of electrons in the first and second *LLs* respectively.

The jumps in the peak position with increasing *B* thus represent a redistribution of electrons between the *LLs*. In a recent experiment by van der Vaart, *et al.*[119], the peaks were actually observed to jump back and forth *in time*. This is shown in Figure 24. Figure 24(a) shows that with two *LLs* occupied a peak that corresponds to *N* electrons in the dot can appear as a double peak. The double peak has a resonance when either $\mu_1(N_1, N_2)$ or $\mu_1(N_1+1, N_2-1)$ aligns with the Fermi energy of the reservoirs. Figure 24(b) shows that the conductance measured as a function of time at a fixed gate voltage switches between two discrete levels. This peak-switching is due to a single electron hopping between the inner and the outer *LL*. At this magnetic field, the time for hopping was on the order of 10 seconds. The tunneling rate between the inner and outer dot is thus incredibly small. This corroborates the point made earlier that the coupling to the inner Landau level is very weak and all of the measurable current is carried by tunneling through the outer *LL*.

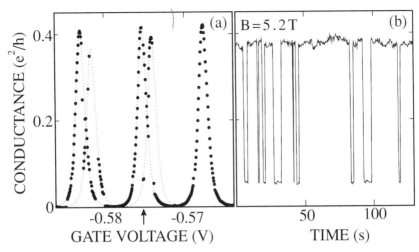

FIGURE 24. (a) Conductance through a quantum dot as a function of gate voltage, measured in a regime where 2 *LLs* are occupied inside the dot. The Coulomb peaks are observed to switch back and forth between two positions. (The dotted lines are a guide to the eye.) (b) Conductance versus time with the gate voltage fixed at the value denoted by the arrow in (a). The switching behavior results from the hopping of a single electron between the 1st and 2nd *LLs*. (From van der Vaart, *et al.*[119].)

C. Open Questions

The model and experiments discussed above indicate that much of the behavior of quantum dots in magnetic fields can be understood based on *LL* quantization and self-consistent electrostatics. Many open questions remain, however. The first is the effect in many-electron dots of Coulomb interactions beyond the Thomas-Fermi approximation. For example, Hartree-Fock models[120,121] yield short-range attractive interactions between electrons of the same spin that lead to larger incompressible regions than in the model above. At higher B, these models predict exotic effects such as an edge reconstruction where the charge density no longer monotonically decreases with increasing radius. Recent experiments by Klein, *et al.*[122] provide evidence for these effects, but more work remains to be done.

Also of potential interest are many-body effects on the tunneling rates of single electrons on and off the dot. If tunneling on the dot requires a complex rearrangement of all other electrons, its rate is predicted to be dramatically suppressed[123]. This "orthogonality catastrophe" may be contributing to the extremely slow tunneling rates between the inner and outer *LL* regions found in the experiment of Figure 24. More experiments are necessary to fully explore these possibilities.

VI. AC TRANSPORT THROUGH QUANTUM DOTS

DC measurements of quantum dots have yielded a remarkable variety of interesting and unexpected phenomena. AC measurements, although still in their infancy, are proving to be similarly exciting. In this section, we describe some early results on AC transport in dots. In Section VI.A, we review the important time scales and discuss the relevant regimes of operation. In Section VI.B, we describe experiments where RF signals are used to move electrons through a dot one at a time – the single-electron *turnstile*[124,13]. In Section VI.C, we discuss experiments at microwave frequencies where *photon − assisted tunneling* on and off a dot is observed[125,126]. Finally, we briefly discuss our theoretical understanding of these systems and look toward the future.

A. Time Scales and Regimes of Operation

Transport processes through a quantum dot encompass a variety of time scales. A listing of these energy/frequency scales is given in Table 1. The first is the single-particle level spacing ΔE, which is about $0.02 - 0.2$ meV for typical lateral dots (see previous sections). The second is the charging energy e^2/C, which is typically $0.2 - 2$ meV. The effects of thermal broadening of the electron energies ($\sim 4k_B T$) determine the observability of these two energy scales. The other characteristic times of the dot are transport times. The first is Γ, the typical time required to tunnel on or off of the dot. This time is set by the transmission coefficient $|t|^2$ of the barriers and by the larger of ΔE or V_{sd}. It can be arbitrarily

TABLE 1. A list of the important energy/frequency scales for transport through quantum dots.

Quantity	Equivalent frequency	Typical parameters
Thermal broadening	$\sim 4 k_B T/h$	10 GHz (at $T = 125$ mk)
Tunneling rate on/off the dot	$\Gamma \sim ([\Delta E \text{ or } eV]/h) \, |t|^2$	0 – 10 GHz
Level spacing (transit time)	$\Delta E/h$	4 GHz – 40 GHz
Charging energy	$(e^2/C)/h$	40 GHz – 400 GHz
Tunneling time	$1/\tau_{tunnel} \sim v_{barr}/L_{barr}$	200 GHz – 1 THz

small for opaque tunnel barriers. The other time scale is the tunneling time, i.e. the actual time spent during tunneling through the barrier. The meaning of such a time is a subject of much controversy[127]. This time is quite fast (~2 ps) for typical barriers (calculated with the Büttiker-Landauer theory[127].)

To access these time scales, AC signals can be applied to the dot, and the effects on the DC characteristics measured. In the experiments reported to date[124-126,13], this has been done by using a capacitor near the sample to couple a high frequency signal from a coaxial cable onto one (or more) of the gate leads. The frequencies of the applied signals have varied from RF (f~1 MHz) to microwave (f~40 GHz). In these experiments, the level spacing of the dots is not an important parameter, and will be neglected in this section.

Ignoring the level spacing, transport can be divided into the regimes shown in Table 2. The first issue is whether the frequency is larger or smaller than the tunneling rate Γ. If $f \ll \Gamma$ then the electrons see an essentially static potential and we are in the adiabatic regime. If $f \gg \Gamma$ then the electron experiences many cycles of the AC signal while it is on the dot. We refer to this as the non-adiabatic regime. The second issue is whether the photon energy hf is greater or less than the thermal smearing of $4k_B T$. If $hf < 4k_B T$, single photon processes are masked by thermal fluctuations, and a classical description is appropriate. If $hf > 4k_B T$, on the other hand, single-photon processes should be observable – we refer to this as the quantum regime. To describe this quantum regime one needs to solve the time-dependent Schrödinger equation for the tunneling electron. In this section, we will describe the three experiments listed in Table 2.

We start by briefly considering the simplest case, the classical adiabatic regime. In this regime, the device behavior can be understood entirely by the DC characteristics. Figure 25 shows an experiment in this regime[128]. Coulomb blockade oscillations are measured when, in addition to the DC voltages, an AC gate voltage of $f = 10MHz$ is applied to one of the gates. The different curves correspond to different amplitudes of the AC signal. To understand these results we note that the AC voltage simply modulates the electrostatic potential of the dot sinusoidally. The result is a Coulomb peak that is, in effect wiggled back and

TABLE 2. Regimes of AC operation of a quantum dot with negligible level spacing. Included in the table are the three experiments discussed in this section.

	Adiabatic ($f \ll \Gamma$)	Non-Adiabatic ($f \gg \Gamma$)
Classical ($hf \ll kT$)	(A) Classical Adiabatic wiggle	(B) Turnstile
Quantum or **Time-dependent** ($hf \gg k_B T$)		(C) Photon- Assisted Tunneling

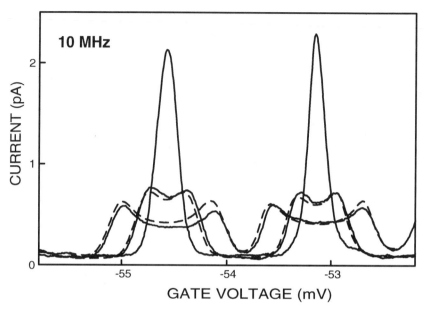

FIGURE 25. Coulomb oscillations of a dot measured with a low frequency (10 MHz) AC signal applied to one of the gates. The narrow peak is for no AC signal; the broadened peaks are for two different AC amplitudes. The dotted line is the expected "adiabatic" result, obtained by convolving the DC data with a sinusoidal AC gate voltage (From Jahuar, *et al.*[128].)

forth by an amount proportional to the amplitude of the AC signal. Figure 25 also shows the expected current (dashed lines), obtained by convolving the DC I-V_g characteristic with a sinusoidal AC signal. The agreement between theory and experiment is excellent. Note that the amplitude of the current at any given V_g is proportional to the time that the oscillating gate voltage spends at that value of V_g. Since a sine wave spends most of its time near its extrema, the result is a broadened current peak that is maximal at its edges.

Low frequency AC voltages can also be applied across the source and drain. The non-linear I-V characteristic of the dot means that the AC voltage will be rectified, with the sign of the rectified current depending on the gate voltage[129].

These experiments demonstrate that RF signals can be coupled onto a quantum dot. However, experiments in this regime are essentially trivial extensions of the DC transport measurements. More interesting possibilities present themselves in the non-adiabatic and quantum regimes, which are discussed in the following two sections.

B. The Single-Electron Turnstile

If the barrier height, and hence the tunneling rate Γ, is made to oscillate by the AC signal, the device can cross over from the adiabatic to the non-adiabatic regime within a given cycle of the AC potential. This is the basis for the quantum dot turnstile[124,130], a device that moves one electron through the dot per AC cycle[131]. The operation of the turnstile is illustrated in Figure 26. Two AC signals are used, with one coupled to the gate controlling the left barrier and the other coupled to the gate controlling the right barrier. The AC voltages applied to these two barriers are 180 degrees out of phase with each other. If the device is perfectly symmetric, the electrostatic potential of the dot is unaffected – the two gates shift the dot potential by equal and opposite amounts. Only the barrier

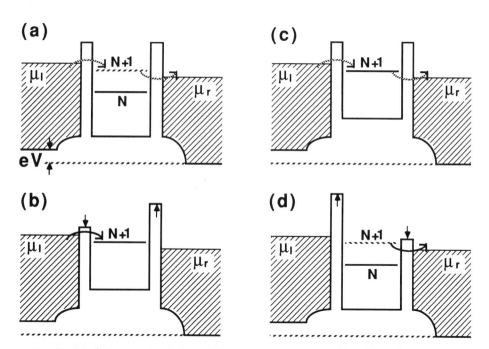

FIGURE 26. Schematic illustration of the four stages of operation of the single-electron turnstile. RF voltages applied to the gates are used to modify the tunneling rates through the barriers. Dashed arrows indicate that the tunneling probability is low; solid arrows indicate that it is high. (a) N electrons occupy the dot. (b) The left barrier is lowered, allowing one electron to tunnel onto the dot. (c) The left barrier is raised; the dot contains $N + 1$ electrons. (d) The right barrier is lowered, allowing the electron to tunnel off. The net result is one electron transferred through the dot per RF cycle. (From Kouwenhoven, *et al.*[124].)

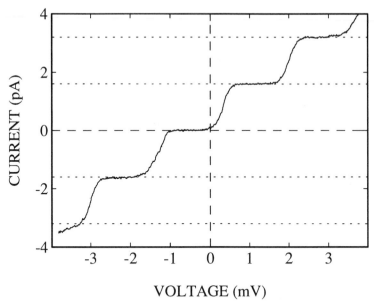

FIGURE 27. Current versus source-drain voltage for a single-electron turnstile operating at 10 MHz. Current steps at multiples of $ef = 1.6$ pA are observed, corresponding to integer numbers of electrons tunneling through the dot per RF cycle. (From Nagamune, *et al.*[13].)

heights, and hence the tunneling rates on and off of the dot, oscillate.

For the operation of the turnstile, the barriers of the dot are adjusted so that the transmission coefficient is very small in the absence of the RF signal. The RF amplitude is adjusted so that tunneling readily occurs during the (negative) peak of the RF signal, and is negligible otherwise. Consider the case where a source-drain bias is as shown in the figure. Energetically, a single electron can tunnel onto the dot from the left lead, and off of the dot to the right. The left barrier is lowered allowing an electron to tunnel onto the dot. The right barrier is high, however, preventing the electron from tunneling off during this part of the cycle. An electron is thus transferred onto the dot. In the second half of the cycle, the right barrier is lowered and the electron tunnels off. The net result is that one electron is transferred through the dot per RF cycle - and a current of $I = ef$ results.

Figure 27 shows an experiment measuring I versus V_{sd} for a dot with oscillating barriers as described above[13]. In this experiment, $f = 10 MHz$, leading to a one-electron-per-cycle current of 1.6 pA. At small V_{sd}, the current is quantized very close to this value. At larger source-drain biases, an integral multiple number of electrons can tunnel onto the dot per cycle and additional plateaus are observed at multiples of 1.6 pA. Superficially, the trace resembles a Coulomb staircase [see Figure 2], but here the height of the plateaus is set by the RF frequency, and not by the tunnel barrier resistances.

Other possibilities for controlling electron flow also exist. If the RF amplitudes are unequal, or if they are not 180 degrees out of phase, the potential of the dot also changes during the RF cycle. In this case, "pumping" of electrons through the dot can occur; the AC signal can transfer electrons through the dot without an applied DC bias voltage, or even against a DC voltage[124].

C. Photon-Assisted Tunneling

At higher frequencies, the photon energy becomes important and we move into the quantum regime. In this regime the effect of the AC potential on electron tunneling can be described in terms of the absorption and emission of photons[132]. For example, an electron may be able to tunnel onto the dot by the absorption of a photon, as is shown in Figure 28. These photon-assisted tunneling

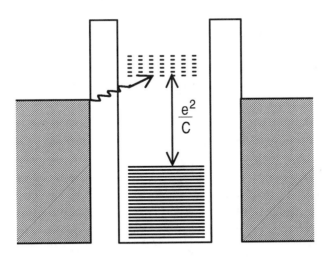

FIGURE 28. Energy level diagram for a quantum dot illustrating photon assisted tunneling. Solid lines are occupied levels, while dotted lines are unoccupied. Shown is the configuration just before a peak in G versus Vg. In the presence of microwaves an electron can overcome the Coulomb gap and tunnel onto the dot via the absorption of a photon, leading to a shoulder on the Coulomb peak (From Kouwenhoven, et al.[126].)

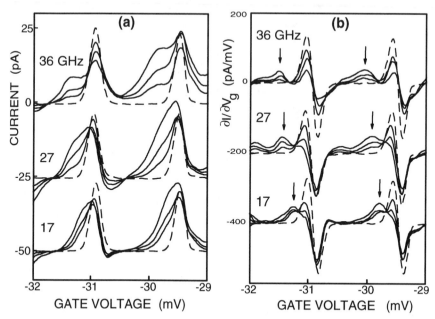

FIGURE 29. (a) Current versus gate voltage V_g for a quantum dot illuminated by microwaves at 3 different frequencies. The dashed curves are without microwaves and the solid lines are for increasing power. A photon-induced shoulder is observed whose position is independent of power but linearly dependent upon frequency. (b) Derivatives dI/dV_g of the data in (a). The arrows indicate the photon-induced features. (From Kouwenhoven, *et al.*[126].)

(PAT) processes strongly affect the DC currents in a quantum dot, as we see below.

Figure 29(a) shows the effect of microwave photons on the Coulomb oscillations. The current without microwaves (dotted) is displayed along with the current in the presence of microwaves at various powers. Results for three different frequencies are shown. The most notable feature is the presence of a shoulder on the left-hand side of the Coulomb-blockade peak. Figure 29(b) shows the derivative, dI/dV_g, of the data in 29(a). The shoulder in Figure 29(a) results in a peak in dI/dV_g in Figure 29(b), as is indicated by the arrows. The position of this peak is independent of the microwave power, but shifts with photon frequency. This shoulder / peak is due to PAT onto the dot, as is schematically illustrated in Figure 28. An electron absorbs a photon when tunneling onto the dot, producing extra current up to hf away from the Coulomb-blockade peak. The fact that the position of this shoulder is independent of the microwave amplitude, but linear in the frequency, is a direct proof of the single-photon origin.

The microwaves can lead to a negative current on the right hand side of the peak. In other words, the current is flowing "uphill" against the applied source drain voltage. This negative photocurrent can be understood if the microwaves couple asymmetrically to the device − in this case if the AC amplitude is larger across the left junction. Just before a peak, transport can only occur when electrons tunnel *onto* the dot by photon absorption. They do this preferentially through the barrier with the larger AC modulation, leading to a net electron current in a particular direction, in this case, in the same direction as the applied DC voltage. Just after a peak, transport can only occur by photon assisted tunneling *off* the dot. Electrons preferentially tunnel off the dot through the barrier, resulting in a net current of the opposite sign.

D. Theory and Future Experiments

These results show that time-dependent studies of transport in quantum dots are possible using microwave techniques. To describe the experiments so far, simple models of photon-assisted transport through quantum dots have been used[125,126,133]. These models are analogous to those developed for superconducting tunnel junctions in the 1960s[132]. In these models, photon absorption/emission changes the transition rates through the tunnel junctions[125]:

$$\tilde{\Gamma}_i(\varepsilon_i) = \sum_{n=-\infty}^{+\infty} J_n^2(e\tilde{V}_i/hf) \cdot \Gamma_i(\varepsilon_i + n \cdot hf). \qquad (13)$$

J_n are the nth order Bessel function, Γ is the transition rate in the absence of microwaves, V_i is the AC amplitude that drops across the $i = 1,2$ barrier, and ε_i is the difference between initial and final states of the many-body system. These calculations contain all the major features observed in the experimental curves, including the shoulders and the negative photocurrent. Some discrepancies remain[126], however, pointing to the need for further work.

It is also interesting that if the sum in Eq. (13) is replaced by an integral over all frequencies, and the Bessel functions replaced by a general weighting function, this approach can describe tunneling assisted by any sort of fluctuations in the environment. These fluctuations could be due to anything from blackbody radiation[134,135], noise[63], or quantum fluctuations in the external circuit in which the dot is embedded[5,136].

These experiments and theory described above really only represent a beginning to the interesting physics that one can explore in this system. The next obvious topic is PAT when quantum levels are important. Photon-assisted resonant tunneling through a single level is well-understood theoretically[137]. The presence of two levels and Coulomb charging has also been studied theoretically[138]. Neither of these regimes has been explored experimentally in dots so far.

Also interesting are the interactions of the photons with other energy/time scales for transport through the dot. For example, if hf matches the quantum level spacing, the microwave signal is in resonance with the classical motion of the electron in the dot. At higher frequencies, the collective excitation frequencies and the tunneling time of the barriers also become relevant. Much of AC transport thus remains unexplored in this interacting, quantum-coherent electron system.

VII. APPLICATIONS AND FUTURE DIRECTIONS

The previous sections have shown how charge and energy quantization are manifested in the electrical properties of individual quantum dots. In Section VII.A, we will consider potential applications of these phenomena. In Section VII.B, we will discuss ways of increasing the operating temperature of quantum dots by reducing their dimensions. Finally, in Section VII.C, we will discuss future directions for research − in particular, the study of coupled quantum dots and dot arrays.

A. Applications

The ability to measure and control current at the single-electron level has a number of potential uses, ranging from metrology to electrometry to computing[139]. In fact, both metal and semiconductor quantum dots are already finding niche applications, though their utility is limited because of the low temperatures required. To broaden their usage, devices must be developed that operate under ambient conditions, i.e. at room temperature. Ways of accomplishing this will be discussed in Section VII.B.

One of the most important Coulomb blockade applications is single-electrometry - the detection of single charges. As discussed already, these devices are very sensitive to small changes in their local electrostatic environment. Sensitivities of $10^{-5} e/Hz^{1/2}$ are possible[140]. In other words, the electrometer can detect a charge e in one second if 10^{-5} of the field lines leaving the charge terminate on the dot. These devices are the electrostatic counterpart to the SQUID, a superconducting device which is sensitive to extremely small magnetic fluxes. There are important differences, however[139]. SQUIDs can be used to measure macroscopic magnetic fields by utilizing flux transformers to couple the macroscopic magnetic field into the SQUID. No such transformer exists to date for electric charge, so the change in the charge over a large object cannot be carefully measured. Nevertheless, as a local electrometer, semiconductor as well as metallic dots may find many uses. Already, they have been used in scientific applications, mainly to monitor the behavior of single electrons in other circuits[24,134,141].

Another application is in the field of metrology. The single electron turnstile, and related devices in metal dots, are being investigated as current standards. They produce a standardized current from a standardized RF frequency, with the conversion factor being the electronic charge e. Accuracies of one part in 10^6 have been obtained in multi-dot metallic circuits[134]. This is still significantly worse than the theoretical limit, most likely due to inadvertent coupling of radiation into the device, leading to unwanted photon-assisted tunneling processes[134]. Various schemes are being investigated to improve the accuracy. These turnstiles would complete the solid-state device "metrology triangle" relating frequencies, currents and voltages[21]. Already, the quantum Hall effect is used to relate current to voltage, and the Josephson effect to relate frequency to voltage. The turnstile would fill in the last leg of the triangle by relating frequencies to currents.

Another application is measurement and regulation of temperature. As discussed in Section III, the Coulomb blockade peak widths are proportional to $k_B T$, and can, once calibrated, be used to measure the temperature of the dot or its surroundings. Even at higher temperatures, where most of the Coulomb structure has been washed out, there are slight nonlinearities in the I-V characteristic that can be used to measure T [71]. Temperature gradients can also be detected, as thermopower measurements of dots have shown[142]. Quantum dots may be able to control the temperature as well as measure it. A quantum dot "refrigerator" that can cool a larger electronic system has been proposed[143]. The idea is to use tunneling through single quantum levels to skim off the hot electrons above E_f and the cool holes below E_f, thereby cooling the electron system.

The experiments discussed in Section VI showed that photon-assisted tunneling over the Coulomb gap can induce DC currents through a quantum dot. This suggests applications for dots as photon detectors in the microwave regime. The tunability of the dot potential relative to the source and drain means that the detector can be frequency-selective. It is even possible for a single photon to lead to a current of many electrons[135]. Photon-detection applications are not limited to the microwave region. For example, a metallic dot operating as a single-electrometer has been utilized to (indirectly) detect visible photons. The dot was fabricated on a semiconductor substrate, and was then used to electrostatically detect the presence of photoexcited electrons within the semiconductor[144].

One can also contemplate electronics applications for these devices – a field sometimes called single-electronics. It is, in principle, possible to perform calculations using quantum dot circuits, based on either charging[1,145] or quantum-coherent phenomena[146], although little experimental work has been done in this direction. Multidot circuits can also serve as static memory elements. This has been tested in the laboratory; for example, a single-electron memory with a hold time of several hours (at millikelvin temperatures) has been demonstrated[141]. One must exercise extreme caution in extrapolating these successes to a useful product, however. The technological barriers to creating complex circuits that work in the real world are enormous. We refer the reader to the chapter by Timp *et al.*[147] for further discussion.

B. Creating Smaller Devices

Increasing the operating temperature of quantum dot devices is clearly a worthy objective - many applications require room temperature operation. In addition, there are subtle effects to be seen at low temperatures, such as the Kondo effect discussed at the end of Section III. Making devices with higher operating temperatures means making the devices smaller. To operate effectively, $e^2/C + \Delta E \sim 5 k_B T$. For room temperature operation, for example, a dot radius of approximately 10 nm is required. Fabricating devices at this size scale is difficult, however. First, lithography is very difficult below 50 nm[47]. Also gates or etched surfaces that are remote (~100 nm) from the 2DEG make reduction below the current 100 nm size range quite difficult. Technological advances, such as in situ regrowth[39], will be necessary to significantly increase operating temperatures in these devices. Silicon devices may offer more promise. Gate oxides can be as thin as 5 nm, making possible very small devices. There have been some promising reports of room temperature operation[148,149] in *Si*, but nothing approaching a controlled technology.

Another approach to getting smaller is to use quantum dots formed by nonlithographic means. For example, natural dots often form during thin film deposition by the growth of islands on a surface. Such dots were used in some of the early studies of charging[17-19], as well as in STM investigations[38]. Ralph *et al.*[150] have recently made a metallic dot device using island growth at a selected location. Charging energies as high as 40 meV were obtained, and at low temperatures the effects of quantum levels were observable. Quantum dots can also be created using chemical techniques. Metal and semiconductor nanocrystals of a desired size can be grown by a variety of techniques and subsequently bound to surfaces using self-assembled monolayers (SAMs)[151,152]. Such clusters have been explored by STM measurements[38], but their incorporation into lithographic structures remains a challenge. The techniques of molecular self-assembly offer particular promise for creating ordered arrays of nanocrystals linked by SAMs. Electrical measurements of such systems would be very exciting, as we discuss below.

C. Future Work – Multi-dot systems

For single dots, most of the open scientific issues were discussed in the previous sections. Here, we turn to a relatively new research direction, interconnected dot systems. If quantum dots are artificial atoms, coupled dots and dot arrays are artificial molecules and solids. Just as in solids, there are two types of effects that can couple the dots. The first is Coulombic – a charge on one dot can shift the electrostatic potential of another. These are analogous to ionic effects in molecules/solids. The second type of effect is quantum-mechanical – an electron can coherently tunnel back and forth between different dots. This is analogous to covalent interactions in molecules and solids.

In the metallic systems, multiple dot circuits have been widely investigated. In these structures, quantum effects are negligible, and only the Coulombic interaction between dots is important. Nevertheless, these Coulombic effects lead to a number of interesting results. They make possible the metallic turnstile[131], where electrons are shifted from one dot to the next during an RF cycle. In 1D arrays, repulsion between electrons in the array leads to the formation of single electron solitons that repel each other and lead to spatially and temporally correlated current flow[153]. In 2D arrays, these solitons are predicted to exhibit interesting phase transitions[154], but the experiments to date are ambiguous.

In semiconductor multiple dots, quantum effects can also be important. Two dots in series – quantum molecules; are being measured by a number of groups. The inter-dot Coulomb interactions have been studied, including the effects of tunneling on the charging energy[155-157]. The effects of the quantum level alignment between the two dots have also been investigated[158-160]. The next goal is to clearly show that coherent transport is occurring between the two dots and that a true molecular state is being formed. Studies of one-dimensional[52] and two-dimensional[161] arrays have also been undertaken, but more work needs to be done to clarify the competing roles of quantum mechanics, Coulomb interactions, and disorder in these systems[162].

We end by noting that, when both Coulomb interactions and quantum tunneling are important, a satisfactory theoretical description of transport in 2D arrays is lacking. Such an array can be viewed as a 2D Hubbard model, widely used to describe, for example, high-Tc superconductors. In dot language, the Hubbard model is a dot array with one (spin-degenerate) quantum level per dot, a charging energy U if two electrons occupy the same dot, and a transmission coefficient T between adjacent dots. When faced with this Hamiltonian, a theorist can only make educated guesses about its behavior. There are many intriguing possibilities, like superconductivity, but no one knows for sure how the system behaves for arbitrary parameters. Only experiments can tell us. It is a tremendous challenge to experimentalists to create dot systems of sufficient quality to address these questions.

ACKNOWLEDGEMENTS

We gratefully acknowledge our (numerous!) collaborators at UC Berkeley, Delft, MIT, Phillips, and the Univ. of Tokyo, who, with the authors, performed most of the work presented here. We also thank our colleagues R. Ashoori, C.W.J. Beenakker, A.A.M. Staring, M. Stopa, and N. Wingreen, who graciously provided figures for this chapter from their published and unpublished work. We thank S. Jauhar for a critical reading of the manuscript and the Office of Naval Research (P.L.M.), the Packard Foundation (P.L.M.), and the Royal Academy of Arts and Sciences (L.P.K.) for their continuing support.

REFERENCES

[1]   Reed, M., *Scientific American* **268**, 118 (1993).

[2]   Kastner, M., *Physics Today* **46**, 24 (1993).

[3]   Grabert, H., and Devoret, M.H., eds., *Single Charge Tunneling*, New York: Plenum Press, 1991.

[4]   Averin, D.V., and Likharev, K.K., *J. Low Temp. Phys.* **62**, 345 (1986); Averin, C.V., and Likharev, K.K., in: *Mesoscopic Phenomena in Solids*, Altshuler, B.L., Lee, P.A., and Webb, R.A., eds., Amsterdam: Elsevier, 1991.

[5]   Ingold, G.-L., and Nazarov, Yu. V., in *Single Charge Tunneling*, Grabert, H., and Devoret, M.H., eds., New York: Plenum Press, 1991.

[6]   Beenakker, C.W.J., *Phys. Rev. B* **44**, 1646 (1991).

[7]   van Houten, H., Beenakker, C.W.J, Staring, A.A.M., *Single Charge Tunneling*, Grabert, H., and Devoret, M.H., eds., New York: Plenum Press, 1991.

[8]   Special issue: **Single Charge Tunneling,** *Zeitschrift für Physik B*, **85**, (1991).

[9]   Special issue: **Few-electron Nanostructures,** *Physica B*, **189**, (1993).

[10] Harmans, C.J.P.M., *Physics World* **5**, 50 (1992).

[11] Likharev, K.K., and Claeson, T., *Scientific American* **266**, 50 (1992).

[12] Devoret, M.H., Esteve, D., and Urbina, C., *Nature* **360**, 547 (1992).

[13] Nagamune, Y., *et al., Appl. Phys. Lett.* **64**, 2379 (1994).

[14] Glazman, L.I., and Shekhter, R.I., *J. Phys.: Condens. Matter* **1**, 5811 (1989).

[15] Kouwenhoven, L.P., *et al., Z. Phys. B* **85**, 367 (1991).

[16] Millikan, R.A., *Phys. Rev.* **32**, 349 (1911).

[17] Gorter, C.J., *Physica* **17**, 777 (1951).

[18] Giaever, I., and Zeller, H.R., *Phys. Rev. Lett.* **20**, 1504 (1968).

[19] Lambe, J., and Jaklevic, R.C., *Phys. Rev. Lett.* **22**, 1371 (1969).

[20] Kulik, I.O., and Shekhter, R.I., *Zh. Eksp. Teor. Fiz.* **68**, 623 (1975); [*Sov. Phys. JETP* **41**, 308 (1975)].

[21] Likharev, K.K., *IEEE Trans. Magn.* **23,** 1142 (1987); *IBM J. Res. Dev.* **32**, 144 (1988).

[22] Mullen, K., *et al.*, *Phys. Rev. B* **37**, 98 (1988); Amman, M., Mullen, K., and Ben-Jacob, E., *J. Appl. Phys.* **65,** 339 (1989).

[23] Fulton, T.A. and Dolan, G.J., *Phys. Rev. Lett.* **59**, 109 (1987).

[24] Lafarge, P., *et al.*, *Z. Phys. B* **85**, 327 (1991).

[25] Fulton, T.A., *et al.*, *Phys. Rev. Lett.* **63**, 1307 (1989).

[26] Geerligs, L.J., *et al.*, *Phys. Rev. Lett.* **65**, 377 (1990).

[27] Tuominen, M.T., *et al.*, *Phys. Rev. Lett.* **69**, 1997 (1992); Hergenrother, J.M., *et al.*, *Phys. Rev. Lett.* **72**, 1742 (1994).

[28] Lafarge, P., *et al.*, *Phys. Rev. Lett.* **70**, 994 (1993); *Nature* **365**, 422 (1993).

[29] Eiles, T.M., Martines, J.M., and Devoret, M.H., *Phys. Rev. Lett.* **70**, 1862 (1993).

[30] Nazarov, Yu. V., *Zh. Eksp. Teor. Fiz.* **95**, 975 (1989); [*Sov. Phys. JETP* **68**, 561 (1989)]; *Pis'ma Zh. Eksp. Teor. Fiz.* **49**, 1349 (1989); [*JETP Lett.* **49**, 126 (1989)].

[31] Devoret, M.H., *et al.*, *Phys. Rev. Lett.* **64**, 1824 (1990).

[32] Girvin, S.M., *et al.*, *Phys. Rev. Lett.* **64**, 3183 (1990).

[33] Flensberg, K., *et al.*, *Phys. Scr.* **T42**, 189 (1992).

[34] Geerligs, L.J., *et al.*, *Europhys. Lett.* **10**, 79 (1989).

[35] Cleland, A.N., Schmidt, J.M., and Clarke, J., *Phys. Rev. Lett.* **64**, 1565 (1990).

[36] Kuzmin, L.S., *et al.*, *Phys. Rev. Lett.* **67**, 1161 (1991).

[37] Eigler, D., Chapter 11 in this volume.

[38] Schönenberger, C., van Houten, H., Donkersloot, H.C., *Europhys. Lett.* **20**, 249 (1992).

[39] Sakaki, H., Chapter 5 in this volume.

[40] Reed, M.A., *et al.*, *Phys. Rev. Lett.* **60**, 535 (1988).

[41] Groshev, A., *Phys. Rev B* **42**, 5895 (1990).

[42] Su, B., *et al.*, *Appl. Phys. Lett.* **58**, 747 (1991); Su, B., *et al.*, *Science N.Y.* **255**, 313 (1992); *Phys. Rev B* **46**, 7644 (1992).

[43] Dellow, M.W., *et al.*, *Phys. Rev. Lett.* **68**, 1754 (1992).

[44] Guéret, P., *et al.*, *Phys. Rev. Lett.* **68**, 1896 (1992).

[45] Tewordt, M., *et al.*, *Solid-State Electronics* **37**, 793 (1994); see also references therein.

[46] Geim, A.K., *et al.*, *Phys. Rev. Lett.* **72**, 2061 (1994).

[47] Tennant, D., Chapter 4 in this volume.

[48] van Wees, B.J., *et al.*, *Phys. Rev. Lett.* **60**, 848 (1988).

[49] Wharam, D.A., *et al.*, *J. Phys. C* **21**, L209 (1988).

[50] Beenakker, C.W.J., and van Houten, H., *Solid State Physics* **44**, 1 (1991). This reviews transport in mesoscopic systems in which single electron charging is not important.

[51] van Wees, B.J., *et al.*, *Phys. Rev. Lett.* **62**, 2523 (1989).

[52] Kouwenhoven, L.P., *et al.*, *Phys. Rev. Lett.* **65**, 361 (1990).

[53] Scott-Thomas, J.H.F., *et al.*, *Phys. Rev. Lett.* **62**, 583 (1989).

[54] van Houten, H., and Beenakker, C.W.J., *Phys. Rev. Lett.* **63**, 1893 (1989).

[55] Meirav, U., *et al.*, *Phys. Rev. B* **40**, 5871 (1989).

[56] Field, S.B., *et al.*, *Phys. Rev. B* **42**, 3523 (1990).

[57] Staring, A.A.M., *et al.*, *High Magnetic Fields in Semiconductor Physics III*, Landwehr, G., ed., Berlin: Springer, 1990.

[58] Kastner, M.A., *Rev. Mod. Phys.* **64**, 849 (1992).

[59] Staring, A.A.M., *et al.*, *Physica B* **175**, 226 (1991); Staring, A.A.M., thesis, Technical University Eindhoven, The Netherlands (1992).

[60] Stopa, M., *Phys. Rev. B* **48**, 18340 (1993).

[61] Macucci, M., Hess, K., and Iafrate, G.J., *Phys. Rev B* **48**, 17354 (1993).

[62] Jovanovic, D., and Leburton, J.P., *Phys. Rev. B* **49**, 7474 (1994).

[63] Vion, D., *et al.*, *J. Appl. Phys.* to be published.

[64]   Meurer, B., Heitmann, D., and Ploog, K., *Phys. Rev. Lett.* **68**, 1371 (1992).

[65]   Hansen, W., *et al.*, *Phys. Rev. Lett.* **62**, 2168 (1989).

[66]   Ashoori, R.C., *et al.*, *Phys. Rev. Lett.* **68**, 3088 (1992); Ashoori, R.C., *et al.*, *Phys. Rev. Lett.* **71**, 613 (1993).

[67]   Korotkov, A.N., Averin, D.V., and Likharev, K.K., *Physica B* **165** & **166**, 927 (1990); Averin, D.V., Korotkov, A.N., and Likharev, K.K., *Phys. Rev. B* **44**, 6199 (1991).

[68]   Meir, Y., Wingreen, N.S., and Lee, P.A., *Phys. Rev. Lett.* **66**, 3048 (1991).

[69]   Averin, D.V., and Nazarov, Yu.V., in *Single Charge Tunneling*, Grabert, H., and Devoret, M.H., eds., New York: Plenum Press, 1991.

[70]   Kouwenhoven, L.P., *et al.*, *Phys. Rev. B* **40**, 8083 (1989).

[71]   Pekola, J.P., *et al.*, *Phys. Rev. Lett.* **73**, 2903 (1994).

[72]   Nazarov, Yu.V., *J. Low Temp. Phys.* **90**, 77 (1993).

[73]   Averin, D.V., *Physica B* **194–196**, 979 (1994).

[74]   Schoeller, H., *Physica B* **194–196**, 1057 (1994).

[75]   Stone, A.D., and Lee, P.A., *Phys. Rev. Lett.* **54**, 1196 (1985).

[76]   Geerligs, L.J., Averin, D.V., and Mooij, J.E., *Phys. Rev. Lett.* **65**, 3037 (1990); Geerligs, L.J., Matters, M., and Mooij, J.E., *Physica B* **194–196**, 1267 (1994).

[77]   Eiles, T.M., *et al.*, *Phys. Rev. Lett.* **69**, 148 (1992).

[78]   Pasquier, C., *et al.*, *Phys. Rev. Lett.* **70**, 69 (1993); and Glattli, D.C., *Physica B* **189**, 88 (1993).

[79]   Glazman, L.I., and Raikh, M.E., *Pis'ma Zh. Eksp. Teor. Fiz.* **47**, 378 (1988); [*JETP Lett.* **47**, 452 (1988)].

[80]   Ng, T.K., and Lee, P.A., *Phys. Rev. Lett.* **61**, 1768 (1988).

[81]   Kawabata, A., *J. Phys. Soc. Jpn.* **60**, 3222 (1991).

[82]   Hershfield, S., Davies, J.H., and Wilkins, J.W., *Phys. Rev. Lett.* **67**, 3720, (1991); *Phys. Rev B* **46**, 7046 (1992).

[83]   Meir, Y., Wingreen, N.S., and Lee, P.A., *Phys.Rev. Lett.* **70**, 2601 (1993).

[84]   Inoshita, T., *et al.*, *Phys. Rev. B* **48**, 14725 (1993).

[85]   Ralph, D.C., and Buhrman, R.A., *Phys. Rev. Lett.* **72**, 3401 (1994).

[86]   Yanson, I.K., *et al.*, *Phys. Rev. Lett.* **74**, 302 (1995).

[87]   Kumar, A., Laux, S.E., Stern, F., *Phys. Rev. B* **42**, 5166 (1990).

[88]   Bryant, G.W., *Phys. Rev. B* **39**, 3145 (1989).

[89]   van der Marel, D., in *Nanostructure Physics and Fabrication*, Reed, M.A., and
       Kirk, W.P., eds., Academic Press, 1989.

[90]   Wang, L., Zhang, J.K., and Bishop, A.R., *Phys. Rev. Lett.* **73**, 585 (1994).

[91]   Merkt, U., Huser, J., Wagner, M., *Phys. Rev. B* **43**, 7320 (1991).

[92]   Pfannkuche, D., *et al.*, *Physica B* **189**, 6 (1993); see also references therein.

[93]   Johnson, N.F., *J. Phys.: Condens. Matter* **7**, 965 (1995).

[94]   Wagner, M., Merkt, U., and Chaplik, A.V., *Phys. Rev. B* **45**, 1951 (1992).

[95]   Su, B., Goldman, V.J., Cunningham, J.E. *Surface Science* **305**, 566 (1994).

[96]   Williamson, J.G., *et al.*, in: *Nanostructures and Mesoscopic Systems*, Reed, M.A.,
       and Kirk, W.P., eds., Academic Press, 1991.

[97]   Foxman, E.B., *et al.*, *Phys. Rev. B* **47**, 10020 (1993).

[98]   Foxman, E.B., *et al.*, *Phys. Rev B* **50**, 14193 (1994).

[99]   Stone, A.D., and Bruus, H., *Physica B* **189**, 43 (1993); see also references therein.

[100] Baranger, H. Chapter 14, Part I in this volume.

[101] Johnson, A.T., *et al.*, *Phys. Rev. Lett.* **69**, 1592 (1992).

[102] Meir, Y., and Wingreen, N.S., *Phys. Rev. Lett.* **68**, 2512 (1992).

[103] Elion, W.J., *et al.*, *Nature* **371**, 594 (1994).

[104] van der Vaart, N.C., *et al.*, *Physica B* **189**, 99 (1993).

[105] Weinmann, D., *et al.*, *Europhys. Lett.* **26**, 467 (1994); Weinmann, D., Hausler, W.,
       Kramer, B., *Phys. Rev. Lett.* **74**, 984 (1995).

[106] Weis, J., *et al.*, *Phys. Rev. Lett.* **71**, 4019 (1993).

[107] Westervelt, R., Chapter 14, Part II in this volume.

[108] Fock, V., *Z. Phys.* **47**, 446 (1928).

[109] Darwin, C.G., *Proc. Cambridge Philos. Soc.* **27**, 86 (1930).

[110] Hawrylak, P., *Phys. Rev. Lett.* **71**, 3347 (1993).

[111] McEuen, P.L., *et al.*, *Phys. Rev. Lett.* **66**, 1926 (1991).

[112] McEuen, P.L., *et al.*, *Phys. Rev. B* **45**, 11419 (1992); McEuen, P.L., *et al.*, *Physica
       B* **189**, 70 (1993).

[113]   Marmorkos, I.K., and Beenakker, C.W.J., *Phys. Rev. B* **46**, 15562 (1992).

[114]   Kinaret, J.M., and Wingreen, N.S. *Phys. Rev. B* **48**, 11113 (1993).

[115]   Evans, A.K., Glazman, L.I., and Shklovskii, B.I., *Phys. Rev. B* **48**, 11120 (1993).

[116]   Stopa, M., (unpublished). Note that in this simulation, the two *LLs* are two *orbital LLs*, $n = 0$ and $n = 1$. The spin splitting is too small to be clearly seen.

[117]   Staring, A.A.M., *et al.*, *Phys. Rev. B* **46**, 12869 (1992).

[118]   Heinzel, T., *et al.*, *Phys. Rev. B* **50**, 15113 (1994).

[119]   van der Vaart, N.C., *et al.*, *Phys. Rev. Lett.* **73**, 320 (1994).

[120]   Yang, S.R. Eric, MacDonald, A.H., and Johnson, M.D., *Phys. Rev. Lett.* **71**, 3194 (1993); MacDonald, A.H., Yang, S.R. Eric, and Johnson, M.D., *Aust. J. Phys.* **46**, 345 (1993).

[121]   de Chamon, C., and Wen, X.G., *Phys. Rev. B.* **49**, 8227 (1994).

[122]   Klein, O., *et al.*, *Phys. Rev. Lett.* **74**, 785 (1995).

[123]   Nazarov, Y.V., and Khaetskii, A.V., *Phys. Rev. B* **49**, 5077 (1994).

[124]   Kouwenhoven, L.P., *et al.*, *Phys. Rev. Lett.* **67**, 1626 (1991); Kouwenhoven, L.P., *et al.*, *Z. Phys. B* **85**, 381 (1991).

[125]   Kouwenhoven, L.P., *et al.*, *Phys. Rev. B* **50**, 2019 (1994).

[126]   Kouwenhoven, L.P., *et al.*, *Phys. Rev. Lett.* **73**, 3443 (1994).

[127]   See, e.g. Landauer, R., and Martin, Th., *Rev. Mod. Phys.* **66**, 217 (1994).

[128]   Jauhar, S., Kouwenhoven, L.P., Orenstein, J., McEuen, P.L., Nagamune, Y., Motohisa, J., and Sakaki, H., (unpublished).

[129]   Weis, J., Haug, R.J., von Klitzing, K., and Ploog, K., (unpublished).

[130]   Odintsov, A.A., *Appl. Phys. Lett.* **58**, 2695 (1991).

[131]   A different kind of turnstile, utilizing three metallic dots in series, predated the quantum dot turnstile. See: Geerligs, L.J., *et al.*, *Phys. Rev. Lett.* **64**, 2691 (1990).

[132]   Tien, P.K., and Gordon, J.R., *Phys. Rev.* **129**, 647 (1963).

[133]   Likharev, K.K., and Devyatov, I.A., *Physica B* **194–196**, 1341 (1994).

[134]   Martinis, J.M., Nahum, M., and Jensen, H.D., *Phys. Rev. Lett.* **72**, 904 (1994), and references therein.

[135]   Hergenrother, J.M., *et al.*, *Physica B* **203**, 327 (1994).

[136]   Hu, G.Y., and O'Connell, R.F., *Phys. Rev. B* **49**, 16505 (1994).

[137]   Johansson, P., *Phys. Rev. B* **41**, 9892 (1990); and references therein.

[138]   Bruder, C., and Schoeller, H., *Phys. Rev. Lett.* **72**, 1076 (1994).

[139]   Averin, D.V., and Likharev, K.K., in *Single Charge Tunneling*, Grabert, H., and Devoret, M.H., eds., New York: Plenum Press, 1991.

[140]   Visscher, E.H., *et al.*, *Appl. Phys. Lett.* **66**, 305 (1994).

[141]   Dresselhaus, P.D., *et al.*, *Phys. Rev. Lett.* **72**, 3226 (1994).

[142]   Staring, A.M.M., *et al.*, *Europhys. Lett.* **22**, 57 (1993).

[143]   Edwards, H.L., Niu, Q., deLozanne, A.L., *Appl. Phys. Lett.* **63**, 1815 (1993).

[144]   Cleland, A.N., *et al.*, *Appl. Phys. Lett.* **61**, 2820 (1992).

[145]   Lent, C.S., Tougaw, P.D., and Porod, W., *Appl. Phys. Lett.* **62**, 714 (1993).

[146]   See for example, Lloyd, S., *Science* **261**, 1569 (1993).

[147]   Timp, G., Howard, R., and Mankiewich, P., Chapter 2 in this volume.

[148]   Yano, K., *et al.*, *IEEE Trans. Elec. Dev.* **41**, 1628 (1994).

[149]   Takahashi, Y., *et al.*, *Elec. Lett.* **31**, 136 (1995).

[150]   Ralph, D.C., Black, C.T., and Tinkham, M., *Phys. Rev. Lett.* **74**, 3241 (1995).

[151]   Brus, L., Chapter 6 in this volume.

[152]   Whitesides, G., and Wilbur, J., Chapter 8 in this volume.

[153]   Delsing, P., in *Single Charge Tunneling*, Grabert, H., and Devoret, M.H., eds., New York: Plenum Press, 1991.

[154]   Mooij, J.E., and Schön, in *Single Charge Tunneling*, Grabert, H., and Devoret, M.H., eds., New York: Plenum Press, 1991.

[155]   Kemerink, M., and Molenkamp, L.W., *Appl. Phys. Lett.* **65**, 1012 (1994).

[156]   Waugh, F.R., Berry, M.J., Mar, D.J., Westervelt, R.M., Campman, K.C., and Gossard, A.C., preprint (1995).

[157]   Blick, R.H., Haug, R.J., Weis, J., Pfannkuche, D., von Klitzing, K., and Eberl, K., preprint (1995).

[158]   Tewordt, M., *et al.*, *Appl. Phys. Lett.* **60**, 595 (1992).

[159]   van der Vaart, N.C., Godijn, S.F., Nazarov, Y.V., Harmans, C.P.J.M., Mooij, J.E., Molenkamp, L.W., and Foxon, C.T., preprint (1995).

[160]   Dixon, D., Kouwenhoven, L.P., McEuen, P.L., Nagamune, Y., Motohisa, J., Sakaki, H., preprint (1995).

[161]   Lenssen, K.-M.H., *et al.*, *Phys. Rev. Lett.* **74**, 454 (1995).

[162]   Stafford, C.A., and Das Sarma, S., *Phys. Rev. Lett.* **72**, 3590 (1994).

Chapter 14

Chaos in Ballistic Nanostructures

Part I - Theory

Harold U. Baranger

Bell Laboratories,
600 Mountain Ave. 1D-230,
Murray Hill, NJ 07974-0636

Part II - Experiment

R.M. Westervelt

Division of Applied Sciences and Department of Physics
Harvard University
Cambridge, Ma. 02139

PART I - THEORY

I. INTRODUCTION

What is the effect of quantum interference on transport in ballistic structures? What are the quantum properties of classically chaotic scattering systems? These are the two questions addressed in this review.

The first question is motivated by the way in which mesoscopic transport physics has developed over the last fifteen years. The main subject of this field[1] is the effects of quantum interference on the transport of electrons in nanostructures. At sufficiently low temperature in small structures, the quantum-mechanical coherence of the electrons causes wave interference phenomena which influence the transmission and hence the conductance. These effects are akin to laser speckle in optical physics[2]. While the starting point of the field is relatively simple – single particle quantum mechanics – the results can be quite surprising because of subtle correlations introduced by the multiple coherent scattering[1,2].

Mesoscopic physics was initially focused on disordered metals in which the motion of the electrons is diffusive[1]. It was shown that quantum interference produces a correction to the classical conductance. The most studied aspect of this correction is its dependence on some external parameter such as the magnetic field. In particular, the average correction – the *weak-localization* effect – and the variance – the *universal conductance fluctuations* – are both sensitive to a weak magnetic field. In magnitude, both effects are of order e^2/h $[(26 k\Omega)^{-1}]$, the natural unit of conductance[1].

More recently, the experimental systems have become cleaner in that the mean free path is larger[1]. This naturally raises the question of interference effects in ballistic nanostructures, nanostructures in which the scattering comes only from specular scattering on the boundary. The main idea is that the interference among the complicated boundary-scattered trajectories gives rise to the same type of transport effects – conductance fluctuations and a correction to the average conductance – that occur in the diffusive regime.

The second question is motivated by developments in the chaos community. Classical chaos in Hamiltonian systems has been intensively investigated in both closed systems and open scattering systems and is well understood for simple systems[3,4]. The defining features of classical chaos are (1) an exponential sensitivity to initial conditions (a small change in the initial conditions of a particle rapidly results in a completely different trajectory) and (2) an exponential proliferation of trajectories (the number of trajectories connecting two points increases exponentially as a function of the path length). In classically chaotic Hamiltonian systems, there is a single constant of the motion – the energy – and so the dynamics cannot usually be treated analytically. It is by now well-established that "almost all" mechanical systems are chaotic: the solvable paradigms that one learns from textbooks are very rare. Because chaotic and non-chaotic dynamics are qualitatively different, these paradigms can be highly misleading.

The connection between classical dynamics and quantum properties is, however, much less well understood; this is the subject of "quantum chaos"[3]. The problem is that the defining features of classical chaos noted above have no direct counterpart in quantum mechanics. The notion of a trajectory for a particle is not natural in quantum mechanics; rather one speaks of the state or wave-function and its time-evolution, governed by the linear Schrödinger equation. Thus the two defining features of classical chaos are apparently necessarily absent from quantum dynamics. However, the quantum properties of classical chaotic systems are often quite different from those of non-chaotic systems. This naturally gives rise to the question how are the quantum properties of a system related to the nature of its classical dynamics, which is the basis for much of the work in "quantum chaos". The deeper question of whether there is real chaos in some form in quantum mechanics continues to be considered but is very controversial at this time.

Most work in "quantum chaos" discusses the connection between classical dynamics and the energy levels of a closed system, either statistics of energy levels or semiclassical prediction of energy levels[3]. From a quantum point of view, it is natural to focus on energy levels since the quantization of energy is perhaps the most dramatic feature of quantum mechanics. Other quantum quantities are important, however, especially for studying systems in which the spectrum is continuous – open systems[4,5]. While the study of energy levels is not possible for such systems, the study of scattering is – the fluctuations of the scattering phase shifts, for instance[5]. The connection between classical dynamics and quantum properties is less well developed for open systems than for closed systems, hence our interest in the question.

In the last five years it has become clear that there is substantial overlap between the answers to these two questions. This overlap stems from the view of transport as scattering from the whole system, a view particularly convenient in the ballistic regime. Through this view, mesoscopic quantum transport is directly connected to classical chaotic scattering. We will use this connection to deduce characteristics of the quantum transport from knowledge of the classical scattering. On the other hand, mesoscopic physics can bring to "quantum-chaos" a variety of new *experimental* systems.

The rest of the paper is organized as follows. We start by reviewing some basic concepts useful for thinking about chaos in nanostructures (Section II, Part I). As a background for the quantum interference effects that are the main subject, in Section III (Part I) classical size effects in nanostructures are reviewed. Section IV (Part I) discusses the quantum transport properties of open ballistic nanostructures. Other systems – closed structures or transport in the tunneling regime, for instance – are briefly discussed in Section V (Part I). Finally, we summarize and conclude in Section VI (Part I).

II. PRELIMINARY CONCEPTS

A. The Coherent Ballistic Regime

There are three important length scales for the phenomena considered in this paper: the Fermi wavelength λ_F, the elastic mean free path l, and the phase coherence length L_ϕ. First, throughout this paper we will be discussing the linear conductance of a degenerate Fermi system, usually a two dimensional electron gas. Because transport in the linear regime occurs at the Fermi surface, the electrons participating in the conduction have a characteristic wavelength λ_F.

Second, the elastic mean free path is caused mainly by impurity scattering. In certain systems – those with a smooth disorder potential – it can be important to distinguish between the total scattering rate and the momentum relaxation rate (transport mean free path) associated with backscattering.

Third, the phase coherence length is conceptually less precise than the first two lengths: it is the length over which electrons retain memory of their phase and so the length over which interference effects occur. The loss of phase coherence is usually associated with inelastic scattering, and so can be controlled by varying the temperature. A quantitative understanding of phase-breaking is clearly crucial to mesoscopic physics. The issue has been studied throughout the history of interference effects in transport[6], and in particular has gained considerable attention in the last few years because of possible dimensionality effects[7].

By comparing these length scales to the size of the system being studied, denoted L, various regimes can be defined. The most familiar is the macroscopic regime – $\lambda_F, l, L_\phi \ll L$ – where the notion of a local conductivity works well. The diffusive regime was first studied in mesoscopic physics: $\lambda_F, l \ll L \ll L_\phi$. In this regime, electrons follow random walks because of scattering from the impurities, and interference among these paths is important[1]. In this paper, we will focus on the *coherent ballistic regime* characterized by $\lambda_F L \ll l, L_\phi$. In this regime, the electrons travel in straight ballistic paths until scattered by the boundary of the structure, and it is the interference among these boundary-scattered paths that is of interest.

One might think that the existence of interference among different paths in a nanostructure would be very difficult to detect. However, there are two knobs available to the experimentalist which make this task much easier: the magnetic field and the wavelength. The presence of a magnetic field changes the phase accumulated along each path: the quantum amplitude is multiplied by $\exp[i \int \mathbf{A} \cdot \mathbf{dl}]$ where $\mathbf{A}$ is the vector potential and the integral is taken along the path. This is the Aharonov-Bohm effect[1]. Likewise, changing the wavelength changes the phase kL along the path. In either case, the phase relations among the different paths varies, and so the interference pattern and conductance vary. It is the sensitivity of the conductance to small magnetic fields or small changes in the energy (wavelength) of the particles that is the hallmark of mesoscopic transport physics.

The material of choice for experimental work in the coherent ballistic regime is the two-dimensional electron gas at the interface between *GaAs* and *AlGaAs*. In this material, the mean free paths are large, and nanostructures can be fabricated; see Part II of this chapter by Westervelt for a review. We will have this system in the back of our minds throughout the theoretical analysis in this paper.

B. Non-interacting Particles

For a physical system with a large number of charged particles, an electron gas in our case, one must always consider the importance of interactions. For conduction in a good metal, it is by now well established that the effects of interactions are much weaker than might originally be expected[8]. Landau's Fermi liquid theory[8] explains this by noting that for non-singular interactions there is one-to-one correspondence between the states of the non-interacting Fermi liquid and an interacting one. Because the lifetime of those states near the Fermi surface is long even in the interacting case, it is appropriate to think in terms of quasi-particles which behave like nearly independent particles. Thus the limit of non-interacting particles is often taken in discussing transport in metals.

Most of the systems to be discussed here are open, in the sense that particles can go from the nanostructure into large metallic regions relatively easily. The conductance of such a system will be large − greater than the quantum of conductance e^2/h − and so the system is a good metal in terms of the nature of the states. Thus Landau's Fermi liquid theory applies, and the transport physics can be largely captured by considering non-interacting particles. We will make this approximation throughout most of this paper; the exception is in the section on transport through nearly closed systems when the conductance is low and so interaction effects must be included.

C. Conductance is Transmission!

A point of view which has proved very useful in thinking about mesoscopic physics has been developed and advocated over the years by Landauer[9], Büttiker[10], and Imry[11], and is summarized by the bumper sticker phrase that is the title of this section: conductance is transmission. The basic idea is to view transport as the response of the system to applied fluxes of particles. To this end, one divides the system conceptually into the sample region − the transport through which one is interested − and "reservoirs" which inject current into leads attached to the sample. Reservoirs are defined essentially by a boundary condition: they emit flux into the lead consistent with their chemical potential and absorb any incident flux. The flow through the sample is characterized by the probability of transmission between each pair of leads attached to the sample. Thus the conductance of the system is related to the scattering properties of the sample region.

This point of view has been used to describe classical transport, quantum transport in one-dimensional wires, transport in multi-lead structures (both quantum and classical), transport in the quantum Hall regime, and more recently time dependent transport[1,12]. In its simplest form, the conductance G through a sample attached to two leads is simply proportional to the transmission through the sample T. For leads of width W which quantum-mechanically support $N = \text{Int}[kW/\pi]$ transverse channels or modes, the conductance summed over incoming (m) and outgoing (n) modes is

$$G = \frac{2e^2}{h} T = \frac{2e^2}{h} \sum_{n=1}^{N} \sum_{m=1}^{N} T_{nm} = \frac{2e^2}{h} Tr\{tt^\dagger\} \qquad (I.1)$$

where t is the matrix of amplitudes between the modes. Generalizing this expression to the multi-lead case, Büttiker showed[10] that the current in lead j, I_j, is related to the voltages applied by reservoirs to the other leads a, V_a, through the transmission probability to go from lead a to lead j, $T^{j,a}$,

$$I_j = \frac{2e^2}{h} \sum_{a=1}^{N} T^{j,a}(V_a - V_j).$$ (I.2)

After applying the appropriate boundary conditions – one lead used to supply the current, another used to sink the current, and zero current in the voltage leads – this equation can be solved for the four-point resistance, defined as the voltage difference between two leads divided by the current. If one measures the voltage between leads j and a when a current goes from lead b to lead k, the general expression for the resistance, $R_{bk,ja}$, is

$$R_{bk,ja} = (h/2e^2)(T^{j,b}T^{a,k} - T^{j,k}T^{a,b})/D,$$ (I.3)

where D is any cofactor of the matrix $d_{ja} = T^{j,a} - N\delta_{ja}$.

The relation between this transmission approach to conductance and the more familiar linear response formalism has been extensively discussed throughout the history of interference contributions to transport. The linear response formalism results in a relation between the conductivity and the current-current correlation function; this is known as the fluctuation-dissipation theorem in classical transport[13] and the Kubo formula in the quantum regime[14]. Several authors have shown that transmission formulas for the conductance can be derived from linear response formalism[15], thus showing that the Landauer-Büttiker and Kubo approaches are essentially equivalent. The tricky point in such a derivation is in treating the leads and reservoirs. Since these appear in the transmission approach as a boundary condition, one must find and argue for the correct boundary condition in the linear response formalism. In fact, the resulting boundary condition is the one that one would naturally use in applying the Kubo formula to the mesoscopic regime, so the picture is entirely consistent.

D. Classical Chaos in Open Systems

Chaos in classical mechanics is defined by the long time properties of classical trajectories: the exponential sensitivity to initial conditions (the Lyapunov exponent) and the exponential growth in the number of trajectories of a given length that connect two typical points (the topological entropy). In an open system, the trajectories spend a finite amount of time in the scattering region, and so these standard definitions of chaos are not appropriate. However, the concept of "transient chaos"[4] has proved useful in many contexts in classical mechanics and in particular has been developed for scattering systems[4,5]. The basic idea is that even though almost all trajectories leave the scattering region at some point, there is an infinite set of trajectories which stay in the scattering region forever. These are, for instance, the periodic orbits which exist in the scattering region and their stable manifolds. For the scattering to be "chaotic", the dynamics on this set must be chaotic in the closed-system infinite-time sense, and the set must have a fractal dimension.

At first sight, characterizing a scattering process by the infinite time properties of a measure zero set of trajectories seems unintuitive. This definition of chaos in an open system is, however, very reasonable. First, consider how scattering occurs in such a system. The incoming particle approaches the set of permanently trapped trajectories, bounces around close to this set for a while, and then gets repelled from the set and ejected from the scattering region since it did not have exactly the right initial conditions to be trapped. With this picture in mind, it seems reasonable that the properties of the permanently trapped trajectories − known as the "strange repeller" − will have a big influence on the scattered trajectories. Second, there has been careful theoretical work[4,5] to connect the more intuitive characteristics of chaotic scattering with the properties of the strange repeller. In particular, the rate at which particles escape from the scattering region − perhaps the most intuitive characteristic of scattering − is directly related to the fractal dimension of the repeller, or equivalently to the difference of the Lyapunov exponent and the topological entropy[4,5].

For the connection between quantum transport and classical chaos, the important classical quantities are the distributions of length and directed area of the classical paths[16-22]. By distribution we mean simply the average over all injected classical trajectories. Such distributions are typically exponential for chaotic scattering[4,5] and power law for non-chaotic scattering[23]. While a sophisticated theory has been developed to show this for the distribution of path length, we give here merely some hand-waving motivation for this well-established property. In chaotic scattering, the particle moves ergodically over the whole energy surface while in the scattering region. Thus, any property which is averaged over the distribution of classical paths will be characterized by a single scale, the average over the energy surface. Because of the presence of a single scale, the distribution is exponential. In contrast, in non-chaotic scattering, the particle moves over only that part of the energy surface consistent with the conserved quantity. So, any distribution will be characterized by multiple scales, one for each value of the conserved quantity. In a situation with multiple scales, it is reasonable that the distribution is a power law.

For any particular structure, one may simply check if the above argument is valid by finding the distributions of length and area through classical simulation. This has been done for a variety of chaotic and non-chaotic structures. All the results for chaotic structures show very good exponential dependence; an example for a stadium cavity is shown in Figure I.1. For non-chaotic structures, the results are clearly not exponential: as an example, the length and area distributions for a rectangular cavity are shown in Figure I.1. The results for non-chaotic structures depend on the particular structure being studied: the properties of non-chaotic structures are not universal, as can be seen, for instance, by comparing the distributions found for the circle[22] with those for the rectangle[19].

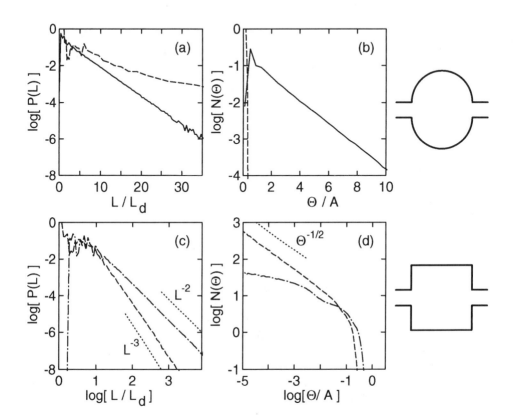

FIGURE I.1. Classical distributions of length [(a),(c)] and effective area [(b),(d)] for the stadium (solid line) and rectangular (dashed line) billiards shown. (a)-(b) Semi-log plot of the distributions: in the stadium both distributions are close to exponential after a short transient region and are very different from the distributions for the rectangle. The exponential distributions are a signature of the single scale characteristic of chaotic systems. (c)-(d) Log-log plot of the distributions in the rectangle show the power-law behavior characteristic of non-chaotic systems in which there is a distribution of scales. Different power laws are indicated by dotted lines. The dash-dotted lines are the two-particle distributions of length or area differences [Eqs. (I.36) and (I.37)]. Note in panel (c) that the angular correlations change the exponent. In panel (d), note the cutoff in both the single and two particle distributions at $\Theta \sim A$. (A is the area of the cavity; L_d is the direct length between the leads.) (From ref. [19].)

III. CLASSICAL SIZE EFFECTS

Before turning to transport in the coherent ballistic regime, it is useful to consider classical transport in ballistic structures. This type of transport occurs when the elastic mean free path is much larger than the sample but the phase-breaking length and wavelength are both smaller than the sample. In this regime, several magnetotransport anomalies are produced by the scattering from the geometry and are thus classical size effects[1]. The unifying theme of these phenomena is that the scattering from the geometry produces a non-equilibrium distribution of the electrons, usually in the sense that the distribution is not isotropic but depends on the direction in which the electron is moving.

A. Bend Resistance

Perhaps the simplest classical magnetotransport anomaly is the resistance caused by forcing a current to turn a corner at a junction between two wires [see Figure I.2]. This "bend resistance" is tremendously enhanced in the ballistic regime because of the strong tendency of the particles to just go straight through the junction[24]. It is straightforward to see this by writing an expression for the bend resistance, R_B, in terms of the transmission coefficients for a symmetric four lead junction. From Eq. (I.3)

$$R_B = (h/2e^2)(T_R T_L - T_F^2)/D \qquad (I.4)$$

where

$$D = (T_R + T_L)[2T_F(T_F + T_R + T_L) + T_L^2 + T_R^2], \qquad (I.5)$$

T_F is the forward transmission, T_R is the probability to turn right, and T_L is the probability to turn left. In fact, because the strong tendency for forward transmission $(T_F > T_R, T_L)$ must be countered by a voltage on the lead directly across from the injecting lead, the bend resistance turns out to be negative[25]. As a magnetic field is applied, the electron paths are bent, and the enhancement of the bend resistance is eliminated as the proportion of straight-through paths decreases.

Even more surprising than the existence of a simple bend resistance is the non-local nature of the resistance caused by a bend[24,25]: in a straight wire with two junctions to other wires, if the current is made to turn a corner at the first junction then a voltage develops across the leads at the second junction [see Figure I.2]. The first junction filters the injected isotropic distribution of electrons, and the second junction responds to this non-equilibrium distribution by developing a voltage drop. One can show explicitly that equilibration of the electron distribution between the two junctions eliminates the effect[26].

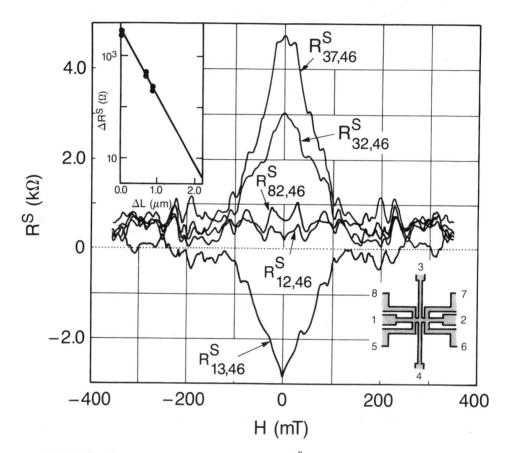

FIGURE I.2. The symmetric magnetoresistance R^S found at 280 mK using different current and voltage leads in the device shown schematically in the lower right. The resistance is measured using different current leads with the same voltage leads (labeled 4 and 6), which are separated by 900 nm. Note (1) the negative bend resistance $R_{13,46}^S$, (2) the increase in the resistance when the current path bends at a junction $R_{12,46}^S \rightarrow R_{32,46}^S \rightarrow R_{37,46}^S$, and (3) the non-local effect of a bend $R_{82,46}^S - R_{12,46}^S$. The inset in the upper left shows the decay of the non-local bend resistance as the distance between the bend in the current path and the voltage lead increase. (From ref. [24].)

B. Collimation

A second example of the connection between ballistic transport and a non-equilibrium distribution of electrons is the presence of collimation in a smooth horn connecting a wire to a two-dimensional region[27]. Collimation is closely related to the presence of an adiabatic invariant[28] and can be explained from either a classical or quantum point of view [see Figure I.3].

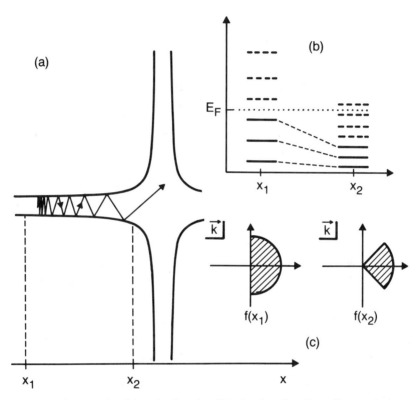

FIGURE I.3. Schematic of the physics of collimation in a junction with rounded corners as shown in panel (a). (b) The threshold energy of the transverse subbands at two places in the structure. For gradual grading, the injected electrons conserve their mode number (dashed lines). Current initially injected in all modes below the Fermi energy (solid lines at x_1) is carried only by low-lying modes near the junction. (c) Distribution functions of classical particles at two places in the structure. The grading rotates the k-vector of the classical particles into the forward direction, as indicated by the trajectory in part (a). Thus particles injected with a hemispheric distribution of k-vectors emerge collimated into a cone. (From ref. [39].)

Classically[27], a particle entering the rounded-corner structure of Figure I.3 with a large transverse momentum, $k_\perp$, has its k-vector rotated by the gradual widening of the wire. Thus, it enters the junction region with a large component parallel to the injecting lead, $k_\parallel$ [see the trajectory in Figure I.3(a)]. So, particles injected with a hemispheric distribution far from the junction reach the junction collimated into a forward cone [Figure I.3(c)]. Quantum-mechanically[29], suppose one injects a wave-packet from the left; what is the behavior of this wave-packet at later times? Figure I.3(b) shows the thresholds for the local transverse subbands at x_1 and x_2. Since at x_1 the wire is narrow, the subbands are widely spaced, while at x_2, the subbands are much closer together. If the widening of the wire is gradual, the electrons will travel adiabatically from 1 to 2, staying in the same mode[28]. Thus, particles injected from a reservoir – for which the current is evenly distributed across all the subbands below E_F [solid lines at x_1 in Figure I.3(b)] – end up in the low-lying transverse subbands as they approach the junction (solid lines at x_2).

Thus, arguing either classically or quantum mechanically, one finds that a non-equilibrium momentum distribution is created in the junction region. "Collimation" refers to this structure in the momentum distribution.

Experimentally, collimation in microstructures has been demonstrated by Molenkamp et al.[30]. They consider a point contact, which acts like a single gradually-widened wire, connected to a large two-dimensional region with a second point contact on the other side. A magnetic field sweeps the electrons emitted from the first contact across the second one and shows that the angular distribution of the electrons is narrower than expected from a diffusive source, hence demonstrating collimation. A more refined experiment by Shepard et al.[31] has mapped out the angular distribution of the intensity injected by a wire, showing both classical collimation and some effects of quantized modes. Some of their data is shown in Figure II.6 of Part II by Westervelt.

Collimation influences magnetotransport because the scattering at the junction [the $T^{j,n}$ in Eq. (I.2)] depends on the momentum distribution. Near the junction, the adiabatic approximation breaks down since the equi-potential contours must turn through 90 degrees. From this point on the electron motion is sensitive to the initial conditions, and the electrons end up in different leads with a wide range of angles. The fact that the scattering at a junction is very different for different angles (or modes in the quantum context)[24,26] implies that the physical result of injecting a non-equilibrium momentum distribution could be quite different from injecting the current in all the modes.

C. Quenching of the Hall Resistance

One physical phenomenon for which collimation is crucial is the suppression of the Hall resistance measured in junctions between very narrow ballistic wires –

a phenomenon known as the quenching of the Hall resistance. The data from the first observation of this phenomenon by Roukes *et al.*[32] is reproduced in Figure I.4. In contrast to the linear dependence of the Hall resistance R_H on magnetic field that is expected in a two-dimensional system (the Lorentz force causing the Hall resistance is linear in B after all), very small values of R_H were observed for an interval of B about zero. This initial observation was later confirmed by several other groups who explored the phenomenon in detail[33-36].

Collimation is present in such simple junctions because of the inevitable rounding of the corners forming the junction. In order to study ballistic transport, the experimental system of choice is the *GaAs/AlGaAs* heterostructure system. In such structures the potential defining the wires is created electrostatically by charges in the dopant layer, at the surface, and on metal gates. Because the distance from these charges to the wire is usually several hundred angstroms, the

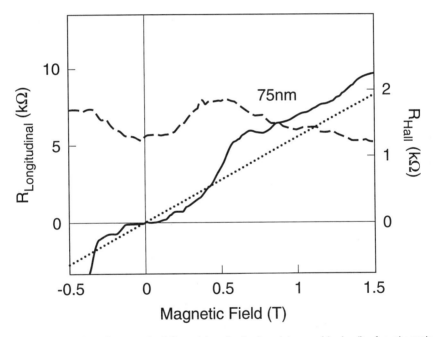

FIGURE I.4. Hall resistance (solid) and longitudinal resistance (dashed) of a six-probe Hall bar composed of 75 nm wide wires. The Hall resistance is suppressed at small magnetic field compared to the two-dimensional value (dotted line). This "quenching" of the Hall resistance is caused by the combination of ballistic transport and rounding of the corners at the junction. $T = 4.2$K. (From ref. [32].)

corners in a junction will be rounded with a radius of curvature of the same order. Such rounding produces a horn-like shape in each of the incoming wires, and hence produces collimation. Other mechanisms for producing a non-isotropic distribution of the current – the essential feature of collimation – have been suggested[37] and may also be present in the experimental structures.

The connection between collimation and the quenching of the Hall resistance was first recognized by Baranger and Stone[29] and demonstrated explicitly in a classical calculation by Beenakker and van Houten[38]. To see this connection, write the Hall resistance for a four-fold symmetric junction using the multiprobe formula (3):

$$R_H = (h/2e^2)(T_R^2 - T_L^2)/D \qquad (I.6)$$

where T_R is the probability to turn right, T_L is the probability to turn left, and D is defined in Eq. (I.5) above. We emphasize that this expression for the resistance is that appropriate to a transport measurement in which currents are applied and the chemical potentials of reservoirs are measured[10]; the behavior of the electrostatic potential in the microstructure (the Hall field, for instance) is a separate topic which we will not discuss here. It is clear that the Hall resistance depends on both the probability to turn a corner and on the difference in the probability to turn left or right. To emphasize this, we rewrite R_H in terms of the relative asymmetry between left-turning and right-turning electrons $\alpha \equiv (T_R - T_L)/(T_R + T_L)$, the total probability to turn a corner $T_{\text{turn}} \equiv T_R + T_L$, and $\tilde{D} \equiv (2e^2/h)D/T_{\text{turn}} = (2e^2/h)[2T_F(T_F + T_{\text{turn}}) + T_{\text{turn}}^2(1 + \alpha^2)/2]$:

$$R_H = \alpha T_{\text{turn}}/\tilde{D}. \qquad (I.7)$$

A detailed analysis[39] showed that collimation influences R_H either by decreasing T_{turn} because intensity is shifted to forward scattering T_F, or by decreasing α because particular paths are selected which minimize the difference between right and left transmission. In most structures both effects are present and contribute to quenching; a large and robust quench seems to require a strong suppression of T_{turn}. Finally, while the basic quenching phenomenon is certainly a classical effect, quantum effects were shown to enhance quenching – leading to a more robust suppression of R_H – and are likely to be important in the experiments[39].

D. The Importance of Short Paths

The importance of short paths in producing magnetotransport anomalies has been dramatically demonstrated in several experiments in which modifications to the geometry of the junction produced large changes in the Hall and bend resistances[34,35]. Data from one experiment[34] is shown as Figure II.4 in Part II by Westervelt. In this case, a negative R_H was produced by introducing a flat wall angled at 45 degrees from which the electrons reflect into the opposite lead. It was shown theoretically[40] that the presence of the 45 degree walls by themselves is *not* sufficient to produce a negative R_H; collimation going into the junction region is required to select the "rebound trajectories". The combination of collimation with certain short paths specific to the sample geometry can explain the main features of the classical magnetotransport anomalies.

E. The Difficulty of Observing Classical Chaos in Nanostructures

The observation of these classical size effects in magnetotransport suggested the possibility of directly observing classical chaos in nanostructures[41]. Since the hallmark of classical chaos is the sensitivity to initial conditions, a direct observation should consist of seeing this sensitivity. In transmission through a four lead junction such sensitivity is apparent in the dependence of the exit lead on the incoming angle or position. By injecting electrons with a particular angle (or mode quantum mechanically), one should produce a strongly fluctuating transmission coefficient into each lead; measuring these fluctuations would constitute a direct measurement of the classical chaos[41].

Unfortunately it has proved very difficult to create the highly controlled initial distributions of electrons that one needs to see these effects. Various geometrical features have been investigated as possible electron selectors – the filtering produced by bends or junctions, the collimation produced by small constrictions, and the refraction produced by a potential well, for instance[42]. While all of these approaches produce a non-isotropic distribution function of the electrons, none of them produces a beam of electrons sufficiently narrow for the classical chaos experiments. *It is ironic that the presence of classical chaos in nanostructures turns out to be more easily visible in the quantum transport properties than in the classical regime.*

There have been several indirect indications of classical chaos in nanostructures. For instance, there are several features of the magnetotransport through junctions or arrays of antidots that are difficult to explain with the simple combination of collimation and short paths – in particular the presence of "plateaus" in $R_H(B)$ at fields below that expected for the quantized Hall effect. The most convincing explanation of these features, put forward by the group of Geisel[43], relies on the presence of chaos and in fact on a mixed phase space connected to the presence of a soft-wall potential in the *GaAs/AlGaAs* system.

IV. QUANTUM TRANSPORT IN OPEN SYSTEMS

We now turn to the main subject of this review, namely coherent transport in systems which are connected by leads to reservoirs. To demonstrate the main phenomena in this regime of transport, and so motivate what needs to be explained, we show the results of a numerical calculation of the transmission through an asymmetrized stadium in Figure I.5 in both the classical and quantum limits[19]. The classical transmitted flux increases linearly as a function of k as the incident flux ($\propto kW/\pi$) increases. The average quantum result also increases

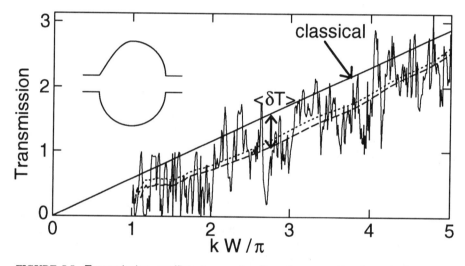

FIGURE I.5. Transmission coefficient as a function of wavevector through the cavity shown. The straight solid line is the classical transmission T_{cl}, the fluctuating solid line is the full quantum transmission T_{qm}, and the dashed (dotted) line is the smoothed T_{qm} at $B=0(BA/\phi_0=0.25)$. Note the following four main effects. (1) The smoothed T_{qm} increases with the same slope as T_{cl}. (2) The smoothed T_{qm} is smaller than T_{cl} by a substantial amount, $\langle\delta T\rangle$, which is sensitive to B and therefore constitutes an average magnetoconductance. (3) T_{qm} fluctuates by an amount of order unity, var$(T) \sim 1$. (4) The spectrum of T_{qm} has a characteristic scale, k_c or B_c, which can be related to classical quantities. (W is the width of the leads, and A is the area of the cavity.) (From ref. [19].)

roughly linearly, but lies below the classical curve, largely because of mode effects from confinement in the lead. Part of this offset is sensitive to a weak magnetic field; this is the weak-localization effect. Perhaps the most striking feature of Figure I.5 is the fine structure in the quantum curve. These are the conductance fluctuations; note that "fluctuations" in this mesoscopic context means a sensitivity to an external parameter such as B or k and does not refer to time-dependent behavior.

The phenomena in Figure I.5 divide roughly into three types. First, there are the effects of having *subbands* in the leads: the shift of the average classical conductance relative to the quantum conductance and the staircase behavior of the transmission as the number of modes (N =Integer $[kW/\pi]$) increases. Second, there is the *shape* of the quantum transport corrections: the characteristic field or wave-vector to which they are sensitive, or more precisely the functional dependence on these parameters. Third, there is their *magnitude*: simply how big these quantum corrections are compared to the classical conductance, characterized for instance by the variance of the fluctuations. While in principle one can ask for an explanation of each bump and wiggle – each "feature" – in the data of Figure I.5, this is on the one hand a very difficult task and on the other hand a not very interesting question – the experimental potentials are not sufficiently well known, after all, to make such a detailed theory worthwhile. Thus we will consider only statistical aspects of the data and theory, treating each type of phenomena noted in Figure I.5 in turn.

While we use the numerical results in Figure I.5 as motivation here, it is important to note that there was experimental motivation for originally undertaking the theory. While carrying out the work on classical size effects discussed above, several experimental groups noticed[33,34] that at lower temperature the magnetotransport curves developed aperiodic fluctuations – the signature of mesoscopic effects. Much of this data remains unpublished because it was not clear how to separate any ballistic interference effects from the known impurity scattering mechanisms for fluctuations. This data did provide, however, an important motivation for the first theoretical studies[17]; systematic experiments required a few more years[45-53].

A. Subbands in Wires

The simplest quantum effect to observe is the quantization of the transverse motion in a wire. The width of a wire is usually the smallest geometrical length in the problem, and thus quantization on this length scale is the first to set in (as temperature is lowered, for instance) and the most robust. Transverse quantization produces modes, and the wires can be thought of as electron waveguides.

The clearest signature of subband effects is in the conductance of a constriction between two large two-dimensional regions. In this two lead case the conductance is simply

$$G = \frac{2e^2}{h} \sum_{m=1}^{N} \left[\sum_{n=1}^{N} T_{nm} \right] \qquad (I.8)$$

where "m" labels the incoming mode. In the absence of backscattering ($\sum_{n=1}^{N} T_{nm} = 1$), the conductance increases abruptly by the quantum of conductance when a new incoming channel opens. Note that the quantum conductance is always less than the classical conductance, $G_{class} = (2e^2/h) kW/\pi$, because $N \equiv$ Integer$[kW/\pi] \le kW/\pi$. This "quantized point contact conductance" was first observed by van Wees *et al.*[54] and Wharam *et al.*[55]; some data from reference [54] is reproduced in Figure I.6. Subsequent work has investigated the sensitivity to temperature and length of the constriction[56-58].

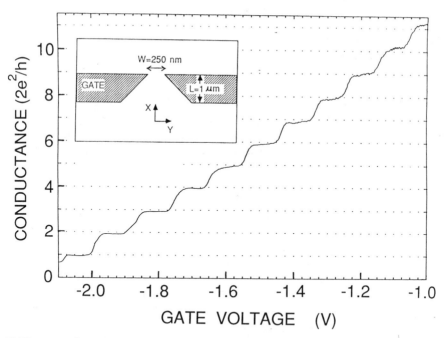

FIGURE I.6. Quantized conductance of the quantum point contact shown schematically in the inset. The conductance was obtained from the measured resistance after subtraction of a constant series resistance of 400 Ω. (From ref. [57].)

The conductance of a point contact has been extensively studied theoretically. Quantization occurs in both very smooth constrictions[28] – the adiabatic limit connected to collimation discussed above – and in the abrupt limit[59]. Longitudinal quantization within the constriction can produce additional structure in the conductance[59,60]. In the experiment, some disorder is present, of course, and so backscattering is not strictly absent. For weak potentials, the rate of backscattering in the constriction is given by the two-dimensional transport mean free path. If the potential is strong but smooth, as is apparently the case in the early experiments[61], the backscattering is dominated by the trapping of an extra mode in the middle of the wire; in this case the length scale for breakdown is the correlation length of the potential not the transport mean free path[62]. An example of a calculated potential profile in this regime is shown in Figure I.7; note that despite the apparent roughness of the potential, the mean free paths are large and quantization of the conductance is good in this case.

Transverse quantization in wires is apparent in more complicated structures as well. One example has already been shown in Figure I.5: the quantum $T(k)$ curve for this cavity lies largely below the classical curve and has some residual staircase structure. Another example is the peaks in the resistance of four-lead junctions associated with threshold singularities[63-66].

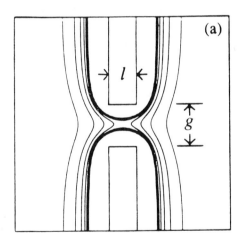 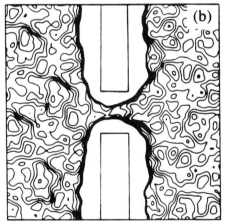

FIGURE I.7. Gate pattern on surface and density of electrons in two-dimensional electron gas for point contacts with $g = 0.3\,\mu m$ and $l = 0.2\,\mu m$. Contours start from 0 and are $4.2\times10^{11}\,cm^{-2}$ apart, corresponding to 1.5 meV; there are two transverse subbands occupied in the constriction. The effect of the random potential is clear by comparing (a), in which there are no random donors, with (b) in which the impurity potential is present. Despite the randomness apparent in (b), the mean-free-path is several microns and the conductance shows good quantization. (From ref. [62].)

B. Shape of Quantum Transport Effects

The main idea of this section is to connect the shape of quantum transport effects – the characteristic scale or functional form of the variation in B or k – with the nature of the classical mechanics, particularly whether it is chaotic or not. The tool for making this connection is semiclassical theory, which has undergone rapid development in the quantum chaos community in recent years[3]. The result[16-22] is a connection between the shape of the quantum effects and the classical distribution functions discussed in the introduction: the qualitative difference in the classical distributions – exponential for chaotic, power law for non-chaotic – produces a qualitative difference in the shape of the quantum transport effects.

1. Basic Semiclassical Theory

Semiclassical theory is an approximation to quantum mechanics valid as $\hbar \to 0$, or in our case as $N \to \infty$. The general result of semiclassical theory is that quantum properties are written as a sum over classical paths of the square root of a classical probability times a phase factor given by the classical action. In the case of transport, the relevant quantity to consider is the transmission amplitudes between modes in the leads. These transmission amplitudes can be written in terms of the Green's function at energy E by simply projecting onto the transverse wavefunctions in the leads:

$$t_{nm} = -i\hbar (v_n v_m)^{\frac{1}{2}} \int dy' \int dy\ \phi_n^*(y') \phi_m(y) \qquad (\text{I.9})$$

$$\times G(L, y'; 0, y; E),$$

where $\phi_m(y)$ is the transverse wavefunction for the incoming mode, $\phi_n(y')$ is for the outgoing mode, $x = 0$ denotes the incoming lead, and $x = L$ denotes the outgoing lead. The intuitive interpretation of the above equations as arriving at the cavity in mode m, propagating inside the cavity, and exiting in mode n is quite straightforward.

These equations constitute an exact starting point which is also the basis of our numerical calculations[67]. As described in detail in reference[39], we obtain the Green's function of a tight-binding Hamiltonian (equivalent to a real-space discretization of the Schrödinger equation) by a recursive algorithm and then do a projection onto the transverse wavefunctions in the leads. This yields, then, the exact transmission amplitude of the discretized problem and the conductance follows from Eq. (I.1).

The semiclassical approximation to the transmission and reflection amplitudes proceeds by making the standard semiclassical approximation[3] to the Green's function, G. Start with the Feynman path integral for the propagator in time. Expand an arbitrary path in this integral to second order about a classical path. Performing the Gaussian integrals yields an expression for the propagator as a sum over classical paths. The Green's function is the Laplace transform of the propagator; performing this transform by stationary phase yields[3]

$$G^{scl}(y';y;E) =$$

$$\frac{2\pi}{(2\pi i\hbar)^{3/2}} \sum_{s(y,y')} \sqrt{D_s} \exp\left[\frac{i}{\hbar} S_s(y',y,E_F) - i\frac{\pi}{2}\mu_s\right], \quad (I.10)$$

where (s) is a classical trajectory at energy E from y in the entrance cross-section to y' in the exit lead. S_s is the action along the path, D_s is the classical probability for following this path (produced essentially by the Jacobian in the Gaussian path integrals, see footnote[68] for an explicit expression) and μ_s is the topological phase[3] given by the number of constant-energy singular points (e.g. caustics).

The semiclassical expression for the transmission amplitudes is obtained by inserting the semiclassical Green's function (I.10) into the expression (I.9) and performing the integrals over the cross-sections by stationary phase. For structures with hard walls, the action is simply k times the length of the path, and the result[17,19] for the transmission amplitude is

$$t_{nm} = -\frac{\sqrt{2\pi i\hbar}}{2W} \sum_{s(\theta',\theta)} \text{sgn}(\theta'\theta)\sqrt{\tilde{A}_s} \exp[ik\tilde{L}_s - i\frac{\pi}{2}\tilde{\mu}_s]. \quad (I.11)$$

The phase factor is $k\tilde{L}_s = S_s/\hbar + ky\sin\theta - ky'\sin\theta'$ plus an additional phase associated with singular points in the classical dynamics[17,19]. The prefactor is $\tilde{A}_s = |(\partial y/\partial\theta')_\theta|/(W\cos\theta')$.

Notice that the semiclassical expression (I.11) has the usual form for a semiclassical approximation: it is a sum over classical paths of the square root of the classical probability times a phase factor related to the action. The most important point is which paths enter the sum: those which enter at (x,y) with fixed angle $\sin\theta = \pm m\pi/kW$ and exit at (x',y') with angle $\sin\theta' = \pm n\pi/kW$. These are exactly the paths that one would expect from a naive semiclassical picture of propagation in a wave guide: the paths which are important are those whose transverse k-vector matches the wavevector of the mode.

Since we are mainly interested in the conductance, it is convenient to write explicitly the semiclassical expression for the total transmission intensity summed over modes:

$$T(k)= \sum_{n,m} |t_{nm}|^2 = \frac{1}{2} \frac{\pi}{kW} \sum_{n,m} \sum_{s} \sum_{u} F_{n,m}^{s,u}(k), \qquad (\text{I}.12)$$

$$F_{n,m}^{s,u}(k) \equiv \sqrt{\tilde{A}_s \tilde{A}_u} \, \exp[ik(\tilde{L}_s - \tilde{L}_u) + i\pi\phi_{s,u}] \qquad (\text{I}.13)$$

where the phase $\phi_{s,u}$ accounts for both the phase produced by singular points and the $sgn(\cdot)$ factor in Eq. (I.11). At this point, one could proceed to evaluate this semiclassical expression directly by summing over the paths numerically. In fact, such direct evaluation has been extensively investigated for closed chaotic systems[69] and has been recently applied to quantum ballistic transport[70]. We prefer, however, to proceed by approximate analytic evaluation of the semiclassical expressions.

2. Average Conductance

Although the most dramatic effect of quantum coherence visible in the data of Figure I.5 is the fluctuations of the conductance, we start our discussion of the shape of interference effects in ballistic billiards by considering the average conductance since it is a simpler quantity to analyze theoretically. For a given ballistic microstructure the average can only be defined by summing over some additional variable (such as wavevector or magnetic field) upon which the conductance is assumed to depend ergodically. Such an average will eliminate all the aperiodic fine structure seen in Figure I.5 leaving any additional smooth dependence on parameters of interest. The main theoretical results for the average conductance reviewed below are, first, that the leading contribution to the average quantum conductance is simply the classical conductance[19,39] and, second, that there is a quantum correction sensitive to the magnetic field whose shape depends on the nature of the classical dynamics[18,19]. Such a correction – analogous to weak-localization in disordered structures[71] – has been observed experimentally in ballistic nanostructures[45-53].

Because the classical transmission coefficient is proportional to kW/π (the outgoing flux is proportional to the incoming flux), we expect a linear contribution to the average quantum transmission; call this slope $\mathbf{T}$. To study magnetic field dependence, we will use an average over k, defined by

$$\langle A \rangle \equiv \lim_{q \to \infty} \frac{1}{q} \int_{q_c}^{q_c+q} dk A(k), \quad q_c W/\pi \gg 1. \qquad (\text{I}.14)$$

Using this procedure to average $T(k)/(kW/\pi)$ over all k, one finds

$$\mathbf{T} \equiv \langle \frac{\pi T(k)}{kW} \rangle = \langle \frac{1}{2} (\frac{\pi}{kW})^2 \sum_{n,m} \sum_s \sum_u F_{n,m}^{s,u}(k) \rangle_k. \qquad (\text{I}.15)$$

As k becomes large, the modes become closely spaced in angle, and the sum over mode number can be converted to an integral over angles, $(\pi/kW) \sum_n \rightarrow \frac{1}{2} \int_{-1}^{1} d(\sin\theta)$. After this conversion, the only k dependence that remains is in the phase factors, $\langle \exp[ik(\tilde{L}_s - \tilde{L}_u)] \rangle$. The average over an infinite range of k means that the length of path u must be *exactly* equal to the length of path s[72]. Both of these paths connect the same modes, m to n, and so must both be chosen from the set of paths which connect incoming angle θ to outgoing angle θ'. How many such paths are there? In a typical structure there will be a *discrete* number of such paths, infinite in the chaotic case, but countable[3]. In the absence of a symmetry which causes a degeneracy in this set, in order for the length of the two paths to be identical, the two paths must be the same:

$$\langle \exp[ik(\tilde{L}_s - \tilde{L}_u)] \rangle = \delta_{s,u}. \qquad (\text{I}.16)$$

Thus the interference terms completely drop out[72] of the expression for $\mathbf{T}$,

$$\mathbf{T} = \frac{1}{2} \int_{-1}^{1} d(\sin\theta) \int_{-1}^{1} d(\sin\theta') \sum_{s(\theta,\theta')} \tilde{A}_s. \qquad (\text{I}.17)$$

Using the definition of $\tilde{A}_s$, one can change variables from $\sin\theta'$ to the transverse initial position y and arrive at the usual expression[38] for the classical probability of transmission

$$\mathbf{T} = \frac{1}{2} \int_{-1}^{1} d(\sin\theta) \int_{0}^{W} \frac{dy}{W} f(y,\theta) \qquad (\text{I}.18)$$

where $f(y,\theta) = 1$ if the trajectory with initial conditions (y,θ) is transmitted and $f(y,\theta) = 0$ otherwise. Thus as expected, the leading order term in the average quantum conductance as the Fermi energy goes to infinity is just the classical conductance[39].

For the quantum corrections to the average conductance, there is one contribution which is quite easy to see: the coherent backscattering of exactly time-reversed paths. To derive this contribution, note that the quantum corrections to the reflection coefficient R can be separated into the terms diagonal in mode number, $\delta R_D \equiv \sum_{n=1}^{N} \delta R_{nn}$ and the off-diagonal terms,

$$\delta R = \frac{1}{2} \frac{\pi}{kW} \left[\sum_{n} \sum_{s \neq u} F_{n,n}^{s,u} + \sum_{n \neq m} \sum_{s \neq u} F_{n,m}^{s,u} \right]. \qquad (I.19)$$

From results on both disordered systems[71,73] and quantum-chaotic scattering[74-76], it is now well-known that a typical diagonal reflection element is twice as large as a typical off-diagonal element when the system is time-reversal invariant. The ratio of mean diagonal to off-diagonal elements is known as the elastic enhancement factor, which is hence equal to 2 in the chaotic case. This is referred to as the coherent backscattering peak. Since we are concerned with the shape of the magnetotransport response, we wish to calculate how this coherent backscattering peak changes as the magnetic field breaks time-reversal invariance.

The argument for the coherent backscattering effect proceeds in much the same way as the calculation of the leading order term T or R (for reflection) above. In calculating $\langle \delta R_D \rangle$, the sum of N reflection elements each with $|\sin\theta| = |\sin\theta'|$ is converted to an integral over angle, $(\pi/kW)\sum_n \to \int d(\sin\theta)$. Then, the only k-dependence is in the exponent so that the average eliminates all paths except those for which $\tilde{L}_s = \tilde{L}_u$ exactly[72]. In the absence of symmetry $\tilde{L}_s = \tilde{L}_u$ only if $s = u$, but with time-reversal symmetry ($B = 0$) $\tilde{L}_s = \tilde{L}_u$ also if u is s time-reversed.

Introducing a magnetic field in general changes both the classical paths traversed and the action along a given path; however, we will be considering low-field effects in which the change in the geometry of the paths is negligible and only the phase difference between time-reversed paths is important. For time-reversed pairs this phase difference (due to the Aharonov-Bohm effect) arises due to the different sign of the "enclosed flux", $(S_s - S_u)/\hbar = 2\Theta_s B/\phi_0$ where $\Theta_s \equiv 2\pi \int_s \vec{A} \cdot \vec{dl}/B$ is the effective "area" enclosed by the path (times 2π) and $\phi_0 = hc/e$. Taking into account this phase change, we obtain

$$\langle \delta R_D(B) \rangle = \frac{1}{2} \int_{-1}^{1} d(\sin\theta) \sum_{\substack{s(\theta,\theta) \\ s(\theta,-\theta)}} \tilde{A}_s e^{i2\Theta_s B/\phi_0} \qquad (I.20)$$

This yields an order unity (k-independent) contribution to the average *quantum* conductance which contains only quantities determined by the *classical* scattering dynamics.

An approximate but plausible assumption about the classical dynamics allows one to evaluate $\langle \delta R_D \rangle$ explicitly. In a chaotic system in which the mixing time for particles within the cavity is much shorter than the escape time, the particles are uniformly distributed in the available phase space. Hence, the probability of scattering out with any angle θ' is just proportional to the projection of the lead on that direction, and one can assume the outgoing distribution is uniform in $\sin\theta'$ for an arbitrary distribution of incoming trajectories. Classical simulations confirm that this is approximately obeyed for the structure in Figure I.5 and improves if the opening to the leads is made smaller. Thus, the sum over back-scattered paths in Eq. (I.20) can be replaced by an average over all $\sin\theta'$, yielding an expression for $\langle \delta R_D(B=0) \rangle$ which is the same as that for **R** [defined in the same way as **T** in Eq. (I.17)]. To obtain the dependence of δR_D on B, group the backscattered paths by their effective area and average over the distribution of this area, $N(\Theta)$,

$$\langle \delta R_D(B) \rangle \sim \int_{-\infty}^{\infty} d\Theta N(\Theta) \exp[i2\Theta B/\phi_0]. \qquad (I.21)$$

Again, to obtain this expression one assumes that the distribution is uniform in $\sin\Theta$ and hence $N(\Theta)$ is independent of θ.

Knowledge of the classical area distribution, reviewed in the introduction, can now be used to obtain explicit predictions for the magnetic field dependence of the quantum conductance. First, consider the chaotic case and assume that the exponential form of the area distribution holds for *all* Θ. Performing the integral in Eq. (I.21) and combining this result with that for the magnitude, one obtains[18] a Lorentzian B-dependence:

$$\langle \delta R_D(B) \rangle \approx \frac{R}{1+(2B/\alpha_{cl}\phi_0)^2}. \qquad (I.22)$$

Note that the field scale can be much smaller than one flux quantum through the area of the cavity since the particle may bounce many times before escaping and enclose an area much larger than the area of the cavity.

Second, consider the non-chaotic case, focusing here on circular and polygo-nal billiards. Since the classical area distribution is power-law, not exponential, one expects that the shape of the average magnetotransport will be different[77]. This is the central conclusion of the semiclassical theory: *the difference in the nature of the classical dynamics in chaotic versus non-chaotic cavities (exponen-tial versus power-law escape) produces a qualitative difference in the shape of the average magnetotransport.* The theoretical argument given above actually deals with only part of the quantum magnetoconductance, namely the coherent backscattering peak. The conclusion is believed to be true for the full

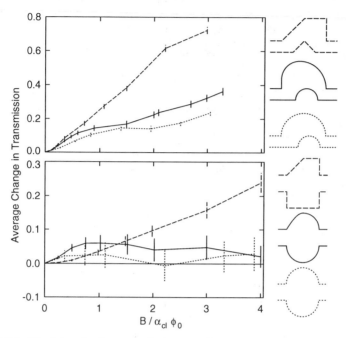

FIGURE I.8. Weak-localization magnitude as a function of magnetic field for the six structures shown. The magnitude is obtained from $\langle T(k,B)-T(k,B=0)\rangle_k$ with $kW/\pi \in [4,11]$. Note the difference between the chaotic and regular structures, as well as the sensitivity to symmetry in the lower panel. α_{cl} is the inverse of the typical area enclosed by classical paths. (From ref. [18].)

conductance; we return to this point in discussing difficulties with the semiclassical theory below.

In order to see whether the semiclassical argument summarized here actually applies to transport through ballistic nanostructures, comparison with both numerical calculations and experiments have been performed. First, with regard to numerical calculations, Figure I.8 shows the average quantum conductance as a function of magnetic field for six different stadium-like and polygonal billiards. The top panel demonstrates the difference between chaotic and regular structures discussed above: the curves for the half-stadia (chaotic) flatten out while that for the half-asymmetric-square (regular) increases linearly (except for very small B where it is quadratic). Thus these numerical results are in agreement with the semiclassical argument connecting the lineshape to the distribution of classical areas. The lack of saturation in some of the chaotic structures is probably due to the bending of the classical paths by the magnetic field (cyclotron radius effects) caused by the small size of the structures. The lower panel of Figure I.8 shows that the weak-localization effect is present even for structures without stoppers. The error bars are larger than in the upper panel because of the greater variation with k produced by the direct paths. However, it is important for experiments that the direct paths do not mask the change in the average conductance and the difference between the chaotic and regular cavities is still clear.

Finally, we comment on the relation between the theoretical results and the experiments[45-53]. As for the theory, extraction of the experimental weak-localization effect from the conductance fluctuations requires an averaging procedure. In experiments this averaging may be achieved from thermal smearing through the Fermi function[45], from explicit energy averaging by varying the potential on a gate[47,48,51], or from measuring the properties of many structures simultaneously[49]. In all cases, the *average* experimental magnetoresistance shows a maximum at $B=0$, in basic agreement with the theory. Of course, because of the conductance fluctuations it is possible to have either a minimum or a maximum at $B=0$ in an *unaveraged* $(T=0)$ magnetoresistance trace.

The most dramatic experiment on the average quantum conductance is that of reference [49] in which a clear difference in lineshape is observed between stadium and circle cavities. Data from this experiment is shown in Figure II.18 of Part II by Westervelt: the magnetoresistance of an array of 48 cavities of each shape. The difference between the rounded Lorentzian shape observed for the stadia and the triangular shape observed for the circles is indeed striking and agrees with the general semiclassical theory outlined above.

Perhaps the most striking feature of this experiment is that the signature of the non-chaotic dynamics in the circles is so clear. After all, the experimental cavities are certainly not perfect circles – there is some surface roughness as well as the smoothly varying disordered potential present in modulation doped heterostructures. To shed some light on this feature of the experiment, calculations were undertaken[49] in disordered cavities with a shape similar to those in the experiment; results are shown in Figure I.9. The top panel shows that the magnetoresistance of the stadium is Lorentzian with or without the disorder, as expected. The bottom panel shows results for circles in three cases: completely ballistic, strong surface roughness, and smooth disorder (solid squares) chosen so that both the transport mean-free-path and the total elastic scattering rate approximately match those in the experiment. In this last case, the magnetoresistance retains the unusual triangular shape, giving one confidence that it is the non-chaotic dynamics in the underlying circle that produces the triangular shape in the experiment. The fact that the experimental magnetoresistance is sharper than the calculation is not understood, but may be connected to the softness of the confining potential in the experimental cavities.

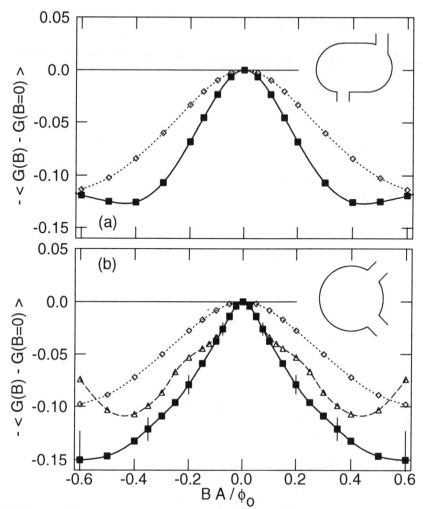

FIGURE I.9. Calculated magnetoconductance $(\times - 1)$ as a function of flux through the geometric area of the cavity for the (a) stadium and (b) circle shown as insets. The lineshape is Lorentzian for the ballistic stadium (solid squares) as well as for a stadium with strong surface roughness scattering (diamonds). The lineshape is more triangular for both the ballistic circle (triangles) and the circle with a weak smooth disordered potential (solid squares), but changes to Lorentzian for strong surface roughness scattering (diamonds). For the disordered potential, the total mean-free-path is approximately 5 times the diameter of the cavity. The similarity of lineshape between this calculation and the experiment [Figure II.18 of Part II by Westervelt] is striking for both structures. (From ref. [49].)

3. *Conductance Fluctuations*

Having discussed the properties of the average quantum conductance, we now turn to fluctuations about this average, the most striking feature of the data in Figure I.5. The shape of these fluctuations is characterized by their power spectrum. The semiclassical treatment of the power spectrum[17,19] was based on the semiclassical approach to S-matrix fluctuations as a function of energy introduced by Gutzwiller[81] and extensively developed by Blümel and Smilansky[16] and Gaspard and Rice[82]. The main conclusion is that the power spectrum is directly related to properties of the classical phase space and, as is the case for the average quantum conductance, this implies a qualitative difference in the form of the spectrum of chaotic and non-chaotic structures.

The fluctuations in the transmission intensity are defined by (in the absence of any symmetries)

$$\delta T \equiv T - \mathbf{T}\frac{kW}{\pi}. \tag{I.23}$$

To characterize the spectrum, it is natural to introduce the correlation function of the fluctuations in k,

$$C(\Delta k) \equiv \langle \delta T(k+\Delta k)\,\delta T(k) \rangle_k. \tag{I.24}$$

The corresponding Fourier power spectrum

$$\hat{C}(x) \equiv \int d(\Delta k)\,C(\Delta k)\,e^{ix\Delta k} \tag{I.25}$$

gives the spectrum of lengths x that are important in producing the k fluctuations.

The semiclassical analysis proceeds by an argument very similar to that for the average conductance above; most of the details will be omitted here in order to focus on the main assumptions and results. First, a certain part of the correlation function can be easily treated: the part analogous to the diagonal part of the reflection coefficient above is the correlation of transmission coefficients between the same modes

$$C_D(\Delta k) \equiv \langle \sum_{n,m} \delta T_{nm}(k+\Delta k)\,\delta T_{nm}(k) \rangle_k. \tag{I.26}$$

Using the semiclassical expression for the transmission amplitudes and replacing the sums over modes by integrals over angles, one finds that the average over k

forces an exact pairing of the four classical paths involved. The resulting exact semiclassical expression for the power spectrum is[19]

$$\hat{C}_D(x) = \frac{\pi}{8} \int_{-1}^{1} d(\sin\theta) \int_{-1}^{1} d(\sin\theta')$$

$$\times \sum_{s(\theta,\theta')} \sum_{\substack{u(\pm\theta,\pm\theta') \\ u \neq s}} \tilde{A}_s \tilde{A}_u \delta(\tilde{L}_s + x - \tilde{L}_u). \qquad (I.27)$$

Note that this expression is independent of k and so is an order unity contribution to the conductance.

In the chaotic case, one can make progress analytically by assuming that (1) the trajectories are uniformly distributed in the sine of the angle, (2) the angular constraints linking trajectories u and s can be ignored, and (3) the constraint $u \neq s$ can be ignored because of the proliferation of long paths. Then, the only remaining constraint on the trajectories is through their lengths. In terms of the classical distribution of lengths, defined by

$$P(L) \equiv \frac{1}{4} \int_{-1}^{1} d(\sin\theta) \int_{-1}^{1} d(\sin\theta') \sum_{u(\theta,\theta')} \tilde{A}_u \delta(L - \tilde{L}_u), \qquad (I.28)$$

the result for the power spectrum is

$$\hat{C}_D(x) \sim \int_0^\infty dL \ P(L+x)P(L). \qquad (I.29)$$

As discussed above, the distribution of lengths is exponential for chaotic billiards for large L; using this form for all lengths, the final result for chaotic billiards is

$$\hat{C}_D(x) \sim e^{-\gamma_{cl}|x|}. \qquad (I.30)$$

This form for the power spectrum implies that the wavevector correlation function is Lorentzian[16],

$$C_D(\Delta k) \sim \frac{1}{1 + (\Delta k/\gamma_{cl})^2}. \qquad (I.31)$$

The argument for the magnetic field correlation function[17,19] is very similar. The correlation function is again defined as an average over k

$$C(\Delta B) \equiv \langle \delta T(k, B + \Delta B)\, \delta T(k, B) \rangle_k. \qquad (I.32)$$

The difference in action in the phase factors comes from the difference in B and can be expanded to first order in field $[S_s(B+\Delta B) - S_u(B+\Delta B) + S_u(B) - S_s(B)]/\hbar = (\Theta_s - \Theta_u)\Delta B/\phi_0]$. One finds exactly analogous to Eq. (I.29)

$$\hat{C}_D(\eta) \sim \int_{-\infty}^{\infty} d\Theta\ N(\Theta + \eta) N(\Theta). \qquad (I.33)$$

Using the known exponential form of the distribution of effective area, $N(\Theta)$, for all values of Θ yields[17]

$$\hat{C}_D(\eta) \sim e^{-\alpha_{cl}|\eta|}\,(1 + \alpha_{cl}|\eta|) \qquad (I.34)$$

for the power spectrum or

$$C_D(\Delta B) \sim \left[\frac{1}{[1 + (\Delta B/\alpha_{cl}\phi_0)^2]} \right]^2 \qquad (I.35)$$

for the correlation function. Note for consistency that the field scale of the fluctuations is twice that of the average quantum correction [Eq. (I.22)] because the relevant phase involves the difference of two "areas" whereas weak-localization involves the sum.

The semiclassical theory of scattering in non-chaotic structures is less well developed than that of chaotic billiards. Nonetheless, based on the results for chaotic billiards, one expects that the spectrum in k or B should be related to the classical distribution of lengths or effective areas for the particles in the cavity. There are at least two relevant differences in the non-chaotic case: the trajectories group into families and there is angular correlation between incident and outgoing angles (the uniformity assumption used successfully in the chaotic case is certainly not valid). Because of these two differences, the classical distributions will *not* be characterized by a single scale and, as discussed in the introduction, these distributions will be power-law rather than exponential.

In the absence of a careful semiclassical treatment of conductance in non-chaotic billiards, a conjecture for the spectrum in the non-chaotic case has been developed[19] by analogy with the chaotic-billiard results. From Eq. (I.27), the power spectrum $\hat{C}_D(x)$ is evidently related to the distribution of two distinct paths with a difference in length of x. In the non-chaotic case, the two distinct paths must come from different families of trajectories, and the constraint that the two paths enter and exit with the same angle must be retained. Thus the conjecture is:

$$\hat{C}_D(x) \propto \int_0^\infty dL \int_0^1 d(\sin\theta)\, P_2(L+x,L,\theta), \qquad (I.36)$$

where $P_2(L+x,L,\theta)$ is the classical distribution for two distinct trajectories at angle $\pm\theta$, one with length L and the other with length $L+x$. In the chaotic case, this two-particle distribution factorizes into the product of length distributions appearing in Eq. (I.29) but in the non-chaotic case P_2 does *not* factorize. The classical integral on the right-hand-side of Eq. (I.36) has been computed numerically for the rectangular billiard, and the result in Figure I.1 shows that it decays more slowly than $P(L)$: $1/x^2$ for large x in contrast to $1/L^3$. In the case of the power spectrum of $T(B)$, the conjecture is that a similar expression holds in terms of the classical distribution for two trajectories at angle θ with effective areas Θ and $\Theta+\eta$:

$$\hat{C}_D(\eta) \propto \int_{-\infty}^\infty d\Theta \int_0^1 d(\sin\theta)\, N_2(\Theta+\eta,\Theta,\theta). \qquad (I.37)$$

The result of classical simulation in Figure I.1 shows that this distribution is roughly constant up to a cutoff which is less than the area of the rectangle, reflecting the strong flux cancellation[19,79,80] in this structure.

In order to test the validity of these semiclassical expressions, we turn first to numerical results and then to experiment. The first prediction to test is the relationship in the chaotic case between the scale of variation of the quantum conductance and the characteristic classical quantity. Figure I.10 shows that γ_{qm}, α_{qm} — obtained by fitting the calculated quantum power spectra to the semiclassical forms — are indeed given by the classical quantities γ_{cl}, α_{cl} — obtained by classical simulation — to high accuracy[17] while they are varied over roughly two orders of magnitude by changing R/W in the two- and four-probe structures shown. *Thus it is possible to predict quantitatively measurable properties of these ballistic quantum conductors from a knowledge of the chaotic classical scattering dynamics.* Indeed, recent experiments[46,47,52] have investigated the variation of γ_{qm} and α_{qm} with the degree of opening of the cavity. They find the correct trends and quantitatively reasonable values.

The second prediction to test by comparison to numerical results is the form of the power spectrum. Figure I.11 shows the power spectra for several stadium and polygonal billiards. The line in panel (a) is an exponential decay using the classical exponent: the agreement between the full quantum numerical results and

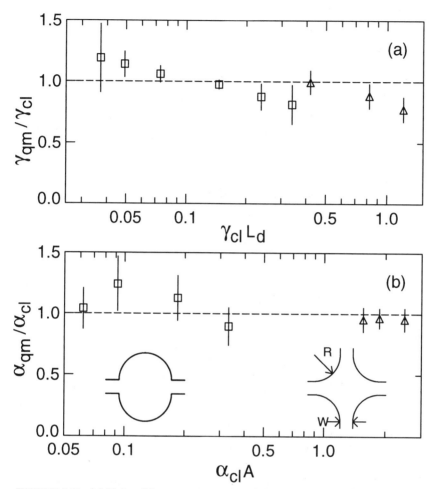

FIGURE I.10. (a) Ratio of the wavevector correlation length (obtained by fitting the power spectrum) to the classical escape rate γ_{cl} as a function of γ_{cl} for both types of structures shown. 4-disk structure (triangles) with $R/W = 1,2,4$, and stadium (squares) with $R/W = 0.5,1,2,4,6,8$. (b) Ratio of magnetic field correlation length (obtained by fitting the power spectrum) to α_{cl}, the exponent of the distribution of effective areas, as a function of α_{cl}. 4-disc structure (triangles) with $R/W = 1,2,4$, and open stadium (squares) with $R/W = 1,2,4,6$. The correlation lengths of the quantum fluctuations agree with the semi-classical prediction over two decades. (A is the area of the cavity; L_d is the direct length between two opposite leads.) (From ref. [17].)

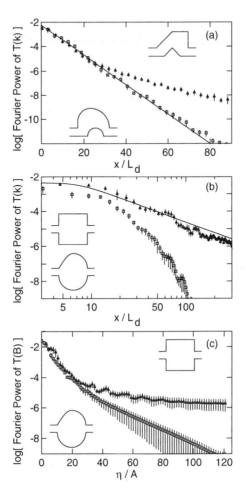

FIGURE I.11. (a) Power spectra of $T(k)$ for the chaotic (squares) and regular (triangles) structures shown. $N = 33$. The regular structure has more power at large frequencies because there are more long trajectories. The line is a fit to the spectrum in the chaotic case: the exponential decay predicted semiclassically holds over 6 decades. (b) Power spectra of $T(k)$ for a second pair of chaotic (squares) and regular (triangles) structures shown on a log-log plot. $N = 21$. The spectrum of the regular structure is consistent with the $1/x^2$ spectrum predicted semiclassically and is certainly very different from the spectrum of the chaotic structure. (c) Power spectra of $T(B)$ for the chaotic (squares) and regular (triangles) structures shown. $N = 6$. The frequency-in-field scale is normalized to the area of the structure, A. The difference between the chaotic and regular structures is apparent in the spectrum of their magneto-fingerprints, and has been studied experimentally[45] [see Figure II.11 of Part II by Westervelt]. (From ref. [19].)

the exponential decay predicted semiclassically is excellent – log $P(x)$ is linear over 6 decades! – when the number of modes is large. The breakdown of semiclassical theory in the few mode limit can be seen numerically[19] and has been studied quantitatively in the context of S-matrix fluctuations[74]. The line in panel (b) is a fit of the spectrum for the rectangle billiard to the conjectured form $1/x^2$; the fit is very good, supporting the conjecture.

All three panels of Figure I.11 show that the non-chaotic structures have more high-frequency power than the chaotic structures: the fluctuations in the non-chaotic case are finer than in the chaotic case[17,19,21,83]. *The difference in the power spectra of chaotic vs. non-chaotic billiards (exponential vs. power-law decay) is explained by the differing classical two-particle distributions of lengths.* Note that one needs at least 2-3 decades of sensitivity to reliably distinguish the spectra of chaotic and non-chaotic billiards.

On the other hand, because the classical distribution of area is bounded for the rectangle [Figure I.1], the conjecture predicts that the spectrum of $T(B)$ should be limited to low frequencies. This is clearly not the case in Figure I.11(c) which looks quite similar to the result for the spectrum of $T(k)$ in panel (a). The reason for the failure of the conjecture Eq. (I.37) is not understood but may well be connected to special properties of the rectangular billiard such as the extreme flux-cancellation effect[19,79,80].

Finally, we note that recent experiments *have* observed different power spectra in ballistic microstructures fabricated with circular vs. stadium-shaped cavities[45,50]. This was the first experimental evidence for a difference between chaotic and non-chaotic behavior in quantum transport. These experiments are discussed in Part II by Westervelt.

4. Difficulties with the Semiclassical Theory

In the above discussion of both the average conductance and conductance fluctuations, the semiclassical theory was used to evaluate only one contribution to the quantum conductance, one that was particularly simple. In the case of the average conductance, for instance, this was the coherent backscattering peak. A semiclassical theory of the complete conductance is problematic at this time. Methods exist for evaluating the full semiclassical conductance [Eq. (I.12)] in both the transmission coefficient approach[84] and in the Kubo formula approach[85-88]. But the result of these methods is physically unappealing and probably incorrect: both approaches predict the absence of any quantum contribution to average conductance (the coherent backscattering contribution is exactly cancelled by other contributions[84]), the Kubo approach does not find any coherent backscattering[86,87], and the transmission approach does not find a periodic orbit contribution to the conductance fluctuations (see the discussion of antidot arrays below). The cause and resolution of these problems are not known at this time. One possibility is that the analytic evaluation of the semiclassical expression (I.12) carried out so far is too crude and that a numerical evaluation of the full semiclassical expression is required[70,89]. Another recent suggestion[88] is that the stationary phase argument used to exactly pair trajectories is in error: there are contributions from pairs of certain trajectories whose phase does not vary rapidly and hence which contribute in spite of having different length or area (for such a pair, for instance, the trajectories may closely follow each other for a long time, of order the Ehrenfest time, before finally separating). A third more radical alternative is that the semiclassical expression (I.12) is intrinsically inadequate for quantum transport and that contributions from non-classical paths are needed[19].

In the absence of a quantitative semiclassical theory of conductance, the argument given above should be regarded as the basis for a conjecture rather than as a demonstration. The conjecture is simply that the shape of the magnetotransport effects – both the average and the power spectrum – will be the same as the shape of the special terms evaluated above. This conjecture is intuitively reasonable since a given property of a chaotic system should be characterized by a single scale which hence should be the same for the full correlation function. As we have seen, the conjecture is supported by considerable numerical evidence.

C. Magnitude of Quantum Transport Effects

Historically, the magnitude of quantum transport effects has been of great interest: in disordered systems, the theoretical result[90,91] that the magnitude is independent of both the strength of the scattering and the size of the system (as long as a single coherent region is considered) – a result known as "universal conductance fluctuations" – attracted a lot of the initial attention to mesoscopic physics[1]. One expects that a similar kind of universality should hold for chaotic systems since disordered materials are one type of chaotic system. In order to show that this is in fact the case, we use random matrix theory to investigate analytically the magnitude of quantum transport effects. The reader may be surprised that the semiclassical theory, so successful in describing the shape, is not extended to treat the magnitude. In fact, because of the difficulties sketched above, we are unable to perform this extension at this time (see [88] however). However, it is interesting to note that the terms which one can evaluate easily with the semiclassical theory – the coherent backscattering peak and the analogous contribution to the fluctuations – give a contribution to the conductance which is independent of the size of the system, see Eqs. (I.20) and (I.27). Thus, the semiclassical theory certainly suggests that "universality" should hold.

Random matrix theory is useful in treating situations in which one knows very little about a particular system other than that it obeys certain symmetries and is somehow "random". The basic idea is to consider the matrix associated with the quantum property in which one is interested, and assume that this matrix is a sample drawn from an ensemble in which all appropriate symmetry constraints are obeyed. Random matrix theory has been most thoroughly developed for Hamiltonians of bounded systems since the discrete set of energy levels that results is one of the more fundamental aspects of quantum mechanics[92]. However, random matrix theory has also been used to treat quantum transport through disordered wires[1] and scattering from complicated potentials[92]. Because of the connection between conductance and transmission, it is this latter work which is related to quantum transport in chaotic cavities. By its very nature, random matrix theory is unable to treat non-chaotic systems; a theory of the magnitude of quantum transport effects in non-chaotic systems will have to wait for a resolution of the semiclassical problems discussed above.

1. Invariant Random S-Matrix Theory

For scattering involving two leads each with N channels, the S matrix describing the quantum scattering problem is

$$S = \begin{bmatrix} r & t' \\ t & r' \end{bmatrix} \qquad (I.38)$$

where r, t are the $N \times N$ reflection and transmission matrices for particles from the left and r', t' for those from the right. Because of current conservation S is unitary, $SS^\dagger = I$, and in the absence of a magnetic field it is symmetric because of time-reversal symmetry, $S = S^T$.

We concentrate on situations where the statistics of the scattering can be described by an ensemble of S-matrices that assign to S an "equal a priori distribution" once the symmetry restrictions have been imposed. In particular, the possibility of "direct" processes, caused by short trajectories and giving rise to a non-vanishing averaged S-matrix, is ruled out; we return to this possibility in a later section. The appropriate ensembles are well-known in classical random matrix theory[92] and are called the Circular Orthogonal Ensemble (COE, $\beta = 1$) in the presence of time-reversal symmetry and the Circular Unitary Ensemble (CUE;$\beta = 2$) in the absence of such symmetry. These ensembles are defined through their invariant measure: the measure on the matrix space which is invariant under the appropriate symmetry operations. To be precise[92] the invariant measure is

$$d\mu(S) = d\mu(S'), \quad S' = U_0 S V_0, \qquad (I.39)$$

where U_0, V_0 are arbitrary fixed unitary matrices in the case of the CUE with the restriction $V_0 = U_0^T$ in the COE. The invariant measure can be written explicitly in several different representations.

In the situation of interest here – transport through a single chaotic cavity – the S-matrix varies as a function of k or B because of the interference effects. The basic assumption – a kind of *ergodic hypothesis* – is that through these fluctuations S covers the matrix space with uniform probability. This should apply to billiards in which the effect of short non-chaotic paths is minimized. In this case, one can model the average over k in which one is interested by an average over the ensemble of S-matrices.

The connection between chaotic scattering and random S-matrix theory was proposed[16] prior to any interest in quantum ballistic transport. Until recently, however, the emphasis in work on random S-matrix theory has been on the eigenphases of S, which are not directly connected to transport because they involve both reflection and transmission. In contrast, several recent studies[93-100] have derived the implications of a random S-matrix theory for quantum transport properties and provided numerical evidence that this theory applies to the class of ballistic microstructures investigated experimentally. In this way experimentally accessible predictions for the quantum transport properties of chaotic billiards are obtained. In what follows, we follow closely the treatment in references [93-95].

An alternative to random S-matrix theory has been pursued by a number of groups in order to find the magnitude of quantum transport effects. In this approach one uses a supersymmetric field theory to model either random matrix behavior or disordered systems. The technique was first used for diffusive systems[101,102] and quantum-chaotic scattering[74] and later applied to ballistic quantum transport[103-106]. With this technique one may in addition calculate the shape of the magnetotransport for chaotic systems[104,106], thus providing a single technique for both the shape and the magnitude.

2. Results for Conductance

We start by deriving the magnitudes $\langle T \rangle$ and var (T) using an integration over the invariant measure (I.39) as the average. One of the pleasant aspects of using random matrix theory is that a great deal is known about these ensembles. In our case, the integrals over both the transmission coefficient and the square of the transmission coefficient exist in the literature[107]. Simply performing the sum over channels, one obtains[93]

$$\langle T \rangle - \frac{N}{2} = -\frac{N}{4N+2} \delta_{1\beta} \qquad \rightarrow -\frac{1}{4} \delta_{1\beta} \qquad (I.40a)$$

$$var(T) = \begin{cases} \dfrac{N(N+1)^2}{(2N+1)^2(2N+3)} & \rightarrow \dfrac{1}{8}, \text{ COE} \\[2em] \dfrac{N^2}{4(4N^2-1)} & \rightarrow \dfrac{1}{16}, \text{ CUE} \end{cases} \qquad (I.40b)$$

where the limit is as $N\rightarrow\infty$. Note that in a classical random scattering model, half of the intensity will be transmitted and half will be reflected; therefore the classical conductance is $N/2$ and Eq. (I.40a) gives the quantum correction to the average conductance.

Several comments concerning these results are in order. (1) Since the number of modes is proportional to the size of the system ($N=$ Integer$[kW/\pi]$), *the fact that one obtains a number independent of N as $N\rightarrow\infty$ shows that the magnitude of quantum transport effects is independent of the size of the system.* This is the analog of the universal conductance fluctuations result for disordered systems. (2) In the large N limit, var(T) in the presence of time-reversal symmetry is twice as large as in the absence of symmetry, as for the diffusive case, demonstrating the universal effect of symmetry. (3) The values obtained in the $N\rightarrow\infty$ limit are the same as those obtained from a random matrix theory for the Hamiltonian[102] in which one assumes that the Hamiltonian of the billiard is described by the Gaussian Ensembles and finds the conductance by coupling the billiard to leads in a random way. In fact, the Hamiltonian and S-matrix models have recently been shown to be equivalent[74,100].

If the pleasant aspect of using random matrix theory is the large base of existing knowledge, the most difficult aspect is justifying why a random matrix theory should apply to the system in which one is interested. There are two approaches to this problem: derive random matrix results from a microscopic theory or compare with numerical results. The former is usually very difficult, but some results exist for energy levels in chaotic systems[108] and for transport in disordered systems[101,102]. Here the second course will be pursued.

The predictions of the random matrix theory are compared to the conductance of a stadium billiard in Figure I.12. In order to increase the likelihood that the ergodic hypothesis implicit in the use of random matrix theory applies, a stadium billiard is considered in which (1) a stopper blocks any direct transmission between the leads, (2) a stopper blocks the whispering gallery trajectories which hug the half-circle part of the stadium, and (3) the stadium is asymmetrized to break all reflection symmetries. The agreement between the energy averages

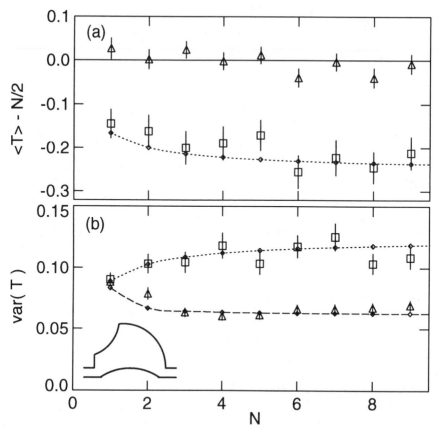

FIGURE I.12. The magnitude of the (a) average quantum correction and (b) conductance fluctuations as a function of the number of modes in the leads, N. The numerical results for $B = 0$ (squares with statistical error bars) agree with the prediction of the COE (dotted line), while those for $B \neq 0$ (triangles) agree with the CUE (dashed line). The inset shows a typical cavity. The numerical results involve averaging over (1) energy at fixed N (50 points), (2) 6 different cavities obtained by changing the stoppers, and (3) 2 magnetic fields for $B \neq 0$ ($BA/\phi_0 = 2, 4$ where A is the area of the cavity). (From ref. [93].)

found numerically and the invariant-measure ensemble averages is excellent.

In one respect, however, the results in Figure I.12 are rather disappointing: the curves are basically flat even in the extreme quantum (small N) regime! In order to find a strong signature of the extreme quantum limit, one must consider more detailed predictions of the random S-matrix theory. In particular, we derive the full distribution of T for small N rather than simply the first two moments of this distribution given in Eq. (I.40). These results are obtained[93,96] by expressing the invariant measure in terms of the eigenvalues of $tt^\dagger$, denoted $\{\mathbf{T}\}$, in terms of which $T = \sum_a \mathbf{T}_a$. Any unitary matrix of the form (I.38) can be written as[109]

$$
S = \begin{bmatrix} v^{(1)} & 0 \\ 0 & v^2 \end{bmatrix} \begin{bmatrix} -\sqrt{1-\mathbf{T}} & \sqrt{\mathbf{T}} \\ \sqrt{\mathbf{T}} & \sqrt{1-\mathbf{T}} \end{bmatrix} \begin{bmatrix} v^{(3)} & 0 \\ 0 & v^{(4)} \end{bmatrix} \tag{I.41}
$$

where $\mathbf{T}$ stands for a $N \times N$ diagonal matrix of the eigenvalues $\{\mathbf{T}\}$ and the $v^{(i)}$ are arbitrary unitary matrices except that $v^{(3)} = (v^{(1)})^T$ and $v^{(4)} = (v^{(2)})^T$ in the presence of time-reversal symmetry. It is straight forward to change variables in the invariant measure to the representation (I.41). Upon doing so one finds[93,96]

$$
d\mu(S) = P_\beta(\{\mathbf{T}\}) \prod_a d\mathbf{T}_a \prod_i d\mu(v^{(i)}) \tag{I.42}
$$

where the joint probability distribution of the $\{\mathbf{T}\}$ is

$$
P_2(\{\mathbf{T}\}) = C_2 \prod_{a<b} |\mathbf{T}_a - \mathbf{T}_b|^2 \tag{I.43a}
$$

$$
P_1(\{\mathbf{T}\}) = C_1 \prod_{a<b} |\mathbf{T}_a - \mathbf{T}_b| \prod_c \frac{1}{\sqrt{\mathbf{T}_c}}, \tag{I.43b}
$$

$d\mu(v^{(i)})$ denotes the invariant, or Haar's, measure on the unitary group, and C_β are N dependent normalization constants[110]. The joint probability distribution of the eigenvalues $\{\mathbf{T}\}$ has the characteristic eigenvalue-repulsion random matrix theories, with the repulsion larger in the unitary case (I.43a) than in the orthogonal case (I.43b).

The distribution of $T = \sum_{a=1}^N \mathbf{T}_a$ follows from Eq. (I.42) by integration over he joint probability distribution,

$$
w(T) = \prod_a \left[\int_0^1 d\mathbf{T}_a \right] P_\beta(\{\mathbf{T}\}) \delta(T - \sum_a \mathbf{T}_a). \tag{I.44}
$$

This integral can be carried out for small N; for instance, in the trivial case $N=1$, $w(T) = 1$ for the CUE and $w(T) = 1/(2\sqrt{T})$ for the COE. For $N = 1-3$ the distributions derived from the random matrix theory are plotted in Figure I.13. *Note the dramatic difference between the CUE and COE in the single mode case, and the difference within each ensemble between $N=1$ and $N=2$.* The results for $N=3$ are close to a Gaussian distribution defined by the two moments given in Eq. (I.40); in the $N \to \infty$ limit, the distribution is known to be Gaussian[111].

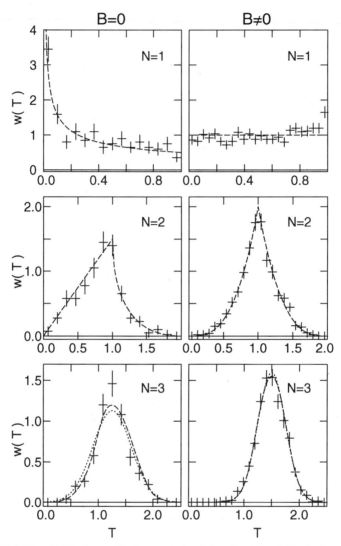

FIGURE I.13. The distribution of the transmission intensity at fixed $N = 1, 2$, or 3 both in the absence (first column) and presence (second column) of a magnetic field. The numerical results (plusses with statistical error bars) are in good agreement with the predictions of the circular ensembles (dashed lines). Note the striking difference between the $N = 1$ and $N = 2$ results and between the $B = 0$ and $B \neq 0$ results for $N = 1$. For $N = 3$, the distribution approaches a Gaussian (dotted lines). The cavities used are the same as those in Figure I.1; for $B \neq 0, BA/\phi_0 = 2, 3, 4$, and 5 were used. (From ref. [93].)

Figure I.13 also shows numerical results for $w(T)$, obtained by making a histogram of values of $T(k,B)$ taken at intervals greater than the correlation lengths. The agreement between the numerical and theoretical results is very good. These effects should be observable in experiments, simply by making a histogram of the experimental data in the same way that the numerical histogram was constructed. Such an observation would provide a clear test of the applicability of random S-matrix theory to experimental microstructures. In fact, an initial experimental measurement of the distribution function has been made[51]. The fact that the effects discussed here were *not* observed leads us directly to the topic of the next section.

3. Phase Breaking

To parallel the discussion of the shape effects above, one should compare the random matrix theory results to both numerical calculations and experiments. In the last section, we saw that the results for the average, variance, and probability density of the conductance are in good agreement with numerical calculations[93]. However, the random-matrix predictions for both the average quantum correction and the variance are larger than the experimental results[49,51]. In addition, the measured probability density is close to a Gaussian distribution[51] when there are two propagating modes per lead ($N=2$), while random-matrix theory predicts a Gaussian distribution only for $N \geq 3$[93].

Inherent in the invariant-measure random matrix theory is the assumption (among others) that one can neglect processes which destroy the coherence of the wave function. It is this assumption that is largely responsible for the discrepancy between theory and experiment mentioned above. To overcome it, a model of phase-breaking must be introduced.

To simulate the effects of phase-breaking events in ballistic transport, several authors[46,94,99] have used a model suggested by Büttiker[112]: in addition to the physical leads 1, 2 attached to reservoirs at chemical potentials μ_1, μ_2, a lead 3 connects the cavity to a phase-randomizing reservoir at μ_3. The basic idea, then, is that there are two types of transmission between reservoirs 1 and 2: direct coherent transmission, and incoherent transmission via reservoir 3. This latter process occurs because a particle entering from lead 1 which exits lead 3 is replaced by a particle from reservoir 3 in order to conserve current. This particle may then proceed to lead 2. The new particle from reservoir 3 is, of course, incoherent with respect to the original particle, and so the third lead introduces phase-breaking.

This model has been discussed extensively for disordered materials[113] and, more recently, for ballistic quantum dots by Marcus *et al.*[46]. A similar model has been used for absorption of microwaves in chaotic-scattering from cavities[75]. Requiring the current in lead 3 to vanish determines μ_3; the two-terminal dimensionless conductance is then

$$g \equiv G/(2e^2/h) = 2 \left[T_{21} + \frac{T_{23} T_{31}}{T_{32} + T_{31}} \right], \qquad (I.45)$$

where T_{ij} is the transmission coefficient from lead j to lead i. N is the number of channels in leads 1 and 2, N_ϕ that in lead 3, and $N_T = 2N + N_\phi$. The N_ϕ channels in lead 3 are physically related to the phase-breaking scattering rate γ_ϕ via the relation $N_\phi / 2N \approx \gamma_\phi / \gamma_{esc}$ where γ_{esc} is the escape rate from the cavity.

We now make the fundamental assumption of an "equal *a priori* distribution" for the *total* $N_T \times N_T$ scattering matrix S. The statistics of the total S-matrix are again given by the Circular Ensembles, but now we treat the statistics of g given in Eq. (I.45). Unfortunately, an analytic treatment of the statistics of g is much more difficult than T_{21}, particularly because of the denominator in the second term (see ref. [99] however). Here, asymptotic results for strong phase-breaking will be presented and then combined with the completely phase-coherent results in an interpolation formula[94].

In the limit $N_\phi \gg 1$, one can obtain[94] results to leading order in $1/N_\phi$ using standard techniques[107] of random matrix theory. The largest contribution to $\langle g \rangle$ is the classical conductance $N/2$; for $N_\phi = 0$ this comes from $\langle T_{21} \rangle$ while for large N_ϕ it comes from the second term in Eq. (I.45) (the incoherent part). The quantum correction to this average is:

$$\delta g = -\frac{N}{N_\phi} + O\left(\frac{1}{N_\phi^2}\right); \qquad (I.46)$$

and the variance is:

$$\mathrm{var}\, g = \left(\frac{N}{N_\phi}\right)^2 \frac{2}{\beta} \left[1 + \frac{2-\beta}{2N}\right] + \cdots . \qquad (I.47)$$

As expected, the magnitude of the quantum corrections decreases as the strength of the phase-breaking increases. That $\mathrm{var}\, g$ decreases with N_ϕ reflects the fact that each S-matrix element fluctuates less as N_T increases. The deviation of the ratio of the variances from 2 is highly unusual; in fact, for $N = 1$ the ratio can be as high as 3.

In order to form a prediction for values of N_ϕ that are not too large, one can combine the asymptotic results with the phase-coherent results in an interpolation formula[94]. For the average conductance, Eqs. (I.40a) and (I.46) suggest

$$\delta g \simeq \frac{N}{2N + N_\phi}.$$ (I.48)

For the fluctuations, one should combine square roots of variances,

$$(\operatorname{var} g)^{1/2} \simeq \left[(\operatorname{var} g)_{N_\phi=0}^{-1/2} + (\operatorname{var} g)_{N_\phi \gg 1}^{-1/2} \right]^{-1} ;$$ (I.49)

the explicit expression for the fluctuations for $\beta = 2$ is

$$\operatorname{var} g \simeq \frac{N^2}{[(4N^2 - 1)^{1/2} + N_\phi]^2}.$$ (I.50)

For $\beta = 1$ a similar, but more complicated, expression holds.

In order to check the accuracy of the interpolation formula, results of random-matrix simulations are presented in Figure I.14. The points are obtained by generating random $N_T \times N_T$ unitary or orthogonal matrices and computing g from Eq. (I.45); $N = 2$ was selected for comparison with the experiment of reference [51]. The insets show that the convergence of the numerical to the asymptotic results (I.46) and (I.47) is rather slow. The curves in the main panels are the interpolation formulae; note that they agree very well with the numerical results.

With these results for phase-breaking in hand, one can now carry out a comparison between the theoretical and experimental magnitudes of these quantum transport effects. Figure I.14 shows the results of the experiment of reference [51] in addition to the random-matrix theory. In this experiment, δg, $\operatorname{var} g$ at $B = 0$, and $\operatorname{var} g$ at non-zero B were measured in a chaotic ballistic cavity for $N = 2$. In addition, an *independent* measure of N_ϕ was obtained through the temperature dependence of the power spectrum of the fluctuations[46,52]. Thus it is possible to compare the theory and experiment with *no free parameters*. Before comparison, however, the variance must be corrected for thermal averaging: convolution over the derivative of the Fermi function produces a reduction of ~0.22–0.38 for a temperature of 50–100 mK. The measured $\operatorname{var} g$ has been increased by the inverse of this reduction factor; the average conductance is not affected since it is already an average effect. The error bars shown result from both the uncertainty in temperature (for the variance) and the experimental fluctuations at small B. The agreement between theory and experiment is very good indeed.

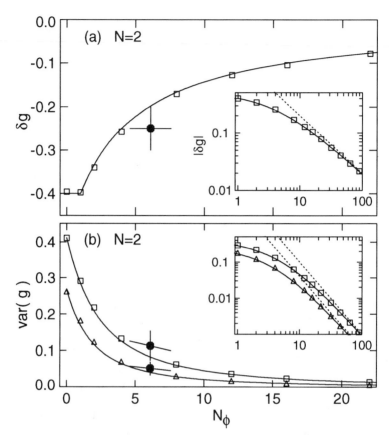

FIGURE I.14. Magnitude of quantum transport effects as a function of the number of phase-breaking channels, N_ϕ, on linear and log-log (insets) scales ($N=2$). (a) The quantum correction to the average conductance. (b) The variance in the orthogonal (squares) and unitary (triangles) cases. Open symbols are numerical results (20,000 matrices used, statistical error is the symbol size). Solid lines are interpolation formulae. Dotted lines are asymptotic results. Solid circles are experimental results of reference [51] corrected for thermal averaging. The interpolation formulae work well, and the agreement between the invariant random-matrix theory and the experiment is good. (From ref. [94].)

The behavior at fixed N_ϕ is relevant to experiments at fixed temperature in which the size of the opening to the cavity is varied. Though δg and $\mathrm{var}\, g$ are nearly independent of N in the perfectly coherent limit – the "universality" of the magnitude – phase-breaking channels cause variation. Thus the universality can be seen only if $N_\phi \ll N$. Theoretical and experimental curves of this type are discussed in Part II by Westervelt [see Figure II.19].

In addition to the mean and variance, the probability density of the conductance, $w(g)$, is experimentally measurable[51] as discussed above. Analytic results for $w(g)$ have been obtained in the case $N = 1$[99], and, on the other hand, it is known that this distribution becomes Gaussian for strong phase-breaking because the transmission amplitudes become uncorrelated[111]. Numerical simulations of the random matrix theory show that the distribution is essentially Gaussian for $N_T \geq 3$[94]. Thus, the theoretical probability density for the estimated $N_\phi = 4 - 8$ of reference [51] is Gaussian, consistent with the experiment. Observation of the interesting non-Gaussian distribution of g obtained theoretically for $N = 1, 2$ and $N_\phi = 0$ requires greatly reduced phase-breaking.

4. Short Paths: Non-invariant Random Matrix Theory

In the work summarized above, direct processes caused by short paths were neglected: the energy averaged, or optical, S-matrix was assumed to vanish. However, short trajectories *are* important in many of the cavities studied numerically[17-19] as well as in all of the experimental structures[45-53]. In fact, for a typical chaotic cavity, one will find that the short paths carry a substantial part of the current between the leads, even though the classical dynamics inside the cavity is completely chaotic: there is no definite connection between chaotic classical dynamics and the absence of short paths[115]. In order to obtain the good agreement between simulation and theory presented above, the absence of short paths was *enforced* in the calculations with two stoppers which block both direct transmission and whispering-gallery trajectories. Thus, the invariant-measure results, though they describe the interference properties which are not sample specific, apply directly to only a small subset of chaotic scattering systems.

Recently, several authors have proposed a model for the common features of typical chaotic scattering processes[95,98,100,114]. In this model, instead of assuming the system has a single characteristic scale as in the invariant-measure case, one includes two time scales: a *prompt* response arising from direct processes, and a time-delayed one arising from the formation of an *equilibrated* intermediate state. This model was introduced in the past in nuclear physics[116] using an information-theoretic approach[117] based on the mathematical development in reference [118]. It has been shown[95,98] that the same theoretical framework is successful in describing transport through a much larger class of quantum dots than considered by the invariant-measure theory.

The ensemble average of S, and hence the prompt component, vanishes when evaluated with the invariant measure. Ensembles in which $\langle S \rangle$ is nonzero contain more information than the invariant-measure ensembles; they are constructed by multiplying the invariant measure by a function of S to give the differential probability $dP_{\langle S \rangle} = p_{\langle S \rangle}(S)\,d\mu(S)$. Any suitable ensemble of S matrices should satisfy the symmetry constraints, should be ergodic, and should consist of S-matrices which are analytic in the upper half of the complex-energy-plane (causality). It can be shown that the probability density known as *Poisson's kernel*[118] satisfies the analyticity-ergodicity requirements[118], and the associated information is less than or equal to that of any other probability density satisfying these requirements for the same $\langle S \rangle$[117]. Thus, Poisson's kernel describes those physical situations in which (a) the system-specific details are irrelevant except for the average S-matrix and (b) the requirements of flux conservation, time-reversal invariance (when applicable), analyticity, and ergodicity must be met.

From the distribution of S, one can, of course, deduce the probability density of the transmission. In some simple cases this can be done analytically[95,98]. In the $N = 1$ case, a detailed comparison between the prediction of Poisson's kernel and numerical results has been carried out for several billiards[95]. In each case the optical S-matrix was extracted directly by energy averaging the numerical data and then used as an input in Poisson's kernel; in this sense the theoretical curves are parameter free. In all cases, the predictions of the information-theoretic model are in excellent agreement with the numerical results.

V. OTHER SYSTEMS

The emphasis in this review is on interference effects in transport through open systems, particularly ballistic cavities. The main reason for emphasizing this aspect of the field is that it is the most easily accessible experimentally and therefore the aspect which has received the most attention from several different groups. There are, however, several other topics in nanostructure physics in which quantum chaos plays a role. Before concluding, we will briefly call the readers attention to four other topics, giving references for further information.

A. Quantum Transport in Arrays of Antidots

In addition to the ballistic cavities discussed above, a system which has received considerable attention is a two-dimensional electron gas with an array of antidots punched into it[120-122]. An "antidot" is a region of the electron gas where the potential is higher than in the immediately surrounding region so that the electron is repelled, rather than attracted as in a quantum "dot". Antidots have been formed in various ways using etching, gates, and illumination with light. An array of antidots usually covers a large area, so that while each antidot individually is a nanostructure, the array as a whole is macroscopic. The pioneering[120] and subsequent[121] work of Weiss *et al.* has addressed both classical and quantum transport through this system.

In the classical regime, peaks in the magnetotransport characteristics are observed when the cyclotron radius is commensurate with the period of the lattice. In the simplest case, this occurs because the electron orbits the antidot without colliding with the other antidots and so is trapped until a rare scattering event (caused, for instance, by residual impurities) causes it to move on. The carriers thus move through the lattice very slowly when the commensurability condition is satisfied, and so the resistance increases.

If the transport is coherent, interference around these trapped trajectories produces a modulation of the resistance, an effect which has been seen experimentally[121]. Such a modulation is very natural from a quantum-chaos point of view[121]: the trapped orbits are periodic orbits of the pure system (antidots but no residual scattering), and one knows since the work of Gutzwiller[3] that quantization of periodic orbits leads to a modulation of the density of states. Since the transport properties and density of states are closely related by the Einstein relation, modulation of the resistance should result. In fact, the connection between periodic orbits and the conductivity has recently been worked out in detail using the Kubo formula[85-88].

B. Quantum Transport in Nearly Closed Systems

As a cavity becomes closed – by pinching off the openings to the leads using gates, for instance – there is a qualitative change in the transport properties. As the conductance dips below e^2/h, the system stops being a good metal, and the non-interacting particle approximation begins to break down. For a nearly closed system – $G \ll e^2/h$ – an extensive theoretical description for the transport properties exists[123]. The main physical effect is the quantization of charge – the number of electrons in the cavity must be an integer. Since for transport, electrons must flow through the cavity, this introduces a new energy scale, e^2/C where C is the capacitance, which is simply the electrostatic energy of the extra electron in the cavity. The suppression of transport because of this charging energy is known as the "Coulomb blockade", and this term is used for transport properties in this regime in general.

When the energetics of the system allow a particle to move across the cavity – because of a sufficiently large applied bias or gate voltage, for instance – the transport proceeds by tunneling from one lead into the dot and then out the other side. In this tunneling process, the overlap of the wavefunction in the cavity with the wavefunctions in the leads is crucial in determining the ease of transport. Thus the conductance is sensitive to the nature of the wavefunction in the cavity near the leads. It was realized by Jalabert et al.[124] that this sensitivity gives rise to a new mesoscopic effect: the interference pattern in the wavefunction can be varied with magnetic field or energy, and the resulting variation in the overlap of the wavefunction with the lead will show up in the conductance.

When the temperature is greater than the typical width of a resonant state in the cavity but less than the mean level separation, one sees[123] peaks in the conductance as a function of gate voltage whose width is controlled by the temperature but whose height is related to the coupling of the state to the leads. The mesoscopic fluctuations should then appear as a fluctuation of the height of the conductance peaks[124]. The characteristic distribution of these conductance peaks has been worked out in several different situations[124-127]. In addition, the sensitivity of the peaks to changes in the shape of the cavity and the correlations of the peak height as a function of an external parameter have been addressed[126]. Very recently, a systematic experimental study of the distribution of Coulomb blockade conductance peak heights was completed[128] which is in good agreement with the theory.

C. Thermodynamics: Orbital Magnetism

Magnetic probes provide information which is complementary to transport and can in addition be used to study isolated systems. The orbital magnetic response of an isolated system is essentially given by the way in which the density of states varies with magnetic field (at fixed particle number, not chemical potential)[1]. It has been recognized for some time that interference effects will influence the orbital magnetic response[129], and there has been a great deal of work on the theory of orbital magnetism in disordered metallic grains and rings[129]. Perhaps best known are the two experiments which probed the "persistent current" in disordered metallic rings[130,131].

More recently, attention has turned to structures in which the transport is ballistic, and there have been experiments on structures patterned in GaAs/AlGaAs heterostructures[132,133]. Because of the relation to the density of states mentioned above, there is a very close connection between the orbital magnetic response and periodic orbits. Theoretically, this has been worked out in detail for fully chaotic systems by several groups for both rings and singly connected structures[134-136]. The result is an expression for the magnetic response in terms of a sum over periodic classical orbits. The theory has also been worked out for the integrable and nearly integrable cases[135,136]. Here because the modulation of the density of states is much larger than in the chaotic case – families of periodic orbits contribute rather than single isolated orbits – the magnetic response is much larger also. This was used[135,136] to explain the magnetic response of GaAs squares observed by Lévy et al.[132] which showed some initially puzzling features in both the magnitude and field scale. Thus the difference between chaotic and non-chaotic dynamics does have experimentally relevant consequences for the thermodynamics of isolated systems.

D. Tunneling Through Quantum Wells

It is natural to try to use tunneling to provide a spectroscopy of the energy levels of an isolated system. Usually when one tries to do this either the system is large so that the spectrum is essentially continuous but interaction effects are weak, or the system is small so that the spectrum is discrete but then the Coulomb blockade effects discussed above must be included. Very recently, a system has been introduced[137] in which both interaction effects are weak and the spectrum is not continuous.

Consider perpendicular transport through a quantum well – a two barrier tunneling problem in the perpendicular direction. In a magnetic field perpendicular to the plane of the well, the energy levels are grouped in Landau levels of high degeneracy spaced by $\hbar \omega_c$: the spectrum is certainly not continuous. Yet because the well is very large, the capacitance is small and charging effects are unimportant. The case of a perpendicular field is, of course, integrable. However, as the field is tilted away from the perpendicular direction, the classical dynamics are no longer integrable; in fact, a transition to fully chaotic dynamics occurs as the tilt angle increases[137-139]. In this tilted field case, the spectrum is not continuous: it is given by the spectrum of an effective two-dimensional problem with an additional degeneracy[138]. Since the actual system remains macroscopic, charging effects continue to be negligible. In this way, one can use tunneling to study the spectrum of an isolated two-dimensional system without Coulomb blockade effects.

Two experimental groups have studied tunneling in this situation[137,138]. Since the connection between classical dynamics and spectra is clearest in the integrable or fully chaotic limits, these are the obvious starting points. The integrable case is, of course, trivial: Bohr-Sommerfeld quantization of a cyclotron orbit immediately gives Landau levels. In the fully chaotic case, Gutzwiller showed that periodic orbits produce a modulation of the density of states[3], and in particular that short periodic orbits produce a modulation on a fairly large energy scale ($\hbar/T$ where T is the period of the orbit). Thus one expects a connection between the tunneling signature and the classical periodic orbits in the well. Such a connection has been demonstrated by Fromhold *et al.*[137]: they showed that the regions of parameter space in which a short periodic orbit exists (determined by classical simulation) correspond to the regions in which a modulation of the tunneling current with the corresponding frequency is seen.

The study of the transition between integrable and fully chaotic behavior undertaken by Müller *et al.*[138] is the first study of such a transition in the nanostructure physics context. This group, as well as an independent theoretical study[139], argues that bifurcations in the classical dynamics as the chaos develops produce a doubling of the number of peaks in the tunneling. They observe such peak doubling experimentally at approximately the parameter values for bifurcations in classical simulations[138]. From both of these experiments it is clear that quantum wells in a tilted magnetic field show great promise for future work on the connection between classical dynamics and quantum properties.

VI. SUMMARY AND FUTURE DIRECTIONS

In this review of chaos in nanostructures, we started by discussing a number of classical size effects: the bend resistance and quenching of the Hall resistance, for instance. These effects first suggested that classical chaos could be relevant to transport in nanostructures. It proved to be difficult, however, to probe classical chaos directly in these structures. On the other hand, it soon became clear that the presence or absence of classical chaos would influence the quantum transport properties. The present theoretical understanding of these properties has been the main focus of this review.

By combining semiclassical analysis with random matrix theory, we have given a fairly complete description of quantum transport through ballistic nanostructures. We have seen that the phenomena are closely analogous to quantum transport phenomena in disordered metals – weak-localization and universal conductance fluctuations exist. The presence of classical chaos enters in determining both the shape of the quantum conductance as a function of k or B and the magnitude of the quantum correction to the conductance. Good agreement between theory and experiment has been demonstrated for both the shape and magnitude.

Concerning quantum transport through open structures, two main theoretical issues remain to be worked out. The first of these is to understand and remove the difficulties with the semiclassical theory noted at the end of Section IV-B. A *quantitative* semiclassical theory is badly needed both to provide a basic understanding and to enable one to investigate new transport phenomena. The second open issue is the behavior of non-chaotic structures. A careful analysis of transport through non-chaotic structures has yet to be carried out. It would be particularly interesting to look at the magnitude of quantum transport effects in such structures since there is numerical evidence[19] that it scales very differently in non-chaotic cavities from chaotic cavities.

One of the most interesting current directions is the widening of the scope of the field: "quantum chaos" in nanostructures is being investigated not only in transport through open systems but in closed, nearly closed, and quantum-well systems (Section V, Part I). As this direction increases in strength, the overlap of mesoscopic and chaos physics should become even more fruitful.

PART II - EXPERIMENT

I. INTRODUCTION

The motion of electrons through conductors is a fundamental part of any electronic circuit. For metals and semiconductors currently used in electronic devices, the Drude model ordinarily applies. Here, we regard electrons as classical particles traveling with the Fermi velocity, which collide with a fixed array of randomly located obstacles, typically atomic impurities or structural defects. The mean free path, l, for momentum relaxation from these collisions is generally much shorter than the dimensions of the conductor, so that electron motion is insensitive to the conductor's shape. The Drude model displays classical chaos: the trajectory of a given electron is exponentially sensitive to the choice of initial conditions. Consequently, the motion is highly unpredictable, and electrons randomly diffuse through the conductor with a well-defined diffusion constant and mobility. The end result of this process is to produce resistance to electrical conduction described by a single macroscopic quantity, the conductivity, σ.

In semiconductor nanostructures at low temperatures, the opposite limit can be achieved: electrons travel as phase-coherent waves with mean free paths much longer than the transverse dimensions of the conductor. When the electron wavelength approximates the width, the conductor acts as an electron waveguide, analogous in many ways to its microwave counterpart. For ideally smooth waveguides, scattering is suppressed, and the electrical resistance is determined by the transmission coefficient through the Landauer formula[1]. A particularly striking example of conductance determined by transmission is the quantization of the conductance through quantum point contacts, $G = 2Ne^2/h$, where N is the number of transverse modes of the electron wave in the constriction at its narrowest point[54,55]. Over the past decade, much experimental and theoretical work has been devoted to the propagation of electron waves in nanostructures with restricted dimensionality[1,71], and the picture of conductors as electronic transmission lines has been developed in detail.

In this chapter we consider the mesoscopic regime of electronic conduction in which the transverse dimensions of the conductor are larger than the electron wavelength, but small enough that quantum confinement and the wave-like nature of electrons are important. An apparent paradox exists for electron waveguides: truly one-dimensional waveguides with transverse dimensions comparable to the wavelength might be expected to exhibit enhanced electron mobility due to the suppression of scattering, however the observed mobility is typically less than for two-dimensional samples. This paradox is resolved by the theoretical prediction that all electron states are localized in long, disordered, one-dimensional conductors at absolute zero, even for small amounts of disorder[140]. In order to achieve good conduction in actual circuits, the width of the waveguide must be increased into the mesoscopic regime. Disordered metals in the mesoscopic regime show novel and important phenomena, including universal conductance fluctuations and weak localization, both extensively studied over the past two decades[1,71]. It is often assumed that the behavior of electrons is more orderly in ballistic structures such as electron waveguides, for which the electron mean free path is longer than the size of the conductor. We will show, below, that the *shape* of the conductor determines the qualitative nature of electron motion in this ballistic mesoscopic regime, and that electron motion is typically chaotic, an example of quantum chaos.

Classical vs. Quantum Chaos

Chaos in classical physics has been extensively studied since the last century, although the term "chaos" was coined recently. Classical chaos refers to the extreme sensitivity commonly found in nonlinear systems to the choice of initial condition. For chaotic trajectories, a small change in initial condition grows exponentially in time, leading to effectively unpredictable behavior – chaos. Classical chaos was originally studied for systems which conserve energy, often in the context of the fundamentals of statistical mechanics and ergodic theory. An important example of a chaotic system is a gas composed of colliding hard spheres. Although the subject is often thought to be esoteric, chaos is certainly more common in everyday life than predictable behavior.

A comparison between regular and chaotic behavior is given in Figure II.1, which illustrates the classical trajectories of particles bouncing inside circular and stadium shaped billiards[141]. The trajectories inside the circular billiard rotate in a predictable manner, because the circle preserves angular momentum, and no

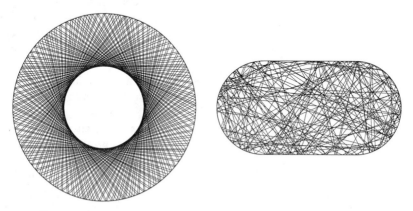

FIGURE II.1. Classical particle trajectories inside circular and stadium shaped billiards[141].

initial condition results in a chaotic trajectory. By contrast, generic choices of initial conditions inside a stadium of this aspect ratio produce chaotic trajectories. The chaotic trajectory shown in Figure II.1 reverses its direction of rotation every three to five bounces, and eventually covers the entire area[142,143]. A different choice of initial condition produces a trajectory which is different in detail, but statistically similar to that shown. Structures with other shapes (e.g., a "lemon" billiard or confocal resonator) typically show a mix of regular and chaotic behavior, with some initial conditions producing regular behavior and others chaotic.

The term "quantum chaos" refers to the quantum mechanical behavior of systems that are chaotic in the classical limit, taken through the correspondence principle by letting $h \rightarrow 0$. Although the term may seem self-contradictory at first, it provides a very useful description of the behavior of quantum systems in the mesoscopic regime. Quantum chaos and the related subject of "spectral rigidity" of energy eigenvalues were studied some years ago in nuclear physics, although the term "chaos" was not used then, and random matrix theory was introduced to study the eigenvalue distributions[3]. More recently, quantum chaos has been experimentally and theoretically studied for highly excited atoms[3,144,145] and molecules where it plays a role in chemical reaction kinetics and dissociation. Diffusion is an example of classical chaos, so universal conductance fluctuations, weak localization, and other mesoscopic phenomena seen in diffusive conductors at low temperatures are, by definition, examples of quantum chaos[1,71]. In this chapter we review recent demonstrations that *ballistic* conductors with shapes that produce classical chaos show corresponding quantum phenomena: conductance fluctuations and coherent backscattering. We focus here on quantum chaos inside quantum dots. Important related work has been done on quantum chaos in open structures: anti-dot lattices[121,122] and tilted square wells in a magnetic field[137,146].

The semi-classical approach described in Part I of this chapter provides a natural way to think about quantum chaos in nanostructures. Let us consider the propagation of electron waves through a ballistic nanostructure, as illustrated in Figure II.2. The triangles are quantum point contacts, which act as antennas to broadcast and receive electron waves in all directions. The propagation of a wavefront can be represented by perpendicular rays through the Huygens construction; each ray corresponds to a classical trajectory. The total transmission from one point contact to the other is obtained by summing over trajectories, as indicated schematically in Figure II.2. Because the rays are simply the classical trajectories, much of the qualitative behavior of the quantum system is determined by its classical counterpart. Quantum effects appear as the interference between pairs of contributing trajectories. For the ideal case in which electron waves are completely phase coherent, all possible trajectories must be summed, including those with arbitrarily long length. Elegant theoretical methods have been developed for this limit[3]. However, in experiments, the phase coherence is

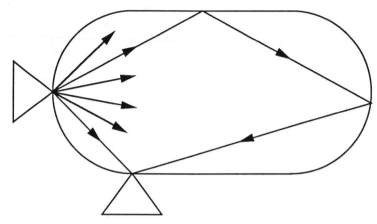

FIGURE II.2. Schematic illustration of interfering electron paths emitted by a quantum point contact.

lost after a characteristic length, l_ϕ, and only the shorter paths contribute to quantum interference, as discussed below. The qualitative characteristics of conduction are nonetheless preserved, as long as l_ϕ is longer than the length over which classically chaotic trajectories diverge.

II. BASIC CONCEPTS

Ballistic Nanostructures

Advances in semiconductor growth technology have led to the production of electron gas samples with nearly ideal characteristics. Electrons trapped at the interface between *GaAs* and *AlGaAs* layers in a heterostructure grown by molecular beam epitaxy (MBE), form a two-dimensional electron gas (2DEG) at low temperature and electron sheet density. Due to the high quality of the interface in the best samples, the electron mean free path can exceed $l = 100$ μm at liquid *He* temperatures[147,148]. The motion of electrons in structures smaller than the mean free path in such samples is ballistic: electrons travel in approximately straight lines between the walls rather than collide with impurity atoms or other imperfections.

Using modern lithographic techniques, one can fabricate structures containing electrons in a wide variety of sizes and shapes. An illustration of one technique for producing a "quantum dot" is given in Figure II.3, which is a scanning electron microscope (SEM) photograph of the top surface of a *GaAs/AlGaAs* heterostructure containing a two-dimensional electron gas located 42 nm below the

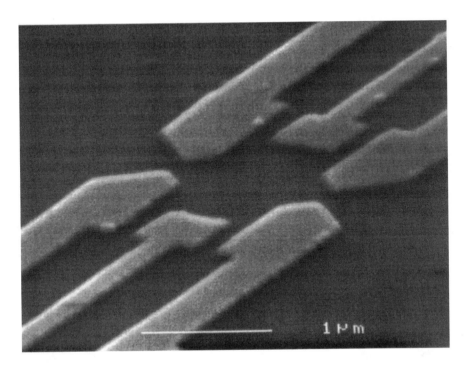

FIGURE II.3. Scanning electron micrograph of quantum dot formed from the two dimensional electron gas located ~50 nm below the surface of the $GaAs/AlGaAs$ heterostructure via electrostatic gates, which appear as the bright areas. By changing the negative gate potential, one can adjust the conductance of the two quantum point contacts as well as the size and shape of the quantum dot.

surface. The bright objects on the surface are $Cr:Au$ gate structures defined via electron beam lithography. When a suitably large negative voltage (typically ~0.3 V) is applied between a gate and the gas, the electron gas beneath is depleted, leading to the formation of a quantum dot underneath the surface in the space between gates. Electrical contact to electrons within the dot is made via a pair of quantum point contacts as shown in Figure II.3. Many other techniques for producing quantum dots and nanostructures of other shapes have been developed for semiconductors, including surface depletion with surfaces profiled by etching, and reduction of the electron mobility via focused ion beam damage[149], each technique having certain advantages.

Ballistic nanostructures show novel and interesting phenomena which can be understood in terms of classical ballistic electron trajectories[1]. In an example particularly relevant to this chapter, Roukes *et. al.*[32] discovered that the Hall effect is "quenched" at low magnetic fields for narrow cross junctions fabricated in a *GaAs/AlGaAs* heterostructure. This effect can be understood via classical electron trajectories[38], as demonstrated by a later experiment by Ford *et. al.*[34], shown in Figure II.4. To test the influence of junction shape on quenching of the Hall effect, a number of cross junctions were fabricated from a 2DEG, one with a central bumper, shown in Figure II.4(a), and one with an injector structure, in Figure II.4(b). The Hall effect arises from the preferential deflection of electrons by the magnetic field into one or the other side lead. Thus, the measured Hall voltage in Figure II.4(a) is enhanced at low magnetic fields, B, above the conventional value, because the central bumper directs electrons into the side leads, as illustrated by the classical trajectory illustrated. In contrast, the Hall voltage in Figure II.4(b) for the injector structure is suppressed at low magnetic fields, because the curvature of the classical trajectory is insufficient to turn into the side leads. These authors observed additional phenomena, including a reversal of the Hall voltage near certain values of magnetic field due to an electron bounce from the corners of the junction[34]. From results such as these, it became clear that classical electron trajectories through ballistic structures could be quite complex. Roukes and Alerhand[41] demonstrated in simulations that ballistic transport through a cross junction is, in fact, chaotic and shape dependent. The influence of classical ballistic electron trajectories has also been studied in other open structures, including anti-dot lattices[120,121,122].

Because electrons travel as phase coherent quantum mechanical waves over substantial distances in semiconductor nanostructures at low temperatures, quantum interference phenomena are important. The electron wavelength for the two-dimensional electron gas in a *GaAs/AlGaAs* heterostructure is relatively long: $\lambda_F = h/(2m_e E_F)^{1/2} \sim 50$ nm, using the electron mass for *GaAs* $m_e = 0.067m_0$ and the typical value $E_F = 10meV$. At low temperatures, the distance an electron wave travels before losing phase coherence can be much longer, greater than 10μm in *GaAs*, determined by the phase coherence time, τ_ϕ, and the trajectory followed by the electron.

In the early 1980s, it was a surprise to learn experimentally that the electron phase coherence length could be much longer than the transport mean free path, l, in diffusive conductors, such as "dirty" elemental metals with atomic impurities and structural defects. Phase coherence in diffusive conductors produces a range of mesoscopic phenomena, including universal conductance fluctuations and weak localization[1,71]. When ballistic semiconductor nanostructures with restricted dimensionality were studied somewhat later, novel phase interference phenomena were also found[1,71], including mesoscopic conductance fluctuations[150,151,152], bend resistance[24], and Aharonov-Bohm oscillations in semiconductor rings[153,154] and inside quantum dots[155].

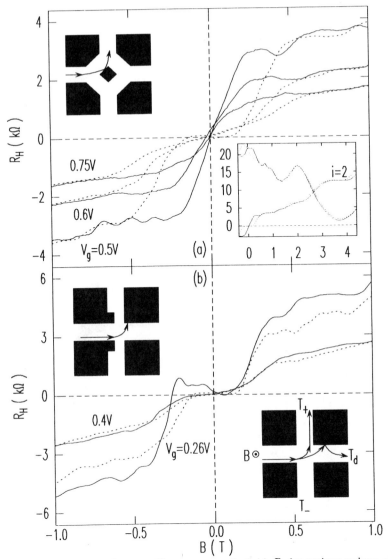

FIGURE II.4. (a) Hall resistance, R_H vs. magnetic field, B, for various values of gate voltage, V_g. The solid and dotted lines are for the widened cross with the central dot (shown in the inset) and a normal cross on the same sample. Inset: High magnetic field data which illustrate tracking of the two curves and the usual quantized Hall effect for V_g = 0.50 V. (b) R_H vs. B for several values of V_g. The solid and dotted lines are for the cross with narrow probes (shown in the inset) and a normal cross (also shown) on the same sample[34].

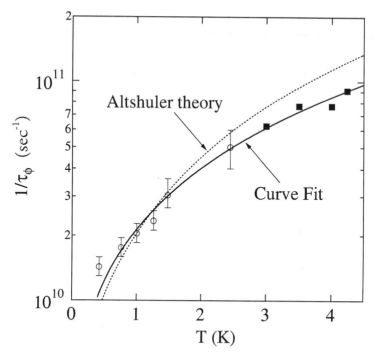

FIGURE II.5. Dephasing rate vs. temperature for $\tau_\phi > \tau_\varepsilon$ (circles) and $\tau_\phi \sim \tau_\varepsilon$ (squares). The results are compared with two-dimensional electron gas theory[159], shown as the dotted curves, using the measured parameter values. A curve fit to the data using the same theory with two adjustable parameters is shown as the solid line (see text)[157]).

The phase coherence time, τ_ϕ, is difficult to measure for the high mobility, two-dimensional electron gas in *GaAs/AlGaAs* heterostructures, because the transport mean free path, l, is sufficiently long at low temperatures that it approaches the dephasing length, and weak localization is suppressed. Recently, τ_ϕ has been measured[156,157] via the weak localization of electrons in long, effectively one-dimensional nanostructures with random scatterers[158], and in quantum dots[52]. Katine *et al.* used a narrow channel, defined in a *GaAs/AlGaAs* heterostructure via electrostatic gates, with electron sheet density $n_s = 4\times10^{11}\text{cm}^{-2}$, elastic mean free path $l_e = 3.8$ μm and elastic scattering time $\tau_e = 13$ psec. Figure II.5 shows the experimental dephasing rate, $1/\tau_\phi$, from Katine *et al.*[157] as data points compared with electron-electron scattering theory, including collective electromagnetic fluctuations[159], shown as dashed and solid curves; the dashed curve is obtained without adjustable parameters, and the solid curve is the result of a fit to the same theory. These data show that electron-electron scattering is the dominant dephasing mechanism in 2DEG samples at liquid *He* temperatures, and demonstrate that the dephasing rate is well described by theory when collective electromagnetic fluctuations (Nyquist noise) are included. As shown in Figure II.5, the ballistic dephasing length, $l_\phi = v_F \tau_\phi$, becomes quite long at low temperatures, reaching $l_\phi \cong 15$ μm at $T = 1$K, and continuing to grow at lower temperatures.

The precise nature of the paths followed by electrons in ballistic nanostructures is an important issue. Semiclassical treatments of this problem commonly assume that electrons travel in straight paths inside nanostructures, and reflect specularly from the walls. These characteristics are necessary to preserve the symmetry of commonly examined structures, such as circular and rectangular quantum billiards. However, experimental semiconductor nanostructures deviate from perfection in ways which can be important for certain shaped structures, particularly those with a high degree of symmetry. First, we consider the specularity of reflections from the walls of ballistic nanostructures, then deviations from straight line trajectories inside.

A common measure of specularity is obtained by assuming that a fraction, p, of reflections are perfectly specular and a fraction $(1-p)$ are perfectly diffuse[160]. Katine *et al.*[157] were able to estimate the degree of specularity, p, for electrostatically confined nanostructures by comparing the measured resistance of a long rectangular channel with theory, obtaining $p \cong 0.95$, in agreement with earlier results of Thornton *et al.*[149]. These data demonstrate that electrostatic confinement of electrons in *GaAs/AlGaAs* heterostructures yields smooth walls which do not deflect electron trajectories away from the specular direction by large angles. Definition of nanostructures, by reducing the mobility via patterned ion beam damage, yields sharper features than electrostatic gates, but a lower degree of specularity[149].

The trajectory of electrons inside semiconductor nanostructures can deviate substantially from straight lines even when the mean free path is considerably longer than the size of the structure. The primary sources of deviation are believed to be small angle scattering due to ionized impurity atoms, and electron density gradients[61,62]. In order to reduce the influence of ionized dopant atoms on the mobility in *GaAs/AlGaAs* heterostructures, the doping layer is typically set back from the electron gas layer by a distance of 20 nm or more. As a result, the electrostatic potential from the doping layer is smoothed spatially, and high wavevector contributions to electron scattering are cut off[161]. The smoothed potential from ionized dopants, together with the electrostatic potential from residual ionized impurity atoms, produces deviations from straight line paths which typically determine the low temperature mobility in two-dimensional electron gases. As a result of small angle scattering, the classical electron trajectories bend continuously by small amounts which add randomly to relax the forward momentum over the transport mean free path, l. Small angle scattering is characterized by the single particle scattering time, τ_s, which is the lifetime of momentum eigenstates of an electron plane wave, typically assumed to be inside a rectangular box[162]. Because these eigenstates are finely spaced, $\Delta p/p \sim \lambda/L$ for boxes of size, L, much larger than the wavelength, λ, the angular deviation associated with τ_s can be quite small. The single particle scattering time can be determined experimentally for a given sample via the amplitude of Shubnikov-de Haas oscillations[163,164,165]. The electron density profile inside semiconductor nanostructures can also create deviations from ideal behavior. Typically, the density profile is not flat, but varies smoothly in a manner determined by the confining potential. Because the Fermi velocity, $v_F \propto n_s^{1/2}$, changes with sheet density, n_s, gradients in electron density bend electron waves like gradients in index of refraction bend light. As a result, electron trajectories curve smoothly rather than follow straight lines.

Whether or not small angle scattering and electron density gradients have a measurable influence on transport statistics depends on the characteristics of the structure. The statistics of electron states in highly symmetric structures such as circular and square billiards can be quite sensitive to small amounts of disorder, and corresponding symmetries in the wavefunctions can be quickly lost. In contrast, the statistics of electron states in structures such as the stadium, with chaotic electron trajectories, are more robust to small amounts of disorder, because symmetry has already been destroyed. Thus, it is relatively easy to produce chaotic structures which correspond well to theory, but quite difficult to achieve ideal behavior experimentally in symmetric structures, such as circles and squares with dimensions much larger than the electron wavelength.

Semiclassical theory does not include the diffraction of electron waves, but the angular patterns of transmission including diffraction are important for experiments on semiconductor nanostructures using quantum point contacts. Typically, the contact contains a relatively small number of transverse modes, so that the contact acts as an antenna radiating electron waves in a multimode pattern. The shape of the contact is also important, both for classical and quantum behavior. The horn shape typically found in point contacts acts to collimate electron trajectories[30]. In an elegant experiment, Shepard et al.[31] measured the angular profile of electrons transmitted through a point contact. The forward transmission coefficient, T_F, is plotted in Figure II.6 vs. angle, for gate voltage biases that progressively open the contact from Figure II.6(a) to II.6(b). As shown, the point contact emits electrons over a considerable angular spread, due to diffraction and to contributions from multiple transverse modes which produce secondary peaks in Figure II.6. These peaks are present even when the contact is nearly pinched off, as in Figure II.6(a), due to the longitudinal profile of the point contact.

Semiclassical calculations of dot conductance, G_{dot}, typically treat the quantum point contacts semiclassically and apply the boundary conditions for electron trajectories inside the leads. Because diffraction in the point contacts is ignored, these calculations are most appropriate for wide contacts with many transverse

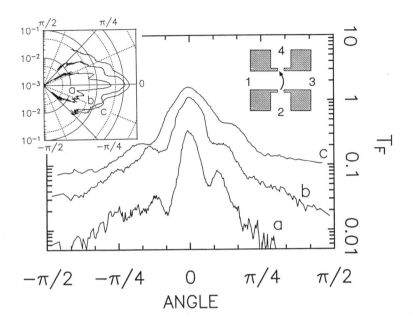

FIGURE II.6. Angular profile of the electron beam emerging from a constriction (shown in the inset) at three biases a, b, and c of the electrostatic gates used to define the structure. The transmission, T_F, is shown as a function of the angle of emission. At these biases the separations, d, between point contacts are 210 nm, 220 nm, and 270 nm, respectively. The corresponding densities within the junction are n_s = (1.8, 2.0, and 2.2) $\times 10^{11} cm^{-2}$, respectively. Right inset: schematic representation of the semiclassical trajectory determining T_F. Left inset: Polar representation of the emerging beam for biases a, b, and c, employing a logarithmic radial coordinate[31].

modes. The semiclassical approach can also be applied to experiments with narrow quantum point contacts for which diffraction is important. Point contact diffraction can be included in an approximate manner by regarding the emitting contact as a transmitting antenna for electron waves, and the receiving contact as receiving antenna, both located along the inside edge of the structure.

III. Shape Dependent Conductance of Ballistic Nanostructures

Quantum dots fabricated from the two dimensional electron gas in *GaAs/AlGaAs* heterostructures provide the experimental means to test the influence of shape on electron transport. The shape of the quantum dots can be chosen to correspond to two-dimensional billiards studied in ergodic theory, so that the behavior of the classical trajectories is well understood for ideal structures. The insets to Figures II.7 show a pair of ballistic quantum dots with quantum point contacts: the white areas are electrostatic gates which define the dot in the two-dimensional electron gas below; the black areas correspond to the shape of the electron gas[45]. The dot in Figure II.7(a) is shaped as a stadium, with an aspect ratio for which all classical trajectories are chaotic, whereas the dot in Figure II.7(b) is circular, with no chaotic classical orbits. Under the conditions of the experiments below, the electron gas inside both dots has the same area, 0.41 μm^2, and their transverse dimensions are a factor ~15 larger than the Fermi wavelength, $\lambda_F \cong 41$ nm. Quantum point contacts are positioned at right angles along the edge of the dots to minimize the influence of direct trajectories between contacts, which do not sense the shape of the dot. In addition, the contacts themselves were shaped to avoid internal standing waves.

Figure II.7 plots the resistance of the stadium-shaped and circular quantum dots measured vs. magnetic field, B, in a dilution refrigerator at $T = 20$ mK[45]. Both resistance traces show two robust features: a resistance peak at $B = 0$, shown in detail in the insets to Figure II.7, and conductance fluctuations away from $B = 0$. We show, below, that the origins of these two features are different. The zero-field peak is due to coherent backscattering (weak localization) associated with the interference of time-reversed pairs of electron trajectories, and conductance fluctuations are due to the interference of forward scattered trajectories. The resistance traces in Figure II.7 accurately reproduce for both signs of magnetic field $(R(B)=R(-B))$, as predicted by time reversal symmetry for two contact resistance measurements $(B \rightarrow -B$ for $t \rightarrow -t)$. Because the resistance is measured via a four probe technique outside the two point contacts, the presence of this symmetry in the data also shows that the influence of the two-dimensional electron gas outside the structure is negligible.

Closer inspection of the data in Figure II.7 reveals characteristic differences between the circle and the stadium. Both the width of the zero field resistance peak and the magnetic field scale for conductance fluctuations are smaller for the circle than for the stadium. Both features indicate that the characteristic areas linking magnetic flux in the circle are larger than in the stadium. This is expected from theory, because electron orbits inside the circle circulate in one direction, whereas orbits inside the stadium reverse direction of rotation every 3 to 5 bounces[20,142]; we return to analysis of the power spectrum of conductance fluctuations below. The dip in the circle magnetoresistance at $B = 0.27T$ [Figure II.7(b)] may be a classical effect due to the bending of electron orbits; at this field the radius of the cyclotron orbit nearly matches that of the circle. For perfectly symmetric circular dots, one might expect to see finely patterned structures in the resistance as a consequence of the circular symmetry. However, the effects of imperfections and the phase coherence time, described above, are

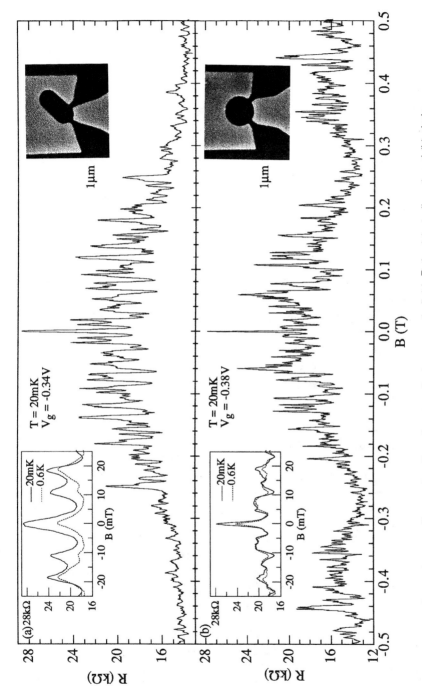

FIGURE II.7. Dot resistance, R, as a function of perpendicular magnetic field, B, for (a) stadium 1 and (b) circle 1, both with $N = 1$ fully transmitted modes in the leads. Insets: Zero-field peaks at 20mK (solid) and 0.6K (dashed), and electron micrographs of devices, with a 1 μm bar for scale[45].

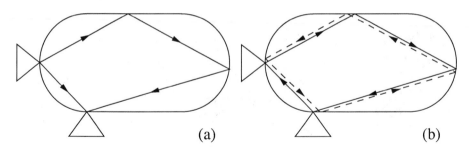

FIGURE II.8. Schematic illustrations of interfering electron trajectories, which give rise to (a) conductance fluctuations and (b) coherent backscattering (see text).

sufficient to disrupt orbits, which sensitively depend on precise geometrical symmetries.

Figure II.8(a) illustrates how conductance fluctuations originate from the interference of a pair of electron trajectories inside the stadium; the locations of the quantum point contacts are indicated by triangles. As shown above in Figure II.6, each spatial mode in the quantum point contact radiates into a range of angles. Electron waves radiating from the same mode in one contact travel along separate paths and interfere in another mode at the second contact to produce conductance fluctuations as the magnetic field is swept. At low magnetic fields, the bending of electron trajectories is relatively small, and the enclosed areas are essentially field independent. Each time one quantum of magnetic flux is added to the area enclosed by a trajectory pair, the conductance executes one complete oscillation. Summing over pairs of trajectories joining the contacts produces the complex conductance fluctuations shown in Figure II.7. Because the interference for a given trajectory pair can be either constructive or destructive, the fluctuations add randomly in the total conductance.

Coherent backscattering is a separate phenomenon, analogous to weak localization in disordered conductors, associated with the interference of time-reversed pairs of paths returning to the emitting contact as illustrated in Fig. II.8(b). For each closed path emitted from and returning to the same spatial mode in one contact (solid line), there exists a time reversed path (dashed line) at zero magnetic field that is identical except for the direction of travel. Because the paths are identical except for the direction of travel, each time-reversed pair interferes coherently to enhance the backscattered amplitude. Unlike the conductance fluctuations above, this interference is always constructive at zero magnetic field; its contribution to the conductance can be obtained by summing over all time reversed pairs. The interference is destroyed when a flux quantum enters the total area enclosed by the time-reversed pair of closed paths, which is twice the area of a single closed path. Therefore, we expect that the characteristic area for coherent backscattering deduced from resistance measurements is approximately twice that for conductance fluctuations. This discussion presents a highly simplified model: a detailed quantitative description of weak localization theory in ballistic nanostructures is given in Part I.

Although a quantitative analysis (presented below) finds characteristic differences, experimental conductance fluctuation data from circles and stadia of the same area can be quite similar. In the past, such similarity has lead to uncertainty about the competing roles of dot shape and disorder. In a recent experiment, Berry *et al.*[48] clearly demonstrated the dominant influence of dot shape on conductance fluctuations and coherent backscattering in ballistic nanostructures by investigating matched pairs of circular dots with (type B) and without (type A) a barrier to circulating orbits, as illustrated in Figure II.9. Other than the presence of the barrier, both shapes are identical. As indicated in Figures II.9(c) and II.9(e), trajectories inside circular dots (type *A*) accumulate large enclosed areas, both for forward scattering between contacts and for coherent backscattering. The barrier in type *B* dots reflects circulating orbits as shown in Figures II.9(d) and II.9(f), and acts to cancel the area linked by magnetic

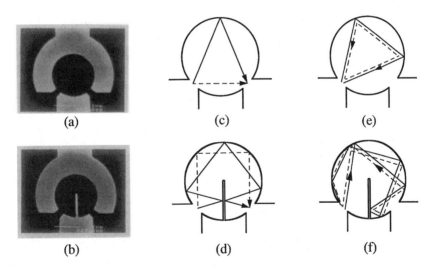

(a)   (c)   (e)

(b)   (d)   (f)

FIGURE II.9. Scanning electron micrographs of (a) large circular shape *A* dot and (b) large circular shape *B* dot with barrier; length bar is 1 μm. In shape *A*, forward-scattered (c) and backscattered (e) trajectories circulate in a single direction; in shape *B* [(d) and (f)], they reverse direction with each rebound off the central bar, partially canceling the accumulated area. Trajectories circulating clockwise (counterclockwise) are shown by the solid (dashed) lines[48].

flux. Thus, we expect that the magnetic field width of the coherent backscattering peak and the characteristic field scale of conductance fluctuations will both be larger for dots with the barrier (type B), compared with circular dots (type A) of the same area. In order to test the influence of device area on the characteristic fields, two pairs of type A and type B dots were fabricated: large, with electron gas area $A_{dot} = 1.6$ μm^2, and small, with $A_{dot} = 0.43$ μm^2.

Figures II.10(a) and II.10(b) plot magnetoresistance data at $T = 0.43K$ for the large and small pairs of dots, respectively; the upper curve in both figures is for type B dots with the barrier to circulating orbits, and is offset for clarity. Within each figure, one finds a striking difference in the width of the zero-field resistance

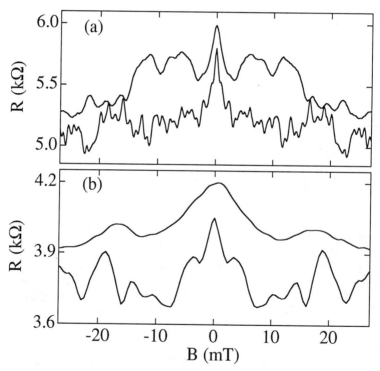

FIGURE II.10. Magnetoresistance data at $T = 0.43K$ for (a) large and (b) small type A and type B quantum dots. In both (a) and (b) the upper trace is type B with the barrier, and the lower is type A, offset for clarity. Note the more rapid fluctuations and sharper zero field peak in type A versus type B and in large versus small dots[48].

peak due to coherent backscattering, and in the characteristic field scale of conductance fluctuations: both are larger for dots with the barrier, as expected. In addition, the magnetic field scales are in proportion to the inverse dot area, as expected. These dramatic differences in the raw data are a clear demonstration of the influence of dot shape on conductance in ballistic nanostructures.

IV. CONDUCTANCE FLUCTUATIONS

The power spectrum of conductance fluctuations versus magnetic field measured for a quantum dot can be used to characterize the distribution of trajectory areas, as described in Part I. For the simple, fully chaotic case, such as the stadium, semiclassical analysis predicts that the distribution of areas decays exponentially with the characteristic area, A_α. The power spectrum, $S(f)$, of conductance fluctuations vs. magnetic frequency, f (cycles/Tesla), is then[17]

$$S(f) = S_0(1 + 2\pi B_\alpha f)\exp(-2\pi B_\alpha f), \qquad (\text{II.1})$$

where B_α is the magnetic field change necessary to increase the flux through the characteristic area, $A_\alpha = \Phi_0/2\pi B_\alpha$, by $\Phi_0/2\pi$, where $\Phi_0 = hc/e = 4.14\text{mT }\mu\text{m}^2$ is the flux quantum for electrons. The power spectrum for chaotic structures decays rapidly with magnetic frequency, because orbits randomly switch their direction of rotation. In contrast, the power spectrum for integrable billiards, such as the circle, possess a non-exponential shape with much more high frequency content, because orbits tend to continue to circulate in the same direction, accumulating area. Numerical results for circular billiards give power spectra that resemble power laws in f [18,19,22,23].

Power spectra of experimental conductance fluctuation data for circular and stadium shaped quantum dots show characteristic differences in agreement with theory. The original comparison made by Marcus *et al.*[45] is shown in Figures II.11(a) and II.11(b), which plot the conductance fluctuation power spectra for stadium-shaped and circular quantum dots for two nominally identical samples. The power spectra were computed from raw data, such as, that in Figure II.7, by first subtracting a smooth background, then averaging the power spectra taken over short overlapping intervals in magnetic field to improve the amplitude resolution and suppress low frequency peaks due to periodic orbits[45]. Fits of the stadium data to Equation 1 are shown as the solid lines. As shown, the power spectra for two stadia are in good agreement with theory for the chaotic case over about three orders of magnitude in power, and the fitted areas, $A_\alpha = 0.19 \mu\text{m}^2$ and $A_\alpha = 0.18 \mu\text{m}^2$, for stadia 1 and 2, respectively, agree quite well. Comparison with the area of the electron gas inside the stadia, $A_{dot} = 0.41 \mu\text{m}^2$, and the typical area of closed trajectories, $\sim A_{dot}/2$, shows that the quantum interference giving rise to conductance fluctuations is dominated by relatively short

paths. Numerical simulations[22,166] for a stadium with contacts in the
same arrangement as in Figure II.11 show that the primary source of conductance fluctuations is interference between one bounce trajectories and direct paths
between contacts. Compared with stadium data, conductance fluctuation power
spectra for circular dots show more high frequency content, corresponding to a
greater proportion of large area trajectories.

The shape dependence of the conductance fluctuation power spectrum is
more clearly shown by comparing the conductance fluctuation power spectra
of the two types of circular dots discussed above open circles
(type A) and circles with a barrier to circulating orbits (type B), as shown in
Figures II.12(a) and II.12(b)[48]. Two pairs of dots were fabricated: large, with
electron gas area $A_{dot} = 1.6 \ \mu m^2$, and small, with $A_{dot} = 0.43 \ \mu m^2$. Corresponding raw data for these structures is shown in Figure II.10, above. The power

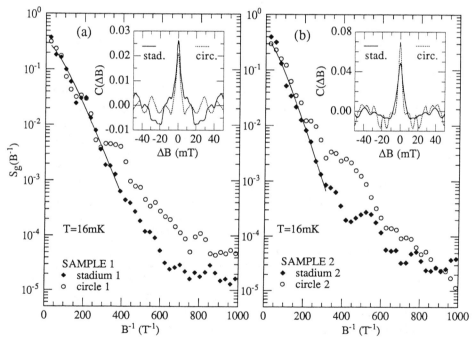

FIGURE II.11. Average power spectra, Sg, of conductance fluctuations, $\delta g(B)$, for a
stadium (solid diamonds) and a circle (open circles) with $N \sim 3$ transverse modes in the
leads. (a) Sample 1. (b) Sample 2. Solid curves are fits of semiclassical theory [Eq.
(II.1)] to stadium data. Insets: Autocorrelations, $C(\Delta B)$, of stadium (solid) and circle
(dashed) for $0.01 T < B < 0.29 T$, with normalization $C(0) = var[g(B)][45]$.

spectrum data for type A and type B dots are shown as squares and triangles, respectively, for large dots in Figure II.12(a), and for small dots in Figure II.12(b). The solid lines are fits to conductance fluctuation theory, Equation (II.1). As shown, chaotic theory fits the data for type B dots (with the barrier) very well, yielding characteristic areas $A_\alpha = 0.30$ μm^2 and $A_\alpha = 0.08$ μm^2 for large and small dots, respectively. The ratio $A_\alpha / A_{dot} = 0.19$ is approximately the same for large and small type B dots; the small value of the ratio is evidence for flux cancellation due to the barrier, as illustrated in Figure II.9. Although a type B

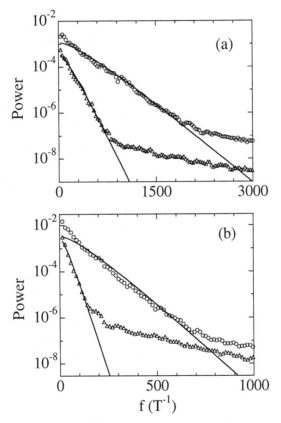

FIGURE II.12. Power spectra of conductance fluctuations for (a) large and (b) small devices versus magnetic frequency, f, averaged over five gate voltages. Data for shape A (circles) and shape B (triangles) are offset for clarity. Curve fits to chaotic theory Eq. (II.1) (solid) agree well with data for shape B (barrier), but deviate for shape A (circular). The tails at high f are a noise effect[48].

dot with ideal geometry geometry would constitute an integrable system without chaotic trajectories, the experimental structure is more rounded and resembles a folded stadium.

In contrast, the power spectra for circular dots (type A) are not well fit by chaotic theory, and correspond to much larger characteristic areas than type B dots with barriers. Examination of Figure II.12 shows that the power spectra for circular dots have a qualitatively different shape: they are concave up, unlike the fits to chaotic theory [Equation (II.1)], shown as solid lines, which are concave down, and fall much faster with magnetic frequency. Although the power spectra for circular dots do not fit chaotic theory well, they typically fall faster with magnetic frequency than power laws predicted by numerical simulations for circular structures. The fitted characteristic areas for circular dots, $A_\alpha = 0.82$ μm^2 and $A_\alpha = 0.25$ μm^2 for large and small dots, respectively, scale with dot area and are both larger by factors $\cong 3$ than the areas for corresponding type B dots with barriers.

The data presented above are primarily taken at low temperatures, $T < 0.5K$, for which the ballistic dephasing length, $l_\phi = v_F \tau \phi$, is considerably larger than the size of the quantum dot, and for which the effects of thermal averaging are small. At higher temperatures, the temperature dependence of the inelastic scattering time, τ_ϕ, leads to a reduction in l_ϕ, and to an increase in the importance of thermal averaging. Both phenomena are covered in detail in Part I of this chapter; we briefly describe the physical picture here.

Inelastic scattering acts to cut off interference between electron trajectories over total lengths longer than the ballistic dephasing length, l_ϕ. As l_ϕ decreases with increasing temperature, this cut off becomes shorter, reduces the number of trajectories contributing to conductance fluctuations, and thereby reduces the total conductance fluctuation amplitude. This physical picture is relatively straightforward, but a full theory of dephasing inside quantum dots has not yet appeared to our knowledge. As described above in Part I, dephasing can be represented by using a simple model in which electrons enter at random a large dephasing reservoir connected to the dot by a point contact with N_ϕ channels[10,46,52,94,167].

Thermal averaging also acts to reduce the size of conductance fluctuations. At temperatures larger than the energy level spacing, $\sim 2E_F/N_e$, with N_e the number of electrons in the dot, and larger than the broadening, $\hbar\gamma \geq 2E_F/N$, due to the residence time, γ^{-1}, inside the dot, transport occurs through a range of energy levels of width $\sim k_B T$ determined by the temperature. This spread in energy results in a spread in wavevector, $\Delta k/k \sim k_B T/2E_F$, which dephases electron waves with different wavevectors along the same trajectory over lengths longer than $\sim 1/\Delta k$. For conductance fluctuations, contributions from different wavevectors add randomly and tend to cancel, so the effect of thermal averaging is to reduce the conductance fluctuation amplitude at higher temperatures. By contrast, contributions to coherent backscattering add with the same sign and do not tend to cancel, so that thermal averaging alone does not reduce the amplitude of the coherent backscattering peak as long as the thermal length, $\hbar v_F / k_B T$, is larger than the sample size.

The experimental dependence of the root mean square conductance fluctuation amplitude δG_{rms} measured for stadium shaped quantum dots is shown versus dot conductance G_{dot} in Figure II.13 for three increasing temperatures, $T = 0.15K$, $T = 0.43K$ and $T = 1.5K$[168]. The dots had the same geometry and size as for Figure II.7, above, and the fluctuation amplitude was obtained by averaging over traces taken on different days with different samples. Because the dot resistance is dominated by that of the quantum point contacts, which add in series at low B[169], the dimensionless conductance, $G_{dot}/(e^2/h)$, is approximately equal to the number of channels in each point contact. Theoretical predictions based on random matrix theory[94] are also plotted in Figure II.13 as the dotted curve at zero temperature ($N_\phi = 0$) and the smooth curves, which are fits to the data at $T = 0.15K$, $T = 0.43K$, and $T = 1.5K$ with $N_\phi = 2.5, N_\phi = 6.6$, and $N_\phi = 16$, respectively. For the sake of simplicity, both dephasing and temperature averaging were represented in the fit by a single empirical parameter, N_ϕ. As shown, random

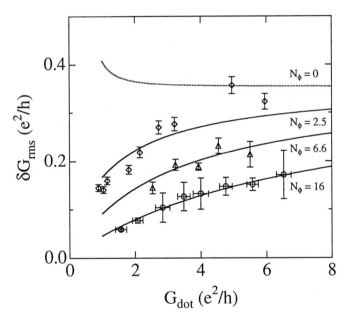

FIGURE II.13. Average root mean square conductance fluctuation amplitude, δG_{rms}, data versus dot conductance, G_{dot}, for three temperatures, $T = 0.15K$ (circles), $T = 0.43K$ (triangles) and $T = 1.5K$ (diamonds), together with fits to random matrix theory, with N_ϕ phase breaking channels[168].

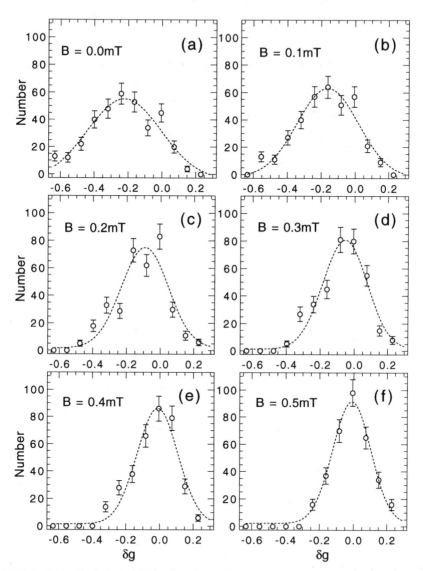

FIGURE II.14. Evolution, (a) to (f), of the distribution of conductance fluctuations, $\delta g(B)$, as time reversal symmetry is broken. Distributions are histograms of shape-distortion conductance fluctuation data at each field (vertical scale is histogram bin count). The dashed lines are Gaussian fits[51].

matrix theory can provide a good fit to the amplitude, δG_{rms}, of conductance fluctuation data. The fitted N_ϕ are larger than one would expect from dephasing alone, consistent with the expected effects of thermal averaging.

The experimental probability distribution, $P(\delta g)$, of conductance fluctuations has recently been measured using a dot with adjustable shape to obtain ensemble averages[51]. The distribution $P(\delta g)$ is predicted by random matrix theory[93] to be Gaussian for dots with a large number of channels in the contacts. For small numbers of channels in the contacts, the distribution is predicted to become asymmetric and to change with magnetic field in a characteristic way as time reversal symmetry is broken, as discussed in Part I of this chapter. Figure II.14 shows the measured probability distribution of conductance fluctuations, $\delta g(B)$, for six increasing values of magnetic field, over which time reversal symmetry is broken by the introduction of magnetic flux inside interfering electron trajectories. The number of channels in each point contact in the experiment is approximately two. As shown by the fits, the distribution remains Gaussian and symmetric over the entire range of magnetic field, as expected for a large number of channels. The difference between experiment and theory is attributed to the effect of phase breaking which effectively adds N_ϕ channels to the contacts[51].

V. COHERENT BACKSCATTERING
AND WEAK LOCALIZATION

The coherent backscattering peak at zero magnetic field is a distinct physical phenomenon due to the constructive interference of time-reversed pairs of trajectories returning to the injecting contact, as described above and in Part I of this chapter. Coherent backscattering in ballistic quantum dots is related to weak localization in diffusively scattering metals. An analog of weak localization also occurs in quantum dots due to the interference of nearby trajectories which contain nearly closed loops[170]. The term coherent backscattering is perhaps more apt when the shortest return paths dominate, and weak localization is more apt when longer interfering paths are important; in this section we use the two terms interchangeably. Coherent backscattering and weak localization are manifested as a robust resistance peak (conductance dip) at zero magnetic field in the low temperature magnetoresistance of ballistic quantum dots [see Figure II.7]. Experiments have shown that the zero field peak persists under changes in conditions which change the sign of the conductance fluctuations: thermal cycling and changes in contact width[48], changes in electron density[47,171], and averaging over an array of nominally identical dots[49]. This robustness argues for a different physical origin for the zero field peak than for the conductance fluctuations, which are also present at zero field. The theory of weak localization and coherent backscattering[18,19] is discussed briefly above and described in detail in Part I of this chapter.

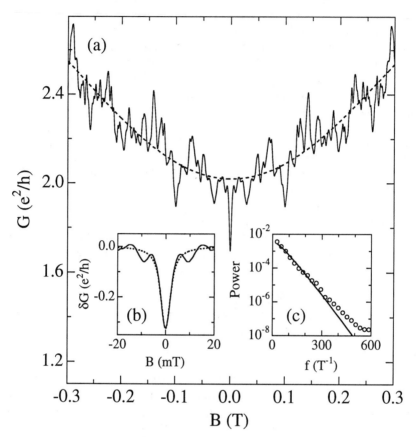

FIGURE II.15. (a) Magnetoconductance at $T = 1.50K$ (solid) shown with a fourth-order polynomial background fit (dashed). (b) Zero-field conductance dip (solid) with a Lorentzian fit (dashed). (c) Averaged power spectrum of conductance fluctuations (circles) with a fit to semiclassical theory[48].

    The area enclosed by time reversed pairs of closed trajectories returning to the injecting contact is approximately twice the area enclosed by interfering trajectories between contacts which give rise to conductance fluctuations, as illustrated by Figure II.8 above. Berry et al.[48] used this property to verify the existence of coherent backscattering in ballistic quantum dots. Figure II.15(a) plots the conductance G_{dot} vs. magnetic field for of a stadium shaped dot nominally identical to the structure used for Figure II.7, above. A higher temperature $T = 1.50K$ was chosen to reduce the relative size of conductance fluctuations. The raw magnetoconductance trace shows a zero field dip as well as conductance fluctuations away from $B = 0$. Theory[18,19] predicts that the zero field conductance dip has a Lorentzian shape in a simple chaotic model; numerical simulations show that the precise form of the conductance dip is dependent on the details of dot shape and contact placement. The data fit a Lorentzian quite well, as shown in the inset of Figure II.15(b), using the half-width at half maximum B_c and the dip height, ΔG_0, as adjustable parameters. The characteristic field, $2\pi B_c$, is just that required to add one flux quantum to the total characteristic area, $A_c = \Phi_0 / 2\pi B_c$, enclosed by the time reversed pair of backscattered trajectories[18,19], which is twice the characteristic area enclosed by a single closed backscattered trajectory. Assuming that the characteristic area enclosed by a single backscattered trajectory is comparable to that enclosed by a pair of interfering forward scattered trajectories, one expects $A_c \cong 2A_\alpha$, where A_α is the characteristic area for conductance fluctuations defined above (Equation 1). In order to test this relationship, A_α was determined by fitting the measured power spectrum with Equation (II.1), as shown in the inset of Figure II.15(c), where theory is the solid line. For this data, $A_c = (0.26\pm0.04)\,\mu m^2$ and $A_\alpha = (0.13 \pm 0.02)\,\mu m^2$, demonstrating the factor of two difference between the characteristic areas for coherent backscattering and conductance fluctuations. Averaging over an ensemble of magnetoresistance data obtained for the same sample at 26 different contact gate voltages gives $\langle A_c \rangle = (0.26\pm0.05)\,\mu m^2$ and $\langle A_\alpha \rangle = (0.133\pm0.003)\,\mu m^2$. Confirmation of the result $A_c \cong 2A_\alpha$ was obtained by Chan et al.[51].

The width of the zero field resistance peak is strongly influenced by dot shape and dot area in a manner consistent with the physical picture given above. Figures II.16(a) through 16(d) show the zero field resistance peaks, together with Lorentzian fits, for the set of large and small circular (type A) and circular with barrier (type B) quantum dots discussed above, taken at $T = 0.43K$. The data are averaged for five different contact widths in order to suppress conductance fluctuations. The ratios of the characteristic areas, A_C, for type A to type B dots obtained from the fitted half width at half maximum B_c are 4.0:1 and 3.1:1 for large and small dots, respectively, demonstrating that the area enclosed by backscattered trajectories is greatly reduced for dots with the barrier to circulating orbits, as expected. The measured areas A_C are also found to be proportional to the dot area within experimental error. Finally, a comparison of the characteristic areas A_C for coherent backscattering and A_α for conductance fluctuations for all four dots yields $A_C / A_\alpha = (2.4 \pm 0.8)$, in agreement with the stadium data above.

Conductance fluctuations are always present to some degree in coherent backscattering data from quantum dots. In order to reduce their importance, Keller et al.[47,171,172] conducted experiments using stadia in which the electron sheet density could be adjusted via a back gate. Magnetoresistance traces were obtained at dilution refrigerator temperatures $T = 50mK$ for a series of carrier densities, and the results averaged over sheet density in order to obtain the mean change in conductance, $\langle \Delta G \rangle$, due to weak localization (coherent backscattering).

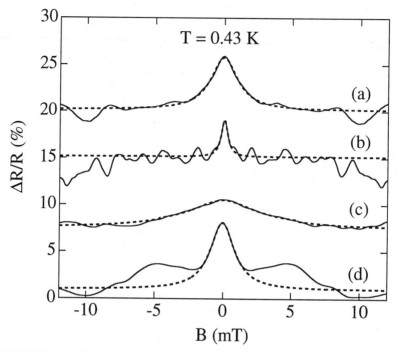

FIGURE II.16. Fractional change in magnetoresistance of (a) large *B*-type, (b) large *A*-type, (c) small *B*-type, and (d) small *A*-type devices averaged over five gate voltages. Data (solid) are offset for clarity and fit to a Lorentzian line shape (dashed)[48].

Figure II.17 plots $\langle \Delta G \rangle$ in units of e^2/h versus normalized magnetic field $B/\Phi_o\, \alpha$ (α is defined above) for a "stomach" shaped stadium illustrated in the inset[172]. As shown, the shape of the zero field conductance dip is in good agreement with both random matrix theory, and with numerical simulations[18,19], demonstrating the existence of the weak localization conductance dip for ensemble averaged data.

In an elegant experiment, Chang *et al.*[49] were able to identify a characteristic change in the weak localization lineshape due to the shape of the quantum dot. This experiment is also important because it demonstrates the existence of shape dependent weak localization (coherent backscattering) in very high mobility samples ($\mu = 1.8 \times 10^6 \, cm^2/V\,sec$) in which small angle scattering does not play an important role. In order to perform true ensemble averaging, two arrays were constructed, one with 48 stadium shaped quantum cavities and one with 48 circular cavities, illustrated in the insets to Figures II.18(a) and II.18(b), respectively. The areas for the circles and stadia were chosen to be the same, equivalent to a diameter of 1.08μm. In this experiment, the small angle scattering length,

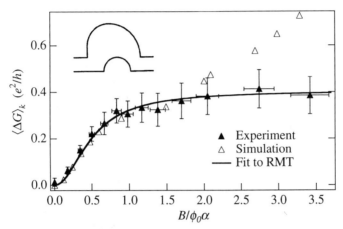

FIGURE II.17. Mean zero-field conductance dip data versus normalized magnetic field, obtained by averaging over different electron density controlled by a back gate, together with fits to random matrix theory and with numerical simulations as indicated[47,172].

$v_F \tau_s \sim 4$ μm, estimated from electron focusing measurements, was larger than the cavity size, and the transport mean free path was 17 μm. As shown in Figures II.18(a) and 18(b), each cavity had two openings, made relatively wide in order for semi-classical theory to be applicable. Figures II.18(a) and II.18(b) plot the measured magnetoresistance at $T = 50$mK for the stadium-shaped and circular cavity arrays, respectively. Both traces show a prominent zero-field resistance peak caused by weak localization in the quantum cavities. The lineshape for the stadium array is approximately Lorentzian, as shown. By contrast, the lineshape for the circular array is approximately triangular. The sharpness near $B = 0$ is evidence for long circulating trajectories, which accumulate relatively large areas before leaving the cavity. Surprisingly, the widths of the two zero field peaks are comparable, probably a consequence of the relatively wide contact openings. Numerical simulations of the lineshape for structures with the same shapes and similar lead configurations fit the experimental data well, as discussed in Part I, confirming the origins of the characteristic difference in weak localization lineshape[49].

At higher temperatures, dephasing of the electron wavefunctions via inelastic scattering acts to reduce the amplitude of the coherent backscattering conductance dip by cutting off interference from trajectories longer than the ballistic dephasing length, $l_\phi = v_F \tau_\phi$. Coherent backscattering is more robust to thermal averaging than conductance fluctuations, because each contribution from interfering backscattered trajectory pairs adds with the same sign. As a result, the zero field

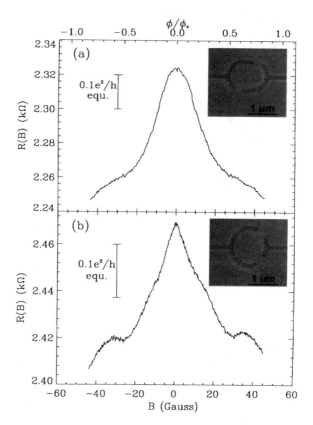

FIGURE II.18. Magnetoresistance for (a) 48 stadium cavities and (b) 48 circle cavities, normalized to a single cavity, at $T = 50$mK. The weak localization peak line shape shows a Lorentzian behavior for the chaotic, stadium cavities. In contrast, the line shape for the nonchaotic, circle cavities shows a triangular shape. The vertical bar indicates the equivalent change in conductance. Insets show electron micrographs of the cavities[49].

conductance peak persists to somewhat higher temperatures than conductance fluctuations. A full theoretical prediction of the temperature dependence for coherent backscattering and conductance fluctuations depends on the details of dot shape and contact placement. As above, one can make progress using a simple model in which dephasing and thermal averaging are represented by an effective number of channels, N_ϕ, opened to a dephasing reservoir.

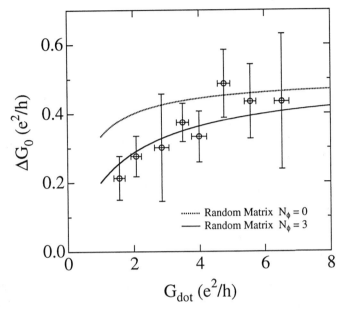

FIGURE II.19. Average amplitude of zero-field conductance dip, ΔG_0, versus dot conductance, G_{dot}, at $T = 1.5$ K, for data from stadium shaped quantum dots, together with random matrix theory at zero temperature and a fit to the data with $N_\phi = 3$ phase breaking channels[168].

The experimental dependence of the amplitude, ΔG_0, of the coherent backscattering dip on dot conductance, G_{dot}, measured for stadium shaped quantum dots with the same geometry and size as for Figure II.7, above, is shown in Figure II.19 at $T = 1.50$K[168]. As above, the dimensionless dot conductance, $G_{dot}/(e^2/h)$, is approximately equal to the number of channels in each point contact. The experimental data were obtained by averaging the zero field conductance dips from data similar to Figure II.7, above. As shown the amplitude ΔG_o increases as the contact width is increased and tends toward the value $\Delta G_o = 0.5 e^2/h$, predicted by random matrix theory for wide contacts with many channels[18,19], as discussed in Part I. Baranger and Mello[94] have also used random matrix theory to calculate the dependence of ΔG_0 on contact width; these results are shown in Figure II.19 as the smooth curves. The dotted curve is for zero temperature ($N_\phi = 0$) and the solid curve is a fit to the experimental data, with the fitted parameter $N_\phi = 3.1$. As shown, random matrix theory can give a good description of the amplitude of the coherent backscattering conductance dip. The fitted value of N_ϕ is smaller than for conductance fluctuations, consistent with a lack of thermal averaging. A careful theoretical analysis that separates the effects of dephasing and thermal averaging will provide additional information.

VI. CONCLUSIONS

The experimental work described above shows that quantum chaos in ballistic microstructures produces mesoscopic phenomena – conductance fluctuations and coherent backscattering – which have previously been associated with diffusive conductors. Mesoscopic manifestations of quantum chaos are commonly present even when the motion of electrons inside the conductor is ballistic, and their characteristics are determined by the shape of the conductor. A practical consequence is that one can expect electron waveguides and cavities in ballistic electronic devices of the future to show chaotic behavior. Most cavity shapes possess chaotic orbits; even those symmetric shapes which do not perfect circles and squares, for example, develop chaotic orbits for even small amounts of disorder. At the simplest level, the role of chaos in ballistic devices is to destroy the memory of initial conditions, just as for diffusive conductors.

At present, experimental work on quantum chaos in ballistic microstructures lags far behind theory. Even the best two-dimensional electron gas samples possess small amounts of disorder that redirect electron trajectories and alter complex orbits which are easily calculated in theory. Statistical properties are accessible in current experiments: as described above, theory and experiment are in good agreement, and the characteristic areas and amplitudes of conductance fluctuations and coherent backscattering resistance peaks can be reliably measured. The coherence time of electrons inside quantum dots is another subject of current interest. Collective contributions to the electron-electron scattering rate depend on device geometry and differ in general from the results for two-dimensional electron gases and one-dimensional wires. The coherence time for electrons inside isolated dots also differs from that for open structures. Single particle energy level statistics are a cornerstone of quantum chaos theory which should be experimentally accessible in the near future. One area where experiment leads theory is the tunneling regime, in which a ballistic quantum dot is connected to its leads by tunnel barriers. Experiments[50] show conductance fluctuations and zero-field resistance peaks for quantum dots in the tunneling regime qualitatively similar to those for open dots. However, current semiclassical theory contains neither tunneling, diffraction, nor collective interactions, all of which are important for closed structures. Direct spatial measurements of electron orbits and electron wavefunctions inside chaotic quantum dots could provide much useful information and directly test quantum chaos theory. In beautiful work, Crommie *et al.*[173] have used scanning tunneling microscopy to measure electron waves inside quantum corrals defined on the surface of a copper crystal by deliberately placing foreign adsorbed atoms. In current work, the corral is not a closed structure, and regular wave patterns are observed, in excellent agreement with semiclassical theory[174]. If a method is found to increase the residence time of electrons inside the corral, direct measurement of chaotic wave patterns may become possible. Measurements of the electron density profile and information about chaotic orbits inside semiconductor quantum dots may also prove possible using movable electrostatic gates with scanning probe techniques.

ACKNOWLEDGEMENTS

In carrying out the research summarized in Part I, H.U. Baranger benefited greatly from working with his collaborators, A.M. Chang, J.H. Davies, R.A. Jalabert, P.A. Mello, A.D Stone, and G. Timp. In addition, I thank H. Bruus, R.V. Jensen, M.W. Keller, C.M. Marcus, H. Mathur, M.L. Roukes, D. Ullmo, and R.M. Westervelt for many valuable conversations and suggestions.

R.M. Westervelt thanks C. Marcus, M. Berry, J. Baskey, J. Katine, and S. Yang for essential contributions to the experimental work, and H.U. Baranger, B. Altshuler, A. D. Stone, and R. Jensen for important theoretical discussions. Robert Westervelt's work was supported in part by the Office of Naval Research under grants N00014-95-1-0104 and N00014-95-1-0886, and the National Science Foundation under grant DMR95-01438.

REFERENCES

[1]   For reviews of mesoscopic physics see: Beenakker, C.W.J., and van Houten, H., in *Solid State Physics*, Vol. **44**, Ehrenreich, H., and Turnbull, D., eds. New York: Academic Press, 1991, pp. 1-228; and Altshuler, B.L., Lee, P.A., and Webb, R.A., *Mesoscopic Phenomena in Solids* Amsterdam: North-Holland, 1991.

[2]   Sheng, P., *Scattering and Localization of Classical Waves in Random Media*, Singapore: World Scientific, 1990.

[3]   For reviews of chaos see: Gutzwiller, M.C., *Chaos in Classical and Quantum Mechanics* , New York: Springer-Verlag, 1990; Giannoni, M.-J., Voros, A., and Zinn-Justin, J., *Chaos and Quantum Physics* , London: Elsevier, 1990.

[4]   For a review of transient chaos see: Tél, T., in *Direction in Chaos*, Vol. **3**, Bai Lin, Hao, ed., Singapore: World Scientific, 1990, pp. 149–211.

[5]   For a review of classical and quantum chaotic scattering see: Smilansky, U., in *Chaos and Quantum Physics*, Giannoni, M.-J., Voros, A., and Zinn-Justin, J., eds.,

New York: North-Holland, 1991, pp. 371–441.

[6]   Concerning phase-breaking, see, for example: Altshuler, B.L., Aronov, A.G., and
      Khmelnitskii, D.E., *Solid State Commun.* **39**, 619 (1981); *J. Phys. C* **15**, 7367
      (1982); Fukuyama, H., and Abrahams, E., *Phys. Rev. B* **27**, 5976 (1983);
      Fukuyama, H., *J. Phys. Soc. Jpn.* **53**, 3299 (1984); Büttiker, M., *Phys. Rev. B* **33**,
      3020 (1986); Datta, S., *Phys. Rev. B* **40**, 5830 (1989); Feng, S., *Phys. Lett. A* **143**,
      400 (1990); Stern, A., Aharonov, Y., and Imry, Y., *Phys. Rev.* **41**, 3436 (1990);
      Kirczenow, G., *Solid State Commun.* **74**, 1051 (1990); D'Amato, J.L., and
      Pastawski, H.M., *Phys. Rev.* **41**, 7441 (1990); Hershfield, S., *Phys. Rev. B* **43**,
      11586 (1991).

[7]   Sivan, U., Imry, Y., and Aronov, A.G., *Europhys. Lett.* **28**, 115 (1994).

[8]   Abrikosov, A.A., *Fundamentals of the Theory of Metals,* Amsterdam: North-
      Holland, 1988.

[9]   Landauer, R., *IBM J. Res. Dev.* **1**, 233 (1957); *Z. Phys. B* **68**, 217 (1987).

[10] Büttiker, M., *Phys. Rev. Lett.* **57**, 1761 (1986).

[11] Imry, Y., in *Perspectives on Condensed Matter Physics*, Grinstein, G., and
      Mazenko, E., eds., Singapore: World Publishing, 1986.

[12] Büttiker, M., Prêtre, A., and Thomas, H., *Phys. Rev. Lett.* **70**, 4114 (1993).

[13] McQuarrie, D.A., *Statistical Mechanics,* New York: Harper and Row, 1976, pp.
      507, 540.

[14] Madelung, O., *Introduction to Solid State Theory,* New York: Springer-Verlag,
      1978, pp. 365–370.

[15] For derivations of transmission formulas for the conductance from the Kubo formula
      see: Economou, E.N., and Soukoulis, C.M., *Phys. Rev. Lett.* **46**, 618 (1981); Fisher,
      D.S., and Lee, P.A., *Phys. Rev. B* **23**, 6851 (1981); Engquist, H.L., and Anderson,
      P.W., *Phys. Rev. B* **24**, 1151 (1981); Langreth, D.C., and Abrahams, E., *Phys. Rev.
      B* **24**, 2978 (1981); Stone, A.D., and Szafer, A., *IBM J. Res. Develop.* **32**, 384
      (1988); Baranger, H.U., and Stone, A.D., *Phys. Rev. B* **40**, 8169 (1989); Kucera, J.,
      and Streda, P., *J. Phys. C* **21**, 4357 (1988); Kucera, J., *Czechoslovak J. Phys.* **41**,
      749 (1991); Shepard, K., *Phys. Rev. B* **43**, 11623 (1991); Azbel, M. Ya., *Phys. Rev.
      B* **46**, 15004 (1992); Fenton, E.W., *Phys. Rev. B* **46**, 3754 (1992); **47**, 10135
      (1993); Nöckel, J.U., Stone, A.D., and Baranger, H.U., *Phys. Rev. B* **48**, 17569
      (1993).

[16] Blümel, R., and Smilansky, U., *Phys. Rev. Lett.* **60**, 477 (1988); **64**, 241 (1990);
      Physica D **36**, 111 (1989).

[17] Jalabert, R.A., Baranger, H.U., and Stone, A.D., *Phys. Rev. Lett.* **65**, 2442 (1990).

[18] Baranger, H.U., Jalabert, R.A., and Stone, A.D., *Phys. Rev. Lett.* **70**, 3876 (1993).

[19] Baranger, H.U., Jalabert, R.A., and Stone, A.D., *Chaos* **3**, 665 (1993).

[20] Jensen, R.V., *Chaos* **1**, 101 (1991).

[21] Lai, Y.-C., *et al.*, *Phys. Rev. Lett.* **68**, 3491 (1992).

[22] Lin, W.A., Delos, J.B., and Jensen, R.V., *Chaos* **3**, 655 (1993).

[23] The distribution in the non-chaotic case is discussed in Meiss, J.D., and Ott, E., *Phys. Rev. Lett.* **55**, 2741 (1985); Bauer, W., and Bertsch, G.F., *Phys. Rev. Lett.* **65**, 2213 (1990); Legrand, O., and Sornette, D., *Physica D* **44**, 229 (1990); Reference [21] and references therein.

[24] Timp, G., *et al.*, *Phys. Rev. Lett.* **60**, 2081 (1988).

[25] Takagaki, Y., *et al.*, *Solid State Commun.* **68**, 1051 (1988); Takagaki, Y., *et al.*, *Solid State Commun.* **71**, 809 (1989); Takagaki, Y., *et al.*, *Solid State Commun.* **75**, 873 (1990).

[26] Baranger, H.U., *Phys. Rev. B* **42**, 11479 (1990).

[27] Beenakker, C.W.J., and van Houten, H., *Phys. Rev. B* **39**, 10445 (1989).

[28] Glazman, L.I., *et al.*, *Pis'ma Ah. Teor. Fiz.* **48**, 218 (1988) [*JETP Lett.* **48**, 238 (1988)].

[29] Baranger, H.U., and Stone, A.D., *Phys. Rev. Lett.* **63**, 414 (1989).

[30] Molenkamp, L.W., *et al.*, *Phys. Rev. B* **41**, 1274 (1990).

[31] Shepard, K.L., Roukes, M.L., and van der Gaag, B.P., *Phys. Rev. Lett.* **68**, 2660 (1992).

[32] Roukes, M.L., *et al.*, *Phys. Rev. Lett.* **59**, 3011 (1987).

[33] Ford, C.J.B., *et al.*, *Phys. Rev. B* **38**, 8518 (1988).

[34] Ford, C.J.B., *et al.*, *Phys. Rev. Lett.* **62**, 2724 (1989).

[35] Chang, A.M., Chang, T.Y., and Baranger, H.U., *Phys. Rev. Lett.* **63**, 1860 (1989).

[36] Roukes, M.L., Scherer, A., and Van der Gaag, B.P., *Phys. Rev. Lett.* **64**, 1154 (1990); Roukes, M.L., *et al.*, in *Science and Engineering of 1- and 0- Dimensional Semiconductors*, Beaumont, S.P., and Sotomayor-Torres, C.M., eds., New York: Plenum, 1990.

[37] Akera, H., and Ando, T., *Phys. Rev. B* **41**, 11967 (1990).

[38] Beenakker, C.W.J., and van Houten, H., *Phys. Rev. Lett.* **63**, 1857 (1989); in *Electronic Properties of Low-Dimensional Semiconductor Structures*, Chamberlain, J.M., Eaves, L., and Portal, J.C., eds., New York: Plenum, 1990.

[39] Baranger, H.U., *et al. Phys. Rev. B* **44**, 10637 (1991).

[40] Baranger, H.U., and Stone, A.D., *Surf. Sci.* **229**, 212 (1990).

[41] Roukes, M.L., and Alerhand, O.L., *Phys. Rev. Lett.* **65**, 1651 (1990).

[42] Sivan, U., *et al.*, *Phys. Rev. B* **41**, 7937 (1990); Spector, J., *et al.*, *Appl. Phys. Lett.* **56**, 967 (1990); **56**, 1290 (1990); **56**, 2433 (1990).

[43] Fleischmann, R., Geisel, T., and Ketzmerick, R., *Phys. Rev. Lett.* **68**, 1367 (1992); *Europhys. Lett.* **25**, 219 (1994).

[44] Timp, G., in *Mesoscopic Phenomena in Solids* , Altshuler, B.L., Lee, P.A., and Webb, R.A., eds., Amsterdam: North Holland, 1991; Kouwenhoven, L.P., private communication (1989); Washburn, S., private communication (1989).

[45] Marcus, C.M., *et al.*, *Phys. Rev. Lett.* **69**, 506 (1992).

[46] Marcus, C.M., *et al.*, *Phys. Rev. B* **48**, 2460 (1993).

[47] Keller, M.W., *et al. Surf. Sci.* **305**, 501 (1994); Keller, M.W., thesis (Yale University 1995).

[48] Berry, M.J., *et al.*, *Phys. Rev. B* **50**, 8857 (1994); Berry, M.J., *et al.*, *Phys. Rev. B* **50**, 17721 (1994); Berry, M.J., *et al.*, *Surf. Sci.* **305**, 480 (1994).

[49] Chang, A.M., *et al.*, *Phys. Rev. Lett.* **73**, 2111 (1994).

[50] Marcus, C.M., *et al.*, *Chaos* **3**, 643 (1993) and *Surf. Sci.* **305**, 480 (1994).

[51] Chan, I.H., *et al.*, *Phys. Rev. Lett.* **74**, 3876 (1995).

[52] Clarke, R.M., *et al.*, preprint (August 1994).

[53] Bird, J.P., *et al.*, *Phys. Rev. B* **50**, 18678 (1994); *Phys. Rev. B* (in press).

[54] van Wees, B.J., *et al.*, *Phys. Rev. Lett.* **60**, 848 (1988).

[55] Wharam, D.A., *et al.*, *J. Phys. C* **21**, L209 (1988).

[56] Timp, G., *et al.*, in *Nanostructure Physics and Fabrication*, Reed, M.A., and Kirk, W.P., eds., New York: Academic Press, 1989, p. 331.

[57] van Wees, B.J., *et al.*, *Phys. Rev. B* **43**, 12431 (1991).

[58] Snider, G.L., *et al.*, *Appl. Phys. Lett.* **59**, 2727 (1991).

[59] Szafer, A., and Stone, A.D., *Phys. Rev. Lett.* **62**, 300 (1989).

[60] Kirczenow, G., *Solid State Commun.* **71**, 469 (1989); *Phys. Rev. Lett.* **62**, 2993 (1989).

[61] Nixon, J.A., and Davies, J.H., *Phys. Rev. B* **41**, 7929 (1990).

[62] Nixon, J.A., Davies, J.H., and Baranger, H.U., *Phys. Rev. B* **43**, 12638 (1991); Laughton, M.J., *et al.*, *Phys. Rev. B* **44**, 1150 (1991).

[63] Ravenhall, D.G., Wyld, H.W., and Schult, R.L., *Phys. Rev. Lett.* **62**, 1780 (1989).

[64] Avishai, Y., and Band, Y.B., *Phys. Rev. Lett.* **62**, 2527 (1989).

[65] Behringer, R., *et al.*, *Phys. Rev. Lett.* **66**, 930 (1991).

[66] Kakuta, T., *et al.*, *Phys. Rev. B* **43**, 14321 (1991).

[67] Lee, P.A., and Fisher, D.S., *Phys. Rev. Lett.* **47**, 882 (1981); Thouless, D.J., and Kirkpatrick, S., *J. Phys. C* **14**, 235 (1981); MacKinnon, A., *Z. Phys. B* **59**, 385 (1985).

[68] The prefactor in the semiclassical expression for the Green's function is $D_s = (v|\cos\theta'|/m)^{-1}|(\partial\theta/\partial y')_y|$ where θ and θ' are the incoming and outgoing angles.

[69] Tomsovic, S., and Heller, E.J., *Phys. Rev. Lett.* **67**, 664 (1991); *Phys. Rev. E* **47**, 282 (1993); and references therein.

[70] Lin, W.A., and Jensen, R.V., preprint (December 1994).

[71] For reviews of weak-localization in disordered systems see: Bergmann, G., *Phys. Rev.* **107**, 1 (1984); Lee, P.A., and Ramakrishnan, T.V., *Rev. Mod. Phys.* **57**, 287 (1985).

[72] The argument to obtain exact semiclassical results can be made in two ways in addition to that given in the text. First, from the the semiclassical approximation to the real space form of the conductance (see Baranger and Stone in ref. [15], one can see that the continuum mode approximation is good up to order $1/N$. This approach provides an alternative way to discuss the limiting procedures leading to the classical results. Second, one can follow the standard semiclassical procedure of evaluating the sum with the Poisson summation formula and then doing the integral by stationary phase. When applied to the sums over modes, this procedure yields the same result as in the text.

[73] Mello, P.A., and Stone, A.D., *Phys. Rev. B* **44**, 3559 (1991).

[74] Lewenkopf, C.H., and Weidenmüller, H.A., *Annals of Phys.* **212**, 53 (1991).

[75] Doron, E., Smilansky, U., and Frenkel, A., *Physica D* **50**,367 (1991); *Phys. Rev. Lett.* **65**, 3072 (1990).

[76] Blümel, R., and Smilansky, U., *Phys. Rev. Lett.* **69**, 217 (1992).

[77] The non-chaotic case is actually considerably more subtle than the glib treatment just given. In particular, one can no longer rely on the uniformity assumption because the distributions may have a significant dependence on the incident angle. So, the whole analytic argument given above becomes dubious, and Eq. (I.20) must be evaluated

by direct simulation, i.e. by injecting classical particles with the appropriate angular distribution and performing the sum over exiting paths with the appropriate phase. The resulting variation of $<\cos(2B\theta/\phi_0)>$ with B is roughly linear for both the polygonal[19] and circular[78] shapes. In the case of the circle, this linear behavior is what one expects from naively using the area distribution in Eq. (I.21)[78]; for polygonal billiards, it is probably associated with the extreme flux cancellation effects which occur in such structures[19,79,80].

[78]   Baranger, H.U., (unpublished).

[79]   DeGennes, P.-G., *Superconductivity of Metals and Alloys*, New York: Addison Wesley, 1966 and 1989, pp. 255–259.

[80]   Beenakker, C.W.J., and van Houten, H., *Phys. Rev. B* **37**, 6544 (1988).

[81]   Gutzwiller, M.C., *Physica D* **7**, 341 (1983).

[82]   Gaspard, P., and Rice, S.A., *J. Chem. Phys.* **90**, 2225 (1989); **90**, 2242 (1989); **90**, 2255 (1989).

[83]   Oakeshott, R.B.S., and MacKinnon, A., *Superlat. and Microstruc.* **11**, 145 (1992).

[84]   Ullmo, D., private communication (1995).

[85]   Wilkinson, M., *J. Phys. A* **20**, 2415 (1987).

[86]   Richter, K., *Europhys. Lett.* **29**, 7 (1995).

[87]   Hackenbroich, G., and von Oppen, F., *Europhys. Lett.* **29**, 151 (1995).

[88]   Argaman, N., preprint (1995).

[89]   Heller, E.J., private communication (1994).

[90]   Lee, P.A., and Stone, A.D., *Phys. Rev. Lett.* **55**, 1622 (1985).

[91]   Altshuler, B.L., *Pis'ma Zh. Eksp. Teor. Fiz.* **41**, 530 (1985) [*JETP Lett.* **41**, 648 (1985)].

[92]   For reviews of random matrix theory see Mehta, M.L., *Random Matrices,* New York: Academic, 1991; Porter, C.E., *Statistical Theories of Spectral Fluctuations,* New York: Academic, 1965.

[93]   Baranger, H.U., and Mello, P.A., *Phys. Rev. Lett.* **73**, 142 (1994).

[94]   Baranger, H.U., and Mello, P.A., *Phys. Rev. B* **51**, 4703 (1995).

[95]   Baranger, H.U., and Mello, P.A., preprint (cond-mat/9502078).

[96]   Jalabert, R.A., Pichard, J.-L., and Beenakker, C.W.J., *Europhys. Lett.* **27**, 255 (1994).

[97]   Jalabert, R.A., and Pichard, J.-L., *J. Phys. I France* **5**, 287 (1995).

[98]   Brouwer, P.W., and Beenakker, C.W.J., *Phys. Rev. B* **50**, 11263 (1994).

[99]   Brouwer, P.W., and Beenakker, C.W.J., *Phys. Rev. B* **51**, 7739 (1995).

[100]   Brouwer, P.W., preprint (cond-mat/9501025).

[101]   Efetov, K.B., *Adv. Phys.* **32**, 53 (1983).

[102]   Iida, S., Weidenmüller, H.A., and Zuk, J.A., *Phys. Rev. Lett.* **64**, 583 (1990); *Annals of Phys.* **200**, 219 (1990).

[103]   Prigodin, V.N., Efetov, K.B., and Iida, S., *Phys. Rev. Lett.* **71**, 1230 (1993).

[104]   Pluhar, A., *et al.*, *Phys. Rev. Lett.* **73**, 2115 (1994).

[105]   Macêdo, A.M.S., preprint (cond-mat/9404039).

[106]   Efetov, K.B., *Phys. Rev. Lett.* **74**, 2299 (1995).

[107]   Mello, P.A., and Seligman, T.H., *Nucl. Phys. A* **344**, 489 (1980); Mello, P.A., *J. Phys. A* **23**, 4061 (1990).

[108]   Berry, M.V., *Proc. R. Soc. Lond. A* **400**, 229 (1985).

[109]   Mello, P.A., and Pichard, J.-L., *J. Phys.* **1**, 493 (1991); see also ref. [73].

[110]   In Eq. (I.42) we have omitted phase factors which reflect the arbitrariness of the decomposition Eq. (I.41) for the CUE case. None of the results reported here are affected; see ref. [109] for a full discussion. Also note that $\prod_i$ runs from $i = 1$ to 2 for the COE and $i = 1$ to 4 for the CUE.

[111]   Politzer, H.D., *Phys. Rev. B* **40**, 11917 (1989).

[112]   Büttiker, M., *Phys. Rev. B* **33**, 3020 (1986).

[113]   Schmidt, B., and Müller-Groeling, A., *Phys. Rev. B* **47**, 12732 (1993) and references therein.

[114]   Doron, E., and Smilansky, U., *Nucl. Phys. A* **545**, C455 (1992).

[115]   Classical chaotic scattering is defined in two equivalent ways. The scattering is chaotic if (1) the dynamics in the fraction of phase space which is trapped for an infinite time is chaotic in the closed system sense, or (2) if as one iterates the scattering map the dynamics is chaotic in the closed system sense as the number of iterations goes to infinity. The connection between scattering and the infinite time limit is very close, as can be seen by the relation between the escape rate and the fractal dimension of the permanently trapped part of phase space. Clearly, neither of these definitions addresses the nature of the short paths in a single pass through the scattering region. In general there will be short non-ergodic paths even in a completely chaotic system.

[116]   Feshbach, H., Porter, C.E., and Weisskopf, V.F., *Phys. Rev.* **96**, 448 (1954); Feshbach, H., "Topics in the Theory of Nuclear Reactions," in *Reaction Dynamics*, New York: Gordon and Breach, 1973.

[117]   Mello, P.A., Pereyra, P., and Seligman, T.H., *Ann. Phys. (N.Y.)* **161**, 254 (1985); Friedman, W.A., and Mello, P.A., *Ann. Phys. (N.Y.)* **161**, 276 (1985).

[118]   Hua, L.K., *Harmonic Analysis of Functions of Several Complex Variables in the Classical Domain*, Providence, RI: Amer. Math. Soc. 1963.

[119]   Agassi, D., Weidenmüller, H.A., and Mantzouranis, G., *Phys. Rep.* **22**, 145 (1975).

[120]   Weiss, D., *et al.*, *Phys. Rev. Lett.* **66**, 2790 (1991).

[121]   Weiss, D., *et al.*, *Phys. Rev. Lett.,* **70**, 4118 (1993); Weiss, D., *et al.*, *Surf. Sci.* **305**, 408 (1994); Weiss, D., and Richter, K., *Physica D* **83**, 290, (1995).

[122]   Schuster, R., *et al.* *Phys. Rev. B* **49**, 8510 (1994).

[123]   For reviews of Coulomb blockade effects see: Grabert, H., and Devoret, M.H., *Single Charge Tunneling*, New York: Plenum Press, 1992: Kastner, M., *Rev. Mod. Phys.* **64**, 849 (1992).

[124]   Jalabert, R.A., Stone, A.D., and Alhassid, Y., *Phys. Rev. Lett.* **68**, 3468 (1992).

[125]   Prigodin, V.N., Efetov, K.B., and Iida, S., *Phys. Rev. Lett.* **71**, 1230 (1993); *Phys. Rev. B* **51**, 17223 (1995).

[126]   Bruus, H., and Stone, A.D., *Phys. Rev. B* **50**, 18275 (1994); *Physica B* **189**, 43 (1993); *Surf. Sci.* **305**, 490 (1994).

[127]   Mucciolo, E.R., Prigodin, V.N., and Altshuler, B.L., *Phys. Rev. B* **51**, 1714 (1995); Alhassid, Y., and Lewenkopf, C.H., preprint (March 1995).

[128]   Chang, A.M., *et al.* preprint (July 1995).

[129]   On orbital magnetism in disordered metals, see, for instance: Oh, S., Zyuzin, A.Yu., and Serota, R.A., *Phys. Rev. B* **44**, 8858 (1991); Altshuler, B.L., *Phys. Rev. B* **47**, 10340 (1993); and references therein.

[130]   Lévy, P.L., *et al.*, *Phys. Rev. Lett.* **64**, 2074 (1990).

[131]   Chandrasekhar, V., *et al.*, *Phys. Rev. Lett.* **67**, 3578 (1991).

[132]   Lévy, L.P., *et al.*, *Physica* (Amsterdam) **189B**, 204 (1993).

[133]   Mailly, D., Chapelier, C., and Benoit, A., *Phys. Rev. Lett.* **70**, 2020 (1993).

[134]   von Oppen, F., and Reidel, E.K., *Phys. Rev. B* **48**, 9170 (1993); Shapiro, B., *Physica* (Amsterdam) **200A**, 498 (1993); Agam, O., *J. Phys. I* (France) **4**, 697 (1994).

[135]   von Oppen, F., *Phys. Rev. B* **50**, 17151 (1994).

[136] Ullmo, D., Richter, K., and Jalabert, R.A., *Phys. Rev. Lett.* **74**, 383 (1995); preprint (July 1995).

[137] Fromhold, T.M., *et al.*, *Phys. Rev. Lett.* **72**, 2608 (1994).

[138] Müller, G., *et al.*, preprint (1995), and private communications.

[139] Shepelyansky, D.L., and Stone, A.D., *Phys. Rev. Lett.* **74**, 2098 (1995).

[140] Mott, N.F., and Davies, E.A., *Electronic Processes in Non-Crystalline Materials* (Clarendon Press, Oxford, 1971).

[141] McDonald and Kaufman, *Phys. Rev. A* **37**, 3067 (1988).

[142] Bennetin and Strelcyn, *Phys. Rev. A* **17**, 773 (1978).

[143] Bunimovic, L.A., *Funkt. Analiz. Jego Prilog.* **8**, 73 (1974).

[144] Iu, C., *et al.*, *Phys. Rev. Lett.* **66**, 145 (1991).

[145] Simons, B.D., *et al.*, *Phys. Rev. Lett.* **71**, 2899 (1993).

[146] Müller, G., Boebinger, G.S., Mathur, H., Pfeiffer, L.N., and West, K.W., preprint (1994).

[147] Pfeiffer, Loren, *et al.*, *Appl. Phys. Lett.* **55**, 1888 (1989).

[148] Spector, J., *et al.*, *Surf. Sci.* **228**, 283 (1990).

[149] Thornton, T.J., *et al.*, *Phys. Rev. Lett.* **63**, 2128 (1989).

[150] Skocpol, W.J., *et al.*, *Phys. Rev. Lett.* **56**, 2865 (1985).

[151] Licini, J.C., *et al.*, *Phys. Rev. Lett.* **55**, 2987 (1985).

[152] Timp, G., *et al.*, *Phys. Rev. Lett.* **59**, 732 (1987a).

[153] Timp, G., *et al.*, *Phys. Rev. Lett.* **58**, 2814 (1987b).

[154] Ford, C.J.B., *et al.*, *J. Phys. C* **21**, L325 (1988).

[155] van Wees, B.J., *et al.*, *Phys. Rev. Lett.* **62**, 2523 (1989).

[156] Kurdak, C., *et al.*, *Phys. Rev. B* **46**, 6846 (1992).

[157] Katine, J.A., *et al.*, in *Proc. 7th Int. Conf. Superlattices, Microstructures, and Microdevices, 1994, Superlattices and Microstructures,* **16**, 211 (1994).

[158] Beenakker, C.W.J., and van Houten, H., *Phys. Rev. B* **38**, 3232 (1988).

[159] Altshuler, B.L., Aronov, A.G., and Khemelnitsky, D.E., *J. Phys. C* **15**, 7367 (1982).

[160] Fuchs, K., *Proc. of the Cambridge Phil. Soc.* **34**, 100 (1938).

[161] Das Sarma, S., and Stern, F., *Phys. Rev. B* **32**, 8442 (1985).

[162] Ando, T., Fowler, A.B., and Stern, F., *Rev. Mod. Phys.* **54**, 437 (1982).

[163] Paalanen, M.A., Tsui, D.C., and Hwang, J.C.M., *Phys. Rev. Lett.* **51**, 2226 (1983).

[164] Harrang, J.P., *et al.*, *Phys. Rev. B* **32**, 8126 (1985).

[165] Coleridge, P.T., Stoner, R., and Fletcher, R., *Phys. Rev. B* **39**, 1120 (1989).

[166] Lin, W.A., and Jensen, R.V., in *Proc. 4th Drexel Symposium on Quantum Nonintegrability,* (1994).

[167] Marcus, C.M., *et al.*, *Semicond. Sci. Technol.* **9**, 1897 (1994b).

[168] Berry, M.J., Katine, J.A., Westervelt, R.M., and Gossard, A.C., to be submitted to Phys. Rev. B (1995).

[169] Kouwenhoven, L.P., *et al.*, *Phys. Rev. B* **40**, 8083 (1989).

[170] Hastings, M.B., Stone, A.D., and Baranger, H.U., submitted to *Phys. Rev. B,* (1995).

[171] Keller, M.W., *et al.*, *Physica B* **1029**, 194–196, 1994.

[172] Keller,M.W., *et al.*, to be published (1995b).

[173] Crommie, M.F., Lutz, C.P., and Eigler, D., *Science* **262**, 218 (1992).

[174] Heller, E.J., *et al.*, *Nature* **369**, 464 (1994).

Chapter 15

Semiconducting and Superconducting Physics and Devices in the *InAs/AlSb* Materials System

Herbert Kroemer and Evelyn Hu

Department of Electrical and Computer Engineering
University of California,
Santa Barbara, CA 93106

I. INTRODUCTION

Advances in the understanding of mesoscopic transport have benefited from the ability to fabricate and characterize structures at the requisite dimensions (≤ 100 nm), having appropriate materials characteristics. For example, the high mobilities of modulation-doped heterostructures, with their resulting long mean free paths, have enabled the observation of quantized conductance, and Aharanov-Bohm oscillations. In addition to the high mobilities, the field-controllable modulation of electron densities has made such semiconductor materials key factors in the observation of mesoscopic phenomena.

Generally, mesoscopic transport is characterized by the persistence of electron phase coherence. On the other hand, superconductivity, too, is a phase-coherent quantum phenomenon, in which the phase of the electron-pair wave function is spatially coherent, even over macroscopic distances. These hitherto separate areas of inquiry, mesoscopic and superconducting transport, have many points of correspondence: the energy scale of the superconductor is set by the gap energy, typically on the order of meVs. The energy scale of the mesoscopic structure is set by a subband spacing, or by a charging energy in small area capacitors, of the same order of magnitude, meVs. Phase coherence is a critical feature of electron transport in both mesoscopic and superconductor structures. Indeed, experiments have already been carried out on mesoscopic superconductor-semiconductor heterostructures, looking at mesoscopic fluctuations in superconducting critical currents[1], or uncovering parity effects in superconducting single electron tunneling transistors[2]. As in all heterostructures, it is not only the quality of the separate heterolayers that is now important; the interface between the materials becomes a critical issue.

For reasons that will be described further in Section I.A, we believe that *InAs-AlSb* structures possess important materials advantages that suit them for studies of mesoscopic transport, and for superconducting-semiconducting studies.

This chapter will review two classes of mesoscopic transport phenomena in quantized *InAs-AlSb* structures: (a) quantized conductance in quasi-one-dimensional ballistic "quantum wire" constrictions; and (b) superconducting weak links in superconductor-semiconductor-superconductor (Sp-Sm-Sp) structures made up from *InAs-AlSb* quantum wells contacted by superconducting electrodes. The two phenomena share not only a common base technology, they also have two important commonalities in the underlying physics: (a) both phenomena rely on maintaining the coherence of the phase of the electron wave functions during the electron flow through the structures; (b) less obviously, both phenomena deal with *ballistic* electron transport. This is self-evident in the case of ballistic transport through constrictions, but it applies also to the transport through Sp-Sm-Sp structures: It has become clear in recent years that the conventional superconducting proximity effect in such structures is drastically modified by phase-coherent multiple *Andreev reflections* of ballistic electrons at the superconductor-semiconductor interfaces (super-semi). Nowhere are those modifications more drastic than in weak-link structures involving *InAs*.

The work carried out so far has developed along two parallel but related paths: one that deals with mesoscopic transport of non-superconducting electrons, and one that examines the interaction of superconducting and non-superconducting electrons, mediated by Andreev reflections across a super-semi interface. Clearly, the next step will involve the realization of a mesoscopic super-semi structure, in this case examining one-dimensional ballistic transport between superconducting contacts. We are on the threshold of being able to realize such structures experimentally, and the *InAs-AlSb* quantum well is an ideal candidate material for these experiments.

Since the common enabling technology and materials for these studies resides in the *InAs-AlSb* heterostructure, we will begin with a short discussion of the salient properties that recommend this material for these studies. This discussion will be amplified in Section IV, with further details of the MBE growth, band line-ups, doping properties, and mobilities of electrons in the *InAs-AlSb* materials. Section II will treat the work carried out in one-dimensional ballistic constrictions fabricated from these materials, while Section III concerns itself with the super-semi weak links comprising *InAs* quantum wells. Since Andreev-reflected electrons play a critical role in determining the transport through the super-semi interface, and inasmuch as many readers, coming from the semiconductor side of mesoscopic transport, are likely to be unfamiliar with Andreev reflections, an introduction to this concept will be part of our presentation. Finally, Section V will attempt to summarize and synthesize, where possible, the results of this chapter.

A. Why *InAs*?

With its low electron effective mass ($m^* \approx 0.023\,m_e$), about one-third that of *GaAs*, *InAs* should, in principle, be one of the most promising semiconductors for mesoscopic studies: For a given quantum confinement geometry, the lower effective mass should lead to larger energy level separations. The low mass also leads to high mobilities: Room-temperature electron mobilities in high-purity *InAs* ($\approx 33,000$ cm^2/Vs) are about four-times those in *GaAs*, which implies a longer mean free path; much of this advantage should carry over to low temperatures. Finally, *InAs* has the unique property that *InAs*-metal contacts, even non-alloyed ones, usually do not form Schottky barriers, and hence are perfectly ohmic down to low temperatures.

Unfortunately, the technology of *InAs* is still much less mature than that of *GaAs*, and as a result, most studies of mesoscopic structures have employed *GaAs* rather than *InAs*. Probably the most important asset of *GaAs* over other III-V semiconductors is the great "crystallographic accident" that a replacement of part or all of the *Ga* atoms by *Al* atoms introduces only a negligible change in the lattice constant, with the result that *GaAs*-$(Al,Ga)As$ interfaces with very low defect densities are easily achieved, forming essentially perfect barriers for the electron confinement in mesoscopic structures.

However, recent research, especially in our own laboratories at UCSB, has shown that high-quality quantum wells can also be achieved with *InAs*, by using nearly lattice-matched *AlSb* or $(Al,Ga)Sb$ as a confinement barrier material. Although the quality of the *InAs*-*AlSb* interfaces has not yet reached the perfection of *GaAs*-$(Al,Ga)As$ interfaces (and perhaps never will), great progress has been made, and impressive results have been achieved with nanostructures based on *InAs*-*AlSb* quantum wells. It is this progress that is the central theme of the present chapter.

B. Band Offsets and Lattice Matching: Why *AlSb* Barriers?

Of the binary III-V compounds, only *GaSb* and *AlSb* have lattice constants that are sufficiently close to that of *InAs* to permit the MBE growth of quantum wells of reasonable width (≥ 10 nm) without the formation of misfit dislocations [see Figure 1].

Given a reasonable lattice match, the dominant reason for an interest in heterostructures is of course the existence of energy gap variations and band offsets at the interfaces. The band offsets within the 6.1Å group have been well-established experimentally. As shown in Figure 2, the *InAs*-*AlSb* pair is characterized by an extraordinarily large conduction band offset of about 1.35 eV at room temperature, first determined by Nakagawa *et al.*[3] and subsequently confirmed by Waldrop *et al.*[4]. Evidently, *AlSb* is well-suited as a barrier material for

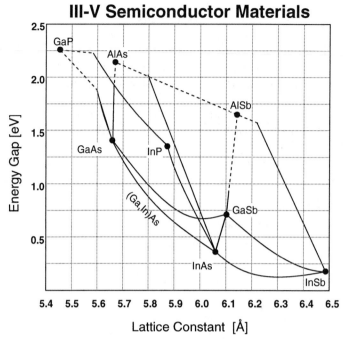

FIGURE 1. The "Map of the World": Energy gap-vs.-lattice constant for the dominant III-V compounds (omitting the wider-gap compounds such as *AlP* and the nitrides).

quantum wells based on *InAs*.

The other member of the 6.1Å group, *GaSb*, is ill-suited as a barrier material, because of the broken-gap band lineup at the *InAs/GaSb* heterojunction, also evident in Figure 2: the top of the *GaSb* valence band lies about 150 meV above the bottom of the *InAs* conduction band[5,6], thus providing an open path between the bands, and making the confinement of electrons in *InAs* by *GaSb* barriers illusory. This last argument does not apply to ternary $(Al, Ga) Sb$ alloys with an *Al* content sufficiently high (>50%) to close the broken gap at the interface, and such alloys are sometimes preferred over binary *AlSb*, because of their improved chemical stability.

The band offsets for *InAs-AlSb* and *InAs-GaSb* are in good agreement with the independently known band offset for *GaSb-AlSb*, $\Delta E_V[AlSb \rightarrow GaSb] \approx 0.35 - 0.40$ eV[7,8], as demanded by of the transitivity postulate of heterojunction band offsets, which states that the three different band offsets combining *InAs*, *GaSb*, and *AlSb* should obey the "closure rule" $\Delta E_V[InAs \rightarrow GaSb] + \Delta E_V[GaSb \rightarrow AlSb] + \Delta E_V[AlSb \rightarrow InAs] = 0$.

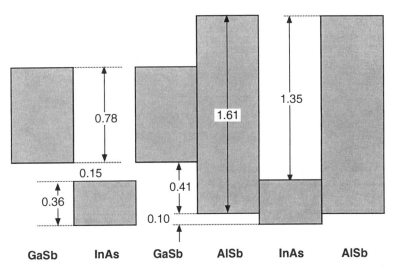

FIGURE 2. *InAs/AlSb/GaSb* Band lineups. The shaded areas represent the forbidden bands; all energies are in eV.

The binary-to-binary lattice match in the *InAs-AlSb* system is not as good as in the *GaAs-AlAs* system, but the mismatch of about 1.3% is sufficiently small that *InAs* quantum wells with *AlSb* barriers with high perfection are easily grown with well widths up to about 15 nm. It would, in principle, be possible to achieve better match by replacing about 10% of the *Sb* atoms in the barriers with *As* atoms. Such alloy barriers have been found useful in conventional FET's[9], but the reasons for using them were related less to lattice matching than to considerations involving valence band offset, which do not apply to mesoscopic transport structures.

A more important question is what substrate to use for epitaxial growth. Ideally, one would like to have a substrate that is both lattice-matched and semi-insulating. Of the three 6.1Å binary compounds themselves, only *GaSb* comes anywhere near this goal: Although not as readily available as semi-insulating *GaAs*, it *is* available in bulk wafer form, and although native stoichiometric defects make undoped *GaSb* invariably *p*-type at room temperature, the material tends to be nearly semi-insulating at low temperatures, the case of interest for mesoscopic transport studies. Of the other two 6.1Å materials, *AlSb* is not available in bulk form at all (it would be unstable against oxidation in any event), and *InAs*, which is available only to a very limited extent, has too narrow an energy gap to make semi-insulating material possible.

In actual practice, however, much of the research on *InAs* quantum wells–including most of our own– has employed far more readily available and far less expensive semi-insulating *GaAs* substrates, despite their 7% lattice mismatch, using various buffer layer schemes to attenuate the high density of misfit-induced threading dislocations in the *InAs*. We will return to this topic in Section IV.

II. QUANTIZED CONDUCTANCE IN ONE-DIMENSIONAL WIRES

A. The Background: Quantized Conductance in *GaAs* Structures

The observation of quantization of conductance in one-dimensional wires has been a signature of quantum-mechanical confinement in those structures. An excellent review of the subject is given by Beenakker and van Houten[10]. A particularly effective means of forming such wires has been through the electrostatic depletion of a high-mobility two-dimensional electron gas beneath a split gate, as shown in Figure 3[11]. The electron gas is 'dynamically two-dimensional'[12]: confined within an electron wavelength of the heterointerface between the $(Al, Ga)As$ and *GaAs*, and free to move in the two spatial directions along that interface. Such a split-gate structure, or constriction, allows a controlled modification of the quantum-mechanical potential that confines the electrons: the gate voltage, V_g, controls the position of the Fermi level with respect to the subbands of the wire.

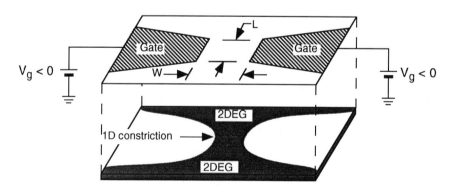

FIGURE 3. Schematic of a split-gate constriction.

Measurements made at $T < 1$K, on split-gate samples fashioned from high-mobility $(2.5\times10^6 - 10^6\,\mathrm{cm}^2/V\cdot s)$ $GaAs - (Al,Ga)As$ substrates grown by molecular beam epitaxy, did indeed demonstrate quantized conductance steps[13, 14]. Subsequently, these split gate constrictions became model systems that allowed the mapping of critical factors in observation of quantum behavior.

The conductance G of the constriction at zero-temperature is given by the Landauer-Büttiker formula[15,16]:

$$G = \frac{2e^2}{h} \sum_{ij=1}^{n} t_{ij}. \qquad (1)$$

Here e is the electron charge, h is Planck's constant, n is the number of occupied one-dimensional (1D) subbands, and t_{ij} is the transmission probability for an electron incident in subband i to exit the constriction in subband j. If the length of the constriction formed by the split gates is less than the mean free path for elastic scattering, $L_e = \hbar k_F \mu/e$, then the electrons may traverse the constriction ballistically, t_{ij} can be replaced by the Kronecker delta δ_{ij}, and the sum in (1) equals n, the number of occupied subbands: $G = 2ne^2/h$. The length scale of interest here presumably is the elastic mean free path L_e, and the structures studied are those for which $L < L_e$, L being a characteristic structural dimension, such as the length of the constriction. In general, for these constrictions, as for the superconducting-semiconducting structures described in Section III, L_e is much smaller than the *inelastic* mean free path, which is the distance over which appreciable de-phasing of the electron waves occurs.

A wealth of information is contained in the transmission factors t_{ij} of Eq.(1), and hence in the precise characteristics of the conductance curves. For example, the position of the conductance plateaus, and their deviation from $2ne^2/h$, provide information on impurity scattering within or near the constriction. Serious factors in the further consideration of such devices in application to practical device technologies are the low temperatures of operation required, and the relative fragility of the quantized conductance behavior.

Achieving Robust Quantized Conductance: The Case for InAs Quantum Wells

The preservation of the initial quantum state of the electron throughout its traverse through the constriction requires the reduction of scattering mechanisms. High-mobility and high-electron-density samples should yield high elastic mean free paths, hence ensure the persistence of quantized conductance over longer constriction lengths. For this reason, much of the work on ballistic electron transport has continued to utilize the $GaAs-(Al,Ga)As$ materials system: the growth of heterostructures of the requisite high mobility and electron density is well understood. Timp has carried out systematic studies of $GaAs/(Al,Ga)As$ constrictions that demonstrate that the quantization will deteriorate as the L_e of the 2DEG becomes smaller, and as the length of the constriction increases[12]. In fact, dramatic deterioration of quantization was observed for constriction lengths almost an order of magnitude *less* than the elastic mean free path of the 2DEG comprising the structures (L of the constriction = 600 nm, and $L_e \sim 7$ μm). Subsequent theoretical analysis suggested that the random potential of poorly-screened impurities in the region of the constriction is responsible for that quantization deterioration[17].

Another influential factor for the robust operation of these constrictions is the energy spacing of the subbands in the confining potential, relative to the thermal energy $k_B T$. The split-gate configuration offers many advantages in ease of fabrication and control of transport parameters. Etched mesa structures can also form 1-dimensional (1D) pathways for the electrons, but bring in the complications of uncontrolled surface states and the possibility of etch-induced damage to the material. However, a limitation of the split-gate geometry is that it provides a gradual, rather than abrupt potential profile, resulting in smaller separation of subband energies. Larger subband energy spacings can be achieved in $GaAs$-$(Al,Ga)As$ split gate geometries by bringing the 2DEG closer to the surface, resulting in the detection of quantized conductance steps up to temperatures as high as 30K[18]. This approach has its limitations, since, as the 2DEG is brought too close to the surface, a degradation of the electron mobility, and hence, elastic mean free path will ensue.

These considerations point to the advantages of forming ballistic constrictions based on $InAs$ quantum wells with a suitable barrier material. The lower effective mass of $InAs$ compared to $GaAs$ should result in larger subband spacings in an $InAs$-based wire, compared to one fabricated in $GaAs$. This in turn should allow persistence of quantized conductance to higher temperatures and larger applied voltages. As discussed in Section I.B, $AlSb$ or $(Al,Ga)Sb$ alloys as the barrier materials with their high conduction band offset relative to $InAs$, will provide excellent electron confinement. Finally, an additional advantage of this material system lies in the mechanism of doping of the quantum well: as will be discussed in Section IV.B, the electron concentration in a not-intentionally doped $InAs$ quantum well may be dominated by electrons from surface states on a $GaSb$ cap layer overlying the $AlSb$ barrier. This allows a close proximity of the 2DEG to the surface with the consequent sharper potential profile and without the trade-off in electron mobility that would affect the 2DEG in a $GaAs$-$(Al,Ga)As$ structure.

B. *InAs-Al(Ga)Sb* Split-Gates: The Basic Structure

The *InAs-Al(Ga)Sb* electron waveguides are fashioned from MBE-grown material similar to that shown in Figure 4. A *GaSb/AlSb* buffer layer about 2 μm thick is grown over a (100) semi-insulating *GaAs* substrate to accommodate the lattice-mismatch between the *GaAs* and subsequently grown layers. The issue of the lattice-mismatch and the possibility of threading dislocations propagating through to the electron channel is undoubtedly of importance. As discussed in Section IV.C, with the help of appropriate buffer layers, we may expect to achieve dislocation densities as low as 10^7/cm^2, giving a dislocation every 3 μm, on the average. At this stage of our investigations, the likelihood of encountering a dislocation within the active region of our electron waveguides is very low, and it is more likely that other intrinsic or process-induced defects are playing a greater role in limiting the performance of our quantum wires.

Subsequent to the growth of the buffer layers, 250 nm of $Al_{0.7}Ga_{0.3}Sb$ is grown to form a chemically stable mesa floor for the fabrication of the waveguides. This is followed by a 10-period *GaSb/AlSb* (2.5 nm/2.5 nm) smoothing superlattice, the 15 nm *InAs* quantum well, a 30.5 nm *AlSb* upper barrier,

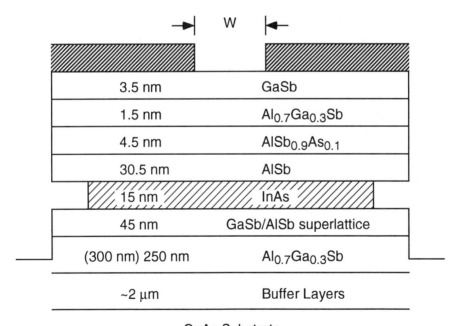

FIGURE 4. Layer structure for *InAs-Al(Ga)Sb* waveguides.

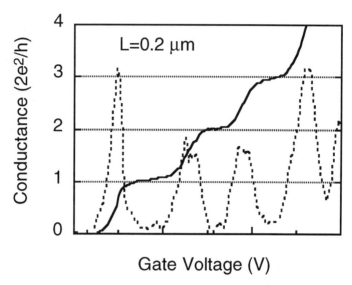

FIGURE 5. Conductance curves (and derivative) of a constriction 120 nm wide and 200 nm long.

4.5 nm of $AlSb_{0.9}As_{0.1}$, followed by 1.5 nm of $Al_{0.7}Ga_{0.3}Sb$, and a final 3.5 nm $GaSb$ capping layer. The electron concentration and mobility at 10K are determined by van der Pauw measurements to be $6.5 \times 10^{11} cm^{-2}$ and $3.8 \times 10^5 cm^2 /V \cdot s$. Mesa isolation of the devices is carried out using selective wet etches[19], and ohmic contacts are formed using non-alloyed Cr/Au metallization. The split-gates are patterned by electron beam lithography, and lifted-off using 30 nm of Au/Pd metal[20].

Two-terminal measurements are carried out in a liquid-helium dewar, at temperatures ranging from 4.2K to >30K. A small bias voltage of 90-180 μV_{rms} is supplied between the source and drain contacts while the current in the waveguide is monitored using a lock-in amplifier. Conductance in the waveguides is measured as a function of the applied gate voltage, and the data are generally taken as the gate voltage is swept in one direction only, toward more negative values. A 'pre-bias' voltage is supplied to the structure prior to the sweep of the gate voltage. The effect and importance of such a pre-bias will be discussed in Section II.D. Figure 5 shows the quantized conductance steps that result for a constriction in the $InAs/AlSb$ material system. The dimensions of the constriction are $W = 120$ nm, $L = 200$ nm, and the data are taken at $T = 4.2$K.

C. Subband Spacings and Temperature Dependence

In determining the temperature limits of quantized conductance, it is useful to examine the *differentiated* conductance, $\partial G/\partial V_g$, versus gate voltage, to more clearly detect the remnant signature of quantized conductance at the highest operating temperatures. Accordingly, Figure 6 shows the differentiated conductance curves at a number of temperatures ranging from $T = 4.2K$ to 40K, pertaining to a constriction 2 μm in length and 0.16 μm in width. A direct analog measurement is made of the differential conductance $\partial G/\partial V_g$ [21]. As shown in Figures 6 (a)-(d), the differential conductance is an oscillatory function of the gate voltage, with oscillations persisting to $T \sim 35K$.

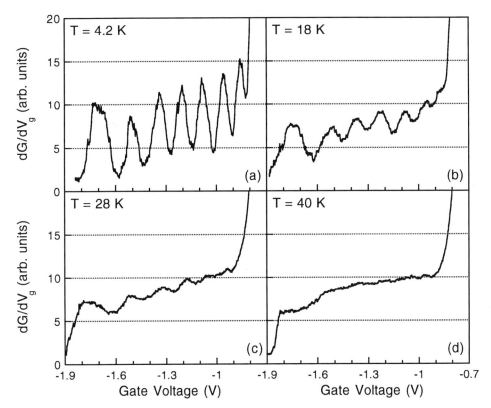

FIGURE 6. Plots of $\partial G/\partial V_g$ vs. V_g for a 2 μm long constriction at several temperatures: (a) 4.2K, (b) 18K, (c) 28K and (d) 40K.

The quality of quantized conductance as a function of temperature should roughly depend on the magnitude of $k_B T$, relative to the subband energies of the waveguide. One way of forming accurate estimates of the subband spacings, $\Delta E_{1D} = E_n - E_{n-1}$ is to use Eq. (1), modified to account for finite (rather than zero) temperature (t_{ij} is set equal to δ_{ij}),

$$G = \frac{2e^2}{h} \sum_n \frac{1}{1 + \exp\left[(E_n - E_F)/k_B T_{eff}\right]}, \qquad (2)$$

where

$$T_{eff} = \sqrt{T^2 + T_0^2}, \qquad (3)$$

is an 'effective temperature.'

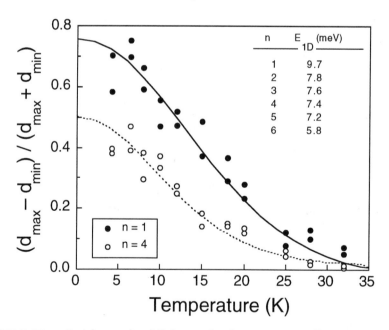

FIGURE 7. The ratio $(d_{max} - d_{min})/(d_{max} + d_{min})$ plotted versus T for $n = 1$ and $n = 4$. The data are indicated by solid and open circles, the best fit curves by the solid and dashed lines.

    In Eq. (3), all effects that would tend to smear the steps, such as reflections due to the gate geometry, scattering from impurities, or heating from gate leakage are subsumed by a temperature-equivalent T_0. Equation (2) predicts a temperature-dependent 'smearing' of the conductance steps, more deleterious for the subbands at higher n with E_n closer to E_F. This prediction can be compared with, and fit to the data of Figure 6, where the modulation of $\partial G/\partial V_g$ diminishes at higher gate voltage, and at higher temperature. The modulation is represented by $(d_{max} - d_{min})/(d_{max} + d_{min})$ in Figure 7 ($d_{min} = \partial G/\partial V_g|_{min}$ and $d_{max} = \partial G/\partial V_g|_{max}$). The modulation is plotted as a function of temperature for subbands corresponding to $n = 1$ and $n = 4$. The solid lines in Figure 7 represent the best fits to the data; ΔE_{1D} and T_0 are the adjustable parameters used to fit the data, and T_0 was found to be 14K. The subband spacings increase with decreasing gate voltage, varying from 5.8 meV for the highest subband, to 9.7 meV for the fundamental mode in the constriction. The rather large value of T_0 points out the presence of factors, other than the direct and simple factor of temperature, that degrade the quantization of the conductance.

    An alternative approach utilizes the controlled variation of an applied magnetic field to help map out the subbands and follows the analysis of Wharam *et al.*[22]. We assume that a lateral parabolic confining potential,

$$V(x) = \frac{1}{2}m^*\omega^2 x^2 + V_0, \qquad (4)$$

describes the electron potential in the most narrow portion of the constriction, with a maximum saddle point potential of V_0. Here, $\omega = [\omega_0^2 + \omega_C^2]^{1/2}$, and ω_C is the cyclotron resonance frequency, eB/m^*. Non-parabolicities are taken into account by using an energy-dependent effective mass[23],

$$m^*(E) = 0.0266m_0[1 + 4.90(E - V_0)], \qquad (5)$$

where E is the energy from the bottom of the lowest confined state in the quantum well, with $E - V_0$ expressed in electron volts. The number of occupied subbands within the confining potential is then given by

$$n = \mathrm{int}\left[(E_F - V_0)/\hbar\omega + \frac{1}{2}\right]. \qquad (6)$$

    These values of n can be determined from the data of Figure 8, which shows the quantized conductance characteristics for a waveguide with $L = 200$ nm and $W = 120$ nm, taken under different values of magnetic field, and for a constant

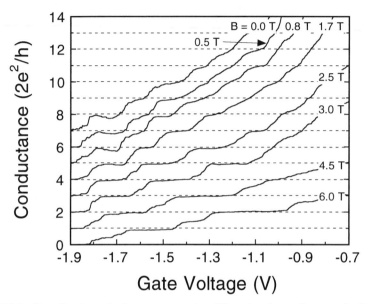

FIGURE 8. Quantized conductance curves for different values of magnetic field. The curves are offset for clarity; $T = 4.2K$.

$T = 0.3K$. The increased density of states and energy level spacings at the higher values of magnetic field result in the wider steps seen in the figure. Equation (6) can then be used to fit those values of n, using eV_0 and $\hbar\omega_0$ as adjustable parameters for the fit. The results are displayed graphically in Figure 9, which plots the number of occupied 1D subbands as a function of inverse magnetic field, for different values of gate voltage. The solid line in the figure is the theoretical fit to the data. As shown in the inset to Figure 9, the subband spacings increase with decreasing gate voltage up to a maximum value of 10 meV, determined for $V_g = -1.48$ V. For gate voltages ≤ -1.5 V, the range of fitting parameters could not be determined, so that we can only estimate an extrapolated subband value of 12 meV for the largest gate voltage, $V_g = -1.74$ V (the middle of the $n = 1$ step, at $B = 0$).

The evidence of quantized conductance in these *InAs* electron waveguides, at temperatures in excess of 30K, seems to confirm the promise of electron transport in this material system. Various estimates of the subband spacings, based on the data, consistently give values of $\Delta E_{1D} \sim 10$ meV, almost an order of magnitude higher than the initially-studied high mobility split-gate structures fabricated in *GaAs*. In absolute terms, however, the temperature of operation and the subband spacing did not exceed those demonstrated in $GaAs - (Al,Ga)As$ waveguides

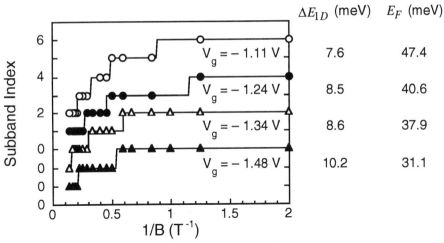

FIGURE 9. Subband index plotted versus inverse magnetic field for various values of gate voltage.

fabricated by Snider *et al.*[18]. Although the 2DEG in that case was placed in closer proximity (25 nm) to the surface, compared to the 40 nm distance used for the *InAs* waveguides, it is clear that we must look further to determine the factors influencing the deterioration of conductance. In particular, we expect that the nature and location of the impurities in the constriction will have an influential role.

D. Degradation of Conductance Characteristics: The Influence of Impurities

1. Controlled Impurity Configurations? – The Influence of Pre-Measurement Biasing

In general, the *InAs* electron waveguides display a strong sensitivity to the precise manner in which the measurements are carried out. This point is illustrated in Figure 10. As the gate voltage is systematically reduced to more negative values, the initial rapid reduction in conductance at V_{t2D} signifies the pinching off of the 2DEG. Further reduction in bias voltage modifies the shape of the confining potential, 'squeezing' the electron channel, and successively depopulating the various subband levels until the 1D region itself is pinched-off at V_{t1D}. If the gate voltage is now increased to zero from its most negative value, the point at which the conductance begins to rapidly rise is shifted to a more negative value $-V'_{t2D}$. That value of V'_{t2D} remains stable through successive cycling of the

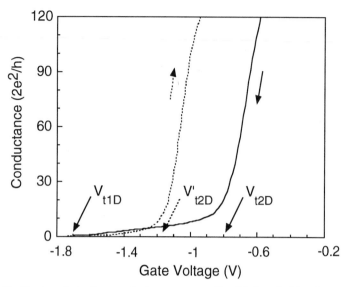

FIGURE 10. Hysteresis in the conductance curves: the 2D threshold voltage of the up-sweep, V'_{t2D}, is shifted negatively with respect to the threshold on the down-sweep, V_{t2D}.

gate voltage between zero and negative values, and V_{t1D} remains the same.

The restoration of the original value of V_{t2D} can be accomplished by the application of a positive voltage, V_p. The larger the value of this pre-measurement bias voltage, the greater the shift of V_{t2D} towards more positive values, as shown in Figure 11. A given value of V_p will produce a reproducible shift in the value of V_{t2D}, and a fixed pre-measurement bias needs to be applied before each measurement of conductance.

Interestingly, the application of the pre-measurement bias not only ensures a reproducibility in the measured characteristics, but also leads to an improvement in the quality of the quantization: Figure 12 shows the conductance characteristics of an electron waveguide for which $W = 120$ nm and $L = 200$ nm, for various values of V_p. The higher value of V_p both increases the number of discernible subbands ($n = 3$ for $V_p = 0$, to $n = 7$ for $V_p = +2V$), and the quality of those steps.

We believe that this dependence of the conductance on the bias history of the split gates results from the charging and discharging of trap states beneath the gates. These trap states would be related to the surface-state donors or the deep near-interface donors discussed in Section IV.B. After the application of a large negative or positive gate bias, the states remain charged until a large bias of the opposite sign is applied. For a given pre-measurement bias, the ensuing values of V_{t2D} and V_{t1D} are extremely repeatable, indicating that the amount of charge

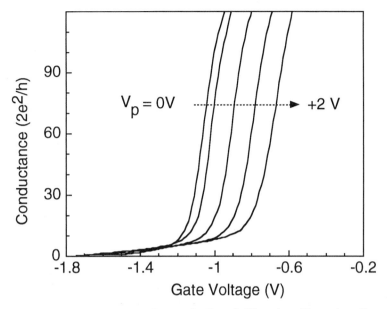

FIGURE 11. Conductance curves after application of different positive gate voltages.

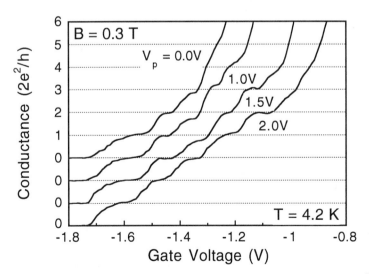

FIGURE 12. Conductance characteristics for various values of pre-bias voltage, V_p. The curves are offset for clarity.

under the gate is well linked to a given value of V_p. We believe that V_p may also influence the configuration of the ionized impurities in the vicinity of the one-dimensional channel, and hence influence the quality of the quantized conductance.

2. How Do Impurities Influence Quantized Conductance?

The transformation of Eq. (1) into $G = 2ne^2/h$ relied on the identity of the transmission coefficients t_{ij} with delta functions. Using the parlance of the transverse modes accessible to the electrons in traversing the waveguide, quantization would seem to depend on the absence of 'mode-mixing', i.e. different transverse modes pass through the waveguide independently. Subsequent calculations have modified this conclusion, finding that quantized conductance is even more robust: direct numerical calculations show that conductance is quantized even if substantial mode-mixing occurs, as long as forward scattering processes dominate[24]. Electrons scattering out of one mode into another can be balanced by electrons scattering via an inverse process[25], leaving the conductance sum of Eq. (1) essentially unchanged. The dominating mechanism of breakdown is back-scattering[26]. This can be direct scattering against impurities situated within the channel; furthermore, calculations have also suggested the importance of long-range impurity potentials which produce quasilocalized states within the channel, providing an 'indirect' mechanism for backscattering[25].

We saw from Section II.C and from the non-zero value of the fitting parameter T_0, that temperature-smearing of the conductance characteristics could not by itself account for the degradation that we observe in those characteristics. In order to better understand the effects of impurities and their limitations to the robust quantization of our waveguides, we will examine the changes in characteristics as we vary the length L of the waveguides.

E. Length-dependence of Quantized Conductance

We examined the conductance curves at $T = 4.2$K of a range of waveguides whose lengths varied from 0.2 μm to 2.0 μm. The data shown in Figure 13 represent the conductance curves for constrictions of 0.6 μm, 1.0 μm, and 2.0 μm, all fabricated on the same substrate material. In all cases $V_p = +2V$ was applied to the samples before the actual conductance measurements were made.

Plotted in Figure 14 are the corresponding heights of each conductance step, G_n, divided by $2ne^2/h$, as a function of the subband index n, for the range of waveguide lengths from 0.2 μm to 2.0 μm. Perfect quantization would be denoted by a straight, horizontal line at $G_n h/2ne^2 = 1$. The step spacings of the shortest device, with $L = 0.6$ μm, are closest to the ideal values. The average values of

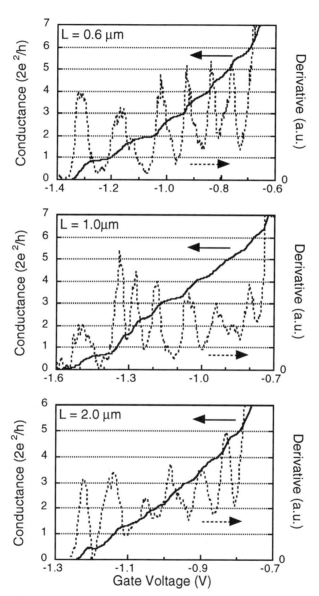

FIGURE 13. Quantized conductance characteristics for constrictions with channel lengths of 0.6 μm, 1.0 μm and 2.0 μm. All constrictions were fabricated from the same wafer; data were taken at 4.2K.

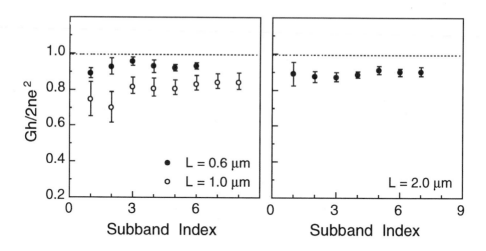

FIGURE 14. Quantization values for constriction lengths of 0.6 μm, 1.0 μm, and 2.0 μm. Perfect quantization would correspond to $G_n h/2ne^2 = 1$.

quantization, in units of $h/2ne^2$, are 0.93 ± 0.05, 0.80 ± 0.11, and 0.89 ± 0.05 for $L = 0.6$ μm, 1.0 μm, and 2.0 μm, respectively. The values for $L = 2.0$ μm are not substantially worse than those for 0.6 μm, and better than the data for $L = 1.0$ μm. We attribute this apparent inconsistency to variations in material quality and impurity configurations over the sample.

The longer constrictions exhibit more variations between nominally identical devices than do the short constrictions, and this may reflect the greater sensitivity of those devices to variations in the impurity configuration. More uniform characteristics for the shorter constrictions were also obtained under 'cycling' of the devices, either through consecutive sweeps of the pre-bias voltage and subsequent device measurements, or through temperature variations from 4.2K to 160K and back. Nevertheless, the value of L_e for our samples is between 4 and 5 μm, and we see evidence of quantized conductance at $T = 4.2$K for waveguides as long as 2 μm. As a point of comparison, we look to the earlier data of Timp et al., taken for *GaAs* waveguides[27]. For that case, $L_e \sim 10$ μm, and the degradation of quantization was evident for constrictions only 0.6 μm long. This latter result has stimulated much interest and examination, for it clearly pointed out that high electron mobilities, and long (2D) elastic mean-free paths were not by themselves sufficient predictors of the robust persistence of quantized conductance. The 'discrepancy' seemed to be well resolved by calculations carried out by Nixon and Davies, which highlighted the effect of the random potential produced by the ionized donors in the *GaAs* − (*Al, Ga*)*As* structure[17].

Using the actual electron waveguide structure utilized by Timp *et al.*, Nixon and Davies calculated a correlation length of 0.2 μm, corresponding to the spatial overlap of the impurity potentials. This can be thought of as producing a 'lumpiness' in the net impurity profile, with a characteristic dimension of 0.2 μm. The variation in the potential can produce quasi-localized states, leading to backscattering that will degrade the quantized conductance. Electrons traversing one-dimensional waveguides that are longer than the correlation length are likely to undergo backscattering and therefore degradation of quantization. The correlation length of the impurity potentials then sets the limit on the maximum waveguide length for which quantized conductance may be observed.

The electron concentration in our *InAs* electron waveguides are comparable to those pertaining to various *GaAs* waveguides. Although the details of the doping of the two dimensional channel are different for our material system, we expect that the nature of the correlated random impurity potentials should be similar to those observed for *GaAs*. Therefore, we should see comparable dependencies of quantization on length, depending on the various mobilities of the samples being studied. However, we observe quantized conductance for waveguides as long as 2 μm, quite a bit larger than the 0.2 μm correlation length calculated by Nixon and Davies. This discrepancy appears not to be associated only with the difference in the materials: Ismail, Washburn, and Lee have also reported quantized conductance steps in one-dimensional waveguides with channel lengths as long as 2 μm[28]. The mobility of the 2DEG of their $GaAs/Al,Ga)As$ devices is comparable to that reported by Timp; an important difference in Ismail's structures is the use of mesa-confinement, rather than the split-gate geometry. What bearing does this have on the limitations on length of the electron waveguides?

Breakdown of Quantized Conductance in InAs Constrictions: Limitations

Nixon and Davies' calculation highlighted an important caveat for these electron waveguides that pertained to the long-range influence of unscreened impurities. Although the spatial separation of dopants from mobile charge carriers (such as is true for modulation doping) provides high carrier mobilities and reduces impurity scattering, the dopants still produce random potential fluctuations in the conduction plane of the 2DEG. Moreover, the carrier concentration within the one-dimensional constriction will be lower than in the 2DEG, hence providing less screening against fluctuations. In fact, earlier researchers have invoked caveats in the facile application of L_e, calculated for the 2DEG, to the estimate of the elastic scattering length in the ballistic constriction itself[12].

However, the variation between our own length-dependent results, and those of other researchers, indicate that factors other than the impurity density and distribution may be important. Recently, Zagoskin *et al.* have looked further into the role of the quasi-localized states which arise from the long-range impurity potentials[29]. These quasi-localized states were shown to be instrumental in electron backscattering through the waveguides, leading to the degradation of quantized conductance. In particular, Zagoskin's results gave a dependence of the density of quasi-localized states that decreases exponentially with the ratio of the 1D subband spacings to the standard deviation of impurity potential fluctuations, $\Delta E_{1D}/\sigma_{imp}$ (σ_{imp} is the standard deviation of the fluctuation potential, $\sigma_{imp}^2 = \langle V_{imp}^2 \rangle$). In other words, the possibility of backscattering is linked to the ratio of the subband energy in the waveguide, to the energy of the potential fluctuations brought about by the charged impurities.

Will the correlation between maximum waveguide length and subband separation prove viable (assuming comparable σ_{imp})? Section II.C revealed that ΔE_{1D} for our structures is ~10 meV. Simulations performed by Snider on Timp's structures give $\Delta E_{1D} = 1.5$ meV for those structures[30]. Snider's large-subband waveguides, fashioned in $GaAs$ had $\Delta E_{1D} \sim 10$ meV. L_e for those structures was ~ 2 μm, and the longest constrictions which exhibited quantized conductance were 0.6 μm ($T = 4.2$K). Finally, Ismail's etched mesa waveguides had values of L_e comparable to that of Timp, but the mesa geometry gave larger subband spacings (~ 10 meV) than the split gate geometry. The comparison between the various electron waveguides is summarized in Table 1:

TABLE 1.

Material	L_e (μm)	ΔE_{1D} (meV)	L_{max} (μm)	Reference
GaAs	10	1.5	0.6	[27]
GaAs	2	10	0.6	[30]
GaAs	10	10	2	[28]
InAs	5	10	2	[21]

More recently, Tarucha *et al.* have seen evidence of quantized conductance in 10 μm long quantum wires formed in $GaAs/(Al,Ga)As$ heterostructures with mobilities of ~ 8×10^6 cm^2/V·s (under illumination), in measurements that were carried out at 1.3K[31]. In this case also, the one-dimensional wire was formed by etching out a mesa past the 2DEG; the shape of the confining potential was controlled by using two side gates, directly adjacent to the 2DEG. It is expected that such a structure would have a more abrupt confining potential than can be easily achieved for split-gate structures, and hence there is the possibility of larger subband splitting.

Finally, the term σ_{imp} is not easily determined, but can be estimated. Using a Thomas-Fermi screening model and simple electrostatics, Davis and Timp gave an estimate of $\sigma_{imp} = 1.6$ meV (comparable to ΔE_{1D}), for the structure used by Timp. Using that as a basis for our own estimate, and accounting for the potential fluctuations arising from deep donors in the barrier layers on either side of the quantum well, and non-uniformities in the *InAs/AlSb* interfaces, σ_{imp} should increase by no more than a factor of two, still less than the 1D subband spacings in our structures[21].

The conclusion is seemingly an obvious one: larger subband spacings produce more robust quantization. However, it is important to remember that the postulated effect of the subband spacing in this case is indirect: through a correlation with the density of quasi-localized states that are responsible for back-scattering, and hence for destruction of quantized conductance. Assuming only small variations of the value of σ_{imp}, methods that will result in the formation of large subband spacings, *whatever* the material system, will allow quantized conductance to be observed for longer constriction lengths: hence the experimental results of Ismail *et al.*[28] and Tarucha *et al.*[31], who achieved large subband splittings in *GaAs*-based heterostructures through deep etch-confinement of the electrons.

F. Future directions

1. Normal Electron Transport: Limitations?

How far can we optimize these structures to realize even greater robustness in the quantized conductance characteristics? The principal limitation in the electronic measurements of these electron waveguides has been the presence of leakage currents. This is related to the large conduction band offset between *InAs* and *AlSb*, and the small bandgap of the *InAs*: the conduction band of *InAs* is separated by only ~300 meV from the *AlSb* valence band [see Figure 2]. Under reverse bias, the interband tunneling distance can become small enough to allow a substantial flow of current from the *AlSb* valence band into the *InAs* quantum well. To some degree, this situation can be mitigated by the addition of *As* to the *AlSb* upper barrier layer, which serves to 'pull down' the valence band of that barrier, and increased the energy separation from the *InAs* conduction band[9].

We noted earlier the beneficial aspects of being able to place the *InAs* quantum well close to the surface, without suffering surface depletion of the conduction electrons: the closer the 2DEG to the surface, the more abrupt the confining potential, and the larger the subband spacing. However, the larger electron conduction in the *InAs* channel, in closer proximity to a *GaSb*-capped surface, also has deleterious consequences for device leakage. The increased electron concentration requires large electric fields under the gate in order to deplete the 2DEG to form the one-dimensional conduction path. The larger electric field and greater band-bending aggravate the problem described in the previous paragraph, and hence the leakage.

Many of these leakage problems are most manifest in the split-gate geometry; this suggests that gated, deep-etched mesa structures, such as those used by Tarucha *et al.*[31], can both circumvent these problems, and produce structures having larger subband spacings. The particular nature of the *InAs*-metal interface (see Section IV.D) may prove particularly beneficial in realizing even higher subband splittings in this material system.

In general, a good 'rule of thumb' suggests that the larger the subband spacing that we can engineer in our quantum structure, the more robust the concomitant quantum phenomena will be. However, the experiments carried out thus far on the *InAs* waveguides highlight the nuances of that 'robust' operation. Higher subband spacings do not necessarily ensure robust high-temperature operation; even at lower temperatures, the otherwise subtle effects of isolated impurities play a major role in these structures. In this case, the higher subband spacings bring greater immunity from the variations in potential resulting from the long-range effects of the impurities. It is also apparent that the independent correlations and results taken from transport studies in these *InAs* materials can be of great use in better understanding and interpreting analogous data from the *GaAs* structures. Both the similarities and subtle differences in the material compositions and manner of doping help to distinguish that which is intrinsic to one-dimensional transport, and that which may be materials– or processing-related.

2. A New Direction: Superconducting Point Contacts

The study of quantized conductance in the *InAs/AlSb* material systems, and the general insights gained by comparison of these results to those in *GaAs* structures of various geometries reveal the power of material-comparative studies of mesoscopic transport. One may then ask, what would ensue by changing, not only the semiconductor material that forms the basis of the mesoscopic structure, but also by changing the normal-metal reservoirs in contact with the mesoscopic conductor, making them superconductors? In fact, this is a question that has been increasingly raised in the past few years, both theoretically and experimentally. In principle a small change (though with numerous experimental challenges), it can illuminate a wide diversity of aspects in the field of mesoscopic transport. Beenakker[32] has concluded that the quantized conductance in this case will be modified, with steps of magnitude $4e^2/h$ rather than $2e^2/h$. Moreover, other calculations, by Furusaki[33], suggest that the critical current may increase in a step-wise fashion depending on both the parameters of the constriction, as well as on the energy gap of the superconducting reservoir.

The controlled variation of the density of highly mobile electrons is an important factor in the selection of materials for these structures, as was true for the normal quantum point contacts. An added, critical concern becomes the clean and reliable coupling of the superconductor to the normal electrons, hence the relatively transmissive interface between superconductor and semiconductor. The advantages of the *InAs-AlSb* material system, and the nature of that coupling, is the subject of Section III.

III. SUPERCONDUCTOR-SEMICONDUCTOR HETEROSTRUCTURES

A. Semiconductor-Coupled Superconducting Weak Links

The idea to combine superconductors and semiconductors in a common device structure dates back to at least the late–'60s, developing quite independently of ideas on phase-coherent ballistic transport. The case of interest to us is what are called semiconductor-coupled superconducting *weak links*. The term *weak links* refers to devices in which two superconducting "banks" are coupled through another *conducting* medium, the critical current through which – if nonzero – is much less than that of the banks. The term *weak link* is used specifically to distinguish those structures from the well-known Josephson *tunnel* junctions, in which the current flow is by Cooper pair tunneling through an *insulating* barrier. For a review of weak links, the reader is referred to the excellent paper by Likharev[34], which gives a very intensive review of the status of weak links up to 1979.

In the case of interest here, the coupling medium is a semiconductor, more specifically, *InAs*. Figure 15 gives an example of a recent (1990+) structure of this kind. An *InAs* quantum well with *AlSb* barriers forms a short (<1 μm) conducting link between two superconducting banks, in this case *Nb*.

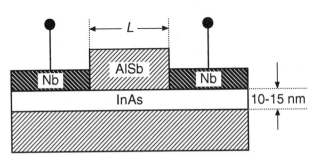

FIGURE 15. Semiconductor-coupled superconducting weak link based on an *InAs-AlSb* quantum well forming a conducting link between two superconducting *Nb* electrodes.

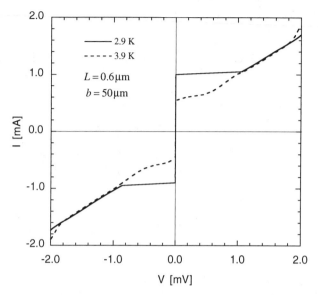

FIGURE 16. Josephson-type I-V characteristics of a device with the structure shown in Figure 15, with 0.6 μm electrode separation, at two temperatures. From Nguyen[19]; see also Kroemer *et al.*[35].

Under favorable conditions, weak links exhibit a pronounced *Josephson effect*, that is, a current-voltage characteristic as in Figure 16, which exhibits a current range inside which a resistance-less *supercurrent* can flow between the two superconducting banks, up to a certain critical current I_c. Only when the latter is exceeded does a voltage appear between the superconducting terminals.

1. Basic Current-Phase and Phase-Voltage Relations– a Tutorial

The basic physics underlying Josephson junctions in general, and weak links in particular, is the following.

(a) As is discussed in many texts on quantum mechanics, on solid-state physics, and on superconductivity itself (see, for example, references [36-39]) the essence of superconductivity is the existence of a common pair wave function for all Cooper pairs in the superconductor. It may be written as:

$$\psi(\mathbf{r})=|\psi(\mathbf{r})|\cdot\exp[i\theta(\mathbf{r})]. \tag{7}$$

Here, the magnitude $|\psi(\mathbf{R})|$ of the pair wave function is related to the local Cooper pair density $N(\mathbf{r})$ via

$$|\psi(\mathbf{r})|^2 = N(\mathbf{r}), \tag{8}$$

and $\theta(\mathbf{r})$ is a phase. The key point is now that this phase is coherent over macroscopic distances, and, in the absence of a current, is the same throughout the superconductor.

(b) In a weak link, two superconductors are coupled through another conducting medium, through which electrons can pass in such a way that the phase of the electrons is somehow preserved in the process. If the phases of the two superconductors are the same, there will be zero net current, but if there is a nonzero phase *difference*

$$\theta_{21}\equiv\theta_2-\theta_1 \tag{9}$$

between the two superconductors, a resistanceless Josephson supercurrent can flow from one superconductor to the other.

The theory shows that in the presence of such a phase-preserving coupling the resulting supercurrent is a function simply of the phase difference θ_{21}: i.e. $I=I(\theta_{21})$. For Josephson *tunnel* junctions the functional relationship, derived in many texts, is simply sinusoidal,

$$I=I_c\cdot\sin(\theta_{21}). \tag{10}$$

Here, for the ordering of the two phases as given in Eq. (9), a positive current designates a flow of Cooper pairs from bank #2 to bank #1. Because the pairs carry a negative charge $(-2e)$, the *electrical* current is in the opposite direction, from bank #1 to bank #2. (Feynman[36] gives incorrect flow directions).

For weak links, Eq. (10) often remains a good first-order approximation, but more complicated relations may occur with additional zeros and multi-valued relations. However, $I(\theta_{21})$ is always an odd function of θ_{21} and, inasmuch as phase differences have a physical meaning only modulo 2π, the $I(\theta_{21})$ relation is necessarily a periodic one with a period 2π: i.e.

$$I(\theta_{21} + 2\pi) = I(\theta_{21}) \qquad (11a)$$

$$I(\theta_{21}) = -I(\theta_{21}) \qquad (11b)$$

Whatever the exact form of $I(\theta_{21})$, this function is the central function in the theory of weak links.

(c) In the absence of a bias voltage between the two superconducting banks, whatever phase difference may be present will not change with time, hence the current will continue to flow– which is why it is called a supercurrent. If an external bias voltage is present, the theory shows that the difference θ_{21} becomes time-dependent according to the simple law:

$$\frac{d}{dt}\theta_{21} = \frac{2e}{\hbar} \cdot V_{21}, \qquad (12)$$

where V_{21} designates the voltage of bank #2 relative to bank #1.

The supercurrent-vs.-phase relation Eq. (11) remains valid in the presence of such a voltage, but the supercurrent now oscillates about zero with the Josephson frequency

$$V_J = \frac{2e}{h} \cdot V_{21}, \qquad (13)$$

where $h = 2$ is Planck's constant. In addition, a time-independent normal current will also flow. If the total *current* rather than the *voltage* is held constant, both the supercurrent, the normal current, and the voltage will oscillate.

2. Some History

Historically, much of the current interest in semiconductor-coupled weak links was stimulated by the 1978 and 1980 papers by Silver et al.[40] and by Clark et al.[41], which proposed the possibility of a *Josephson Field Effect Transistor* (JOFET). The JOFET is a *three*-terminal FET-like device based on semiconductor-coupled Josephson junctions (JJ's). The central idea was simple. One of the most characteristic features of semiconductors is that their electron concentration is not a fixed quantity, but can be modulated by a gate electrode. Hence, in a semiconductor-coupled JJ, it should be possible to modulate the current-voltage characteristics by a gate electrode acting on the semiconductor, leading to an FET-like current-voltage characteristic. (The work by Clark *et al.* actually goes back to Clark's 1971 Ph.D. dissertation at the University of London, but Clark did not publish his work until after the appearance of the paper by Silver *et al.*)

The experimental realization of true JOFET's took a number of years. Throughout much of the '80s, the materials choice was influenced more by technological maturity than by physics preferences. In fact, the first demonstration of a working JOFET, by Nishino *et al.*, employed *Si*[42,43], hardly a material ideal for JOFET applications. Shortly following the *Si* work, Ivanov and Claeson demonstrated true JOFETs in *GaAs*[44,45]. This was also the first such work employing the 2D electron gas at a heterostructure interface, the familiar *GaAs*-(*Al,Ga*)*As* hetero-interface. Indium alloy contacts served as the superconducting contacts. *GaAs* is still not ideal for this purpose, principally because it suffers from the difficulty of making truly barrier-free ohmic contacts.

The importance of this contact barrier problem was recognized already by Silver *et al.*, who proposed the use of *InAs* as the preferred semiconductor. The first *InAs*-based JOFET was demonstrated in 1985 by Takayanagi *et al.*[46], and much subsequent work, not only on JOFET's but on semiconductor-coupled weak links in general, has emphasized *InAs*. This has culminated in what must be the best true three-terminal JOFET's to date, by the Japanese NTT group, which employed *InAs* quantum wells, with confinement barriers made from (*Ga,In*)*As*/(*Al,In*)*As*[47-49] [See Figure 17].

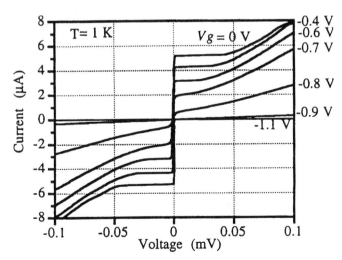

FIGURE 17. Current-voltage characteristics of an experimental *InAs*-channel JOFET, from Akazaki *et al.*[49].

It is not the purpose of this chapter to give a detailed review of these early developments. The interested reader is referred to the papers cited, where complete references to related work can be found. In fact, work on true (three-terminal) JOFET's forms only a small fraction of the overall recent work on semiconductor-coupled weak links that was stimulated by the original JOFET proposal. None of the JOFETs actually reported to date have shown the kind of performance that offers promise for practical applications. Their most severe problem is that the obtainable drain-to-source voltage swings are typically much less ($\ll$ 1 mV) than the gate voltage swings ($\gg$ 1 mV) required to achieve significant drain current changes. These devices therefore have painfully low voltage gains, which appear to be inherent in their physics. Even the best JOFETs reported to date[47-49] just barely achieve a voltage gain of unity under optimal loading conditions. To achieve net power gain, a necessity for both amplifying and logic devices, the low voltage gains would have to be made up for by compensating large current gains, a non-trivial task in practical circuits. It remains to be seen whether or not future developments will overcome this problem. As it stands now, a more likely application of JOFETs is as current-routing switches in superconducting networks, drawing on the fact that a JOFET is an FET with a true zero-resistance on-state, something no pure semiconductor device can offer.

The remainder of this chapter deals exclusively with two-terminal semiconductor-coupled weak links.

3. Enter InAs-AlSb Quantum Wells

In the use of *InAs* quantum wells with confinement barriers made from $(Ga,In)As/(Al,In)As$, a severe lattice mismatch $\approx 4\%$ remains between the *InAs* and the barrier material. To avoid the formation of misfit dislocations during growth, the *InAs* quantum well must be kept very narrow (the NTT group used 4 nm). Although recent developments in strained-layer epitaxy have largely overcome the technological problems associated with the built-in strain, there are at least two fundamental problems associated with the very narrow wells: (a) the mobility in narrow wells is significantly reduced by interface roughness scattering[50,51]; (b) the large quantization energies associated with narrow wells, combined with the relatively low height of $(Ga,In)As$ barriers, limit the electron concentration achievable in the wells. The desire for lattice-matched barriers of greater height naturally leads again to the use of *AlSb* barriers, just as in the case of *InAs* for quantum wires, discussed in Section II. We initiated research towards super-semi structures employing such quantum wells in 1989, and in 1990 reported a first *InAs-AlSb-Nb* superconducting weak link, which showed high current densities even for a remarkably wide electrode spacing of 0.6 μm[52]. Our research since then is what the present chapter will discuss.

B. Andreev Reflections

1. The Premise: Semiconductor-Coupled Weak Links as "Clean" Weak Links

The basic weak-link physics discussed above holds independently of the nature of the mechanism that preserves the phase of the electrons. In the semiconductor-coupled weak links of interest here, the mean free path of the electrons tends to be larger than the inter-electrode separation, in which case the mechanism for the phase transfer tends to be dominated by the phase-coherent flow of *ballistic* electrons between the banks. In the jargon of superconductivity, such weak links are called "clean" weak links, in contrast to the more common "dirty" weak links extensively studied in the past, in which the electron transport is diffusive. Unfortunately, much of the literature on weak links, including Likharev's classical review[34], is still dominated by considerations of dirty weak links.

As we shall see, the ballistic nature of the transport is responsible for some of the most remarkable properties of clean semiconductor-coupled weak links. It therefore forms another bond between the two fields of quantized conductance in ballistic constrictions, and of superconducting weak links. The principal difference between the case of superconducting electrodes and normal-metal electrodes is that in the latter case the phases of the individual electrons are independent of each other, while in the former case the electrons enter the semiconductor already with a common and well-defined phase, determined by the phase of the Cooper pairs in the superconductor.

Furthermore, the mean free path that matters for the phase transfer is not the elastic mean free path that determines the low-field mobility, but the *inelastic* mean free path that is responsible for any de-phasing of the electron waves, and which is typically much longer than the elastic mean free path. For example, in impurity scattering, the phase of the scattered wave is coherent with the phase of the incident wave, and while such scattering may create a chaotic wave front, this does not constitute phase-incoherence in the sense of weak-link theory: there is still a fixed phase relation between any two points in the wave field.

Given an inelastic mean free path much longer that the inter-electrode spacing, the dominant phase-altering process for the electrons becomes the scattering, not inside the semiconductor, but at the semi-super interface between the electrons in the semiconductor and those in the superconductor. Now, electron-electron scattering is normally a phase-destroying process. However, at a super-semi interface, at sufficiently low temperatures, the only electrons available for participation in scattering on the superconductor side are the Cooper pairs. But, as we saw earlier, the Cooper pairs all have the same well-defined phase. As a result, the scattering interaction of electrons in the semiconductor with electrons in the superconductor becomes itself a phase-coherent process. It is universally referred to as Andreev scattering or, more commonly, as *Andreev reflections* (AR's), in honor of the man who discovered the possibility of such a process in 1964[53].

Although postulated over thirty years ago, Andreev reflections have received major attention only during the last few years, when it became clear that their understanding is central to the understanding of clean-limit weak links. As a result of this belated recognition, they have not yet found their way into current textbooks on superconductivity. Even the 1979 weak-link review by Likharev mentions Andreev reflections only in passing. In fact, Likharev, on page 132 of his paper[34], explicitly lists a number of experimental observations that are not consistent with the then-existing theoretical understanding, all of which find their explanation via Andreev scattering. We therefore provide here the necessary background on this topic.

2. Basic Concept

The basic idea behind Andreev reflections is simple. Consider a semi-super interface between a degenerately doped semiconductor and a superconductor. As shown in Figure 18(a), a superconducting energy gap has opened up on the super-conductor side. If now an electron with an energy E *above* the Fermi level (but still inside this gap) is incident on the interface from the semiconductor side, the absence of single-particle states within the gap prevents that electron from enter-ing the superconductor as a *single* electron, and one might expect this electron to be reflected, and the electrical resistance to current flow across the interface actu-ally to *increase* at the onset of superconductivity in the metal.

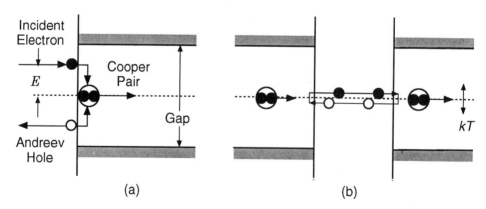

FIGURE 18. Andreev reflections. (a): Basic concept. (b): Persistent current flow by multi-ple Andreev reflections.

However, the electron may pair up with a second electron at the same energy *E below* the Fermi level, forming a Cooper pair, which *can* enter the superconductor, causing a *doubling* of the current compared to that in the absence of superconductivity, rather than a suppression. The electron removed from the semiconductor below the Fermi level leaves behind a hole in the Fermi sea. The generally accepted jargon associated with this phenomenon is to say that the incident electron is *reflected as a hole*.

In a semiconductor with a large mean free path for the electrons ($\approx 3\,\mu m$ in our structures), the Andreev hole left behind at the interface has a large mean free path itself, roughly equal to that of the original electron. The theory shows that the hole travels back into the semiconductor along a trajectory that essentially retraces the trajectory of the original incident electron. If its mean free path is sufficiently large, the hole will eventually reach the opposite superconducting electrode. In the absence of any bias across the structure, the energy of the hole is still within the superconducting gap on that side. Such a hole cannot enter the superconductor, but it can be annihilated by breaking up a Cooper pair inside the other superconductor: One of the electrons of the pair annihilates the hole, the other electron takes up the annihilation energy, and is injected into the semiconductor as a ballistic electron *above* the Fermi level, at an energy E exactly equal to that of the initial electron. This process, illustrated in Figure 18(b), can evidently be repeated: the Andreev reflections act as what we might call a "Cooper pair pump" annihilating Cooper pairs on one side, and re-creating them on the other. If *all* reflections of electrons and holes were Andreev reflections rather than "ordinary" reflections, and if there were no other kinds of scattering processes, the result would be a persistent current.

However, perturbations are always present, and we would expect simply an enhancement of the conductivity by a factor equal to the number of ballistic traverses before an unfavorable reflection or collision event randomize either the electron or the hole flow in this chain reaction.

Also, even if no unfavorable reflection and collision events took place, the diagrams in Figure 18(b) show only half the story: For each state with a given direction of arrows there exists another state with all current flows reversed. If both states of such a pair are occupied, their currents will cancel. To understand how a stable supercurrent can arise, we must go beyond the pure ballistic particle picture of Figure 18(b), and must take into account the wave properties of the unpaired electrons and holes, and of the Cooper pairs. This leads us back to the idea that supercurrents depend on the phase difference between the pair wave functions in the two superconducting banks[32,54].

3. Stable Supercurrents as a Phase-Coherence Phenomenon

Waves have phase, and even in the absence of any scattering events, the simple current-carrying state illustrated schematically by Figure 18(b) is a quantum-mechanically allowed stationary state only if the round-trip phase shift along the electron-hole loop is an integer multiple of 2π,

$$\Delta\phi_{RT} = 2\pi i. \tag{14}$$

For every value of i, there will actually be two states, corresponding to opposite directions of the flow arrows in Figure 18(b).

Up to a point, the above condition is exactly the same as for the bound states in an "ordinary" one-dimensional semiconductor quantum well. These, too, are states for which the round trip phase changes are the different multiples of 2π. In fact, with regard to the spatial confinement of the unpaired electrons and holes inside the semiconductor portion of the structure, the stationary states may indeed be viewed as a new kind of bound state[32,54], the difference being that the "heterojunction" barriers are now formed, not by the conventional energy gap of another semiconductor, but by the superconducting energy gap of the two superconducting electrodes.

However, there are two decisive differences. The first is that for an Andreev bound state confined by superconducting energy gap barriers, the phase on one of the two traverses is carried by an electron, on the other traverse by a hole. This does not change the requirement that the *total* round-trip phase change be a multiple of 2π, but it means that these kinds of bound states actually carry a current across the semiconductor, in contrast to the current-less conventional bound states in a $GaAs-(Al,Ga)As$ quantum well. The two states belonging to a given i belong to opposite directions of that current flow.

A second consequence is the following. As in a conventional quantum well with barriers of finite height, the round-trip phase shift contains a contribution from the reflections at the two superconductor barriers. In a semiconductor quantum well, these contributions simply represent the finite penetration of the wave function into the barrier, and they are responsible for lowering the bound state energies with decreasing barrier height.

But in the case of Andreev reflections there is an *additional* phase shift at each bank, equal in magnitude to the phase of the Cooper pair wave function in that bank, but with a sign depending on whether an electron or a hole is reflected. When an electron is reflected at a superconductor with phase θ, the wave function of the hole resulting from the reflection acquires an additional phase shift by $(-\theta)$. This behavior can be readily understood by realizing that the Andreev reflection of an incident electron creates an additional Cooper pair with phase θ, and the phase shift $(-\theta)$ of the reflected hole simply compensates for the phase of the new Cooper pair.

Conversely, if a hole is reflected, the wave function of the resulting electron acquires the phase $(+\theta)$, with a similar interpretation. What matters for the overall Andreev bound states is, of course, the net round-trip phase shift. If the two superconducting banks have the same phase, the phase shifts by $(\pm \theta)$ at the two banks cancel, but if there is a phase difference between the two banks, it will make a contribution

$$\Delta\phi = \pm\theta_{21} \qquad (15)$$

to the round-trip phase shift with the following sign rule: if, in Figure 18(b), the left-hand bank is bank #1, the minus-sign applies, otherwise the plus sign applies.

If we compare this sign rule with the sign rule for Josephson supercurrents stated in Section III.A. – see Eq. (10) and the text following it – we may re-phrase our new rule as follows: if the net ballistic electron flow in the Andreev loop in Figure 18(b) is in the same direction as the Josephson supercurrent in a weak link with the same phase difference θ_{21}, then the net phase shift $\Delta\phi$ is positive; otherwise, $\Delta\phi$ is negative.

In order to retain the round-trip condition Eq. (14) in the presence of the phase shift contribution $\Delta\phi$, the latter must be compensated for by an opposite change in the phase shift contribution associated with the ballistic flight through the semiconductor itself. That ballistic phase contribution depends, of course, on the energy of the ballistic electron. Hence, the introduction of a phase difference between the two superconductors leads to changes of the energies of the Andreev bound states. A positive barrier contribution to the round trip phase shift requires a lowering of the ballistic phase contribution, and hence a lowering of the bound-state energy, while a negative contribution raises the latter.

At this point, the sign difference in Eq. (15) becomes decisive: it means that, in the presence of a nonzero phase difference θ_{21} the energies of the bound states will depend on the direction of current flow in each state, in such a way that the states with a current flow in the direction proper for a Josephson supercurrent will have a lower energy and therefore a higher thermal occupation probability, than those with a current flow in the opposite direction. Hence, in this case there will be a thermodynamically stable net current flow, even in the presence of scattering events.

Recall finally that a time-independent phase difference θ_{21} corresponds to zero bias voltage. Hence the stable current is a true zero-resistance supercurrent, with a certain maximum value, the critical current, for some particular value of the phase difference θ_{21}.

So long as the superconducting banks remain superconducting, thereby permitting phase-coherent Andreev reflections, the critical current will never be rigorously zero. But it will usually drop precipitously with increasing temperature, becoming "astronomically small" once the temperature increases above a certain *de-facto* critical temperature, which may be far below the critical temperature of the banks. Any currents in a "real" experiment will then exceed the critical current, and the net effect will be almost the same as if the *de-facto* critical temperature were a well-defined sharp critical temperature associated with the heterostructure, albeit one well below that of the superconducting banks themselves.

The reader wishing to go beyond our purely qualitative description of AR-induced supercurrents must be referred to the relevant literature. A good point of departure is a 1991 paper by van Wees *et al.*[55]. By neglecting all "real-world complications" and stripping the problem down to its essentials (assuming one-dimensional current flow, equal Fermi velocities in the two materials, etc.), the paper gives a remarkably transparent description of the basic nature of the super-current. One unique aspect of van Wees' paper is a discussion of the changes in the central $I(\theta_{21})$ characteristic when there is a small bias applied between the normal-conductor "reservoir" and the two superconducting banks.

For a more detailed and quantitative theory of the supercurrents Andreev-reflection-based supercurrents – a formidable undertaking– the interested reader is referred to the work by Schüssler and Kümmel[56]. Readers wishing to go back further to the early fundamental work following the original paper of Andreev are referred particularly to the 1972 paper of Bardeen and Johnson[57], to which all subsequent authors are indebted.

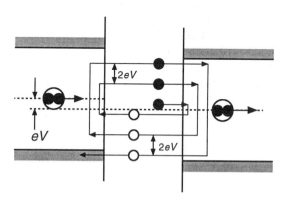

FIGURE 19. Andreev "spiral" in the presence of a non-zero bias voltage.

C. Non-Zero Bias

Consider now what happens when the current through the structure exceeds the critical current, and a bias voltage develops across the semiconductor. We again start with a purely ballistic particle description as in Figure 18(b), ignoring temporarily the phase of the electrons.

In the presence of a non-zero bias, the Andreev loop in the ballistic particle picture of Figure 18(b) ceases to be a closed loop, and develops into a spiral, as shown in Figure 19. The number of electron/hole traverses then becomes necessarily finite, decreasing with increasing bias voltage.

Whenever, with increasing bias, the number of traverses decreases by one, the differential conductance also decreases in step-like fashion, with the steps occurring at voltages roughly equal to the integer fractions of the superconducting gap voltage. Such steps are readily observed experimentally. Figure 20 shows an

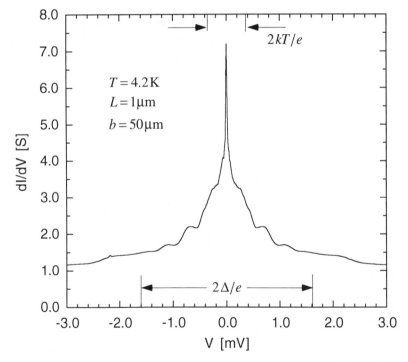

FIGURE 20. Zero-bias spike, and sub-harmonic gap structure in the differential conductance dI/dV vs. voltage at 4.2K for a device with 1 μm electrode spacing. From Nguyen *et al.*[58].

example of the differential conductance at 4.2K as a function of bias voltage for one of our samples with a relatively large 1 μm electrode spacing, and hence, no superconductivity in the temperature range investigated (>1.4K). (The sample *does* show a pronounced enhancement of the zero-bias conductivity, to which we will return presently.)

The existence of such a sub-harmonic-gap structure in the dI/dV characteristic is generally considered the key "fingerprint" evidence for the presence of pronounced multiple ARs[59,60]; it is the most direct evidence that the electron transport in the kind of Sp-Sm-Sp structures discussed here is governed by multiple ARs.

It is instructive to carry this argument to very small bias voltages ($e|V| \ll k_B T$). If there were no competing scattering processes, each *initial* electron would undergo a total of $\Delta/e|V|$ traverses, where Δ is the superconducting gap, and hence lead to a total charge transfer of $\Delta/e|V|$ electrons, diverging as the bias voltage goes to zero! However, at the same time, the total number of *available* initial electrons becomes restricted by the number of states available within the semiconductor, in the energy interval of width $e|V|$ separating the Fermi levels of the two superconductors. This number decreases linearly with decreasing $|V|$ as $V \to 0$. As a result, the two voltage dependences cancel out, and the current saturates at some nonzero value as $V \to 0$, leading to an I-V characteristic with a Josephson-like discontinuous step at $V = 0$.

In reality, there will always be some scattering event truncating the loop, and the current will start dropping with decreasing voltage already at some nonzero voltage. The sharp and narrow zero-bias conductance spike simply reflects that sharp drop of the current with decreasing bias. The larger the number of traverses before a competing scattering event takes place, the smaller will be the voltage at which the current will start to fall off, and the taller will be the conductance spike around zero bias. Evidently, the zero-bias conductance spike is a non-equilibrium phenomenon in which each electron picks up an energy much larger than the potential difference between the superconducting banks, even though this energy is not necessarily larger than $k_B T$. Conventional "linear" conductivity theory, which views linear conductance as a near-equilibrium phenomenon, is inapplicable here.

Just as for the supercurrent, a detailed and quantitative theory of the dissipative portion of I-V characteristics again requires going beyond the simple ballistic particle picture of Figure 19, and taking the phase of the electron waves into account. The need for a full quantum-mechanical treatment of this problem is apparent already from the data in Figure 20. Those data evidently show a zero-bias conductance enhancement by about a factor 7 relative to the conductance at large bias voltages. But about one-half of this enhancement takes place over a bias range much narrower than $\Delta/7e$, the range one might expect from the ballistic argument given above. Many other samples studied by us show a similar behavior.

Once again, the reader wishing to go beyond our purely qualitative description must be referred to the relevant literature, especially to a series of papers by Kümmel and his collaborators[61-65].

Historically, the first paper interpreting actual data on super-semi devices in terms of AR's appears to have been a key 1990 paper by Kleinsasser *et al.*[66]. Those authors studied the I-V characteristics of superconducting contacts to (*Ga,In*)*As*, and found a significant enhancement, by about a factor 2, of the zero-bias conductance, which they interpreted as due to Andreev reflections. Ironically, their interpretation actually *disagreed* with the then-accepted theories about the affect of AR's which predicted, not a single peak at zero bias, but a pair of peaks at bias voltages separated by the energy gap voltage of the superconductor. It was subsequently pointed out by van Wees *et al.*[67] that a single peak should indeed be expected as a result of the phase conjugation between electrons and Andreev-reflected holes. Our own later work was greatly influenced by those two papers, which led to the realization that the current transport in our structures was heavily influenced by Andreev reflections[68]. Much of the remainder of the present chapter deals with work done since then.

D. Selected Recent Results on *InAs/AlSb*
Quantum Wells-Coupled Weak Links

1. The Structures: Single-Gap Devices
vs. Multi-Gap Grating Structures

The selected data reported above, in the context of our discussion of Andreev reflections, were all obtained on the single-gap device structure shown schematically already in Figure 15, consisting of an *InAs* quantum well, typically 15 nm wide, with barriers made from straight *AlSb*, an (*Al,Ga*) *Sb* alloy or an *AlSb-GaSb* superlattice. In those structures, the barriers are selectively removed by chemical etching, and replaced with *Nb*, deposited either by electron beam (e-beam) evaporation or by sputtering. Typical inter-electrode separations ranged from 1 μm (done by optical lithography) to below 0.6 μm, done by e-beam lithography or laser holography. Figure 16, shown earlier, displayed the I-V characteristic of one of the better Josephson weak links obtained in this way.

In our first weak link made by this technology[52], an unsuccessful attempt was made to place a gate over the *AlSb* barrier separating the two *Nb* electrodes, and to achieve a true JOFET, but the gate electrodes (and their bonding pads) proved to be too leaky and the attempts were abandoned pending the development of a better technology. Apparently, others encountered similar problems. The title of a more recent paper by Maemoto *et al.*[69] promises *"Superconducting Transistors,"* but the paper contains no data to live up to that promise beyond a statement that "only a slight modulation ... has been observed."

More recently, we found it useful to go beyond a single-gap device geometry, and to study series-connected periodic arrays, prepared by laser holography. Figure 21 shows schematically such a series-connected array involving (≈ 300) gaps. The initial reason for going to such arrays was to be able to generate narrower gaps than what is achievable by conventional optical lithography, without having to use e-beam lithography. But the grating structure has advantages of its own: (a) working with a series connection of a large number of devices improves the voltage sensitivity of the measurements in the range of near-vanishing resistance; and (b) the narrow superconducting *Nb* grating lines interfere less than large-area *Nb* electrodes with studies of magnetic field effects on the properties of the devices. Electron sheet concentrations in these devices, obtained by modulation doping with *Te*[70], are typically around $8 \times 10^{12}/\text{cm}^2$ with low-temperature mobilities around 90,000 $\text{cm}^2/V \cdot s$.

Following the optical patterning by a combination of laser holography and conventional optical lithography, the *InAs* surface is again selectively exposed by etching, just as for the single-gap devices. The structures are then sputter-cleaned, followed by the immediate sputter-deposition of *Nb*, without breaking vacuum. The *Nb* is finally lifted everywhere except where it contacts the *InAs* directly,

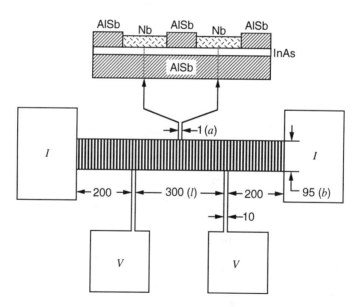

FIGURE 21. Overall layout (bottom) of *Nb* grating structure, along with (top) a schematic cross-section through a pair of *Nb* lines separated by a narrow stripe of *InAs-AlSb* quantum well. All dimensions are in micrometers.

yielding a parallel array of a large number of *Nb* lines. Typical dimensions for the devices are as indicated in Figure 21, which shows the dimension of our first sample. It had over 700 *Nb* lines, each with a length of 95 μm, a period of 1 μm, and a gap between the *Nb* lines of about 0.4 μm. The gap-to-*Nb* ratio of the array depends on processing details; devices with gaps in the range 0.25 μm to 0.4 μm were prepared and studied. As part of the processing, narrow *Ti-Au* side arms for voltage measurements were applied about 300 μm apart, and current-supplying *Ti-Au* end contact pads were applied approximately 200 μm past the voltage side arms. The parameters of this first sample remained typical for subsequent samples, the most important variation being a narrower gap between the *Nb* lines in some of the subsequent samples, down to about 0.25 μm.

Up to a point, these devices might be viewed as simple series connection of a large number (300 or more) individual *Nb-InAs-Nb* Josephson junctions, each having a lateral width equal to the length of the *Nb* lines. However, the analogy is imperfect. The single-event Andreev reflection probability is likely to be less than 100%. Hence, it must be expected that a fraction of the electron flow through the device does not pass through any given *Nb* stripe, but passes ballistically underneath. This implies the presence of an additional coupling between successive diodes. The magnitude of this effect, and its consequences, are not clear.

During our work it also became clear that these grating devices show a wealth of interesting physical phenomena, not only when superconductivity is achieved, but also when the inter-electrode spacing is still too large or the temperature is still too high to yield a zero-resistance state. We already showed in Figure 20 the rich structure in the differential conductance characteristics of a typical 1 μm-gap sample at 4.2K, with its conductance steps at integer fractions of the superconducting gap voltage, and especially the very sharp conductance spike at zero bias. We briefly discuss here two aspects of this precursor spike, its temperature dependence, and its sensitivity to magnetic fields.

2. Temperature Dependence

One of the most important properties of any weak link is its dependence on temperature, starting with the question as to the effective "critical temperature" of the device itself. The critical temperature is defined as the temperature at which the device becomes superconducting, as evidenced by a Josephson-type I-V characteristic as in Figure 16. The technology of these structures is still at a immature stage, with huge sample-to-sample variations, and no single property shows this variation more than the effective critical temperature. Many devices simply never became superconducting at all in the temperature range investigated by us ($T > 1.4$K), but a separate study of two such samples at lower temperatures (~ 0.1K)[71] showed superconductivity even in those devices, suggesting that even "poor" devices become superconducting at sufficiently low temperature.

However, all but the poorest samples show an increase in their zero-bias differential conductance as soon as the *Nb* banks become superconducting (7K-9K), increasing continuously with decreasing temperature. What varies greatly from device to device is the *rate* of this increase. An important aspect of this conductance increase is that it is inevitably non-ohmic: i.e. the I-V characteristics become non-linear, with a conductance spike at zero bias, qualitatively similar to the spike shown in Figure 20. Usually the spike is accompanied by more or less pronounced Andreev satellites as in that figure.

For bias voltages significantly larger than the superconducting gap voltage of the *Nb* electrodes (≈ 3 mV), the I-V characteristics become linear again, but with the current shifted upward relative to a perfectly ohmic characteristic by a finite amount. This shift, commonly referred to as an *excess current*, is simply one-half of the integral over the zero-bias conductance spike relative to the differential conductance at large bias voltages.

In all samples in which the zero-bias resistance ultimately vanishes with decreasing temperature, this vanishing always takes the form of a *continuous* decrease in which the resistance becomes lower and lower until it finally reaches values below the noise limit of the measuring setup. We never saw any evidence for a discontinuous transition as for clean type-I superconductors. We therefore view this conductance peak as a *precursor* of true superconductivity, even in samples where superconductivity is not reached in the temperature range investigated.

One of the implications of this behavior is that it is, strictly speaking, not possible to give a well-defined critical temperature for the onset of superconductivity in these structures; the measured value always depends (roughly logarithmically) on the sensitivity of the measurement employed. This sensitivity problem is not trivial. Because of the strong non-linearity of the I-V characteristics, measurements of the height of the zero-bias conductance spike require small measuring currents. (We typically used 2 µA.) Together with the low resistance, this leads to the need to measure very small voltages. One of the advantages of the use of multi-gap grating structures, which are essentially series-connections of a large number of individual devices, is the large enhancement of the sensitivity with which the vanishing of the resistance at the onset of superconductivity can be studied.

According to Gunsenheimer et al.[64,65], the temperature dependence of the zero-bias conductance should be essentially that of the mean free path for *inelastic* scattering, the process responsible for the phase-breaking that terminates the Andreev loop of electrons and holes of Figure 19. The exact mechanism that causes the phase breaking that is responsible for the re-appearance of a nonzero resistance is at present not yet understood, hence such measurements should be very revealing as to that mechanism. However, actual measurements of the temperature-dependence of the zero-bias resistance have shown a bewildering variety of behavior, even on samples from the same MBE growth, and processed one-at-a-time by *nominally* identical procedures but by different individuals.

As an example, Figure 22 shows an Arrhenius plot of the temperature dependence of the zero-bias resistance for what is probably our best sample to date[72]. The data evidently do not fall on a straight line, and hence cannot be characterized by a single temperature independent activation energy. In the temperature range below 5K, the data can be fitted to an activation energy of about 17 meV, whereas in the range 5.4K to 6.4K, the slope of the curve corresponds to an activation energy of about 8.4 meV. Yet, another sample from the same wafer as the sample of Figure 22, studied by Harris *et al.*[73], showed a linear Arrhenius plot over a wide resistance range, but with a much lower activation energy of only about 3.4 meV.

Neither set of data is compatible with a simple phonon mechanism, as suggested by Gunsenheimer[64,65]. The latter would yield a T^3 power law, just like the low-temperature Debye specific heat, much weaker than what is observed. Evidently, additional research is called for – and is in fact going on.

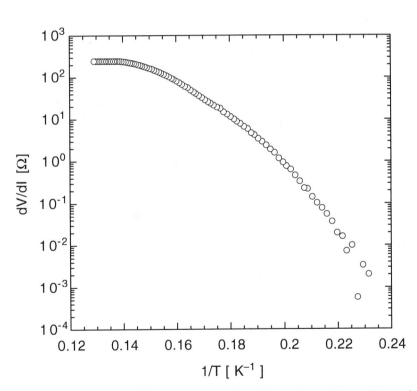

FIGURE 22. Arrhenius plot of the temperature dependence of zero-bias resistance for a grating sample[72].

3. Magnetic-Field Dependence

Another important property of any weak link is its high sensitivity to magnetic fields. In fact, superconductive quantum interference devices (SQUID's) made up from pairs of Josephson junctions (JJ's), and capable of measuring extremely small magnetic fields, constitute one of the most important present-day applications of JJ's. We would therefore expect this sensitivity to carry over to the *InAs*-coupled weak links of interest here – both below and above the onset of superconductivity.

As an example of a field-dependence above the onset of superconductivity, we have studied the field dependence of the zero-bias conductance spike shown earlier, in Figure 20. As we indicated already in Section III.C, this central spike is much narrower than one would expect from the simple ballistic Andreev Loop argument illustrated in Figure 19, which ignored the role of the phase of the wave function inside such a spatial loop. Now it is well known that phase effects in superconductivity – and in mesoscopic transport in general – are very sensitive to shifts introduced by the magnetic vector potential of even very weak magnetic fields. Hence, to the extent that the conductance spike in Figure 20 may be viewed as a phase-coherent precursor of true superconductivity, we would expect it to exhibit a strong magnetic field sensitivity. This is indeed the case, even though the details of this dependence differ drastically from the textbook-type behavior of "ordinary" JJ's. In Figure 23 we show the dependence of the height of the spike in Figure 20 as a function of a *nominal* magnetic field (i.e., the field at "infinity") perpendicular to the sample plane[19,35].

Evidently, *part* of the conductance enhancement is quenched by very small magnetic fields ($< 1 \, \mu T$), while the remainder drops off much more slowly with increasing magnetic field, a behavior very different from the textbook pattern characteristic of the magnetic field dependence of conventional JJ's. We interpret this behavior in terms of contributions from both coherent and incoherent AR's to the overall conductance enhancement, with the coherent contribution being quenched by fields corresponding to flux on the order of a single flux quantum threading through the gap of the device, while the incoherent contribution remains almost unaffected. In all samples that showed a vanishing resistance at all, a nonzero resistance was restored by weak magnetic fields applied perpendicularly to the sample plane via a suitable small solenoid.

Again, this is a problem best studied with a grating structure, not only because of the sensitivity argument given earlier, but also because in such a geometry the magnetic flux passing through each grating cell is better defined. A first example of such a measurement was shown in reference [35]. In Figure 24 we show the data for a more recent example[74], which is believed to be more typical of the phenomena occurring. [It is the same sample as in Figure 22.] Above certain field strengths, a non-zero resistance re-appears across the sample, increasing gradually with increasing field, but – and this is the key point – containing a strong

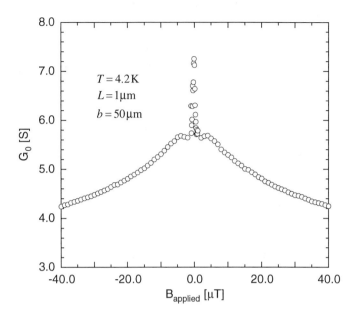

FIGURE 23. Decrease of the height of the zero-bias conductance peak of Figure 20 with increasing magnetic field. A very sharp drop is followed by a much slower decrease, suggesting a transition from phase-coherent to incoherent multiple Andreev reflections. Note the suppressed zero of the vertical axis.

oscillatory component.

The gradual emergence of a *non-periodic* resistance component with increasing field probably has a very simple explanation: The magnetic flux that penetrates through the space between the *Nb* lines is presumably present in the form of quantized vortices. Because of the high structural perfection of the high-mobility *InAs-AlSb* quantum wells, the flux pinning in the *InAs* is probably very weak, which would lead to a finite resistance of the structure due to transverse flux motion between the *Nb* "rails". The entire structure should then behave similarly to a type-II superconductor with very weak flux pinning.

The oscillations are puzzling, though. The smallness of the fields involved and the periodicity in *B* indicate a quantum interference phenomenon, presumably involving the quantized flux vortices just mentioned. However, the overall behavior does not resemble the "Fraunhofer diffraction pattern" observed in conventional JJ's, discussed in many textbooks, which predicts a Josephson critical current that varies with magnetic field according to:

$$I_C(\Phi) = I_C(0) \cdot \frac{\sin(\pi\Phi/\Phi_Q)}{(\pi\Phi/\Phi_Q)} = I_C(0) \cdot \frac{\sin(\pi B/B_Q)}{(\pi B/B_Q)} . \qquad (16)$$

Here, Φ is the magnetic flux passing through the area between the two superconducting electrodes, $\Phi_Q = \pi h/e$ is the "canonical" superconductive flux quantum, B is the applied magnetic field, and

$$B_Q = \frac{\Phi_Q}{ab} = \frac{\pi\hbar}{eab} \qquad (17)$$

is the value of the applied magnetic field that would correspond to *one* superconductive flux quantum per unit cell of the structure. For the dimensions of the sample of Figure 24 (a = 0.936 μm, b = 95 μm), we should have $B_Q = 23.6\,\mu T$.

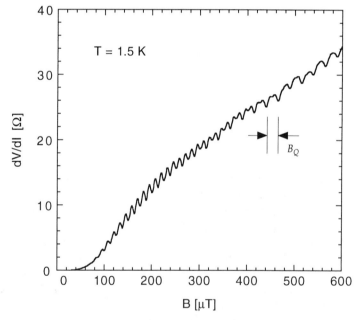

FIGURE 24. Onset of nonzero differential resistance with increasing applied magnetic field. The field strengths are nominal ones, ignoring the lateral flux expulsion from the grating due to the superconducting *Nb* grating lines. From Thomas *et al.*[74].

Not only do the data in Figure 24 not resemble the shape demanded by Eq. (16), but – and perhaps more importantly – the periodicity is different: A close inspection of Figure 24 shows that the period actually doubles between about 300 and 500 μT, with every second minimum gradually dropping out with increasing field. At high fields, the period is approximately 25.6 μT, which is sufficiently close to the theoretical value just given to suggest that the remaining discrepancy might be the result of lateral flux expulsion from the grating.

What remains puzzling is that at low fields, the period is one-half the "canonical" value. The data suggest that the flux, at least at low fields, is not uniformly distributed over the grating cells, but the details are once again not yet understood. In fact, in one sample, reported in an earlier paper, in which the oscillations were studied at fields only up to 100 μT, we observed a much larger discrepancy, by about a factor 5[19,35]. However, such a large discrepancy was seen only in that particular sample.

IV. THE *InAs-AlSb* MATERIALS SYSTEM: TECHNOLOGY AND BASIC PROPERTIES

A. MBE Growth

The MBE technology of both *InAs* and *AlSb* has been worked out to the point that the growth of such structures is now fairly routine. Following earlier pioneering work of the IBM group of Chang *et al.*, first on *InAs/GaSb*[6], and subsequently on *InAs/AlSb*[75], Tuttle *et al.* started, in late-1985, to investigate the properties of *InAs/AlSb* quantum wells systematically and in considerable depth[76-78]. As with other III-V semiconductors, RHEED observations and oscillations are important diagnostic tools. They were extensively studied by Subbanna *et al.*[79], in a paper that formed the backbone of much subsequent work on improving the technology. Extensive subsequent contributions and further refinements were made by Nguyen *et al.*[70]. The reader interested in the MBE technology of these structures is referred to the papers cited.

An important "nuisance property" of *GaSb* and *AlSb* is that *Si* and *Sn*, the two preferred donors in the arsenides, act as acceptors in *GaSb* and *AlSb*, which necessitates the use of column-VI elements as donors, such as *S*, *Se*, and *Te*. Rather than using the elements themselves, we, and others, have found it preferable to follow an idea of McLean *et al.*[80] to use them in the form of a volatile compound such as *PbTe*. The *PbTe* vaporizes as a *PbTe* molecule without dissociating, but splits up on the *AlSb* surface, with a significant fraction of the *Te* being incorporated into the growing crystal, and the *Pb*, which has a fairly high vapor pressure, evaporating[81].

A final important problem is the need to protect the surface of the *AlSb* top barrier from oxidizing. Similar to the *GaAs-AlAs* system, the chemical stability of

AlSb is greatly enhanced by working with $(Al, Ga) Sb$ alloy. Even a relatively small *Ga* addition of 10% greatly enhances the stability, and there does not appear to be any need to go beyond 20%. Associated with such alloying is a reduction in energy gap. This leads to a reduction of the electron barrier, and – probably more important – to a reduction in the *residual* gap at the interface, defined as the energy difference between the bottom of the *InAs* conduction band and the *AlSb* valence band. This can lead to leakage problems in gated structures.

A simple approach for protecting the *AlSb* is by capping the *AlSb* layer with a very thin layer of either *GaSb* or *InAs*. Even a few nanometers of such capping provide complete protection, except, of course, at the edges of any mesa structures, where the *AlSb* is exposed, and must be protected by other means if long-term chemical stability of the structure is required. For short-term experimental purposes (a few months) no such protection appears to be necessary. As we shall see presently, the choice of the cap has a strong effect on the electron concentration in the wells.

B. Doping, Defects, and Mobilities

The doping properties of the *InAs-AlSb* quantum wells are dominated by the fact that these wells are extraordinarily deep, reaching almost to the bottom of the energy gap of the *AlSb* barriers. As a result, even energetically very deep donors in the barrier will drain any available electrons into the well, even donors at a significant distance from the well. Consequently, even not-intentionally-doped wells tend to show relatively high electron sheet concentrations, on the order of 3×10^{11} electrons per cm^2 and more. This has both desirable and undesirable consequences, with different implications for ballistic transport in quantum wires and for superconducting weak links.

To understand the doping properties of these wells, (at least) three distinct sources of electrons must be considered: intentionally added (*Te*) donors; surface-state donors; and deep donors.

1. Conventional (Te) Donors, and Modulation Doping

Because of the high barriers, relatively high electron concentrations are easily achieved by placing the donors into the barriers rather than into the wells, an important consideration for super-semi structures, which call for high electron concentrations. This *modulation doping* technique has the advantage of reduced impurity scattering due to the spatial separation of the ionized impurities from the electrons, which in turn leads to enhanced mobilities, at least relative to bulk-doped wells. In this way, room temperature mobilities equal to the high-purity bulk limit of about 33,000 cm^2/V·s are readily achieved even for electron sheet concentrations around $1 \times 10^{12} \mathrm{cm}^{-2}$, corresponding to volume concentrations approaching $10^{18} \mathrm{cm}^{-3}$. Low-temperature mobilities up to $1 \times 10^6 \mathrm{cm}^2/\mathrm{V \cdot s}$ have also been achieved at similar electron concentrations[19].

By increasing the doping levels, much higher electron concentrations can be obtained – a topic of great interest for superconducting weak links – albeit at some loss in mobility[70].

2. Surface-State Donors

One important source of electrons, at least for wells with *GaSb*-capped top barriers, is a very high density of surface states ($\gg 10^{12}/\ \mathrm{cm}^2$) that is present on the surface of *GaSb*, at an energy about 0.85 eV below the conduction band edge of the *AlSb*[82]. Under flatband conditions, this is still about 0.5 eV above the bottom of the *InAs* well. As a result, electrons from the surface states will drain into the well until the electric field introduced into the barrier by the charge

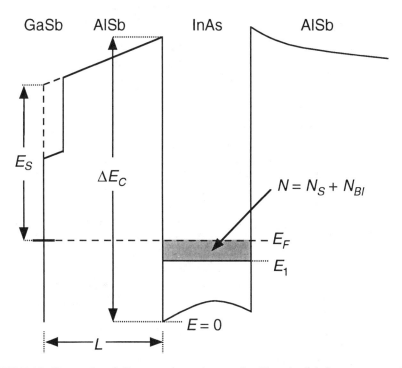

FIGURE 25. Energy band diagram of quantum well, with potential drop across surface barrier due to electron transfer from surface states to the quantum well. From Nguyen *et al.*[82].

transfer has pulled the surface state down to the same energy as the Fermi level inside the well [see Figure 25][82]. The resulting charge transfer increases with decreasing barrier thickness. Well-controlled electron sheet concentrations in the range from about $3 \times 10^{11} \text{cm}^{-2}$ to more than $1.5 \times 10^{12} \text{cm}^{-2}$ have been achieved by a controlled variation of the overall barrier thickness. In fact, in not-intentionally-doped wells with sufficiently thin GaSb-capped top barriers, the electron concentrations in the wells are dominated by electrons from surface states on the GaSb cap used to protect the AlSb.

The surface state appears to be specific to the GaSb cap. If an InAs cap is used instead, the electron concentrations are much lower, corresponding to a surface pinning at least 0.35 meV lower than on GaSb. The exact value apparently depends on the chemical state of the InAs surface, including, specifically, the degree of contamination, with Sb[83]. By working with multiple caps, like an InAs cap over a GaSb cap, and then selectively removing the upper cap, interesting lateral structuring can be achieved[84].

3. Deep Near-Interface Donors

By using sufficiently thick top barriers and/or an InAs cap rather than a GaSb cap, electron sheet concentration in the range $2\text{-}3 \times 10^{11} \text{cm}^{-2}$ are readily achieved in not-intentionally-doped wells, as desired for quantized-conductance quantum wire experiments.

These lower-limit concentrations are still much higher than what is observed in typical GaAs-(Al,Ga)As heterostructures, indicating the presence of an additional source of electrons. The concentration is far too high to be attributable to background donors in the InAs; the data call for a high concentration of donors either associated with the interface itself, or inside the barrier, but very close to the interface. A quantitative analysis of the temperature dependence of the residual electron concentrations[85] calls for a donor level less than 50 meV above the bottom of the bulk conduction band of InAs, which implies a level below the bottom of the quantum well, and below the Fermi level in the well at the observed electron concentrations. But in this case only a small fraction of the donors will be ionized, calling for a donor concentration much higher than the observed electron concentrations, on the order of about 3×10^{12} donors per cm^2 per well.

The nature of these donors is not at all clear. Kroemer proposed that the donors are not ordinary point defects at all, but are Tamm states at the InAs-AlSb interface[85], inherent to the band structure of that interface. More recently, Shen et al.[86] proposed that the donors are very deep bulk donor states associated with AlSb anti-site defects, that is, Al atoms on Sb sites. With Al having two valence electrons less than Sb, one might naively expect that such defects would act as a double acceptor, but the authors argue that in the vicinity of a quantum well that is even deeper than the energy levels associated with the anti-sites, the defects, rather than accepting one or two electrons from the *AlSb* valence band, can act as "false-valence" donors by giving off one of their electrons into the *InAs* well.

A critical discussion of the nature of this "mystery donor" would go beyond the scope of this chapter. Evidently, more research on this crucial question is needed, especially in the context of the use of *InAs-AlSb* quantum wells for quantized conductance studies in ballistic constrictions. For one, it might sometimes be desirable to be able to work with lower electron concentrations than are currently readily achievable. More importantly, remote impurity scattering by those donors almost certainly plays a role in limiting the low-temperature mobilities to values significantly less than those in *GaAs*, a point we will take up presently. The strong hysteresis effects observed in the studies of quantized conductance in ballistic constrictions, discussed extensively in Section II.D, also indicate the presence of a deep defect of some sort, acting as a trapping center for electrons.

4. Electron Mobilities

Room-temperature mobilities in the range $30,000 - 33,000$ cm^2/V·s, four times the value for *GaAs*, are routinely achieved in "good" *InAs-AlSb* quantum well samples with typical well widths and electron sheet concentrations (around 15 nm and 1.5×10^{12}/ cm^2). However, what matters for our purposes are the low-temperature mobilities, from about 30K down, especially around and below 4.2K. At a given temperature, these mobilities still vary strongly with electron concentration, well width, and that elusive parameter called "sample quality." Inasmuch as the highest mobilities constitute an existence proof of what *can* be achieved, we concentrate on those here.

The highest low-temperature (4.2K) mobility observed to date, $980,000$ cm^2/V·s[19], was obtained in a 15 nm wide well that was modulation-doped to an electron sheet concentration of 1.5×10^{12}/cm^2. Most of our studies used that well width. For narrower wells the mobilities decrease, due to interface roughness scattering[50,51]. For wider wells, mobilities decrease too for the following reason. Shubnikow-de-Haas measurements indicate that in the samples with the highest mobilities only the lowest 2D subband is occupied[19]. But with increasing well width, the energy of the second 2D subband soon drops below the Fermi level causing that subband to become occupied. This, in turn, leads to a mobility reduction due to inter-subband scattering. An additional problem with wider wells is the formation of dislocations due to the lattice mismatch between the *InAs* and the *AlSb* barriers.

Given a well width of 15 nm, the mobilities tend to peak near an electron concentration of 1.5×10^{12}/cm^2, dropping off for both higher and lower concentrations. *Increasing* the electron concentration leads to electron spillover into the second subband, just as an increase in well width would do, with the same mobility reduction consequences. Still, remarkably high mobilities are achievable even for the high electron concentrations that are desirable for weak links. In two recent samples with an electron sheet concentration of $5.5 \times 10^{12}/\text{cm}^2$ and $6.9 \times 10^{12}/\text{cm}^2$, we have obtained mobilities as high as 220,000 and 196,000 cm^2/V·s[87].

With *decreasing* electron concentration, the Fermi velocity in the degenerate 2D electron gas decreases. This, in turn, increases impurity scattering, just as impurity scattering increases with decreasing temperature in a non-degenerate electron gas. In principle, in this low-concentration range, mobilities similar to those in *GaAs* should be attainable, like values in excess of 10^6 cm^2/V·s for electron concentrations far below $10^{12}/\text{cm}^2$. However, in this range the mobilities in *InAs-AlSb* quantum wells tend to drop to values an order of magnitude or more below state-of-the-art *GaAs* values. We believe that these discrepancies are not of a fundamental nature, but simply reflect a state-of-the-art for the new materials system which, while vastly improved over the last few years, is still far behind that for the more mature *GaAs* system. Presumably, the "mystery donor" referred to above plays an important role in limiting the mobility. Evidently, a lot of research remains to be done.

C. Substrate Choice and Dislocations

As we stated at the end of Section I.B, the most "natural" substrate for *InAs/AlSb* quantum wells would be *GaSb*, because of its close lattice match to both *InAs* and *AlSb*. In fact, with the lattice constant of *GaSb* being in between those of *InAs* and *AlSb*, a certain amount of beneficial strain compensation would be present in the *InAs/AlSb* structures themselves. Although *GaSb* is not a perfect semi-insulating material even at the low temperatures of interest, essentially perfect isolation is readily achieved by growing *AlSb* buffer layers between the *GaSb* substrate and the *InAs* quantum wells.

Nevertheless, most of our work (and much of the work of others in this field) has employed semi-insulating *GaAs* substrates, working with more or less elaborate buffer layer schemes (see, for example reference [70]) to attenuate the high density of threading dislocations in the *InAs* that are inevitably induced by the large 7% lattice mismatch between *GaAs* and the two 6.1Å semiconductors. In the early stages of our work, there appeared to be little incentive to work with hard-to-obtain and expensive *GaSb* substrates until after other growth problems had been solved that were more urgent than the dislocation problem. As the quality of the layers improved, periodic comparison growths on *GaSb* substrates *did* reveal some improvements in the quality of layers grown on *GaSb* substrates (see again reference [70]), but the improvements were not sufficiently dramatic to create a strong incentive to use such substrates routinely. This does not necessarily mean that dislocations are "benign" defects; it may simply indicate that, at the present stage of technology development, defects other than dislocations are still more important in limiting the quality of the layers.

Furthermore, there is indeed some question about the urgency of eliminating threading dislocations below what is readily achieved by suitable buffer layer schemes. A detailed discussion of the problems and prospects of dislocation attenuation by such schemes would go far beyond the scope of the present chapter, but the basic argument is simple[88]. Experience with the growth of *GaAs* on *Si* indicates that suitable buffer layer schemes should be readily able to attenuate the threading dislocation density at least to 10^7 dislocations per cm^2 – possibly below $10^6/cm^2$. Even at a density of $10^7/cm^2$, there is, on the average, only one dislocation per 10 μm^2. (We are not claiming to have achieved such values in our work; we do not, in fact, not have any recent data on the dislocation count in our growths.)

Consider now a quantum wire with a wire dimensions of $0.05 \mu m \times 1 \mu m = 0.05 \mu m^2$. There is only a 0.5% probability that any given device will have a dislocation directly within the active area of the wire, in which case the device would presumably fail outright. While this might be a failure rate too high for future integrated circuits, it hardly matters in today's research on single devices. Only a small fraction of the devices will fail for *that* reason. More relevant than outright failure due to a dislocation actually penetrating the active area of the wire is the gradual degradation of the electrical properties due to the strain field of dislocations in the vicinity, even several micrometers away. This may indeed be a problem, but we suspect that at the present stage of the technology other defects still play a more important role in limiting the performance of our quantum wire structures. Controlled experiments comparing growths on *GaSb* and *GaAs* substrates should be performed to resolve that question.

In the superconducting weak links we have described in Section III, the area per cell is sufficiently large that the presence of many dislocations per cell is guaranteed. While the impressive performance of these structures suggests that dislocation effects do not play a large role, the possibility cannot be ruled out that much better performance could be achieved if dislocations were essentially absent. Because of the large area of these devices, the dislocation density would have to be reduced by several orders of magnitude. At present it appears difficult to see how such a massive reduction could be achieved by improved buffer layer schemes; this is an area where the use of a lattice-matched substrate such as *GaSb* may indeed be necessary.

D. The *InAs*-Metal Interface

An important parameter in *InAs* contact physics is the band lineup of the *InAs* relative to the Fermi level of the metal. Although strong evidence has existed for many years that the Fermi level pins inside the *InAs* conduction band, there does not seem to exist a unique pinning energy that is characteristic of *InAs*. Instead, the energy appears to depend strongly on both the contact metal and the processing technology.

The earliest data on this topic appear to be those in the extensive 1964 set of Schottky barrier measurements by Mead and Spitzer[89]. These authors report that gold Schottky barriers on *n*-type *InAs* are ohmic down to low temperatures, indicating at most a very shallow electron barrier. In fact, for *p*-type *InAs*, C-V measurements showed a hole barrier height of 0.47 eV at 77K, to an interface Fermi level position about 100 meV above the *InAs* conduction band edge. It was subsequently shown by Walpole and Nill[90] that such C-V measurements systematically under-estimate the hole barrier height, but the authors were not able to give a numerical estimate of the correction. Hence, the 100 meV estimate is almost certainly only a lower limit, with the true value being even higher. Whatever the exact value, an interface Fermi level *somewhere* inside the *InAs* conduction band is undoubtedly present – at least for the interface processing techniques that are commonly employed in making these contacts. This last qualifier is important, because it appears that the interface Fermi energy is not a fixed number, but depends on the processing details[91].

Unfortunately, super-semi contacts are far more sensitive to residual interface barriers than conventional "ohmic" contacts intended for room-temperature operation. Even for interface energies safely inside the conduction band, the quantum-mechanical reflections at the interface still depend on the exact value of that energy. The same is true for the electron concentration on the *InAs* side of the metal-to-*InAs* contact, leading to variations in the superconducting properties of *InAs*-coupled weak links.

In fact, actual data on superconductor-*InAs* contacts clearly show that the low-temperature transport properties of these contacts are bedeviled by processing variations, much of them not yet under control. It was reported already by Silver *et al.*[40], that the properties of *Pb*-to-*InAs* contacts depend drastically on whether the *Pb* was applied by evaporation ("bad") or by electroplating ("good"). Our own work since 1989 using *Nb*-to-*InAs* contacts has only heightened this concern. The I-V characteristics of *InAs*-coupled superconducting weak links are consistently found to depend very sensitively on processing details. For example, e-beam evaporated *Nb* works much less well than sputter-deposited *Nb*[19]. In fact, even for a given technique, as yet uncontrolled processing variations from sample to sample are still common.

V. CONCLUSIONS

The ability to engineer heterostructures such as the *InAs/AlSb* material system has given us access to materials having deeply confined electrons with high mobilities, and hence large mean free paths – elastic and inelastic. The ability to fabricate structures from those materials, with critical dimensions that are substantially smaller than those mean free paths, has allowed us to examine the effects of coherent electron propagation, and given us the seeds of new device ideas, as well as more powerful tools to probe heterointerfaces and impurities on a local scale. Our ability to understand and control heterointerfaces between different materials, such as superconductors and semiconductors, has opened up the possibility of studying coherent phenomena and interactions in a way that has not been previously possible. Those initial studies already reveal the natural commonalities of energy and length scales that pertain to these mesoscopic superconducting structures.

The optimization of the diverse material properties that make those studies possible is critical and demanding. The work described in this chapter emanated from two seemingly disparate vantage points: studies of 1-dimensional transport in constrictions fabricated in the *InAs-AlSb* material had as one motivation the formation of more 'robust' electron waveguides, with quantized conductance apparent at higher temperatures, longer lengths and larger applied bias. Studies of the super-semi structures were initially motivated by a desire to form a three-terminal superconducting device, to achieve gate-modulatable superconductivity, but have subsequently taken on a path of their own.

It is almost certainly unwise to judge these 'devices' – or *any* drastically new technology – by their ability to compete with well-developed entrenched technology. History shows that the applications of new technology are almost always dominated by new applications that were made possible by the new technology, not by pre-existing applications where the new technology could do something a little better that could already be done pretty well. Hence, the primary value of each set of studies may in fact not lie in realization of either more robust electron waveguides, or in JOFETs, but in the knowledge gained through the melding of the two areas – and in the part that this knowledge that will play in whatever new devices based on mesoscopic principles will emerge in the years and decades to come.

ACKNOWLEDGEMENTS

Our work was supported by the Office of Naval Research and by the National Science Foundation. The NSF support was under the auspices of QUEST, an NSF Science and Technology Center on Quantized Electronic Structures. We are also deeply indebted to Dr. Chanh Nguyen and Dr. Steve Koester, who actually performed most of the experimental work at UCSB described in this chapter, as well as our other collaborators who helped to lay the foundation of understanding of the materials growth and processing, device design and characterization. Special thanks go to John English, without whom our MBE Laboratory, where all our samples were prepared, would be unthinkable. Thanks also go to Mike Rooks and the e-beam writing facilities at the Cornell National Nanofabrication Facility, and our own e-beam facilities at UCSB.

REFERENCES

[1]   Takayanagi, H., Hansen, J.B., and Nitta, J., *Physica B* **203**, 291–297 (1994).

[2]   Hergenrother, J.M., *et al.*, *Physica B* **203**, 327–339 (1994).

[3]   Nakagawa, A., Kroemer, H., and English, J.H., *Appl. Phys. Lett.* **54** , 1893–1895 (1989).

[4]   Waldrop, J.R., *et al.*, *J. Vac. Sci. Technol. B* **10**, 1773–1776 (1992).

[5]   Sakaki, H., *et al.*, *Appl. Phys. Lett.* **31**, 211–213 (1977).

[6]   Chang, L.L., and Esaki, L., *Surf. Sci.* **98**, 70–89 (1980).

[7]   Gualtieri, G.J., *et al. Appl. Phys. Lett.* **49**, 1037–1039 (1986).

[8]   Cebulla, U., *et al.*, *Phys. Rev. B* **37**, 6278–6284 (1988).

[9]   Bolognesi, C.R., *et al.*, *IEEE Elect. Dev. Lett.* **14**, 13–15 (1993).

[10] Beenakker, C.W.J., and van Houten, H., *Solid State Physics* **44**, 1 (1991).

[11] Thornton, T.J., *et al.*, *Phys. Rev. Lett.* **56**, 1198–1201 (1986).

[12] Timp, G., in *Nanostructured Systems, Semiconductors and Semimetals* **35**, Reed, M.A., ed., New York: Academic Press, 1990, pp. 113–190.

[13] Wharam, D.A., *et al.*, *J. Phys. C* **21**, L209–214 (1988).

[14] van Wees, B.J., *et al.*, *Phys. Rev. Lett.* **60**, 848–850 (1988).

[15] Landauer, R., "Transport as a consequence of incident carrier flux," in *International Conference on Localization, Interaction and Transport Phenomena*, Braunschweig, Germany, 1984, Kramer, B., Bergmann, G., and Bruynseraede, Y., eds., Springer-Verlag, pp. 38–50.

[16] Büttiker, M., *Phys. Rev. Lett.* **57**, 1761–1764 (1986).

[17] Nixon, J.A., Davies, J.H., and Baranger, H.U., *Phys. Rev. B* **43** , 12638–12641 (1991).

[18] Snider, G.L., *et al.*, *Appl. Phys. Lett.* **59**, 2727–2729 (1991).

[19] Nguyen, C., "Current Transport in *InAs-AlSb* Quantum Well Structures with Super-conducting Niobium Electrodes", Ph.D. Dissertation, UCSB, 1993 (unpublished).

[20]   Rooks, M.J., *et al.*, *J. Vac. Sci. Technol. B* **9**, 2856–2860 (1991).

[21]   Koester, S.J., Brar, B., Bolognesi, C.R., Caine, E.J., Patlach, A., Hu, E.L., Kroemer, H., and Rooks, M.J., *Phys. Rev. B.*, submitted 1995.

[22]   Wharam, D.A., *et al.*, *Phys. Rev. B* **39**, 6283–6286 (1989).

[23]   Yang, M.J., *et al.*, *Phys. Rev. B* **47**, 6807–6810 (1993).

[24]   Brataas, A., and Chao, K.A., *Mod. Phys. Lett. B* **7**, 1021–1027 (1993).

[25]   Laughton, M.J., *et al.*, *Phys. Rev. B* **44**, 1150–1153 (1991).

[26]   Glazman, L.I., and Jonson, M., *Phys. Rev. B* **44**, 3810–3820 (1991).

[27]   Timp, G., *et al.*, "When isn't the conductance of an electron waveguide quantized?", *International Symposium on Nanostructure Physics and Fabrication*, College Station, TX, 1989, Reed, M.A., and Kirk, W.P., eds., Academic Press, pp. 331–345.

[28]   Ismail, K., Washburn, S., and Lee, K.Y., *Appl. Phys. Lett.* **59**, 1998–2000 (1991).

[29]   Zagoskin, A.M., *et al.*, *J. Phys. Condensed Matter* **7**, 6253–6270 (1995).

[30]   Snider, G.L., "Design, fabrication, and analysis of split-gate ballistic constrictions," Ph.D. Dissertation, UCSB, 1991 (unpublished).

[31]   Tarucha, S., *et al.*, "Random potential scattering and mutual Coulomb interaction in long quantum wires," *2nd International Workshop on Quantum Functional Devices*, Matsue, Japan, 1995, FED–143, pp. 8–11.

[32]   Beenakker, C.W.J., and van Houten, H., "The superconducting quantum point contact," *International Symposium on Nanostructures and Mesoscopic Systems*, Santa Fe, NM, 1991, Kirk, W.P, and Reed, M.A., eds., Academic Press, pp. 481–497.

[33]   Furusaki, A., Takayanagi, H., and Tsukuda, M., *Phys. Rev. Lett.* **67**, 132–135 (1991).

[34]   Likharev, K.K., *Revs. Mod. Phys.* **51**, 101–158 (1979).

[35]   Kroemer, H., *et al.*, *Physica B* **203**, 298–306 (1994); (*Proc. NATO Advanced Research Workshop on Mesoscopic Superconductivity*, Karlsruhe, 1994).

[36]   Feynman, R.P., Leighton, R.B., and Sands, M., *The Feynman Lectures on Physics; Vol. 3: Quantum Mechanics*, Reading: Addison-Wesley, 1965. See Section 21–9.

[37]   Kittel, C., *Introduction to Solid State Physics*, New York: Wiley, 1986.

[38]   Tinkham, M., *Introduction to Superconductivity*, New York: McGraw-Hill, 1975.

[39]   de Gennes, P.G., *Superconductivity of Metals and Alloys*, New York: Benjamin, 1966.

[40] Silver, A.H., *et al.*, "Superconductor-Semiconductor Device Research," *Future Trends in Superconductive Electronics*, Charlottesville, VA, 1978, Deaver, J.B.S., Falco, C.M., Harris, H.H., and Wolf, S.A., eds., *Am. Inst. Phys. Conf. Ser., vol. 44*, Am. Inst. Physics, pp. 364–379.

[41] Clark, T.D., Prance, R.J., and Grassie, A.D.C., *J. Appl. Phys.* **51**, 2736–2743 (1980).

[42] Nishino, T., *et al.*, *IEEE Elect. Dev. Lett.* **EDL–6**, 297–299 (1985).

[43] Nishino, T., *et al. IEEE Elect. Dev. Lett.* **10**, 61–63 (1989).

[44] Ivanov, Z., Claeson, T., and Andersson, T., *IEEE Trans. Mag.* **MAG–23**, 711–713 (1987); (*Proc. Applied Superconductivity Conference*, 1986).

[45] Ivanov, Z., Claeson, T., and Andersson, T., *Jpn. J. Appl. Phys.* **26** Supplement 3, DP31–32 (1987); (*Proc. 18th Internat. Conf. on Low Temperature Physics*, Kyoto, 1987).

[46] Takayanagi, H., and Kawakami, T., "Planar-Type *InAs*-Coupled Three-Terminal Superconducting Devices," *Internat. Electron Devices Meeting*, Washington, D.C., 1985, IEDM Digest, IEEE, pp. 98–101.

[47] Takayanagi, H., *et al.*, *Jpn. J. Appl. Phys.* **34**, 1391–1395 (1995).

[48] Akazaki, T., Nitta, J., and Takayanagi, H., *IEEE Trans. Applied Supercond.* **5**, 2887–2891 (1995).

[49] Akazaki, T., Takayanagi, H., Nitta, J., and Enoki, T., *Appl. Phys. Lett.*, submitted 1995.

[50] Bolognesi, C.R., Kroemer, H., and English, J.H., *J. Vac. Sci. Technol. B* **10**, 877–879 (1992).

[51] Bolognesi, C.R., Kroemer, H., and English, J.H., *Appl. Phys. Lett.* **61**, 213–215 (1992).

[52] Nguyen, C., *et al.*, *Appl. Phys. Lett.* **57**, 87–89 (1990).

[53] Andreev, A.F., *Sov. Phys. JETP* **19**, 1228–1231 (1964).

[54] van Houten, H., and Beenakker, C.W.J., *Physica B* **175**, 187–197 (1991).

[55] van Wees, B.J., Lenssen, K.-M.H., and Harmans, C.J.P.M., *Phys. Rev. B* **44**, 470–473 (1991).

[56] Schüssler, U., and Kümmel, R., *Phys. Rev. B* **47**, 2754–2759 (1993).

[57] Bardeen, J., and Johnson, J.L., *Phys. Rev. B* **5**, 72–78 (1972).

[58] Nguyen, C., Kroemer, H., and Hu, E.L., *Appl. Phys. Lett.* **65**, 103–105 (1994).

[59] Blonder, G.E., Tinkham, M., and Klapwijk, T.M., *Phys. Rev. B* **25**, 4515–4532 (1982).

[60] Klapwijk, T.M., Blonder, G.E., and Tinkham, M., *Physica B&C* **109/110** , 1657–1664 (1982).

[61] Kümmel, R., and Senftinger, W., *Z. Physik B* **59**, 275–281 (1985).

[62] Kümmel, R., Gunsenheimer, U., and Nicolsky, R., *Phys. Rev. B* **42**, 3992–4009 (1990).

[63] Gunsenheimer, U., Schüssler, U., and Kümmel, R., *Phys. Rev. B* **49**, 6111–6125 (1994).

[64] Gunsenheimer, U., and Zaikin, A.D., *Phys. Rev. B* **50**, 6317–6331 (1994).

[65] Gunsenheimer, U.G., "Josephson Ströme und dissipativer Ladungs transport in mesoskopischen Supraleiter-Normalleiter-Supraleiter- Kontakten," Ph.D. Dissertation, Würzburg, 1994.

[66] Kleinsasser, A.W., *et al.*, *Appl. Phys. Lett.* **57**, 1811–1813 (1990).

[67] van Wees, B.J., *et al.*, *Phys. Rev. Lett.* **69**, 510–513 (1992).

[68] Nguyen, C., Kroemer, H., and Hu, E.L., *Phys. Rev. Lett.* **69**, 2847–2850 (1992).

[69] Maemoto, T., *et al.*, *Jpn. J. Appl. Phys.* **33**, 7201–7209 (1994).

[70] Nguyen, C., *et al.*, *J. Electron. Mat.* **22**, 255–258, (1993).

[71] Chaudhuri, S., Crocker, B., Bagwell, P., Thomas, M., and Kroemer, H., "A study of *InAs-AlSb*-coupled weak links at temperatures below 1.4K," (unpublished).

[72] Thomas, M., Blank, R., Wong, K., Kroemer, H., and Hu, E.L., "Temperature Dependence of the Precursor to Superconductivity in *InAs*-Coupled Superconducting Weak Link Arrays," (unpublished).

[73] Harris, J., and Yuh, E., "Temperature Dependence of the Zero-Bias Resistance of a Series Array of *InAs*-Coupled Superconducting Weak Links," (personal communication).

[74] Thomas, M., Wong, K., and Kroemer, H., "Flux-Periodic Resistance Oscillations in a Series Array of Superconducting Weak Links of *InAs-AlSb* Quantum Wells with *Nb* Electrodes," (unpublished).

[75] Chang, C.-A., *et al.*, *J. Vac. Sci. Technol. B* **2**, 214–216 (1984).

[76] Tuttle, G., Kroemer, H., and English, J.H., *J. Appl. Phys.* **65**, 5239–5342 (1989).

[77] Tuttle, G., Kroemer, H., and English, J., "Electron Transport in *InAs/AlSb* Quantum Wells: Interface Sequencing Effects," III-V Heterostructures for Electronic/Photonic Devices, San Diego, 1989, Tu, C., Mattera. V.D., and Gossard, A.C., eds., *MRS Symposia Proceedings, vol. 145*, Materials Research Society, pp. 393–398.

[78]   Tuttle, G., Kroemer, H., and English, J.H., *J. Appl. Phys.* **67**, 3032–3037 (1990).

[79]   Subbanna, S., *et al.*, *J. Vac. Sci. Technol. B* **7**, 289–295 (1989).

[80]   McLean, T.D., *et al.*, *J. Vac. Sci. Technol. B* **4**, 601 (1986).

[81]   Subbanna, S., Tuttle, G., and Kroemer, H., *J. Electron. Mat.* **17**, 297–303 (1988).

[82]   Nguyen, C., *et al.*, *Appl. Phys. Lett.* **60**, 1854–1856 (1992).

[83]   Nguyen, C., *et al.*, *J. Vac. Sci. Technol. B* **11**, 1706–1709 (1993).

[84]   Nguyen, C., *et al.*, *Appl. Phys. Lett.* **63**, 2251–2253 (1993).

[85]   Kroemer, H., Nguyen, C., and Brar, B., *J. Vac. Sci. Technol. B* **10**, 1769–1772 (1992).

[86]   Shen, J., *et al.*, *J. Appl. Phys.* **77**, 1576–1581 (1995).

[87]   Thomas, M., and Blank, R., "Improved Electron Mobilities in Heavily Modulation-Doped *InAs-AlSb* Quantum Wells," personal communication.

[88]   Kroemer, H., Liu, T.-Y., and Petroff, P.M., *J. Cryst. Growth* **95**, 96–102 (1989).

[89]   Mead, C.A., and Spitzer, W.G., *Phys. Rev.* **134**, 713–716 (1964).

[90]   Walpole, J.N., and Nill, K.W., *J. Appl. Phys.* **42**, 5609–5617 (1971).

[91]   Brillson, L.J., *et al.*, *J. Vac. Sci. Technol. B* **4**, 919–923 (1986).

Index